Annual Review of Nuclear and Particle Science
Volume 32, 1982

CONTENTS

ANNUAL REVIEWS INC. is a nonprofit scientific publisher established to promote the advancement of the sciences. Beginning in 1932 with the *Annual Review of Biochemistry*, the Company has pursued as its principal function the publication of high quality, reasonably priced *Annual Review* volumes. The volumes are organized by Editors and Editorial Committees who invite qualified authors to contribute critical articles reviewing significant developments within each major discipline. The Editor-in-Chief invites those interested in serving as future Editorial Committee members to communicate directly with him. Annual Reviews Inc. is administered by a Board of Directors, whose members serve without compensation.

1982 Board of Directors, Annual Reviews Inc.

Dr. J. Murray Luck, Founder and Director Emeritus of Annual Reviews Inc.
Professor Emeritus of Chemistry, Stanford University
Dr. Joshua Lederberg, President of Annual Reviews Inc.
President, The Rockefeller University
Dr. James E. Howell, Vice President of Annual Reviews Inc.
Professor of Economics, Stanford University
Dr. William O. Baker, *Retired Chairman of the Board, Bell Laboratories*
Dr. Robert W. Berliner, *Dean, Yale University School of Medicine*
Dr. Winslow R. Briggs, *Director, Carnegie Institute of Washington, Stanford*
Dr. Sidney D. Drell, *Deputy Director, Stanford Linear Accelerator Center*
Dr. Eugene Garfield, *President, Institute for Scientific Information*
Dr. Conyers Herring, *Professor of Applied Physics, Stanford University*
Dr. D. E. Koshland, Jr., *Professor of Biochemistry, University of California, Berkeley*
Dr. Gardner Lindzey, *Director, Center for Advanced Study in the Behavioral Sciences, Stanford*
Dr. William D. McElroy, *Professor of Biology, University of California, San Diego*
Dr. William F. Miller, *President, SRI International*
Dr. Esmond E. Snell, *Professor of Microbiology and Chemistry, University of Texas, Austin*
Dr. Harriet A. Zuckerman, *Professor of Sociology, Columbia University*

Management of Annual Reviews Inc.

John S. McNeil, Publisher and Secretary-Treasurer
Dr. Alister Brass, Editor-in-Chief
Mickey G. Hamilton, Promotion Manager
Donald S. Svedeman, Business Manager

ANNUAL REVIEWS OF
Anthropology
Astronomy and Astrophysics
Biochemistry
Biophysics and Bioengineering
Earth and Planetary Sciences
Ecology and Systematics
Energy
Entomology
Fluid Mechanics
Genetics
Immunology
Materials Science
Medicine

Microbiology
Neuroscience
Nuclear and Particle Science
Nutrition
Pharmacology and Toxicology
Physical Chemistry
Physiology
Phytopathology
Plant Physiology
Psychology
Public Health
Sociology

SPECIAL PUBLICATIONS
Annual Reviews Reprints:
Cell Membranes, 1975–1977
Cell Membranes, 1978–1980
Immunology, 1977–1979

Excitement and Fascination
of Science, Vols. 1 and 2

History of Entomology

Intelligence and Affectivity,
by Jean Piaget

Telescopes for the 1980s

For the convenience of readers, a detachable order form/envelope is bound into the back of this volume.

J.H.Walker

ANNUAL REVIEW OF
NUCLEAR AND
PARTICLE SCIENCE

EDITORIAL COMMITTEE (1982)

CHARLES BALTAY

GORDON A. BAYM

JAMES D. BJORKEN

GERALD T. GARVEY

HARRY E. GOVE

ERNEST M. HENLEY

J. D. JACKSON

ROY F. SCHWITTERS

Responsible for the organization of Volume 32
(Editorial Committee, 1980)

CHARLES BALTAY

GORDON A. BAYM

HARRY E. GOVE

ERNEST M. HENLEY

J. D. JACKSON

ROY F. SCHWITTERS

G. M. TEMMER

JAMES D. BJORKEN

Production Editor   MARGOT PLATT
Indexing Coordinator   MARY GLASS

ANNUAL REVIEW OF NUCLEAR AND PARTICLE SCIENCE

Volume 32, 1982

J. D. JACKSON, *Editor*
University of California, Berkeley

HARRY E. GOVE, *Associate Editor*
University of Rochester

ROY F. SCHWITTERS, *Associate Editor*
Harvard University

ANNUAL REVIEWS INC.   4139 EL CAMINO WAY   PALO ALTO, CALIFORNIA 94306 USA

ANNUAL REVIEWS INC.
Palo Alto, California, USA

COPYRIGHT © 1982 BY ANNUAL REVIEWS INC., PALO ALTO, CALIFORNIA, USA. ALL RIGHTS RESERVED. The appearance of the code at the bottom of the first page of an article in this serial indicates the copyright owner's consent that copies of the article may be made for personal or internal use, or for the personal or internal use of specific clients. This consent is given on the condition, however, that the copier pay the stated per-copy fee of $2.00 per article through the Copyright Clearance Center, Inc. (21 Congress Street, Salem, MA 01970) for copying beyond that permitted by Sections 107 or 108 of the US Copyright Law. The per-copy fee of $2.00 per article also applies to the copying, under the stated conditions, of articles published in any *Annual Review* serial before January 1, 1978. Individual readers, and nonprofit libraries acting for them, are permitted to make a single copy of an article without charge for use in research or teaching. This consent does not extend to other kinds of copying, such as copying for general distribution, for advertising or promotional purposes, for creating new collective works, or for resale. For such uses, written permission is required. Write to Permissions Dept., Annual Reviews Inc., 4139 El Camino Way, Palo Alto, CA 94306 USA.

International Standard Serial Number: 0163-8998
International Standard Book Number: 0-8243-1532-4
Library of Congress Catalog Card Number: 53-995

Annual Review and publication titles are registered trademarks of Annual Reviews Inc.

Annual Reviews Inc. and the Editors of its publications assume no responsibility for the statements expressed by the contributors to this *Review*.

PRINTED AND BOUND IN THE UNITED STATES OF AMERICA

SOME RELATED ARTICLES IN OTHER *ANNUAL REVIEWS*

From the *Annual Review of Astronomy and Astrophysics*, Volume 20 (1982):

The Universe: Past and Present Reflections, Fred Hoyle

Spectra of Cosmic X-Ray Sources, Stephen S. Holt and Richard McCray

From the *Annual Review of Earth and Planetary Sciences*, Volume 10 (1982):

Early Days in University Geophysics, J. Tuzo Wilson

Pre-Mesozoic Paleomagnetism and Plate Tectonics, Rob Van der Voo

Halley's Comet, Ray L. Newburn, Jr., and Donald K. Yeomans

From the *Annual Review of Fluid Mechanics*, Volume 14 (1982):

Strongly Nonlinear Waves, L. W. Schwartz and J. D. Fenton

Gravity Currents in the Laboratory, Atmosphere, and Ocean, John E. Simpson

The Strange Attractor Theory of Turbulence, Oscar E. Lanford III

From the *Annual Review of Materials Science*, Volume 12 (1982):

Synchrotron Radiation Topography, M. Kuriyama, W. J. Boettinger, and G. G. Cohen

Recombination-Enhanced Reactions in Semiconductors, D. V. Lang

From the *Annual Review of Physical Chemistry*, Volume 33 (1982):

The Way It Was, Joseph E. Mayer

Polyacetylene, $(CH)_x$: The Prototype Conducting Polymer, S. Etemad, A. J. Heeger, and A. G. MacDiarmid

Solid State NMR of Biological Systems, Stanley J. Opella

Ann. Rev. Nucl. Part. Sci. 1982. 32:1–34

BEAM-FOIL SPECTROSCOPY[1]

H. G. Berry
Physics Division, Argonne National Laboratory,
Argonne, Illinois 60439

M. Hass
Department of Nuclear Physics,
Weizmann Institute of Science, Rehovot, Israel

CONTENTS

INTRODUCTION

The technique of beam-foil spectroscopy was developed in the late 1960s: most experiments used Van de Graaff type accelerators to produce fast atomic ions whose atomic properties could be studied from

[1] The US Government has the right to retain a nonexclusive, royalty-free license in and to any copyright covering this paper. This work was supported by the US Department of Energy, Office of Basic Energy Sciences, contract W-31-109-Eng-38.

1

their fluorescence after transmission through thin carbon foils. The *Proceedings of the First Beam-Foil Conference* (Bashkin 1968) describe most of the early work. Two principal advantages of such a light source are clear from the earliest experiments: first, the highly efficient production of excited states of highly stripped ions; and second, the highly localized excitation of these states at the carbon foil, which allows simple time resolution of better than 10^{-9} second. By the straight-forward technique of measuring the light intensity emitted by the ionic beam as a function of distance down beam from a thin foil, one can determine decay rates of individual spectral lines. These results are of principal importance both in checks of theoretical work on transition probabilities in atoms and ions and in providing oscillator strengths needed for determining the relative abundance of the atomic elements in the sun and stars.

These two applications provided the main impetus for developing beam-foil spectroscopy in the early 1970s. However, the beam-foil source was also found useful in studying the spectra of ionized systems, allowing classifications, particularly in highly excited systems such as multiply excited states and states of high angular momentum. Atomic alignment in the source led to fine and hyperfine structure studies using quantum-beat and level-crossing measurements. Also, studies of the solid foil-fast ion interaction involve measurements of both light yields and scattered particle distributions and have developed more recently into measurements of the surface interaction under clean conditions of ultra high vacuum.

All these applications and developments are reviewed in the various conference proceedings (Martinson et al 1970, Bashkin 1973, Sellin & Pegg 1976, Désesquelles 1979, Knystautas & Drouin 1982). Other useful reviews of all or part of this field are by I. Martinson (1981) on spectra and lifetimes, H. J. Andrä (1974) on quantum-beat phenomena, H. G. Berry (1977), and in *Beam Foil Spectroscopy*, edited by S. Bashkin (1976). In this review, we briefly summarize the lifetime and spectroscopy measurements. We treat in more detail the alignment and excitation aspects of the beam-foil source, particularly in the applications to nuclear magnetic moment measurements.

LIFETIME MEASUREMENTS

From the first experiments (Kay 1963, Bashkin 1964) showing the electronic decay in flight of fast ions, it was clear that the beam-foil technique could directly determine lifetimes of excited states in atoms and ions. Thus, for nitrogen ions of 3-MeV energy, a typical electronic decay time of 1 ns corresponds to a decay of 0.64 cm. The decay rate from a state of population $N_i(t)$ at time t is

$$\frac{dN_i}{dt} = \left(-\sum_j A_{ij}\right) N_i(t) = \frac{-N_i(t)}{\tau} \qquad 1.$$

where the state i decays to states j with transition probabilities A_{ij} and τ is the lifetime of the state i. If the population $N_i(t)$ is determined solely by the impulsive collision in the foil, $t \leq 0$, then the resulting light yield to each lower state j will be an exponential decay with a rate corresponding to the upper-state lifetime τ_i:

$$I_{ij}(t) = I_{ij}(0)\, \exp\left(\frac{-t}{\tau_i}\right). \qquad 2.$$

However, the beam-foil collision is not state-selective, and populates most levels of a given ion. Thus, the population $N_i(t)$ is due both to its initial population in the collision $N_i(0)$, and to cascading from other more excited levels during the period $t > 0$ up to the time of observation, and we must write:

$$\frac{dN_i}{dt} = -\sum_j (A_{ij}) N_i(t) + \sigma_i Q(t) + \sum_k A_{ki} N_k(t), \qquad 3.$$

where the source Q is generally taken as a delta function at $t = 0$. This set of coupled differential equations can be solved, if all initial populations are known, in terms of the transition probabilities as a sum of exponentials:

$$I_{ij}(t) = I_{ij}(0)\, \exp\left(\frac{-t}{\tau_i}\right) + \sum_k P_k^{(i)} \cdot \exp\left(\frac{-t}{\tau_k}\right). \qquad 4.$$

In many cases the decay is dominated by the exponential decay from the initial population, e.g. for resonance-type transitions in light ions. This is the basis for multiexponential fitting of the observed data using nonlinear minimum χ^2 computer programs where the secondary exponentials are often averages of many of the dominant cascading factors into the state of interest. Frequently, approximate calculations of the mean-lives of the cascading levels, together with reasonable assumptions on the various excitation populations, become valuable inputs for such parametric solutions. Livingston et al (1980) considered some cases in detail, and showed that up to four exponentials can be usefully fit to beam-foil decay data. Numerous studies of multiexponential fitting programs for decay-type data have been made. The reader is referred to the work of Pinnington et al (1974, 1979) for further information.

Cascading can be used through coincidence measurements of two successive photons within a system of three levels to produce a single exponential decay. However, unlike most excited nuclear systems, many decay branches usually occur for excited electronic states and these rapidly reduce the single-to-noise ratios. Thus, only one or two cases have been attempted (Masterson & Stoner 1973), and even then with results strongly limited by statistics. Even in light sources of high yields of excited states, lifetime precisions are limited to a few percent, very little better than most multiexponentially fitted beam-foil data (Camhy-Val et al 1970).

Curtis (1974, Curtis et al 1971) introduced several other techniques for using information about cascading decay schemes to reduce the number of arbitrary fit parameters in a multiexponential decay curve. A review and bibliography of these lifetime analysis methods has been published (Curtis 1976). Here we just briefly sketch his method of "joint analysis of cascade-related decay curves" (or ANDC). We can rewrite Equation 3 as

$$\tau_i \frac{dI_{ij}}{dt} = -I_{ij}(t) + \sum_k \xi_k^{(i)} I_{ki}(t). \qquad 5.$$

Or, in integral form for the time limits (t_I, t_F),

$$\tau_i[I_{ij}(t_F) - I_{ij}(t_I)] = -\int_{t_I}^{t_F} I_{ij}\, dt + \sum_k \xi_k^{(i)} \int_{t_I}^{t_F} I_{ki}\, dt. \qquad 6.$$

Both Equations 5 and 6 are of the linear form

$$\sum_n \xi_n x_n(t_I, t_F) = 0, \qquad 7.$$

where $\xi_0 = 1$, $\xi_1 = \tau_i$, $\xi_{n \geq 2} = \xi_k^{(i)}$ are the unknown parameters.

Hence, measurements of the decay curves of the primary decay and all direct cascades, even though they themselves are cascaded, are sufficient to determine the lifetime τ_i and the additional efficiencies $\xi_k^{(i)}$. In practice, it is sufficient to measure all dominant cascades, and most results have been obtained with only a single cascading term.

In Figures 1 and 2 we show examples of a successful ANDC fit and a comparison with multiexponential fitting and theory. These indicate the precision of most beam-foil lifetime measurements to be about 5% when cascades are carefully taken into account. The improvements in such analyses of lifetime data are shown in Figure 3, where measurements d–f take cascades into account and compare well with the best theoretical value indicated by the arrow. Pinnington & Gosselin (1979)

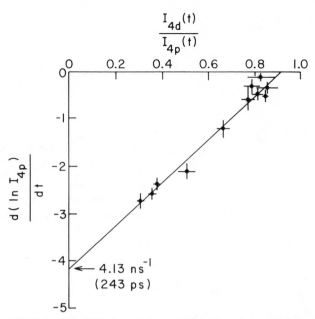

Figure 1 ANDC fit to Kr VIII data of the transitions $4s_{1/2}$-$4p_{3/2}$ and $4p_{3/2}$-$4d_{5/2}$. The intercept of the fitted line with the y-axis determines the reciprocal lifetime of the $4p_{3/2}$ level. The good fit indicates that all significant cascades pass through $4d_{5/2}$ (Livingston et al 1980).

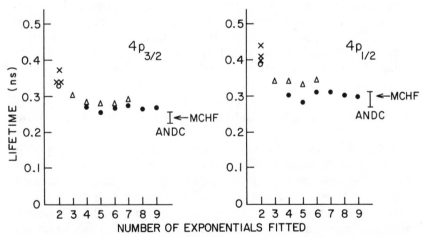

Figure 2 Lifetime results for the $4p_{1/2}$ and $4p_{3/2}$ states obtained by multiexponential fitting of the decay data for the individual levels. Results are shown for two to nine fitted exponential components, including zero (*open circles*), one (*triangles*), or two (*closed circles*) short-lived (growing-in) cascades. For comparison, correlated-decay (ANDC) results are indicated, as are the lifetimes derived from the multiconfiguration Hartree-Fock calculation of Froese-Fischer (1977) (Livingston et al 1980).

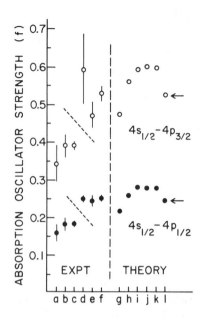

Figure 3 Comparison of theory and experiment for the 4s-4p transitions of Kr VIII (Livingston et al 1980). See Livingston et al (1980) for detailed references. The experimental measurements above the dashed line took cascades into account in detail.

and Younger & Wiese (1978) also discussed the conditions needed for accurate resolution of multiexponential decay curves both by ANDC techniques and by simple exponential fitting techniques. It is now thought unlikely that most beam-foil lifetime inaccuracies are produced by line blending, as suggested earlier (Bashkin 1979).

Good agreement of beam-foil lifetimes with theory can now be obtained for most transitions in light ions, as given for example in nitrogen by Dietrich & Leavitt (1979), in aluminum by Ceyzeriat et al (1979) and Baudinet-Robinet et al (1979), and in more highly ionized systems such as in silicon and phosphorus by Träbert & Heckmann (1980).

Although beam-foil lifetime measurements continue to be used in obtaining elemental abundances from astrophysical data [see, for example, H. Nussbaumer (1979)], the rapid increase in calculational techniques for atomic transition probabilities provides a useful testing of both experiment and theory. In particular, relativistic effects in heavier atomic systems can be tested almost solely by beam-foil measurements. Desclaux (1979) and Luc-Koenig (1979) compare the *ab initio* and parametric potential methods for such data. However, electron correlation corrections to Hartree-Fock calculations generally dominate the relativistic effects, as is shown in several calculations by Froese-Fischer (1977).

SPECTRA

Introduction

Most early beam-foil spectra were plagued with large linewidths, which allowed observations only of simple spectra of light ions where line blending would not cause severe problems. These linewidths were due in part to the low luminosity of the light source ($\sim 10^{13}$ ions per second in the observation region), but principally to the Doppler shifts from the high ion velocities. For observation at an angle θ to a beam velocity v, the observed wavelength is

$$\lambda = \gamma\lambda_0(1 + \beta \cos \theta), \qquad\qquad\qquad 8.$$

where $\beta = v/c$, $\gamma = (1 - \beta^2)^{-1/2}$, and λ_0 is the wavelength emitted by a stationary source. Even at $90°$ to the beam axis, the second-order Doppler shift, $\Delta\lambda = (\gamma - 1)\lambda_0$, due to time dilatation, is nonzero. Also, most collection devices subtend a nonzero solid angle to the beam, which leads to a Doppler width $\delta\lambda = \gamma\lambda_0\beta \sin \theta \cdot \delta\theta$; this typically is a few ångströms in the visible region. Some measurements have been made of light emitted paraxial to the beam to minimize $\delta\lambda$ (Dufay 1970, Bakken & Jordan 1970), but this technique results in a loss of the time resolution needed for many experiments.

Stoner & Leavitt (1971, 1973), and Bergkvist (1976) in a more general treatment, showed how to greatly reduce the line broadening using refocussing techniques. Resolutions, $\delta\lambda/\lambda$, are then $(1.0 - 5.0) \times 10^{-4}$. Thus, with careful analysis of line shapes and other Doppler shift factors, wavelength precisions of the order of 10^{-5} can be attained in most beam-foil spectra (De Serio et al 1981). This precision is equivalent to that of most other sources of spectra of highly stripped ions.

Other advantages of the beam-foil source, especially in spectral analysis, are the lack of impurities in the beam after momentum analysis of the ions and the lack of external perturbations (e.g. collisions, Stark effects) as excited states of the beam decay in a vacuum.

Multiply Excited States

Since significant inner-shell electron excitation occurs in the foil collision process (e.g. Betz 1972), the capture of one or more electrons at the final surface often leads to the production of multiply excited states, with energies far above the first ionization potential of a particular ion.

The beam-foil source remains essentially the only light source that populates in profusion the multiply excited states of ions. Bickel et al (1969) and Buchet et al (1969) were first able to classify transitions in doubly excited lithium from beam-foil spectra. The quartet states of this lithium isoelectronic sequence have been studied in detail using the beam-foil source, and many of the results are reviewed by Berry (1975). More recent work has benefited greatly from the calculations of Bunge et al (1982, Bunge & Bunge 1978), especially in the three-electron systems of low ionization stages.

Most doubly excited states autoionize rapidly (in 10^{-14} s) through auger electron emission. The first measurements by Sellin et al (1972) showed the strong excitation of these levels. Rødbro et al (1979) and Schneider et al (1981) review the recent spectral analyses made in the three-, four-, and five-electron doubly excited systems. It should be

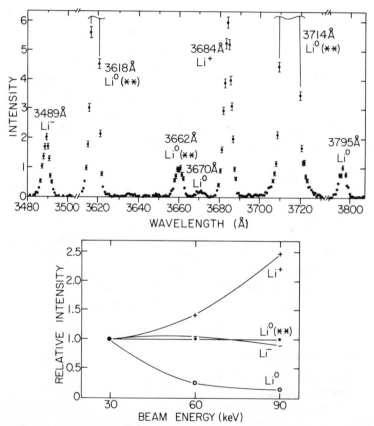

Figure 4 Lithium spectrum (*upper half*), 90-keV beam energy, 1.90 mm from the foil, used to obtain relative light yields from Li^-, Li^{0**}, and Li^+ excited states as a function of beam energy (*lower half*). The yields of Li^{0**} have been normalized to 1.0 at each energy (Brooks et al 1980).

noted that superposed on the electron spectra are the convoy electrons of velocity close to that of the beam, and hence observed close to the forward direction (Harrison & Lucas 1970, Sellin 1979, Gladieux & Chateau-Thierry 1981). A recent experiment (Vager et al 1982) suggests that many of these "convoy electrons" may be due to field dissociation of higly excited Rydberg atoms.

Long-lived excited states of negative ions are generally multiply excited where two or more outer electrons share the net core charge of $+1$. Such states of high spin, which cannot then autoionize rapidly, have been predicted in several negative ions (Bunge 1982). In Figure 4 we show the spectrum of negative lithium, the first transition observed in a negative ion (Brooks et al 1980, Mannervik et al 1980). The identification was verified by observing the photoemission during acceleration of the negative ions in an electric field (Figure 5).

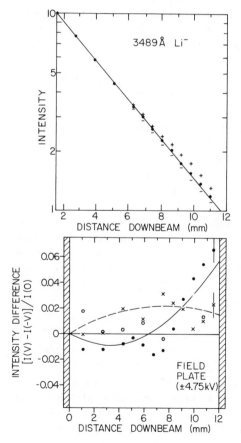

Figure 5 Decay curves of the Li⁻ decay in zero-field (*closed circles*), in accelerating field (*plusses*), and in retarding field (*dash*). In the lower half are shown the differences of the field-affected decay curves of 3489 Å (*closed circles*), 3684 Å (*crosses*), and 3714 Å (*open circles*) at 70 keV. The theoretical curves are given by a solid line for Li⁻ (3489 Å) and a dashed line for Li⁺ (3684 Å) (Brooks et al 1980).

Fine structure measurements of these simple multiexcited systems have led to precision tests of relativistic corrections in *ab initio* atomic structure calculations (Cheng et al 1978). We show an example in Figure 6 for the three-electron system (Livingston & Berry 1978). The four-electron system has also been studied (Berry et al 1982).

Relativistic Atomic Structure

Most relativistic effects are strong functions of Z, the nuclear charge, since the stronger electromagnetic fields yield higher average electron velocities. Thus, the highly stripped heavy ions produced in beam-foil collisions provide a good testing ground for such theories. The status of relativistic atomic theory and experimental tests in general was recently reviewed (Berry et al 1980b).

One-electron tests of QED range from the separated oscillatory field measurements in hydrogen (Lundeen & Pipkin 1981) at a precision of 2×10^{-5}, to the quenching measurements in Ar^{17+} by Gould & Marrus (1978) with a precision of 3%. A new measurement in Cl^{16+} shows agreement with the calculations of Mohr (1975) for the $2s_{1/2}$-$2p_{1/2}$ separation to $\pm0.7\%$ (D. E. Murnick, private communication).

There exist many precision measurements of the 2s-2p transition wavelengths in two-electron systems ranging from Li II (Bacis & Berry 1974) to Fe XXV (Buchet et al 1981). The experiments and theory are reviewed by DeSerio et al (1981), who show that two-electron QED effects can be accounted for (see Figure 7).

In the three-electron system, the resonance transitions $1s^2 2s\ ^2S_{1/2} = 1s^2 2p\ ^2P_{1/2,3/2}$ have been measured and compared successfully with theory along the isoelectronic sequence to Ni XXVI (Berry et al 1980a).

Heavy atomic systems with simple valence-shell structures provide tests of both *ab initio* and semiempirical calculations of atomic structures. The Na I isoelectronic sequence has been studied using both Hartree-Fock wavefunctions (Kim & Cheng 1978, Weiss 1977) and parametric expansions (Edlén 1979, Curtis 1982). The heavier Cu I and Zn I isoelectronic sequences are also of interest in tokamak plasmas (Drawin 1981) in the observation of resonance lines of impurity ions. Both beam-foil (Livingston et al 1980, Pinnington et al 1979) and vacuum spark measurements show systematic deviations from relativistic Hartree-Fock calculations (Cheng & Kim 1978)

Deslattes (1980) discussed the status of calculations relative to inner-shell vacancies in heavy ions. Improved resolution of better than 1 eV in 100 keV for x-ray wavelengths is now feasible. The application of these high resolution x-ray techniques to spectra of one- and

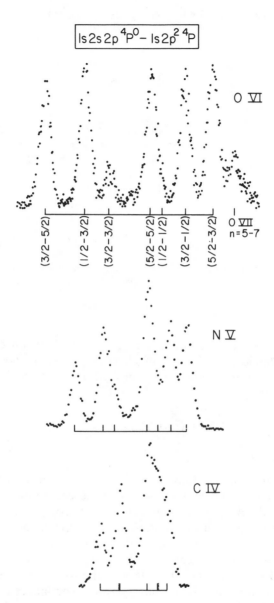

Figure 6 Fast-ion spectra showing resolved fine structure in three-electron ^4P states (Livingston & Berry 1978).

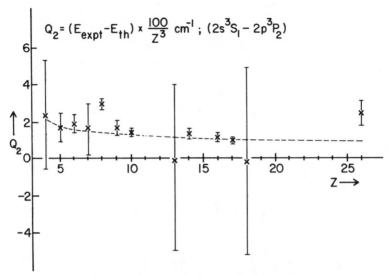

Figure 7   Experiment minus theory for 1s2s $^3S_1$-1s2p $^3P_2$ energy difference (scaled by Z^3) versus Z. The dashed curve shows a calculation of the two-electron Lamb shift correction (DeSerio et al 1981).

two-electron heavy ions is in progress (J. P. Briand, private communication) and should provide tests of QED as precise as those presently made using laser techniques on the ground state of hydrogen (Wieman & Hänsch 1980).

ALIGNMENT AND EXCITATION

General Introduction

The interaction of fast ions in solids is still an area of very active research. Bohr (1948) initially described the energy loss processes in terms of a dipole polarization model. The dynamic nature of this polarization has been further understood through both observations and calculations of the systematics of fast molecular ions passing through solids. In particular, the temporal variations in the solid can be probed by a second projectile travelling through the solid, separated by a few ångströms from its initial molecular partner. The induced polarization wake (e.g. Vager & Gemmell 1976, Vager 1981) significantly and measurably changes the particle trajectories (e.g. Gaillard et al 1977, Kanter et al 1979). For most beam-foil experiments, the ions reach an excitation equilibrium in the foil, and the major points of interest become the energy loss in passage, the mean charge of the projectile, and its state of excitation. Clearly, the latter two may be affected by the final surface interaction. However, since neither can be measured easily within a solid, they can generally be considered as experimental parameters determined by a beam ion, its velocity, and the foil as a whole.

However, several interesting experiments are able to distinguish between the surface and bulk interactions. Charge state effects are minimal (Chateau-Thierry & Gladieux 1976) since all materials give close to the same charge state. However, when the times between "collisions" in the solid can be increased so that "equilibrium" is not attained, even excited state distributions can be dramatically changed. Thus, for example, Okorokov (1965) predicted that for a crystal in the channeling condition, excitation at successive crystal planes in resonance with specific atomic transitions in the moving ion can occur. This produces resonant excitation as a function of the ion velocity and was initially observed by Datz et al (1978). However, such resonances were not observed directly through light yields from excited states (Berry et al 1974b).

For an amorphous or polycrystalline foil, the bulk of the solid excites the moving ion, whose electron cloud on leaving the foil is cylindrically symmetric with respect to the ion velocity direction. If this direction is taken as the z axis, then we say that the ion can be *aligned* in this direction. In terms of the orbital angular momentum projections m_L, the cross sections from excitation, $\sigma(m_L)$, must obey $\sigma(m_L) = \sigma(-m_L)$, but may be a function of $|m_L|$. Clearly, for S states $(L = 0)$, no alignment is possible and the states are spherically symmetric, and for P states $(L = 1)$, the *alignment* can be defined by a single parameter:

$$A'(^1P) = \frac{\sigma(m_L = 0) - \sigma(m_L = 1)}{\sigma(m_L = 0) + \sigma(m_L = 1)}. \qquad 9.$$

In general, light emission from an aligned state is linearly polarized (Percival & Seaton 1958, Fano & Macek 1973) and characterizes excitation of cylindrical symmetry, as in the normal beam-foil case.

An interesting breakage of this cylindrical symmetry is the foil surface itself, which imposes a definite direction on the symmetry axis. Thus $+z$ is not the same as $-z$, since the ion feels a directed force at the surface boundary. This allows the production of states of nondefinite parity with respect to reflections in the plane of the foil surface, say the $z = 0$ plane. Eck (1973) showed that these states, which are equivalent to a linear superposition of states of well-defined but opposite parity, can easily be observed for the excitation of hydrogen to the 2s and 2p states. The measurements by Sellin et al (1973) and by Gaupp et al (1974) verified Eck's predictions. The lack of parity, of course, is equivalent to the production of an atom with an electric dipole moment, where the center of charge of the electron cloud does not coincide with that of the nucleus. In Figure 8 are shown the results of Gabrielse (1981) for the size of the electric dipole moment and the velocity of the captured electron cloud relative to the proton velocity at the foil surface. Both appear to have broad maxima near a proton energy of 200 keV. This energy (velocity) dependence is not yet understood.

The cylindrical symmetry of the beam-foil process can be further broken by tilting the surface itself relative to the beam direction (Berry et al 1974a). Excitation parameters dependent on the surface properties will, in general, be a function of this foil tilt angle α. They can thus be distinguished, at least in part, from bulk properties of the foil. Just as in the cylindrical case considered above, we can envisage the directed surface force on the outgoing atom to affect the properties, most particularly the shape, of the electron cloud about the moving nucleus. The interaction system has now lost its cylindrical symmetry. Its symmetry can be defined by two vectors: the outgoing surface normal, $\hat{\mathbf{n}}$, and the direction of the moving ion velocity, $\hat{\mathbf{z}}$. However, there still exists reflection symmetry in the plane containing $\hat{\mathbf{n}}$ and $\hat{\mathbf{z}}$. This sym-

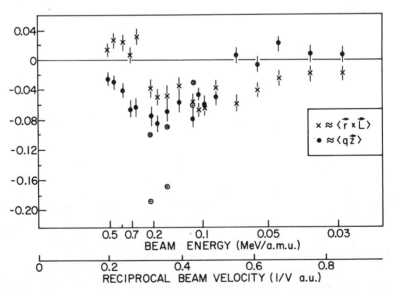

Figure 8   The electric dipole moment of the $n = 2$ hydrogen state, $\langle q\,\mathbf{z}\rangle$, and the relative electron velocity, or linear momentum, $\mathbf{r} \times \mathbf{L}$, after foil excitation of protons (adapted from Gabrielse 1981). The vertical scale is in atomic units.

metry is the same as that of a two-particle scattering event with central forces when the two particles are detected in coincidence. Fano & Macek (1973) discussed such geometries for atomic scattering, with light emission, and we adopt most of their results in the brief development below.

Measurement of Alignment and Orientation

We use the notion that any charge distribution can be expanded in terms of electromagnetic multipole moments and we use components that are directly proportional to a spherical tensor expansion of the excitation density matrix (as used, for example, by Ellis 1973). For the case of an atomic state of definite parity and with reflection symmetry (in say the x–z plane of a standard Cartesian system), we have an electrostatic charge E0, a single-component magnetic dipole M1, a three-component electric quadrupole E2, plus higher-order terms. The more general case of relaxing the parity condition was discussed by Gabrielse (1980, Gabrielse & Band 1977). The dipole gives rise to the atomic *orientation*, O_1^c, and the quadrupole components describe the *alignment* of the atom, A_0^c, A_1^c, A_2^c, where

$$O_1^c = \frac{\langle L_y \rangle}{L(L+1)} = \frac{-K}{3\sqrt{2}} \rho_1^1 \qquad\qquad 10.$$

$$A_0^c = \frac{\langle 3L_z^2 - L^2 \rangle}{L(L+1)} = \frac{K}{10} [(2L-1)(2L+3)]^{1/2} \rho_0^2 \qquad\qquad 11.$$

$$A_1^c = \frac{\langle L_x L_z + L_z L_x \rangle}{L(L+1)} = \frac{K}{5\sqrt{6}i} [(2L-1)(2L+3)]^{1/2} \rho_1^2 \qquad\qquad 12.$$

$$A_2^c = \frac{\langle L_x^2 - L_y^2 \rangle}{L(L+1)} = \frac{-K}{5\sqrt{6}} [(2L-1)(2L+3)]^{1/2} \rho_2^2, \qquad\qquad 13.$$

with

$$K = \frac{2(2S+1)(2L+1)^{1/2}}{[L(L+1)]^{1/2}}, \qquad\qquad 14.$$

where the standard density matrix components ρ_q^k are defined by Ellis (1973), the state has an electron spin angular momentum S, an orbital angular momentum L, and the interaction has been assumed to be independent of S.

Since the photon of deexcitation carries away angular momentum, whose value is defined by its polarization, we can use the analysis of the observed polarization to obtain information on the atomic orientation and alignment. Stokes (1852) showed that all the information can be measured in terms of four numbers, now called Stokes parameters, that can be defined in terms of the total light yields $I(\psi)$ viewed through a polarizer at angles ψ in the plane perpendicular to the light propagation axis:

$$I = (\text{total intensity}) = I(0°) + I(90°) \tag{15.}$$

$$M = I(0°) - I(90°) \tag{16.}$$

$$C = I(45°) - I(-45°) \tag{17.}$$

and the net circular polarization

$$S = I_{\text{rhc}} - I_{\text{lhc}}. \tag{18.}$$

The relative Stokes parameters for light emitted in the direction (θ, Φ) measure the alignment and orientation parameters through

$$\frac{M}{I} = \tfrac{3}{4}R_2(1 - F)^{-1}[A_0^c \sin^2 \theta + A_1^c \sin 2\theta \cos \Phi$$
$$+ A_2^c(1 + \cos^2 \theta) \cos 2\Phi] \tag{19.}$$

$$\frac{C}{I} = \tfrac{3}{2}R_2(1 - F)^{-1}(A_1^c \sin \theta \sin \Phi - A_2^c \cos \theta \sin 2\Phi) \tag{20.}$$

$$\frac{S}{I} = \tfrac{3}{4}R_1(1 - F)^{-1}O_1^c \sin \theta \sin \Phi, \tag{21.}$$

where

$$F = \tfrac{3}{4}R_2[A_0^c (\cos^2 \theta - \tfrac{1}{3}) + A_1^c \sin 2\theta \cos \Phi + A_2^c \sin^2 \theta \cos 2\Phi] \tag{22.}$$

and

$$R_k = \frac{\begin{Bmatrix} L_f & L & 1 \\ k & 1 & L \end{Bmatrix}}{\begin{Bmatrix} L & L & 1 \\ k & 1 & L \end{Bmatrix}}. \tag{23.}$$

The four alignment and orientation parameters A_0^c, A_1^c, A_2^c, and O_1^c can be derived algebraically from single measurements of S/I and C/I and two measurements of M/I at different observation angles (θ, Φ) (Berry et al 1974a, 1977b). Alternatively, Andrä et al (1976) showed how the same results may be obtained by single measurements, including the application of a magnetic field. A useful polarimeter was described by Berry et al (1977a).

For unresolved fine and hyperfine structures, the above expressions (19–22) must be multiplied by the time-independent depolarization factors Q_k ($k = 1$ for S/I, and $k = 2$ for M/I, C/I, and F) for time-averaged optical observations (Berry 1977).

Quantum beats may be observed in time-resolved measurements of unresolved fine and hyperfine structures. The factors Q_k are then replaced by time-dependent modulation factors as described in detail, for example, by Andrä (1974) and by Berry (1977).

POLARIZED LIGHT FROM THIN FOIL EXCITATION

Many observations of light yields from perpendicular and tilted-foil excitation show that the alignment is generally small and less than 10% (see Liu et al 1971, 1974, Church et al 1974, 1975, Hight et al 1977). Only for light ions such as helium and lithium does the alignment reach 20% (e.g. Hight et al 1977, Gay et al 1981, Andrä 1974). This is true for all beam energies, and very little dependence on beam energy is observed (Schectman et al 1980, Gay et al 1981).

The orientation by tilted foils can be significantly higher, as observed by Berry et al (1974a), Liu et al (1974), Brooks & Pinnington (1978, 1980), Pedrazzini et al (1978), and Burns et al (1979). In Figure 9, we

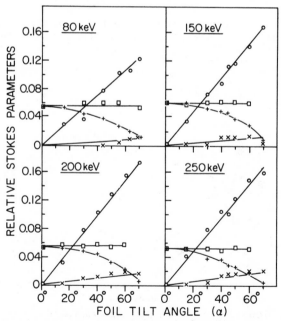

Figure 9 The Stokes parameters of the ^{4}He I, 3889 Å, 2s ^{3}S-3p ^{3}P transition for four different beam energies, as a function of foil tilt angle (Brooks et al 1982): S/I (*circles*); C/I (*crosses*); M/I (0°) (*plusses*); M/I (90°) (*squares*).

show some examples of Stokes parameter measurements from thin foils. The angle dependences are typical of most excited states. The orientation (S/I) increases almost linearly with angle and is always in the same sense, such that the ion is rotating into the nearer foil surface, as shown in Figure 10. As discussed by Schröder & Kupfer (1976) and Burgdörfer & Gabriel (1979), this can be attributed to enhanced electron interaction and pickup with the greater electron density at lower relative velocities with respect to the moving ion. Only this qualitative agreement with the observed orientation has been found as also with the

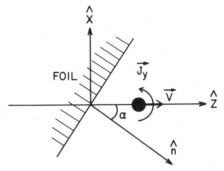

Figure 10 The tilted-foil geometry. All measurements of atomic orientation $\mathbf{J}_y$ give a null or positive value, such that $\mathbf{J}_y$ is up, out of the plane in the geometry shown.

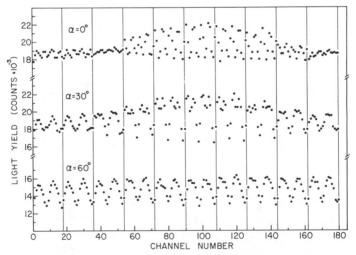

Figure 11 Polarization data for a set of foil tilt angles, each taken over one beat length of $J = 1, 2$ of the 3889 Å transition of ^{4}He I. The data consist of an 18-step rotation (20 degrees/step) of a phase plate at 10 equidistant positions along a single quantum beat (Brooks et al 1982).

tilt-angle dependence α, which has been predicted by several theories (e.g. Eck 1974, Band 1975, 1976, Herman 1975, Lombardi 1975, Lewis & Silver 1975). The most successful theory of Band (1975, 1976) is probably not applicable for large tilt angles $\alpha \gtrsim 10°$, when his expansion parameters become too large (M. Dehaes, 1981, private communication). The alignment parameters generally remain small at increased angles, but can have both signs, depending on α and energy; see, for example, results for hydrogen (Winter 1979) and for helium (Schectman et al 1980).

The first alignment measurements were the quantum-beat observations (Andrä 1970), which provide measurement of atomic fine and hyperfine structures of the moving ions, and several results have been

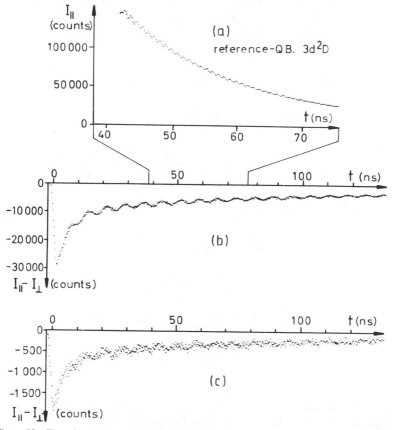

Figure 12 Experimental result of the fs-quantum-beat measurement of the 6d²D term of Li I. (*a*) Reference fs-beat (autocalibration) of the 3d ²D term. (*b*) Sum of the difference $I_{\parallel} - I_{\perp}$ of 30 runs. (*c*) The difference $I_{\parallel} - I_{\perp}$ for a single run (Wangler et al 1981).

obtained in light ions using both zero-field excitation (e.g. Andrä 1974) and magnetic field Zeeman quantum beats (e.g. Liu et al 1971, Gaillard et al 1973). We show two examples of time-resolved quantum beats in Figures 11 and 12.

POLARIZED LIGHT FROM SURFACE EXCITATION

Andrä (1975, Andrä et al 1977) first reported that similar optical polarization can be observed from ions scattered at grazing incidence from surfaces. The interaction of these ions leaving the surface has the same symmetry as that of ions leaving the final surface of the tilted foil. However, these experiments showed a greater circular polarization component, or atomic orientation. This was presumably because of the much greater effective tilt angle ($\alpha \approx 85°$) compared with most tilted-foil

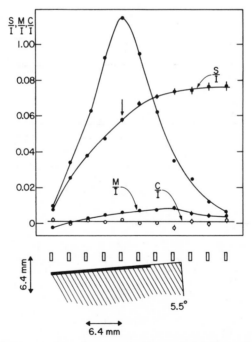

Figure 13 The lower part shows the geometry for measuring the spatial dependence of the Stokes parameters. Observations at the rectangular detection windows shown gave the Stokes parameter values I, M/I, C/I, and S/I for the Ar II 4610 Å transition shown in the upper figure, the abscissa scale being the same for both parts of the figure (Berry et al 1977b). The target is shown below, the heavy line denoting the collision area.

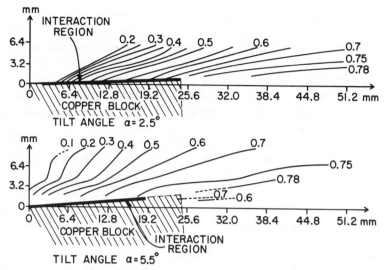

Figure 14 Contour plots of equal circular polarization S/I of Ar II 4610 Å for tilt angles of 2.5° and 5.5° of a copper surface. The incident Ar$^+$ energy was 1.0 MeV (Berry et al 1977b).

experiments of $\alpha \lesssim 70°$. More detailed angular measurements showed the basic interactions in the two cases to be similar. Thus, Berry et al (1977b, c) showed that the total light yields from surface and foil excitation are similar. Qualitative agreement was also noted in the angular dependence of the atomic orientation, which always had the same sign, as indicated above; and the alignment was small (Andrä et al 1976). Figures 13 and 14 indicate some of these results.

Further measurements in improved vacua have shown some dependences, particularly of the produced orientation, on the smoothness and cleanliness of the surface (Tolk et al 1977, 1978, 1979, Andrä et al 1979). Measurements made with clean, single crystals also stress the surface character of the interaction and compare in more detail the tilted-surface and tilted-foil excitations (Tolk et al 1979, 1981).

NUCLEAR ALIGNMENT AND ORIENTATION

The alignment and orientation produced at the thin-foil or surface collision is in terms of the electronic orbital angular momentum **L**, while the interaction is presumed to be independent of electron spin **S** and the nuclear spin **I** (see Ellis 1973, 1977 for discussion of these conditions). The observation of quantum beats is evidence of the periodic transfer of this asymmetry to one or both of the two spin systems. This is just the depolarization factor Q^k as discussed above. The electron and nuclear spins gain the asymmetry lost by the time-averaged orbital.

Since the atomic alignment is usually small and of varying sign, the net transfer to the nuclei through all the electronic excited states in a beam is likely to be very small. The orientation production is a different story: the orientation of all excited states (and probably ground states) is in the same direction for both tilted-foil and grazing-surface excitation, and it also becomes very large for large tilt angles.

The first observation of nuclear orientation was made in the hyperfine quantum beats of ^{14}N IV by Berry et al (1975). However, a clear demonstration that the total time-averaged nuclear orientation in a beam is significant was shown by Andrä et al (1977). They showed that a perpendicular foil excitation of a nitrogen beam, which had previously undergone a grazing-incidence surface excitation, can produce circularly polarized quantum beats. This orientation cannot occur in the foil (a cylindrically symmetric excitation) but is in the nucleus before the electron pickup at the final foil surface. We show the results of this experiment in Figure 15. The quantum beats, which are "time-

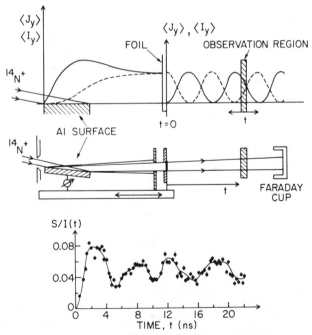

Figure 15 Time-averaged production of a nuclear spin-polarized ion beam and its use for a hfs-quantum beat measurement (Andrä et al 1979).

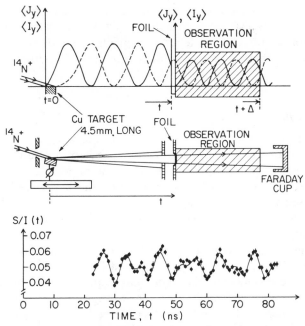

Figure 16 Ground term hfs-quantum beats via transient orientation transfer to the nucleus as observed by time-averaged optical detection (Andrä et al 1979).

averaged" in the time *between* surface and foil excitation, indicate the transfer of nuclear orientation to electronic spatial orientation. This is the opposite of the single foil transfer of electronic to nuclear orientation.

Observation of electronic ground-state orientation was verified in the similar measurement indicated in Figure 16 (Andrä et al 1979). The quantum beats are now "time-averaged" in the time *after foil excitation*. The quantum beats are due predominantly to the $2p^2\ ^3P$ ground state of ^{14}N II, but are perturbed by contributions from ground states and long-lived states in other nitrogen ions (see also below).

NUCLEAR MAGNETIC MOMENT MEASUREMENTS

The same hyperfine interaction that produces nuclear orientation (as discussed above) can also produce a *precession* of the nuclear spin around the resultant angular momentum $\mathbf{F} = \mathbf{I} + \mathbf{J}$. For an excited nuclear state, such a precession will result in a rotation of the angular correlation of decay γ rays. This rotation of the angular correlation serves as a novel method to measure magnetic moments of excited nuclear levels. It also serves as a manifestation that *atomic* orientation persists for deeply bound electronic configurations of highly stripped ions for which there is no simple method of observing circular polarization.

Introduction

The nature of the hyperfine interaction in free ions recoiling at high velocity out of solid targets has been established in recent years as a largely static interaction in all regions of the periodic table that have been investigated (Goldring 1981). Clearly, the hyperfine interaction can be used to study the magnetic moments of nuclear levels. These magnetic moments played a crucial role in establishing the structure and internal organization of nuclei. In recent years, states of high angular momentum have come under particular scrutiny, and, again, measurements of magnetic moments are involved in order to determine the nature of these states and the underlying pattern of nuclear structure. The development of techniques for measuring nuclear magnetic moments has therefore recently been concerned to a large extent with problems related to high spin states. Several general reviews of these techniques exist (e.g. Goldring & Hass 1982, Niv et al 1981) and we limit ourselves here to a discussion of tilted-foil methods.

The application of the intense hyperfine magnetic fields present in an ensemble of highly stripped fast ions recoiling into gas (or vacuum) to the measurement of nuclear magnetic moments (Goldring 1981) has one severe intrinsic limitation. Since these magnetic fields do not, in general, have a preferred direction of polarization, only the absolute value of the particular nuclear moment can be determined. However, the sign of the -g-factor is an important property since, even by itself, it can often reveal the nature of the underlying nuclear structure. Also, the measured quantity in such an experiment is the attenuation of the angular correlation of the emitted γ rays, which may be difficult to detect for short lifetimes. For the purposes of determining signs of nuclear moments and for improved experimental sensitivity, it is desirable to establish mechanisms that can polarize the atomic configurations and at the same time maintain the attractive features, especially for high-spin states, of the recoil-into-gas technique. The tilted (multi) foil and the transient field techniques (Benczer-Koller et al 1980) provide these desired characteristics.

The first tilted-foil measurements showing the applicability of this technique for measuring the signs of nuclear g-values were in ^{18}O and

[16]O (Hass et al 1977, Goldring et all 1977). Since then, the more sensitive multifoil technique has been established (Dafni et al 1981, Broude et al 1981).

Theoretical Development

The standard geometry for g-value measurements via the hyperfine interaction produced in tilted foils is shown in Figure 17. In the presence of oriented hyperfine fields there is a net rotation of the angular distribution of the decay γ rays, in addition to an attenuation. This rotation is both a signature of polarized atomic fields and a measure of the magnitude and sign of the nuclear magnetic moment.

In the nuclear experiments, we shall ignore the initial atomic alignment (A_0^c, A_1^c, A_2^c, which are small) and assume only an orientation O_1^c (which can be large). The orientation is generally defined (Hass et al 1977) as

$$p_J = \frac{\langle J_y \rangle}{\sqrt{J(J+1)}} = O_1^c \langle J \rangle^{1/2} = -\rho_1^1 \frac{\sqrt{2(2L+1)}}{3}$$ 24.

using Equation 10 and neglecting electron spin ($J \equiv L$).

Transforming the y axis to a new z axis, one can describe the initial alignment of a nuclear level of angular momentum I, by nuclear

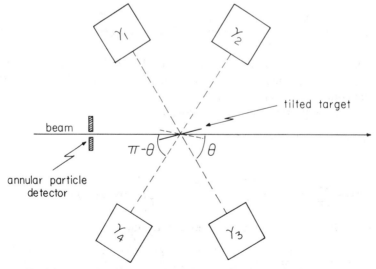

Figure 17 Schematic drawing of an experiment to measure a small precession of the angular distribution of decay γ rays. The γ counters are placed at symmetric angles with respect to the beam axis. The direction of polarization is perpendicular to the detector plane.

statistical tensor components $\rho_q^k(I, t=0) = \rho_{-q}^k(I, t=0)$ with even k, q. This yields an angular distribution of the decay γ rays of the form

$$W(\theta, \Phi) = \sum_{k,q} U_k \rho_q^k Y_q^k(\theta, \varphi),$$ 25.

with U_k depending upon the nuclear de-excitation process. The hyperfine interaction modifies the angular distribution through the coupling of the electronic and nuclear statistical tensors:

$$\rho_q^k(I, t) = \sum_{k',q'} G_{qq'}^{kk'}(t) \rho_{q'}^{k'}(I, t=0),$$ 26.

where

$$G_{qq'}^{kk'}(t) = \sum_{\substack{FF' \\ k_e=0,1}} \exp(-i\omega_{FF'}t) \, (2F+1) \, (2F'+1)[(2k+1)(2k'+1)$$

$$(2k_e+1)]^{1/2} \times \langle k'q'k_eq_e|kq\rangle \cdot \rho_0^{k_e}(J, t=0)/\rho_0^0(J, t=0)$$

$$\times \begin{Bmatrix} J & I & F \\ J & I & F' \\ 0 & k' & k \end{Bmatrix} \begin{Bmatrix} J & I & F \\ J & I & F' \\ k_e & k' & k \end{Bmatrix}.$$ 27.

Here

$$\omega_{FF'} = \frac{a\{F(F+1) - F'(F'+1)\}}{2\hbar}, \qquad a = -g\mu_N H_0.$$ 28.

The only nonvanishing components of the electronic tensor $\rho_0^{k_e}(J, t=0)$ are for $k_e = 0, 1$, which thus gives only diagonal elements $k' = k$, and the nondiagonal elements $k' = k \pm 1$.

The diagonal elements can be written

$$G_{qq}^{kk}(t) = G^k(t) - iqH^k(t),$$ 29.

where

$$G^k(t) = \sum_{FF'} \frac{(2F+1)(2F'+1)}{2J+1} \begin{Bmatrix} F & F' & k \\ I & I & J \end{Bmatrix}^2 \cos(\omega_{FF'}t)$$ 30.

is the well-known attenuation coefficient, and

$$H^k(t) = \frac{3p_J}{k(k+1)\sqrt{J(J+1)}} \sum_{FF'} \frac{(2F+1)(2F'+1)}{2J+1} \frac{F(F+1) - F'(F'+1)}{2}$$

$$\times \begin{Bmatrix} F & F' & k \\ I & I & J \end{Bmatrix} \sin(\omega_{FF'}t).$$ 31.

This term is proportional to the initial orientation and is responsible, as shown below, for the rotation of the angular distribution.

The nondiagonal terms with $k' = k + 1$ have contributions only from $k_e = 1$. In a typical experimental arrangement, where the γ detectors are positioned in the plane defined by the beam and the normal to the foil (see Figures 17 and 18), they have no effect on the angular distribution since then $\theta = 90°$ and $Y_k^q(\theta = 90°)$ vanishes for odd k and even q. This term leads to the nuclear orientation observed by Andrä et al (1977) and discussed above.

For a multifoil array such as that in Figure 18, we can assume a total of N equidistant foils of total transit time T, and the nuclear statistical tensor becomes

$$\rho_q^k(I, T) = \sum_{k'} G_{qq}^{(N)kk'}(T)\rho_q^{k'}(t = 0), \qquad\qquad 32.$$

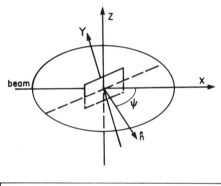

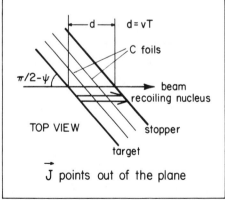

Figure 18 The geometry of the tilted-foil arrangement. The electronic polarization is along the ζ direction and the γ detectors are in the x–y plane (*top*). Schematic view of the multifoil arrangement (*bottom*).

where $G_{qq}^{(N)kk'}(T)$ describes a product of single-foil interactions as in Equation 27. Dafni et al (1981) considered the limiting case of $N \to \infty$, but finite T, so that $t \to 0$ at each foil. They conclude $G^{(N)}(T)$ has only diagonal terms and can be written as

$$G_{qq}^{(N)kk'}(T) \xrightarrow[N \to \infty]{} \exp(-iqp_J\omega_J T) \qquad\qquad 33.$$

with $\omega_J = a\sqrt{J(J+1)}/\hbar$. In a typical experimental arrangement, the γ detectors are placed at a polar angle $\theta = 90°$, and the angular distribution can be written [with $Y_q^k(\theta, \varphi) + V_{kq}P_{kq}(\cos \theta)\exp(iq\varphi)$]

$$W(\theta = 90°, \Phi, T) = \sum_{k, q \text{ even}} U_k \rho_q^k(I, 0) V_{kq} P_{kq}(0) \cdot G_{qq}^{(N)kk}(T) \exp(iq\varphi)$$

$$\simeq \sum_{k, q} U_k \rho_q^k(I, 0) V_{kq} P_{kq}(0) G_k\left(\frac{T}{\sqrt{N}}\right)$$

$$\exp[iq(\Phi - p_J\omega_J T)^2]$$

$$\xrightarrow[N \to \infty]{} W(\theta = 90°, \varphi - p_J\omega_J T, 0), \qquad 34.$$

which is just a classical precession with a frequency $p_J\omega_J$ independent of the nuclear spin I. We would like to note that under these conditions the nuclear *orientation* is negligible.

The first-order correction to the pure precession is a superimposed attenuation, but the attenuation is reduced owing to the scaling of the time by the factor $1/\sqrt{N}$. This is analogous to attenuation in gas, with N referring to the number of ion-gas collisions or the transient field case, where N refers to the number of ion-solid collisions. In the transient field experiments, N is very large, the attenuation can be completely neglected, and the measured effect is just the classical precession.

It has been assumed so far that the ions emerge from each polarizing foil in a state with a unique value of J and H_J, and moreover, that these values do not change from foil to foil. The actual situation is much more complex, and the ions emerge with a broad distribution of charge states, ionic configurations, and hyperfine frequencies. The main conclusions do not change, however, in the limit of very large N, and the resulting expression for the matrix elements is simply an averaging process:

$$G_{qq}^{(N)kk}(T) \approx \left[1 - \frac{k(k+1)}{3}\langle\omega_J^2\rangle\frac{T^2}{N}\right]\exp(-iq\langle p_J\omega_J\rangle T). \qquad 35.$$

The last expression reflects an important advantage of the multicollision techniques. The precession does not depend on the details of the frequency distribution or on the nuclear spin, and all that is needed to relate the precession to the nuclear g-factor is a single ionic parameter, namely the effective field $\langle p_J H_J\rangle^*$. The first term corresponds to the well-known Abragam-Pound attenuation in gas with the time T/N replacing the mean time between collisions, τ_c.

The tilted multifoil array (or the transient field technique) transforms the hyperfine perturbation in a striking way: instead of a complex isotropic perturbation with a small (and also complex) superimposed precession, it gives rise to a regular precession in a constant field. Experimentally one benefits from this situation in a two-fold way: the accumulation of many small precessions creates a large rotation that is relatively easy to detect, and the reduction in attenuation preserves the original anisotropy and with it the experimental sensitivity to this rotation.

Experiment

The first tilted-foil g-factor measurements (Hass et al 1977, Goldring et al 1977) established an asymmetry using the scheme shown in Figure 17, where the foil plane can be flipped from right to left of the beam axis. Thus if

$$\rho_{ij} = \left[\left(\frac{W_i^R}{W_i^L} \right) \left(\frac{W_j^L}{W_j^R} \right) \right]^{1/2} \qquad i = 1, \ldots, 4 \qquad 36.$$

where W_i^R etc are counts in detector i, with "right" tilt. The precession angle $\Delta\varphi$ is

$$\Delta\varphi = \frac{1}{W} \frac{dW}{d\theta} \bigg|_{\theta=\theta_0} \frac{1-\rho}{1+\rho},$$

Table 1 Tilted-foil results; $\varepsilon = (1-\rho)/(1+\rho) = 1/W \cdot dW/d\theta \cdot \Delta\Phi^a$

Nucleus	Level	Mean-life τ(ps)	Foil	Tilt angle Ψ	ε	$\Delta\Phi$ (mrad)	Alignment p
^{16}O	3_1^-	27	C	70°	+0.0080(15)	−2.6(5)	0.02
^{18}O	2_1^+	3.6	C	70°	−0.0045(21)	+1.7(8)	0.02
^{40}Ca	3_1^-	68	C	70°	+0.018(4)	−8.6(19)	0.06
^{44}Ca	2_1^+	4.2	C	70°	−0.010(4)	+2.8(13)	0.06
^{146}Nd	2_1^+	38	C	70°	+0.031(10)	−16(5)	0.10
^{148}Nd	2_1^+	144	C	70°	+0.026(6)	−25(6)	0.10
^{148}Sm	2_1^+	10.5	C	70°	+0.043(8)	−17(3)	0.10
^{150}Sm	2_1^+	70	C	70°	+0.023(4)	−19(3)	0.10
^{150}Sm	2_1^+	70	Sm	70°	+0.005(5)	−4(4)	~0.015
^{150}Sm	2_1^+	70	Au	70°	+0.012(9)	−10(8)	~0.04
^{110}Pd	2_1^+	62	C	65°	+0.010(4)	−12(4)	~0.03

[a] The results correspond to a recoil velocity of $v/c \approx 0.01$ except for the ^{18}O and ^{110}Pd cases, which correspond to $v/c \approx 0.03$.

where $\rho = (\rho_{14}\rho_{32})^{1/2}$ and θ_0 is an angle for large slope in the angular distribution $W(\theta)$.

Niv et al (1979) applied the technique to calcium and found the asymmetries (0.06) to be larger than 0.02. In Table 1 we summarize all the asymmetry results for these tilted-foil measurements (Goldring 1981).

The Multifoil Technique

The general ideas of the multi-tilted-foil method were tested in a Coulomb excitation experiment using the 2_1^+ level of ^{110}Pd and ^{150}Sm at recoil velocities of $v/c = 0.03$ (Dafni et al 1981). The multifoil concept is technically realized by inserting a stack of thin (5–10 μg cm^{-2}) carbon foils in between the target and stopper foils of a conventional plunger device used in standard time-differential measurements. Mechanical details of the multifoil assembly used in this stage of experiments are described by Dafni et al (1981). The immediate conclusion drawn from that experiment was that the measured asymmetry is indeed large and that the sensitivity and usefulness of the tilted-foil technique are appreciably increased by the introduction of multifoil assemblies.

Broude et al (1981) used the multifoil technique to measure a g-factor hitherto unknown; i.e. they measured the g-factor of the 8^+ level in ^{84}Sr reached in a (^{12}C,4n) reaction. The g-factor of the 8_1^+, $\tau = 626$ ns isomeric level in ^{86}Sr was measured before to be $g = -0.241(15)$, with the time-dependent perturbed angular distribution (TDPAD) technique, which confirms the ($\nu g_{9/2}^{-2}$) nature of this level. In the ^{84}Sr nucleus, with four neutrons away from the closed $N = 50$ shell, the 8^+ level has a mean-life of only 210 ps and a similar measurement of its g-factor is impossible; it is, however, well suited for the multifoil technique. Such a measurement can distinguish between a single-particle nature and the collective rotational pattern found in the lighter Sr isotopes. By setting the target-stopper distance at $T \geq 50$ ps, one can observe the γ rays only from the 8^+ level decay without contaminating precessions from the short-lived feeding states.

Measurements were carried out with stacks of 8 foils at distances equivalent to interaction times of 50 ps and 100 ps, and for 50 ps with no carbon foils. The results for the various transitions depopulating the 8^+ level yield $g(8^+) < 0$ to better than six standard deviations and thus confirm the neutron single-particle nature of this level. Comparing the ^{84}Sr precession to attenuation and precession data for ^{80}Se (2^+) and ^{77}Se ($5/2^+$) in a time-integral measurement at similar recoil velocities, one can determine the parameters $\langle H_J \rangle$ and $\langle P_J \rangle$ and derive the

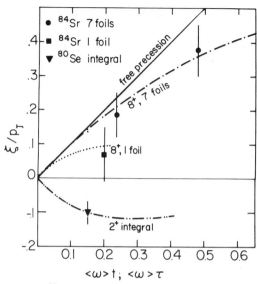

Figure 19 Results of the ^{84}Sr multifoil experiment (Broude et al 1981), in terms of the parameter ξ/p (roughly proportional to the measured asymmetry) for the ^{84}Sr (8_1^+) and ^{80}Se (2_1^+) levels. The improvement in the sensitivity of the measurement and the approach to a classical precession regime achieved by the introduction of several foils are apparent. The result for ^{80}Se (2_2^+) corresponds to a positive g-factor.

magnitude of the g-factor of the 8^+ level in ^{84}Sr. Figure 19 presents the results in terms of the parameter ξ defined by Dafni et al (1981). The lines are drawn using a model calculation with $J = 5/2$ and the best fit value of $g(^{84}$Sr $[8^+]) = -0.15(7)$. The error on the magnitude is substantially increased compared to the error on the sign; nevertheless it is evident that the result is close to the g-factor value of the 8^+ level in ^{86}Sr, which indicates a similar $(\nu g_{9/2})^{-n}$ configuration.

CONCLUSIONS

The applications of beam-foil spectroscopy have expanded greatly in the last several years. We have only been able to give a brief outline here of recent developments, particularly in atomic physics and the use of lifetime and wavelength measurements in testing basic relativistic atomic theory. These measurements are also relevant to the testing of magnetic and inertial fusion devices where diagnosis techniques for determining the plasma parameters often depend on the properties of highly stripped ions. The new developments of multicharged ion sources, for example, of completely stripped neon and argon, should lead to atomic spectroscopy of much more highly stripped heavy ions than so far possible. The related technique of bombarding a gas target with highly stripped fast heavy ions may make feasible improved precision in measuring the wavelengths of slow but highly stripped recoil ions (Beyer et al 1980). The production of alignment in beam-foil excitation has been treated in more detail, especially in its applications to nuclear magnetic moment measurements. This new area is developing rapidly with the promising multifoil technique.

ACKNOWLEDGMENT

One of the authors (H. G. B.) wishes to thank members of the Laboratory of Aimé Cotton, Orsay, for their kind hospitality during a stay there when most of this review was written.

Literature Cited

Andrä, H. J. 1970. *Phys. Rev. Lett.* 25:325
Andrä, H. J. 1974, *Phys. Scr.* 9:257
Andrä, H. J. 1975, *Phys. Lett.* 54A:315
Andrä, H. J., Plöhn, H. J., Gaupp, A., Fröhling, R. 1977. *Z. Phys. A* 281:15
Andrä, H. J., Fröhling, R., Plöhn, H. J., Silver, J. D. 1976. *Phys. Rev. Lett.* 37:212
Andrä, H. J., Fröhling, R., Plöhn, H. J., Winter, H., Wittmann, W. 1979. *J. Phys. C* 1:275
Bacis, R., Berry, H. G. 1974. *Phys. Rev. A* 10:466
Bakken, G. S., Jordan J. A. Jr., 1970. *Nucl. Instrum. Methods* 90:181
Band, Y. 1975, *Phys. Rev. Lett.* 35:1272
Band, Y. 1976. *Phys. Rev. A* 13:2061
Bashkin, S. 1964. *Nucl. Instrum. Meth* ʰs 28:80
Bashkin, S., ed. 1968. *Beam Foil Spectroscopy*, Vols. 1, 20 New York Gordon & Breach
Bashkin, S., ed. 1973. 3rd Int. BFS Conf., Tucson. *Nucl. Instrum. Methods* 110
Bashkin, S., ed. 1976. *Beam Foil Spectroscopy*. Berlin:Springer
Bashkin, S. 1979. *J. Phys. C* 1:125
Baudinet-Robinet, Y., Dumont, P. D., Garnir, H. P., Biémont, E., Grevesse, N. 1979. *J. Phys. C* 1:175
Benczer-Koller, N., Hass, M., Sak, J. 1980. *Ann. Rev. Nucl. Part. Sci.* 30:53
Bergkvist, K. E. 1976. *J. Opt. Soc. Am.* 66:837
Berry, H. G. 1975. *Phys. Scr.* 12:5
Berry, H. G. 1977. *Rep. Prog. Phys.* 40:155
Berry, H. G., Curtis, L. J., Ellis, D. G., Schectman, R. M. 1974a. *Phys. Rev. Lett.* 32:751

Berry, H. G., Gemmell, D. S., Holland, R. E., Poizat, J. C., Remillieux, J., Worthington, J. N. 1974b *Phys. Lett.* 49A:123
Berry, H. G. Curtis, L. J., Ellis, D. G., Schectman, R. M. 1975. *Phys. Rev. Lett.* 35:274
Berry, H. G., Gabrielse, G., Livingston, A. E. 1977a. *Appl. Opt.* 16:3200
Berry, H. G., Gabrielse, G., Livingston, A. E. 1977b. *Phys. Rev. A* 16:1915
Berry, H. G., Gabrielse, G., Livingston, A. E., Schectman, R. M., Désesquelles, J. 1977c. *Phys. Rev. Lett.* 38:1473
Berry, H. G., De Serio, R., Livingston, A. E. 1980a. *Phys. Rev. A* 22:998
Berry, H. G., Cheng, K. T., Johnson, W. R., Kim, Y. K. 1980b. *Proc. Workshop Found. Relativistic Theory Atom. Struct.*, ANL-80-126. Argonne, Natl. Lab., Ill.
Berry, H. G., Brooks, R. L., Cheng, K. T., Hardis, J. E., Ray, W. 1982. *Phys. Scr.*225:391
Betz, H. D. 1972. *Rev. Mod. Phys.* 44:465
Beyer, H. F., Schartner, K. H., Folkmann, F. 1980. *J. Phys. B* 13:2459
Bickel, W. S., Bergström, I., Buchta, R., Lundin, L., Martinson, I. 1969. *Phys. Rev.* 178:118
Bohr, N. 1948. *Mat. Fys. Medd. Dan. Vid. Selsk.* 18:8
Brooks, R. L., Pinnington, E. H. 1978. *Phys. Rev. A* 18:1454
Brooks, R. L., Pinnington, E. H. 1980. *Phys. Rev. A* 22:529
Brooks, R. L., Hardis, J. E., Berry, H. G., Curtis, L. J., Cheng, K. T., Ray. W. J. 1980. *Phys. Rev. Lett.* 45:1318
Brooks, R. L., Berry, H. G., Pinnington, E. H. 1982. *Phys. Rev. A* 25:161–68

Broude, C., Dafni, E., Goldring, G., Goldberg, M. B., Hass, M., Zemel, A. 1981. *Phys. Lett.* 105B:119

Buchet, J. P., Denis, A., Désesquelles, J., Dufay, M. 1969. *Phys. Lett.* 28A:529

Buchet, J. P., Buchet-Poulizac, M. C., Denis, A., Désesquelles, J., Druetta, M., Grandin, J. P., Husson, X. 1981. *Phys. Rev. A* 23:3354

Bunge, C. F. 1982. See Knystautas & Drouin 1982

Bunge, C. F., Bunge, A. V. 1978. *Phys. Rev. A* 17:816

Bunge, C. F., Galan, M., Jauregui, R. 1982. See Knystautas & Drouin 1982

Burgdörfer, J., Gabriel, H. 1979. *J. Phys. C* 1:315

Burns, D. J., Hight, R. D., Greene, C. H. 1979. *Phys. Rev. A* 20:1468

Camhy-Val, C., Dumont, A. M., Dreux, M., Vitry, R. 1970. *Phys. Lett.* 32A:233

Ceyzeriat, P., Buchet, J. P., Buchet-Poulizac, M. C., Désesquelles, J., Druetta, M. 1979. *J. Phys. C* 1:171

Chateau-Thierry, A., Gladieux, A., Delaunay, B. 1976. *Nucl. Instrum. Methods* 132:553

Cheng, K. T., Kim, Y. K. 1978. *At. Data Nucl. Data Tables* 22:547

Cheng, K. T., Desclaux, J. P., Kim, Y. K. 1978. *J. Phys. B* 11:L359

Church, D. A., Kolbe, W., Michel, M. C., Hadeishi, T. 1974. *Phys. Rev. Lett.* 33:565

Church, D. A., Michel, M. C., Kolbe, W. 1975. *Phys. Rev. Lett.* 34:1140

Curtis, L. J. 1974. *J. Opt. Soc. Am.* 64:495

Curtis, L. J. 1976. *See* Bashkin 1976, p. 63

Curtis, L. J. 1982. See Knystautas & Drouin 1982

Curtis, L. J., Berry, H. G., Bromander, J. 1971. *Phys. Lett.* 34A:169

Dafni, E., Goldring, G., Hass, M., Kistner, O. C., Niv, Y., Zemel, A. 1981. *Phys. Rev. C.* In press

Datz, S., Moak, C. D., Crawford, O. H., Krause, H. F., Dittner, P. F., Gomez del Campo, J., Biggerstaff J. A., Miller, P. D., Hvelplund, P., Knudsen, H. 1978. *Phys. Rev. Lett.* 40:843

Desclaux, J. P. 1979. *J. Phys. C* 1:109

De Serio, R., Berry, H. G., Brooks, R. L., Hardis, J., Livingston, A. E., Hinterlong, S. 1981. *Phys. Rev. A* 24:1872

Deslattes, R. D. 1980. See Berry et al 1980b, p. 12

Désesquelles, J., ed. 1979. 5th Int. BFS Conf., Lyon. *J. Phys. C* 40:Suppl. 1

Dietrich, D. D., Leavitt, J. A. 1979. *J. Phys. C* 1:169

Drawin, H. W. 1981. *Phys. Scr.* 24:622

Dufay, M. 1970. *Nucl. Instrum. Methods* 90:15

Eck, T. G. 1973. *Phys. Rev. Lett.* 31:270

Eck, T. G. 1974. *Phys. Rev. Lett.* 33:1055

Edlén, B. 1979. *Phys. Scr.* 19:255

Ellis, D. G. 1973. *J. Opt. Soc. Am.* 63:1322

Ellis, D. G. 1977. *J. Phys. B* 10:2301

Fano, U., Macek, J. H. 1973. *Rev. Mod. Phys.* 45:553

Froese-Fischer, C. 1977. *J. Phys. B* 10:1241

Gabrielse, G. 1980. *Phys. Rev. A* 22:138

Gabrielse, G. 1981. *Phys. Rev. A* 23:775

Gabrielse, G., Band, Y. 1977. *Phys. Rev. Lett.* 39:697

Gaillard, M. L., Carré, M., Berry, H. G., Lombardi, M. 1973. *Nucl. Instrum. Methods* 110:273

Gaillard, M. J., Poizat, J. C., Ratkowski, A., Remillieux, J., Auzas, M. 1977. *Phys. Rev. A* 16:2323

Gaupp, A., Andrä, H. J., Macek, J. 1974. *Phys. Rev. Lett.* 32:268

Gay, T. J., Berry, H. G., De Serio, R., Garnir, H. P., Schectman, R. M., Schaffel, N., Hight, R. D., Burns, D. J. 1981. *Phys. Rev. A* 23:1745

Gladieux, A., Chateau-Thierry, A. 1981. *Phys. Rev. Lett.* 47:786

Goldring, G. 1981. In *Heavy Ion Collisions*, ed., R. Bock. Amsterdam: North Holland. In press

Goldring, G., Hass, M. 1982. In *Heavy Ion Science*, ed., D. A. Bromley. New York: Plenum. In press

Goldring, G., Niv, Y., Wolfson, Y., Zemel, A. 1977. *Phys. Rev. Lett.* 38:221

Gould, H., Marrus, R. 1978. *Phys. Rev. Lett.* 41:1457

Harrison, K. G., Lucas, M. W. 1970. *Phys. Lett.* 33A:142

Hass, M., Brennan, J. M., King, H. T., Saylor, T. K., Kalish, R. 1977. *Phys. Rev. Lett.* 38:218

Herman, R. 1975. *Phys. Rev. Lett.* 35:1626

Hight, R. D., Schectman, R. M., Berry, H. G., Gabrielse, G., Gay, T. J. 1977. *Phys. Rev. A* 16:1805

Kanter, E. P., Cooney, P. J., Gemmell, D. S., Groeneveld, K. O., Pietsch W. J., Ratkowski, A. J., Vager, Z., Zabransky, B. J. 1979. *Phys. Rev. A* 20:834

Kay, L. 1963. *Phys. Lett.* 5:36

Kim, Y. K., Cheng, K. T. 1978. *J. Opt. Soc. Am.* 68:836

Knystautas, E., Drouin, R., eds. 1982., 6th Int. BFS Conf., Quebec. *Nucl. Instrum. Methods.* In press

Lewis, E. J., Silver, J. D. 1975. *J. Phys. B* 8:2697

Livingston, A. E., Berry, H. G. 1978. *Phys. Rev. A* 17:1966

Livingston, A. E., Curtis, L. J., Schectman, R. M., Berry, H. G. 1980. *Phys. Rev. A* 21:771
Liu, C. H., Bashkin, S., Bickel, W. S., Hadeishi, T. 1971. *Phys. Rev. Lett.* 26:222
Liu, C. H., Bashkin, S., Church, D. A. 1974. *Phys. Rev. Lett.* 33:993
Lombardi, M. 1975. *Phys. Rev. Lett.* 35:1172
Luc-Koenig, E. 1979. *J. Phys. C* 1:115
Lundeen, S. R., Pipkin, F. M. 1981. *Phys. Rev. Lett.* 46:232
Mannervik, S., Astner, G., Kisielinski, M. 1980. *J. Phys. B* 14:L441
Martinson, I. 1981. *Phys. Scr.* 23:126
Martinson, I., Bromander, J., Berry, H. G., eds., 1970. Proc. 2nd Int. BFS Conf., Lysekil. *Nucl. Instrum. Methods* 90
Masterson, K. D., Stoner, J. O. 1973. *Nucl. Instrum Methods* 110:441
Mohr, P. J. 1975. *Phys. Rev. Lett.* 34:1050
Niv, Y., Hass, M., Zemel, A., Goldring, G. 1979. *Phys. Rev. Lett.* 43:326
Niv, Y., Hass, M., Zemel, A. 1981. *Hyperfine Interact.* 9:181
Nussbaumer, H. 1979. *J. Phys. C* 1:102
Okorokov, V. V. 1965. *J. É. T. P. Lett.* 2:111
Pedrazzini, G. J., Christiansen, P. G., Ross, J. E., Gardiner, R. B., Liu, C. H. 1978. *J. Phys. B* 11:3939
Percival, I. C., Seaton, M. J. 1958. *Philos. Trans. R. Soc. A* 251:113
Pinnington, E. H., Gosselin, R. N. 1979. *J. Phys. C* 1:149
Pinnington, E. H., Livingston, A. E., Kernahan, J. A. 1974. *Phys. Rev. A* 9:1004
Pinnington, E. H., Gosselin, R. N., O'Neill, J. A., Kernahan, J. A., Donnelly, K. E., Brooks, R. L. 1979. *Phys. Scr.* 20:151
Rødbro, M., Bruch R., Bisgaard, P. 1979. *J. Phys. B* 12:2413
Schectman, R. M., Hight, R. D., Chen, S. T., Curtis, L. J., Berry, H. G., Gay, T. J., De Serio, R. 1980. *Phys. Rev. A* 22:1591
Schneider, D., Bruch, R., Butscher, W., Schwarz, W. H. E. 1981. *Phys. Rev. A* 24:1223
Schröder, H., Kupfer, E. 1976. *Z. Phys. A* 249:13
Sellin. I. A. 1979. *J. Phys. C* 1:225
Sellin, I. A., Pegg, D. J., eds. 1976. 4th Int. BFS Conf., Gatlinburg, *Beam-Foil Spectroscopy*, Vols. 1, 2. New York: Plenum
Sellin, I. A., Pegg, D. J., Griffin, P. M., Smith, W. W. 1972. *Phys. Rev. Lett.* 28:1229
Sellin, I. A., Mowat, J. R., Peterson, R. S., Griffin, P. M., Lambert, R., Haselton, H. H. 1973. *Phys. Rev. Lett.* 31:1335
Stokes, G. G. 1852. *Trans. Cambridge Philos. Soc.* 9:399
Stoner, J. O. Jr., Leavitt, J. A. 1971. *Appl. Phys. Lett.* 18:477
Stoner, J. O. Jr., Leavitt, J. A. 1973. *Opt. Acta* 20:435
Tolk, N. H., Tully, J. C., Heiland, W., White, C. W., eds. 1977. *Inelastic Ion-Surface Collisions*, New York: Academic
Tolk, N. H., Tully, J. C., Krans, J. S., Heiland, W., Neff, S. H. 1978. *Phys. Rev. Lett.* 41:643
Tolk, N. H., Tully, J. C., Krans, J. S., Heiland, W., Neff, S. H. 1979. *Phys. Rev. Lett.* 42:1475
Tolk, N. H., Feldman, L. C., Kraus, J. S., Tully, J. C., Hass, M., Niv, Y., Temmer, G. M. 1981. *Phys. Rev. Lett.* 47:487
Träbert, E., Heckmann, P. H. 1980. *Phys. Scr.* 22:489
Vager, Z., Gemmell, D. S. 1976. *Phys. Rev. Lett.* 37:1352
Vager, Z. 1981. *Ann. Isr. Phys. Soc.* 4:139
Vager, Z., Zabransky, B. J., Schneider, D., Kanter, E. P., Zhuang, G. Y., Gemmell, D. S. 1982, *Phys. Rev. Lett.* 48:588
Wangler, J., Henker, L., Wittmann, W., Plöhn, H. J., Andrä, H. J. 1981. *Z. Phys.* 299:23
Weiss, A. W. 1977. *J. Quant. Spectrosc. Radiat. Transfer* 18:481
Wieman, C., Hänsch, T. W. 1980. *Phys. Rev. A* 22:192
Winter, H. 1979. *J. Phys. C* 1:307
Younger, S. M., Wiese, W. L. 1978. *Phys. Rev. A* 17:1944, 2366

Ann. Rev. Nucl. Part. Sci. 1982. 32:35–64
Copyright © 1982 by Annual Reviews Inc. All rights reserved

STATISTICAL SPECTROSCOPY[1]

J. B. French and V. K. B. Kota[2]

Department of Physics and Astronomy,
University of Rochester, Rochester, New York 14627

CONTENTS

1. SPECTROSCOPY AND STATISTICAL MECHANICS

Two landmarks in statistical spectroscopy are Bethe's (1936) derivation
of the nuclear-level density, and Wigner's introduction (1955) of
Hamiltonian matrix ensembles in order to study, among other things,
some features of the spectral (energy-level and strength) fluctuations.
These complementary developments, whose results are of permanent
value, are remarkable in their economy, relying as they do on general
arguments and making use of relatively little input information. For the
level-density the number of active particles, the angular momenta, and
the single-particle energies are taken lightly into account, though the

[1] Supported in part by the US Department of Energy.
[2] On leave from Physical Research Laboratory, Ahmedabad 380009, India.

35

0163-8998/82/1201-0035$02.00

interactions are ignored (except implicitly as they contribute to the single-particle energies); combinatorial statistical mechanics generates the level density; and the central limit theorem (CLT) decomposes it by angular momentum. In Wigner's theory the states are treated as abstract vectors, and the Hamiltonian as an ensemble of H matrices in which no attention is paid to any specific features of H except that it should be time-reversal invariant (TRI), the ensembles being then of real symmetric matrices. There is an implicit assumption of ergodic behavior so that the results of ensemble averaging should apply to individual spectra.

It is usually taken for granted that statistical calculations like these could only be valid at high excitation energies, say above the neutron threshold. Detailed conventional spectroscopy on the other hand could only be valid at relatively low energies [even though the model Hamiltonian might have a wide spectrum span, e.g. ~ 100 MeV for $(ds)^{12}$]. This is because only a small segment of the single-particle spectrum can be dealt with (5 MeV in the ds shell). The concerns in these calculations are with solutions of the equations of motion in the restricted model space defined by the single-particle orbits and the applicable exact symmetries — either exact solutions, as with shell-model calculations, or approximate ones as with random phase approximation (RPA), Hartree-Fock (HF) calculations, etc. The model spaces may be of fairly large dimensionality (currently up to a few thousand), the Hamiltonians used are specified in detail, the calculations usually involve the construction of many-particle wave functions, and completely detailed answers are given to questions asked. On comparing the two procedures, one is led to wonder whether some essential aspects of nuclear structure have been missed in the statistical calculations, whose simplicity might then be deceptive. Or have some general principles, which would justify the statistical procedure, been ignored in the spectroscopic, whose complexity is then unreasonable? Or is it simply that one must use one procedure at high energies and the other at low?

The subject of statistical spectroscopy considers these questions. It works in the model spaces of conventional spectroscopy but, since it does not deal directly with equations of motion, it is not limited in the dimensionality of these spaces. For the smoothed level density it uses statistical methods based on spectral averaging (i.e. averaging along the spectrum of the given model Hamiltonian); for the fluctuations, a two-point correlation function and its corresponding fluctuation measures derive from an averaging over a Hamiltonian ensemble closely related to Wigner's ensemble. We thus make contacts with both calculations described above. One significant difference, however, is that the combinatorial feature of Bethe's calculation, which works even for unbounded single-particle spectra, has no real counterpart for interacting particles; hence we bound the spectrum, as in conventional spectroscopy, use the CLT for both the level-density and its symmetry decompositions, and then extend the spectrum by a method of partitioning. All of this works because:

1. There is, in the model space, a *macroscopic simplicity* deriving from the action of the CLT. The smoothed eigenvalue density is close to a characteristic form, usually Gaussian, describable therefore in terms of a small number of low-order Hamiltonian traces (moments).

2. There is a *microscopic simplicity* corresponding to a remarkable spectral rigidity (Dyson & Mehta 1963), which extends over the whole spectrum and ensures that the fluctuations are small and for the most part carry little information.

3. There is indeed, as implied above, usually a *sharp separation* between the secular behavior of the spectrum and its fluctuations, so that the two can be treated separately and by different methods. This separation also arises from CLT action.

4. There is a *propagation of information* (i.e. of traces) throughout the set of model subspaces defined by N (the number of single-particle states) and the symmetries that label the subspaces. This enables us to express either exactly or approximately, depending upon the symmetries involved, the many-particle traces as linear combinations of the few-particle "input" traces.

5. The ensembles that one uses have a *strong ergodic behavior.*

Just as with the partition function in conventional statistical mechanics, parametric differentiation of the level density, given as a function of the system parameters, yields information about other quantities (transition strengths and electromagnetic moments, for example) because the two functions are related by Laplace transformation. The CLT arises from the direct-product nature of the model spaces. If we could ignore the interactions (and the much less important Pauli blocking effects), the energies would be additive and the m-particle eigenvalue density an m-fold convolution of the one-particle density, which would rapidly become Gaussian as m increases; $m \sim 6$ would give a decent Gaussian. No such simple argument applies for interacting particles. But Gaussian spectra are found by averaging over an appropriate H ensemble, as well as by direct evaluation of the third and fourth cumulants of "realistic" model Hs; and of course they are found also in shell-model calculations.

The spectral rigidity may be regarded as representing an extended version of the von Neumann–Wigner (1929) level repulsion, which, in its ensemble version, may be demonstrated even for two-dimensional matrices. Thus the eigenvalue spacing S for a Hermitian matrix H with $H_{11} = A$, $H_{22} = B$, and $H_{12} = C + iD$ is

$$S = E_> - E_< = [(A - B)^2 + 4(C^2 + D^2)]^{1/2}. \qquad\qquad 1.$$

Repulsion arises because generating an ensemble by choosing A, B, C, and D independently generally requires, for a zero spacing, the simultaneous vanishing of three quantities. If H is real symmetric $(D = 0)$, we need only a double vanishing so that level-repulsion effects, and spectral rigidity in general, should be larger for non-time-reversal-invariant (NTRI) interactions than for TRI.

The economy in the level-density calculations (in their extended form, with interactions) arises as always from the action of a CLT. As we continue to add particles the effective convolution represented by the CLT (there is not a true convolution for interacting particles) generates an eigenvalue density that becomes steadily smoother; it is described therefore by low-order Hamiltonian moments. The information not carried by these moments is filtered away. The smoothing does not extend directly to the fluctuations because the level spacing is simultaneously decreasing and the fluctuations, which then examine things on a smaller and smaller scale, escape the direct action of the CLT. It is this that gives the decoupling between the density and fluctuations called for above. It implies also that the process of spectral averaging is rapidly convergent only "to within fluctuations." Finally it should be clear that, to the extent that the experimental data about fluctuations agree well with the results given by Wigner's ensemble, they carry no information except about time-reversal invariance.

We can hardly expect that contemporary problems can be handled in ways as economical as those of Bethe and Wigner (and the whole subject would be far less interesting if they could be). In any spectroscopy, statistical or otherwise, bounding the single-particle spectrum at ε_N, automatically introduces the group U(N) of unitary transformations among the single-particle states; this follows from the fact that the $\binom{N}{m}$ states for m fermions generate an irreducible representation of U(N), its column structure being $[m]$. But at the same time there enters the whole array of U(N) subgroups, some of which represent the projection into the model space of the "exact" symmetries (J, T) while others are fairly good model symmetries connected, for example, with rotational and pairing effects. Beyond these there are badly broken symmetries, not necessarily interesting in themselves but often useful in statistical spectroscopy. We remark also that the subject has the capacity to respond to all of the "spectroscopic" questions about nuclear structure (those that one could imagine answering via some immensely extended shell model); but the responses to some of them (especially some pertaining to the ground-state region) may be inadequate. Those questions would call for a synthesis of statistical and nonstatistical methods.

The next section we devote to a simple treatment of spectrally averaged phenomena, giving some elementary applications to binding energies, single-state occupancies, spin cut-off factors, and so forth. In Section 3 we pay more attention to phenomena arising from the finer [U(N) subgroup] partitioning of the model space. In Section 4 we deal with energy-level fluctuations, paying special attention to the comparison of theory and experiment and the implications of the agreement that is found. We also describe connections between fluctuations and other phenomena.

Our purpose in this article, which may serve as an introduction to more extensive reviews, is to make clear the principles and, without technical detail, something of the methods, results, possibilities, and limitations of statistical spectroscopy.

2. SIMPLE AVERAGES AND THEIR APPLICATIONS

2.1 *Information and its Simple Propagation*

We have an $\binom{N}{m}$-dimensional model space formed by distributing m particles over N single-particle states corresponding to some set of orbits, spherical or otherwise. The choice of orbits and the Hamiltonian H that acts in the space is dictated, as in conventional spectroscopy, by experience, the phenomena of interest, the complexity that can be tolerated, and (for H) theoretical treatments of the nucleon-nucleon interaction. We can, however, anticipate dealing by statistical methods with indefinitely large model spaces.

It is easy to produce a many-particle spectrum that is far from Gaussian; $Q \cdot Q$ for example, with Q the quadrupole operator, gives in large spaces a χ^2_5 spectrum. Nonetheless detailed calculations with reasonable Hs give good to excellent Gaussian spectra. Leaving an explanation of this until later, let us accept for the present that our spectra are Gaussian (thereby incidentally ignoring corrections to Gaussian, which in practice we would take account of). Thus we need only calculate the spectral centroid $\mathscr{E}(m)$ (the average of the eigenvalues) and variance $\sigma^2(m)$. An essential result for the trace of a k-body operator $G(k)$ over the m-particle states is that

$$\langle G(k)\rangle^m = \binom{m}{k}\langle G(k)\rangle^k; \qquad \langle\langle G(k)\rangle\rangle^m = \binom{N-k}{N-m}\langle\langle G(k)\rangle\rangle^k. \qquad 2.$$

Here we have written the trace as $\langle\langle G(k)\rangle\rangle^m$, and its renormalized version (the average eigenvalue) as $\langle G(k)\rangle^m [= d^{-1}(m)\langle\langle G(k)\rangle\rangle^m$, where $d(m)$ is the dimensionality of the space; in the present case $d(m) = \binom{N}{m}]$. This is a simple example of the propagation of a piece of information, a trace, from a space in which the operator is defined to a more complicated one. Equation 2 can be easily derived, for example, by a combinatorial argument. It is more useful for us to recall that with fermions the only independent U(N) scalar is the number operator n, so that $\binom{n}{k}$, which obviously has maximum "particle rank" k (a k-body operator has rank k) but vanishes in spaces with fewer than k particles, is the only independent unitary-scalar k-body operator. The result (Equation 2), which for $m = k$ must be an identity, then follows from the scalar nature of the trace operation, under which only the unitary-scalar part of G survives.

Since, for example, H^2 has mixed particle ranks, ≤ 4 for a two-body H, we should extend Equation 2 to operators G with *maximum* rank u. But then $\langle G\rangle^m$ is a polynomial of order u in m, expressible therefore in terms of its values on any set of $(u+1)$ m values, say $m = t_i$, $i = 1, 2, \ldots, (u+1)$. Thus

$$\langle G\rangle^m = \sum_{j=0}^{u} \prod_{i \neq j} \frac{m - t_i}{t_j - t_i} \langle G\rangle^{t_j} \qquad 3.$$

where, to simplify the calculation of the input traces, we choose the t_j subspaces to be as simple as possible. For $H^2(2)$, the square of a two-body operator, we could take $t = 0, 1, 2, N-1, N$ and then the spectral variance for a $(0+1+2)$-body H is

$$\sigma^2(m) = \binom{m}{2}\binom{N-m}{2}\binom{N-2}{2}^{-1}\left[\sigma^2(2) - \frac{2(N-3)(m-2)}{(N-1)(m-1)}\sigma^2(1)\right.$$
$$\left. + \frac{2(m-2)}{(N-1)(N-m-1)}\sigma^2(N-1)\right]. \qquad 4.$$

The last two terms in Equation 4 vanish if the single-particle states as well as the single-hole states are degenerate (as they would be if, for example, the model space were generated by a single spherical orbit). Note that we have gone from the second moment to the second cumulant by invoking the fact that Equation 3 applies equally well to any homogenous expression of order u in the moments of H.

An alternative procedure whose ingredients are more physically significant and that, unlike Equation 3, has a natural extension to U(N) subgroups follows from the further decomposition of $G(k)$ according to the irreducible representations (irreps) of U(N) . For this we should recognize that since $\psi(k)$ belongs to $[k]$, its adjoint belongs to $[N - k]$, so that $G(k) \sim \psi(k)\psi^\dagger(k)$ belongs to a set of one- or two-columned irreps $[N - \nu, \nu]$ where the "unitary rank" $\nu = 0, 1, \ldots, k$. Since the νth irrep can also be generated in a ν-body space, a little thought will show that (Vincent 1967, Chang et al 1971)

$$G(k) = \sum_{\nu=0}^{k} G^\nu(k) = \sum \binom{n - \nu}{k - \nu} \mathcal{G}^\nu(\nu), \qquad 5.$$

where the $\mathcal{G}^\nu(\nu) \equiv \mathcal{G}^\nu$, a ν-body operator of unitary rank ν that derives from $G(k)$, would vanish on further contractions. $\mathcal{G}^0$, which identifies the fully contracted form of $G(k)$, has been given above as $\langle G(k)\rangle^k$. For the interaction Hamiltonian, $H = \sum \varepsilon_i n_i - \frac{1}{4}\sum W_{ijkl} a_i^\dagger a_j^\dagger a_k a_l$, the $\nu = 1$ part is

$$H^{\nu=1} = \sum_i \xi_i n_i \qquad 6.$$

$$= \sum_i \left\{ \xi_i(1) + (n - 1)(N - 2)^{-1}\left[\sum_j W_{ijij} - N^{-1}\sum_{jk} W_{jkjk}\right]\right\} n_i,$$

where $\xi_i(1) = \varepsilon_i - N^{-1}\sum \varepsilon_j$ is the centered single-particle energy and ξ_i is then its renormalized equivalent. (The difference between the two is the Hartree-Fock-like "induced" single-particle energy.) Since the fermion anticommutation rules are invariant under the hole $\leftrightarrow$ particle transformation $a_\mu \leftrightarrow a_\mu^\dagger$ ("holes are also fermions"), it follows that a hermitian contracted operator $\mathcal{G}^\nu$ must transform into itself under this operation. This is so since the commutator terms that would normally appear must necessarily vanish because they belong to a lower unitary rank. The precise result is $\mathcal{G}^\nu \leftrightarrow (-1)^{\nu(\nu-1)/2}\mathcal{G}^\nu$ where the phase derives from the commutation of two sets of ν fermions each. But now, since the product of two tensors of different rank cannot contain a unitary scalar, we have

$$\langle F(k')G(k)\rangle^m = \sum_{\nu=0}^{k_<} \binom{m - \nu}{k' - \nu}\binom{m - \nu}{k - \nu}\binom{N - m}{\nu}\binom{N - \nu}{\nu}^{-1}\binom{m}{\nu}$$

$$\times \langle \mathcal{F}^\nu(\nu)\mathcal{G}^\nu(\nu)\rangle^\nu \qquad 7.$$

where we have used Equations 3 and 5 with the set ($t = 0, \ldots, \nu$; $N - \nu + 1, \ldots, N$), only a single member of which, for each ν, can contribute. Note in particular that, just as the *linear* trace of a k-body operator propagates in a simple way by Equation 2 (or by Equation 7 if we take $F(k') = 1$), so does the *quadratic* trace of a unitarily irreducible operator.

Here we have simple examples of the propagation of information throughout the set of model spaces, the m-particle traces being linear combinations of those in simpler spaces. This yields a great, indeed essential, simplification in the calculations. More important, when combined with the CLT it describes a filtering away of most of the information carried by the operators and gives the statistical "economy" that we have been looking for. As discussed in the next section, the relationship between symmetry and statistical behavior, displayed especially by Equation 7 when coupled with the action of the CLT, is fundamental. Also, the simple result of Equation 2 gives the m-particle trace as a two-particle trace amplified by the weight with which the two-particle space is contained in the m-particle one. This pretty picture of a complicated process also applies in much more general circumstances.

2.2 *Model-Space Geometry, Norms, and Correlations*

Just as in more conventional statistical mechanics with the partition function, we can calculate essentially all further quantities of interest by parametric differentiation on the eigenvalue density. Thus first-order perturbation theory gives for $K(E)$, the expectation value of an operator K in the Hamiltonian eigenstate $\Psi(E)$, assumed nondegenerate for an inessential simplicity,

$$K(E) \equiv [\Psi(E) K \Psi(E)] = \left\{ \frac{\partial E_\alpha}{\partial \alpha} \right\}_{\alpha=0} \qquad 8.$$

where E_α ($\xrightarrow{\alpha \to 0} E$) is an eigenvalue of the "deformed" Hamiltonian $H_\alpha = H + \alpha K$. Extending this to a parametric derivative of the eigenvalue density (rather than of the eigenvalue itself) gives the energy distribution of the expectation value.

It is standard to think of a probability density in terms of its *location* (fixed by the centroid), *scale* (by the translation-invariant variance), and *shape* (by a set of translation- and scale-invariant parameters, such as the cumulants of order $\nu \geq 3$, or the $\nu \geq 3$ polynomial moments defined by a reference shape). For our spectra, discrete with fluctuations, it is better to describe the higher-order parameters (labeling those "excitations" of the reference spectrum which have wavelength $\sim$ spacing) as *fluctuation* parameters, and agree then that "shape" implies "smoothed to within fluctuations." That being understood, let us assume that the level density, which we write as $I(x) = d \times \rho(x)$ so that $\int \rho(x)\, \mathrm{d}x = 1$, has when smoothed a characteristic CLT shape that to first order in α is preserved under the deformation. Then, for the expectation value $K(E)$, we need consider only the consequences of the change in the centroid $\mathscr{E}$ and the variance σ^2, as $H \to H_\alpha$. The centroid variation obviously leads to a constant $K(E)$, and that of the variance to a linear E dependence. Formally, since

$$\frac{\partial \mathscr{E}_\alpha}{\partial \alpha} = \frac{\partial \langle H + \alpha K \rangle^m}{\partial \alpha} = \langle K \rangle^m; \quad \left\{ \frac{\partial \sigma_\alpha^2}{\partial \alpha} \right\}_{\alpha=0} = 2 \langle K(H - \mathscr{E}) \rangle^m, \qquad 9.$$

we have that

$$K(E) \xrightarrow{\ \text{CLT}\ } \langle K \rangle^m + \langle K(H - \mathscr{E}) \rangle^m (E - \mathscr{E}) / \sigma^2 \qquad 10.$$

so that expectation values are characteristically linear in the energy. Similar results apply also to the distribution of the "strengths," the squares of the matrix elements connecting two eigenstates via a transition operator (Draayer et al 1977).

The basic result displayed in Equation 10 has a geometrical interpretation, the geometry being that generated by assigning to an operator G the norm $\|G\| = \{ \langle (G - \langle G \rangle)^\dagger (G - \langle G \rangle) \rangle^m \}^{1/2}$. This is a proper norm, which moreover is compatible with the unitary norm for the states. Thus, weighed by this norm, sums and products of operators as well as the result of operating on a state are never unexpectedly large. Now, defining the (traceless) unit operator $g = \{G - \langle G \rangle\} / \|G\|$, we have

$$k(E) \xrightarrow{\ \text{CLT}\ } \frac{(\mathbf{k} \cdot \mathbf{h})(E - \mathscr{E})}{\sigma} \qquad 11.$$

where $(\mathbf{k} \cdot \mathbf{h})$ can be interpreted in geometrical terms (for Hermitian operators) as the cosine of the angle between the centered operators, or in statistical terms as the correlation coefficient between them.

The usefulness of the geometry introduced here depends crucially on the CLT convergence. Since, to within a dimensionality factor, $\|G\|^2 \sim \sum \lambda_G^2$ where the λ_G^2 are the eigenvalues of $G^\dagger G$ (a hermitian G has eigenvalues λ_G), we see that $\|G\|$ is a "democratic" norm, assigning equal weight to all eigenvalues. However, only a small part of a model space (especially an unpartitioned one) is of direct physical interest, and thus this choice of the norm seems a priori unreasonable. But the CLT convergence, since it fixes the low-lying eigenvalues in terms of a few global quantities, permits an extrapolation from the centroid (which typically lies very far above the states that are well treated) down into the physical domain, and thereby makes our geometry an effective one.

As indicated above $\zeta_{K-H} = (\mathbf{k} \cdot \mathbf{h})$, which is easily calculable by a simple extension of Equation 4, or by Equation 7, is a true correlation coefficient (formally between the matrix elements in any basis or, if the operators commute, between the eigenvalues). It therefore takes on values in the range $(-1, 1)$; the extremes indicate that then the operators are effectively multiples of each other. Thus, for a simple example, the fact that $Q \cdot Q$ is strongly anticorrelated with reasonable Hs $(\zeta \approx -0.5)$ immediately tells us that, to within fluctuations, its expectation value will be large near the ground state and decrease with slope $\zeta \|Q \cdot Q\|/\sigma$ as the excitation energy increases. There are many other uses of these correlation coefficients, some of which we discuss in the next subsection.

When we have a rapid convergence to a characteristic form it is good, as we have just done, to discuss the asymptotic limiting case (Equation 10). But we must also be able to calculate corrections. Since $\langle G \rangle^m = \int G(x)\rho(x)\,dx$ we see that $\rho(x) = \langle \delta(H-x) \rangle^m$ and then

$$K(x) = \rho^{-1}(x)\langle K\,\delta(H-x) \rangle^m = \sum_{\nu=0}^{d} \langle KP_\nu(H) \rangle^m P_\nu(x) \qquad 12.$$

where we expanded $\delta(H-x)$ in terms of the orthonormal polynomials defined by $\rho(x)$ as a weight function. The number of such polynomials is strictly d, the dimensionality, but we can just as well take $d = \infty$ and use continuous instead of lattice polynomials. Since $P_0(x) = 1$ and $P_1(x) = (x - \mathscr{E})/\sigma$, the first two terms of Equation 12 reproduce the CLT result. In practical cases it is good to take one or two terms beyond the CLT limit, these giving quadratic (and cubic) curvature corrections to the expectation value and, by their size, allowing us to assess the convergence rate.

2.3   Simple Applications

2.3.1   ENERGIES   Our spectral averaging has generated a smoothed level distribution. There is no way to undo the smoothing and recapture thereby the discrete spectrum E_r $(r = 1, 2, \ldots, d)$ but we can produce a plausible "fluctuation-free" spectrum $\hat{E}_r$ based on approximating the exact (staircase) distribution function $F(x) = \int_{-\infty}^{x} \rho(z)\,dz$ by a smooth curve passing through the center of each step. Then if we know the spectrum to be nondegenerate, its smoothed version follows (Ratcliff 1971, Langhoff 1980) by solving $F(\hat{E}_r) = [r - \frac{1}{2}]/d$, with a simple modification if the spectrum has known degeneracies. The low-lying energies, those satisfying $(\mathscr{E} - E_r) \gg \sigma$, in a large-$d$ Gaussian spectrum are given approximately by $\hat{E}_r \approx \mathscr{E} - \{2 \ln d/[r - \frac{1}{2}]\}^{1/2}\sigma$, and are easily calculated via the propagation equations. For model spaces that are not too large (say $d \leq 10^4$) this gives a quite useful binding-energy ($r = 1$) value, and of course its variation with particle number. For larger model spaces and for the low-lying spectrum itself we would need the extensions described in the next section, which give the binding energies of light nuclei with an error ~ 2–3 MeV (Chang & Zuker 1972, Lougheed & Wong 1975).

2.3.2 OCCUPANCIES AND OTHER EXPECTATION VALUES   Since expectation values have a polynomial rather than an exponential form, high accuracy, to within fluctuations, can be anticipated even in the ground-state region. The unitary decomposition, and Equation 7 that derives from it, break up the scalar product into a sum of scalar products of irreducible tensors. A particularly simple but important application of this is to the expectation value of one-body operators K. Since in the correlation trace $\langle K(H - \mathscr{E})\rangle^m$, $(H - \mathscr{E})$ has no $\nu = 0$ part (being traceless) and K has no $\nu = 2$ part (being one-body), the trace depends on $H^{\nu=1}$, given by Equation 6. Thus the expectation value in the CLT limit depends on the Hartree-Fock-like shifts of the one-body energies caused by the interaction, and may be used to determine them, for example from the occupancies derived from single-nucleon-transfer experiments; in that case K is the number operator for a single-particle state or orbit. (Potbhare & Pandya 1976, Kota & Potbhare 1979). By the same token one can derive results for the fluctuation-free one-body density matrix.

The J decomposition of the level density is determined by the CLT applied to the random variable J_z. It is expressed, as in Bethe's calculation, by the "spin-cut-off factor," which is the (locally averaged) expectation value of J_z^2. Because J_z^2 is only weakly correlated with H (typically $\zeta \sim -0.15$ in ds-shell nuclei) the spin cut-off factor varies only slowly with excitation energy and therefore, by Equation 7, has approximately the value $\frac{1}{3}m(N - m)(N - 1)^{-1}N^{-1}\sum_r(2j_r + 1)j_r(j_r + 1)$ where the sum is over the (spherical) orbits defining the model space. More precise results follow from Equations 12, or 13 below, as do forms for the isospin and spin-isospin cut-off parameters and the various level dimensionalities as well. And of course expectation values have many other uses not explicitly connected with symmetries or with statistical behavior: in calculating sum-rule quantities for transitions and excitations, in studying what parts of Hamiltonians are most effective at various energies, and so forth.

2.3.3 COMPARISON AND ANALYSIS OF HAMILTONIANS   Since the effective Hamiltonian in a given model space is by no means precisely known, one must ask to what extent and in what ways do various given forms for H agree or disagree with each other. For any two Hs the first procedure is to use Equation 7 to evaluate the norms and correlation coefficient as a function of particle number (and ideally of other exact quantum numbers as well). For example (Potbhare 1977) between two of the conventional "realistic" interactions BK-12.5 (Kuo and Brown 1965) and KLS-R (Kahana et al 1969) we find both in $(ds)^8$ and $(ds)^8_{T=0}$ that $\zeta = 0.99$ so that these interactions could differ significantly in these spaces only in scale; that could be checked from their norms. On the other hand, they differ considerably from the MSDI interaction (Arvieu & Moszkowski 1966) with reasonable parameter choices; the correlations are respectively .89, .87 in $(ds)^8$ and .92, .90 in the zero-isospin subspace. To understand the differences we could in the same way analyze the Hamiltonians in terms of various components ($Q \cdot Q$, pairing, etc) as they operate in the many-particle spaces. A different decomposition of H, valuable for example in studying symmetry breaking, is into G tensors where G is a U(N) subgroup. This decomposition is orthogonal with respect to the U(N) norm as well as to norms supplied by the exact symmetries that are conserved by the G tensors.

3.   PARTITIONS AND FINER AVERAGES

3.1   *Averages for Fixed Symmetries*

The simple theory of the last section gives useful approximate results for many phenomena and indicates some general features of complex systems; for example that expectation values usually are roughly linear with energy and vary in a simple way with the number of active particles. But the assumption of Gaussian shape will be inadequate for large model spaces (CLT convergence is always slowest in the spectral extremes). By calculation of one or two higher moments, we can accommodate small departures from Gaussian shape, but this is not adequate for large ones. We must instead take advantage of the array of U(N) subgroups that come into play along with U(N) when the single-particle spectrum is truncated at ε_N, in order to partition the model space into subspaces and hence the density into partial densities better represented by Gaussians (perhaps with low-moment corrections for the low-lying densities). At the same time we gain information about the goodness of the model symmetries [e.g. SU(3) rotational and symplectic pairing] defined by these subgroups, and where in the spectrum the various irreps are located. Similarly we learn about the way in which ground state energies, expectation values, correlation coefficients, etc vary with the "good" symmetries (angular momentum, isospin), which in our model space are also defined by U(N) subgroups. Information propagation, throughout the lattice of subgroup representations becomes now much more interesting and useful than in the simple case of the previous section. Such propagation has no counterpart for subspace decompositions that are not symmetry defined, which illustrates again the intimate connection between symmetries and statistical behaviour (Parikh 1978).

When we partition the space by $(m) = \sum_\Gamma (m, \Gamma)$ where Γ (which may be a multiple index) labels the subspaces, the moments decompose similarly, $\langle\langle H^p \rangle\rangle^m = \sum_\Gamma \langle\langle H^p \rangle\rangle^{m, \Gamma}$ and similarly for the densities $I_m(x) = \sum_\Gamma I_{m, \Gamma}(x)$ with $I = d \times \rho$ as above. Decompositions with respect to exact symmetries are of course always interesting; the isospin case is simple, the angular momentum case quite difficult. These two are in fact not really analogous, for all the states of a given isospin belong to a single irrep [of U($\frac{1}{2}N$) × U(2)], whereas the states of given J form a set of equivalent irreps. A broken symmetry is interesting if it is fairly good (with states having an intensity say $\geq 90\%$ in a single irrep). The relative contributions of the various irreps in a given energy region are then a matter of real interest. Approximate intensities may be read off from the values of the (m, Γ) densities, though better methods are available.

Even a badly broken symmetry can be useful as long as the centroids are well separated and the individual variances considerably smaller than the overall "scalar" variance. For such a symmetry makes a rough partition of the energy range into partially overlapping domains, in each of which one or a few irreps may be dominant. This effectively reduces the size of the model space that contributes at a given energy, a result particularly important near the ground state, which must be located as accurately as possible since the excitation energies are specified with respect to it.

If our main interest is with accurate methods of working in very large spaces, the simplest partitioning is by configurations, in which N is partitioned as $\sum N_r$, usually in terms of spherical orbits, and similarly for m. The alternative procedure represented by Equations 5–7 for dealing with the scalar [U(N)] case extends immediately to the present partitioning, which involves a "direct-sum" subgroup of U(N). Since the single-particle energies generate large centroid splittings, all our needs are satisfied.

Leaving aside these practical matters, let us consider two more general aspects of complex spectroscopic systems, the first of which is displayed by the form of the expectation value $K(E)$ of an operator K. Using, in the domain about the centroid of the (m, Γ) subspace, the orthonormal polynomials generated by $I_\Gamma(x)$ as a weight function, we find the remarkable result

$$K(E) = \sum_\Gamma \left\{ \frac{I_\Gamma(E)}{I(E)} \right\} K(E:\Gamma);$$

$$K(E:\Gamma) = \sum_\nu \left[\langle K P_\nu^{(\Gamma)}(H) \rangle^{m,\Gamma} \right] \times P_\nu^{(\Gamma)}(E)$$

13.

At first sight this is exactly what we would expect since the first factor is the obvious branching ratio, while $K(E:\Gamma)$ appears to be the expectation value, as in Equation 12. However, in evaluating $\langle K P_\nu^{(\Gamma)}(H) \rangle^{m,\Gamma}$ we must, when the Γ subspace is not defined by a good symmetry, take account of the excitations to and between the other subspaces. Thus by a formally simple device the equation, while maintaining its form and its linearity in the I_Γ, accommodates the whole hierarchy of interactions between subspaces; this fact has on occasion been seriously misunderstood (see for example Brody et al 1972). Once again, of course, as long as we have rapid convergence only low-order nonlinearities in H are encountered.

In dealing with trace propagation among the $U(N)$ spaces we encountered the $U(N)$-scalar k-body operator

$$\binom{n}{k} = \sum_\alpha \psi_\alpha(k)\psi_\alpha^\dagger(k) \equiv e(k).$$

As indicated, it is also the operator representation of the trace of the k-body density matrix [n is the number operator, $\psi_\alpha(k)$ is the state operator that, acting on the particle vacuum, generates $\Psi_\alpha(k)$, and the summation extends over all k-particle states]. Similarly when we have a subgroup decomposition $U(N) \supset V$ (there could also be intervening subgroups), we encounter the k-body V-scalars $e(k, \Lambda)$; these are density-matrix trace operators defined like $\binom{n}{k}$ above but with α taken only over the (k, Λ) subspace. A little manipulation will show that

$$\langle\langle G(k) \rangle\rangle^{m,\Gamma} = \langle\langle \bar{e}(N-m, \Gamma_c)G(k) \rangle\rangle^k$$

$$= \sum_\Lambda \langle\langle \bar{e}(N-m, \Gamma_c)G(k) \rangle\rangle^{k,\Lambda},$$

14.

where $(N-m, \Gamma_c)$ is the subspace complementary to (m, Γ) and, for any operator F, $\bar{F}$ is its "hole ↔ particle adjoint" defined by $a_\mu \leftrightarrow a_\mu^\dagger (\mu = 1, \ldots, N)$. Since, for $U(N)$,

$$e(N-m) = \binom{n}{N-m}$$

so that

$$\bar{e}(N-m) = \begin{pmatrix} N-n \\ N-m \end{pmatrix},$$

Equation 14 properly reproduces Equation 2. The basic equation (14), which has an easy extension for operators G of mixed particle rank and which as it stands is valid for arbitrary decompositions (k, Λ) (not necessarily defined by symmetries), is in form analogous to a potential problem with $\bar{e}$, a "surface Green's function," carrying information from the k-particle to the m-particle surface. When (k, Λ) is defined by a $U(N)$ subgroup, as we have taken for granted, then $\bar{e}$ is a V scalar. But by a theorem of Weyl (1946, see also Quesne 1975) all such scalars are representable as polynomials in a finite set of basic scalars forming the "integrity basis" for the group structure involved. For $U(N)$ itself the basis is one dimensional for fermions or bosons, while for the isospin group we have two basic scalars n and T^2. Note then that in the latter case the fixed-isospin traces propagate just as simply as do the complete m-particle traces, because all the polynomial scalars (being "Casimir representable") act as constants in every (m, T) subspace. This simple propagation occurs in certain other circumstances also, and then it is easy to construct the $\bar{e}$ operators, which, for fixed k, act as projection operators. But in still others the scalars are not completely expressible in terms of Casimir operators. Then the situation is much more complicated (Quesne 1976). We go no further into these matters except to say that there are easy ways to find the size and nature of the integrity basis (French & Draayer 1979) but no easy ways as yet to make use of it in the non-Casimir case). There are expansions of the traces in powers of the single-state occupancies that for low orders and for arbitrary groups generalize the $U(N)$ embedding result referred to at the end of Section 2.1. There are extensions of the propagation concept for dealing with information carried not by group scalars but by more general tensors. The basic Equation 3 is valid also for bosons, as are corresponding symmetry extensions (but not, for example, Equation 4, which uses hole $\leftrightarrow$ particle arguments having no relevance for bosons).

3.2   Further Applications

The discussion following Equation 13 makes it clear that the decomposition,

$$\sigma^2(f) = \sum_{f'} \sigma^2(f, f'),$$

of the spectral variance into partial variances is particularly significant in studying symmetry admixtures. Here $d^{-1}(f')\sigma^2(f, f') = d^{-1}(f)\sigma^2(f', f)$ is the mean-square matrix element connecting the f and f' subspaces (Parikh & Wong 1972). The partial variances, which propagate by laws that are often quite complicated, have been derived for and used in studying various features of Wigner's SU(4) symmetry (Hecht & Draayer 1974, Chakraborty et al 1980), pairing symmetries (Quesne & Spitz 1974, 1978, Nissimov et al 1972), and IBA symmetries (Kota 1981).

The orthogonal polynomial expansion of Equation 13 has been used for a theory of the effective interaction (Chang 1978, Chang & Draayer 1979); its convergence properties are closely related to the CLT action in the model space, which has no counterpart in the conventional perturbation expansions. There are many other uses for the polynomial expansion.

Finally, we stress that with partitioning we come to a theory of level densities for interacting particles that apparently can be extended to model spaces of arbitrary dimensionality (Chang et al 1971, Haq & Wong 1980, Grimes 1980). Since configurations carry definite parities, the fixed-parity decomposition of the level density follows at the same time.

4. FLUCTUATIONS

4.1 *Comparison of Random-Matrix Models with Experiment*

The problems with fluctuations are how to characterize them and how to deduce from them the information they carry. In principle the latter analysis might yield values or limits for certain physical parameters; or it could happen that the fluctuation patterns are essentially parameter-free, corresponding then to a general law whose nature and origin are of some fundamental interest.

These complicated problems have not been directly attacked. Instead, analytically tractable ensembles, featureless except for a specification of their behavior under rotations and time reversal, have been introduced and various measures (e.g. variances of kth-nearest-neighbor spacing distributions) calculated and compared with experiment. We should first ask:

1. How well do ensemble results agree with experiment? If the agreement is good, then

2. What mechanisms, generated by the Hamiltonian, would affect the fluctuations, and what limits on their magnitudes are imposed by the agreement?

3. What is the source of the agreement? Why are the fluctuations not affected by the special features of the Hamiltonian that exhibit themselves in particular nuclei?

We deal first with energy-level fluctuations, beginning with an ensemble of $d = 2$ (two-dimensional) matrices:

$$H = \begin{Bmatrix} A + \lambda & C + i\alpha D \\ C - i\alpha D & B - \lambda \end{Bmatrix} = \begin{Bmatrix} X_1 + X_2 + \lambda & X_3 + i\alpha X_4 \\ X_3 - i\alpha X_4 & X_1 - X_2 - \lambda \end{Bmatrix} \quad 15.$$

where α and λ are real constants, which may be taken nonnegative, while the X_i are real independent random variables of type $G(0, v^2)$, i.e. Gaussian with zero centroid and variance v^2. It follows then that A, B, C, D are real independent, the first two of type $G(0, 2v^2)$ and the latter two $G(0, v^2)$. A nonzero λ defines a nonrandom component of H. For $\lambda = 0$ the probability density, with respect to the volume element ΠdH_{ij} supplied by the distinct matrix elements, is proportional to $\exp[-(\mathrm{Tr}\ H^2/v^2)]$, defining therefore an ensemble that is invariant under orthogonal transformations: $\alpha = 0$ gives the $d = 2$ Gaussian orthogonal ensemble (GOE), each of whose members can be diagonalized by such a transformation; $\alpha = 1$ defines the corresponding unitary ensemble (GUE), invariant under the larger group of unitary transformations, whose members in general are not diagonalizable by an orthogonal transformation. The trace $T = 2X_1$ and the spacing S, given as in Equation 1 by $S^2 = 4[(X_2 + \lambda)^2 + X_3^2 + \alpha^2 X_4^2]$ are independent. T is of type $G(0, 4v^2)$ while an elementary calculation (e.g. via convolutions) gives for S, which takes on values $x \geq 0$,

$$\rho_S(x) = \frac{x}{4v^3 \alpha \sqrt{2\pi}} \exp\left(\frac{-\lambda^2}{2} - \frac{x^2}{8v^2}\right) \int_0^x I_0\left[\frac{\lambda (x^2 - z^2)^{1/2}}{2v}\right]$$

$$\times \exp\left[\frac{(\alpha^2 - 1)z^2}{8\alpha^2 v^2}\right] dz. \quad 16.$$

Our major interest is with $\lambda = 0$ and then (Favro & McDonald 1967)

$$\rho_s(x) = \frac{x}{4v^2 \sqrt{1 - \alpha^2}} \exp\left(\frac{-x^2}{8v^2}\right) \mathrm{erf}\left[\left(\frac{1 - \alpha^2}{8\alpha^2 v^2}\right)^{1/2} x\right]. \quad 17.$$

In the small and large argument limits, the error function can be approximated to give, respectively,

$$\rho_S(x) \simeq \frac{x^2}{4\sqrt{2\pi}\alpha v^3} \exp\left(\frac{-x^2}{8v^2}\right)\left[1 - \frac{(1-\alpha^2)x^2}{24\alpha^2 v^2} + \ldots\right] \qquad 17a.$$

$$\rho_S(x) \simeq \frac{x}{4v^2(1-\alpha^2)^{1/2}} \exp\left(\frac{-x^2}{8v^2}\right). \qquad 17b.$$

We then find, for $\alpha \leq 1$ a level repulsion that is "linear" $[\exp(x^2/8v^2)\rho \propto x]$ for large x and quadratic for small x, the former then being precise for GOE ($\alpha = 0$) and the latter for GUE ($\alpha = 1$). The mean spacing (D) and the variance (σ^2) are

$$\bar{S}_\alpha = D_\alpha = (8/\pi)^{1/2} v\left[(1-\alpha^2)^{-1/2} \tan^{-1}\frac{(1-\alpha^2)^{1/2}}{\alpha} + \alpha\right]$$

$$\sigma_\alpha^2 = \frac{4v^2(2+\alpha^2)}{D_\alpha^2} - 1 \xrightarrow{\alpha \ll 1} \left(\frac{4}{\pi}-1\right) - \frac{2\alpha^2}{\pi} = \sigma_0^2 - \frac{4\alpha^2 v^2}{D_0^2}$$

18.

For arbitrary dimensionality d, dropping the nonrandom component, we write H as the sum of independent real symmetric and imaginary antisymmetric parts[3]:

$$H_{ij}(\alpha) = H_{ij}(S) + i\alpha H_{ij}(A); \qquad \bar{H}_{ij}(S) = \bar{H}_{ij}(A) = 0;$$

$$\bar{H}_{ij}^2(S) = (1 + \delta_{ij})v^2; \qquad \bar{H}_{ij}^2(A) = (1 - \delta_{ij})v^2$$

19.

The technique used above for $d = 2$ is obviously not available in general, there being no analog of Equation 1. The classical theory of energy-level fluctuations, introduced by Wigner (1955) and developed by many people but especially by Dyson & Mehta (Mehta 1967), deals with the problem for the special cases[4] $\alpha = 0$ and 1. The invariance properties of these ensembles represent a realization of the statistical-mechanical doctrine of equal a priori probabilities, or, from the information-theory standpoint, of "minimal information." Both of these descriptions are tempered by the fact that TRI is good for $\alpha = 0$ and irrelevant ("completely broken") for $\alpha = 1$. Exploiting these properties yields the joint probability eigenvalue distributions for these ensembles (and a Gaussian "symplectic" ensemble (GSE) that we need not consider here).

---

[3] Parenthetically we remark that both parts of H could be replaced by "random-sign" ensembles (with $H_{ij} = \pm 1$), which in principle we could generate by coin tossing. The two-point fluctuation properties would be unchanged (as is obvious from Equation 24 in Section 4.2).

[4] Very recently A. Pandey and M. L. Mehta (private communication) gave exact results for arbitrary α.

The real difficulties then lie with the evaluation of measures. This classical theory has been referred to by Lieb & Mattis (1966) as follows: ". . . it is the one-dimensional theory par excellence! Not only does it have immediate usefulness and validity for real physical systems but, from the mathematical point of view, it has given rise to profound results and makes use of the deepest theorems of analysis."

We avoid the difficulties of the classical theory here by relying on a much simpler and more generally applicable approximate theory. This is based on the use of moment methods with a certain binary correlation approximation first used in Wigner's (1955) calculation of the eigenvalue density. But let us begin with our first question, namely the agreement between experiment and the results of the GOE model appropriate for TRI systems.

The fact that there is a rough general agreement between experiment and the simplest predictions of the GOE has been known for decades. About a dozen years ago, with a dramatic increase in the quality of the data, it became feasible to make use of more demanding measures (Garg 1972). But only during the past year (Haq et al 1982) was a decisive test of the GOE model made — by combining the available high quality data (about 1400 levels from about 30 nuclei) and using new kinds of measures for the analysis. The first measure used for this test was a spectrally averaged version of the Dyson-Mehta Δ_3 (n), which essentially measures the mean-square deviation of an n-member run of levels from the levels of the best-fit uniform spectrum. Analytic results for the Δ_3 expectation value were supplemented by Monte-Carlo calculations of its variance, the latter replacing an inadequate asymptotic (large n) calculation. The Δ_3 autocorrelation function, needed to evaluate the variance of the spectrally averaged measure, has also been derived from Monte-Carlo calculations. The agreement found for Δ_3 (which measures two-point fluctuations) and for its variance and distribution function (which could have contributions from higher-order fluctuations) shows that Wigner's GOE model fits remarkably well, leaving little if any room, with presently available data, for the detection of significant discrepancies (see Figure 1).

4.2   Time-Reversal Invariance and Symmetry-Breaking Effects

We have answered our first question above, but the ensembles required to deal with the other two violate the symmetries invoked *ab initio* with the classical theory. For the second question, we use in Equation 19 a nonzero α, appropriate to the violation of TRI. With $d = 2$ small α gives

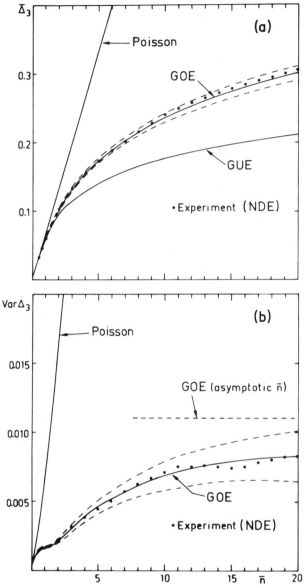

Figure 1 Comparison of the Dyson-Mehta Δ_3 statistic with results from 30 experimental sequences (27 from neutron resonances with $A \geq 110$, and 3 from proton resonances with $A = 44, 48$). Spectrally averaged values of Δ_3, and its variance ($\bar{\Delta}_3$, Var Δ_3) are calculated for each nucleus and weighted averages taken over the set of nuclei considered (the "nuclear data ensemble," NDE). Results are given as functions of $\bar{n}$, the number of levels defining Δ_3. Dashed lines correspond to one standard deviation about GOE values. Poisson and unitary ensemble (GUE) results are also indicated. Taken with permission from Haq et al (1982).

small deviations from the GOE prediction as in Equation 18, but for large d the relevant parameter is not α but $d^{1/2}\alpha$ so that the GOE $\rightarrow$ GUE transition goes very much faster with α and indeed is discontinuous in the large-d limit. This comes about because, with fixed v and $\alpha = 0$, the spectrum span increases with dimensionality only as $d^{1/2}$ so that the levels are pushed together by a factor of $d^{1/2}$. But then the imaginary matrix elements entering when $\alpha > 0$ become $d^{1/2}$ times as effective, at least in the small-α limit. Moreover, since they generate complex admixtures of the real wave function amplitudes, the effects of the imaginary matrix elements should show up in the fluctuation measures (see the discussion following Equation 1). We have here an analogue of a phase transition, the statistical chaos changing very rapidly with increasing α from GOE to GUE type, which has indeed quantitatively different properties. It should be clear that no corresponding GUE $\rightarrow$ GOE transition is to be expected (in the neighborhood of $\alpha = 1$) for in that case we are dealing with admixtures of complex amplitudes generated by real matrix elements. The difference clearly arises from the fact that the orthogonal transformations form a subgroup of the unitary group, the rapid change with α then deriving from the breaking of a subgroup symmetry. Moreover, these simple arguments suggest that we have stable chaos except in the transition region; thus we might expect the fluctuation measures [expressed in local spacing units so that the $H(\alpha)$ normalization is irrelevant] to be of GUE type, independent of α, for all $\alpha \gg d^{-1/2}$.

These arguments rely on a considerable extrapolation from the $d = 2$ results. A much more complete demonstration of the rapid transition (J. B. French, V. K. B. Kota, and A. Pandey, to be published) follows from a simple perturbation argument showing that the variance of the kth-nearest-neighbor spacing distribution ($k = 0$ is nearest neighbor), measured in terms of the local spacing D, satisfies the equation (a special case of which follows from Equation 18):

$$\left(\frac{\partial \sigma_\alpha^2(k)}{\partial(\alpha^2)}\right)_{\alpha = 0} = -\frac{4v^2}{D_0^2}$$

and therefore, to order α^2,

$$\sigma_\alpha^2(k) = \sigma_0^2(k) - \frac{4\alpha^2 v^2}{D_0^2} = \sigma_0^2(k) - 4\Lambda(\alpha). \qquad 20.$$

This indicates that $\Lambda^{1/2}(\alpha)$, which varies over the spectrum (in a manner expected from the simple argument above) and has the value $(1/\pi)$ $d^{1/2}\alpha$ at the center (with the normalization $v^2 d = 1$), is the transition parameter. Its value $\Lambda^{1/2} \simeq 1$ marks the completion of the GOE-GUE transition, not only for the nearest-neighbor spacing but for spacings of higher order and for all other two-point measures as well. This is because (French et al 1978) the $\sigma^2(k)$ essentially determine the two-point correlation function itself. To be rigorous we should use the two-point measures $\Sigma^2(k)$, the variances of the number of levels in an interval of length kD, but the two are closely related: $\Sigma^2(k) \simeq \frac{1}{6} + \sigma^2(k-1)$. We have here a demonstration of the fact that the level repulsion effects are not at all restricted to adjacent levels (which is all we could be certain of from the $d = 2$ results) but in fact extend over the whole spectral range. Indeed the most striking feature of complicated spectra is the long-range rigidity, first derived by Dyson & Mehta (1963), who described spectra as "semicrystalline," and demonstrated by the behavior of $\Sigma^2(r) \simeq (2/\beta\pi^2) \ln \beta\pi r$ ($\beta = 1$, 2, 4 for GOE, GUE, GSE) rather than $\Sigma^2(r) \simeq r$, which would characterize a Poisson process. This rigidity has been accurately verified up to $r = 20$ (Figure 1) by the behavior of $\Sigma^2(r) \simeq (2/\beta\pi^2) \ln \beta\pi r$ ($\beta = 1$, 2, 4 for GOE, GUE, GSE) rather than $\Sigma^2(r) \simeq r$, which would characterize a Poisson would be of great interest.

For the more general (nonperturbative) theory, used to replace the classical theory, consider first Wigner's (1955) derivation of the ensemble-averaged GOE ($\alpha = 0$) eigenvalue density. Its odd moments $M_{2\nu+1}$ vanish since the basic distributions are symmetric about zero. The even moments are

$$M_{2\nu} = \overline{\langle H^{2\nu}\rangle^m} \xrightarrow{d\to\infty} t_\nu\{\overline{\langle H^2\rangle^m}\}^\nu = t_\nu(dv^2)^\nu; \qquad t_\nu = \frac{1}{\nu+1}\binom{2\nu}{\nu}. \qquad 21.$$

The logic involved here is that $\overline{HOH}$ is smaller by order d^{-1} than $\overline{HHO}$, for almost all traceless operators O, since only diagonal matrix elements of O contribute to the first while all contribute to the second. Thus $M_{2\nu}$ is determined by counting the *unlinked* binary correlations in $\langle H^{2\nu}\rangle^m$; e.g. pairings (1-2, 3-4) and (1-4, 2-3) contribute to M_4, but not (1-3, 2-4). The counting here gives the "Catalan number" t_ν, in contrast to the $(2\nu-1)!!$ familiar for Gaussian processes in which all binary pairings contribute. The t_ν define Wigner's "semicircular" density. In fact, a simple extension to the case of arbitrary α gives

$$\bar{\rho}_\alpha(x) = \frac{[4dv^2(1+\alpha^2) - x^2]^{1/2}}{2\pi[dv^2(1+\alpha^2)]} \xrightarrow{dv^2(1+\alpha^2)=1} \frac{1}{2\pi}[4-x^2]^{1/2}$$

$$= \frac{1}{\pi}\sin\psi(x); \qquad \psi(x) = \pi - \cos^{-1}(x/2), \qquad 22.$$

where in the last form we adopted a normalization giving $\bar{\rho}_\alpha = \bar{\rho}$, a semicircle of radius 2. Considered as a weight function, $\bar{\rho}(x)$ defines a set of orthonormal Chebyshev polynomials, valid for $-2 \leq x \leq 2$,

$$v_\zeta(x) = (-1)^\zeta \frac{\sin(\zeta+1)\psi(x)}{\sin \psi(x)} = \sum_s (-1)^s \binom{\zeta - s}{s} x^{\zeta - 2s}. \qquad 23.$$

Turning now to the fluctuations, it is not surprising that there is a representation of the two-point correlation function in terms of a superposition of random Chebyshev-polynomial "excitations" of the semicircular density. The derivation proceeds by an extension of that used for the eigenvalue density. Writing $S^\rho(x, y) = \overline{\rho(x)\rho(y)} - \bar{\rho}(x)\bar{\rho}(y)$ as the density correlation function, $\Sigma^2_{p,q}$ as its moments, and $S^F(x, y)$ as its doubly integrated form, we have (Mon & French 1975, J. B. French, V. K. B. Kota, and A. Pandey, to be published)

$$\Sigma^2_{p,q}(\alpha) = \int S^\rho(x, y) x^p y^q \, dx \, dy = \overline{\langle H^p \rangle^m \langle H^q \rangle^m} - \overline{\langle H^p \rangle^m}\,\overline{\langle H^q \rangle^m}$$

$$= \sum_\zeta \underbrace{\overline{\langle H^p_\alpha \rangle^m \langle H^q_\alpha \rangle^m}}_{\zeta} = \sum_\zeta u^p_\zeta u^q_\zeta \underbrace{\overline{\langle H^\zeta_\alpha \rangle^m \langle H^\zeta_\alpha \rangle^m}}_{\zeta}$$

$$= \frac{1}{d^2} \sum_\zeta u^p_\zeta u^q_\zeta \zeta A_\zeta(\alpha), \qquad 24.$$

where

$$u^p_\zeta = \binom{p}{\dfrac{p-\zeta}{2}}; \quad A_\zeta(\alpha) = 1 + \eta^\zeta,$$

with

$$\eta = \frac{1-\alpha^2}{1+\alpha^2}.$$

We have first written the moment in terms of traces, then recognized that binary correlations can involve two Hs in the same trace ("internally correlated") or one in each trace ("cross correlated"), and decomposed the moment according to ζ, the number of cross-correlated pairs, which corresponds also to the inverse wavelength of the excitation. For fixed ζ, we could have generated the cross-correlated product by starting with $p = q = \zeta$; its evaluation is straightforward. We should then insert into the first factor $(p - \zeta)/2$ correlated pairs in such a way that each pair contracts only over internally correlated pairs. A tricky combinatorial argument gives u^p_ζ, the number of ways of doing this.

Similarly for the second factor. The $\zeta = 0$ term does not contribute and because the number of modes is bounded we have a natural cut-off, $\zeta_{max} \simeq d$.

The moments (Equation 24) yield for the correlation function

$$S_\alpha^F(x, y) \simeq \frac{1}{d^2} \bar{\rho}(x)\bar{\rho}(y) \sum_{\zeta=1}^{d} A_\zeta \zeta^{-1} v_{\zeta-1}(x) v_{\zeta-1}(y)$$

$$\rightarrow S_{GUE}^F + \frac{1}{4\pi^2 d^2} \ln\left\{ \frac{1 + \eta'^2 - 2\eta' \cos[\psi(x) + \psi(y)]}{1 + \eta'^2 - 2\eta' \cos[\psi(x) - \psi(y)]} \right\}; \quad 25.$$

where $\eta' = \eta \exp(-\tau/d)$. In order to extend the ζ summation to ∞, its evaluation being then elementary, we have introduced a cut-off factor $(\exp -\tau/d)^\zeta$, which combined with the η^ζ of A_ζ (Equation 24) gives $(\eta')^\zeta$. The value of $\tau(\sim 1)$ can be fixed, for example, by using Equation 20. Then all the two-point measures follow immediately. As an example, we have for the number variances $\Sigma_\alpha^2(r)$, with r defined by the interval (x, y),

$$\Sigma_\alpha^2(r) = d^2[S^F(x, x) + S^F(y, y) - 2S^F(x, y)]$$

$$= \Sigma_{GUE}^2(r) + \frac{1}{2\pi^2} \ln\left[1 + \frac{r^2 \eta'}{4\pi^2 d^2 \bar{\rho}^4(x)(1 - \eta')^2} \right], \quad 26.$$

which is plotted for $r = 1$ in Figure 2 and compared there with Monte-Carlo calculations. The agreement is within statistical error for the entire range $(0-\infty)$ of α and confirms the semiquantitative arguments, given at the beginning of this section, for the transition generated by varying α.

Returning now to the second question at the beginning of this section, we observe that the agreement of GOE $(\alpha = 0)$ predictions with experimental results, coupled with the rapid GOE $\rightarrow$ GUE transition of the measures as α increases, should give us an upper limit to the magnitude of the NTRI interaction. Specifically it should give a limit for the transition parameter $\Lambda^{1/2}(\alpha)$, the RMS matrix element of the NTRI interaction in the appropriate GOE model measured in units of the local spacing D. Combining Equation 26 with the Haq et al (1982) analysis of all the data (divided into four groups with the average D varying from ~ 10 to ~ 100 eV) gives for this matrix element a limit $\lesssim 0.2 \, D$.

What we now must fix in order to complete the determination of α are the values of the TRI GOE parameters v^2 appropriate to the individual nuclei considered. The interpretation of v as an RMS matrix element is unsatisfactory, being unstable with respect to extensions of the model

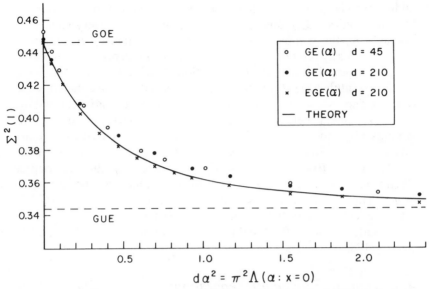

Figure 2   The Σ^2 (1) transition curve of Equation 26 compared with Monte-Carlo results for 50-member ensembles, $GE(\alpha)$, of 45-and 210-dimensional $H(\alpha)$ matrices (Equation 19), and for 50-member 210-dimensional embedded ensembles, $EGE(\alpha)$, involving two-body interactions in a four-particle space. The theoretical GUE value is 0.344. For $d = 210$ and $d\alpha^2$ beyond the range indicated we have found 0.347, 0.344, 0.340 for $d\alpha^2 = 4.8$, 210 (GUE) and ∞. The embedded ensembles give 0.342, 0.346, 0.342, 0.341, 0.346, 0.347 for $d\alpha^2 = 4.8$, 9.8, 60, 210, 735, ∞. Observe that the transition is effectively completed at $\Lambda(\alpha) \simeq 1$. The quality of the agreement is as expected from the sample errors. Taken from J. B. French, V. K. B. Kota, and A. Pandey (to be published).

space. Instead we recognize $2v^2$ as the ensemble variance (and hence by ergodicity the spectral variance) of the expectation value of the *two-body* interaction $\mathcal{H}$. This is stable and calculable by a double polynomial expansion of the strength which represents a natural extension of Equation 13 (Draayer et al 1977). Therefore we must choose for each nucleus a reasonable model space, effective interaction $\mathcal{H}$, and single-particle energies ε_i (which generate the stability) and carry out the spectral calculation by partitioning with respect to configuration and angular momentum. This nontrivial calculation is not yet completed.

Our third question above asks how an almost featureless fluctuation model can give results in agreement with experiment. The natural way to answer this is to supplement the Hamiltonian ensemble with nonrandom terms that take explicit account of the dominant terms in H. This was done to a considerable extent by Pandey (1981), who, by using Stieltjes-transform methods, demonstrated more or less explicitly that adding even large operators to the ensemble has little or no effect on the local fluctuations. Perhaps this is because, for almost every (traceless) operator K, the complete set of its orthogonal images generates an ensemble closely related to the GOE; it has the same distributions of the matrix elements but with weak correlations ($\sim d^{-1}$, d^{-2}) between them. To the extent that the correlations can be ignored, we have then the sum of two independent GOEs, which is itself a GOE, the variances adding.

The argument here is strengthened by the fact that one of its variants gives the rapid GOE $\rightarrow$ GUE transition because the unitary images of K form a "weakly correlated GUE." We can write $H(\alpha) = (1 - \alpha)H(\text{GOE}) + \alpha H(\text{GUE})$ and argue then that, as long as $d^{1/2}\alpha \gg 1$, the second term dominates the first, which may be very much larger. Thus the diagonalizing transformations convert $(1 - \alpha)H(\text{GOE})$ into an effective GUE that combines with $\alpha H(\text{GUE})$. Similar arguments and results extend to the breaking of model symmetries (Dyson 1962, Pandey 1981).

4.3 Density-Fluctuation Separation: Gaussian Density and Ergodic Behavior

Wilson (1979) noted: "In general, events distinguished by a great disparity in size have little influence on one another; they do not communicate and so the phenomena associated with each scale can be treated independently." We have taken for granted that the eigenvalue density and its fluctuations can be treated separately. We should understand this better. We remark first that our random-matrix models are a priori unrealistic and incompatible with our treatment of the density itself, for they ignore the constraints required to specify a two-body interaction and hence describe only simultaneous interactions between all particles, for which there can clearly be no CLT. That the ensemble density is semicircular rather than Gaussian (or equivalently that its moments are fixed by counting unlinked binary correlations rather than all of them) is a sign of that. A much more realistic model generates a random interaction in the two-particle space (perhaps supplemented by a one-body term) and uses it in the m-particle space, resulting then in an "embedded" ensemble (EGOE, for example) in which the particle rank of the interaction is specified. A recalculation of the eigenvalue density (Mon & French 1975), completely parallel with that above for the GOE, yields an asymptotic ($m \rightarrow \infty$, $N \rightarrow \infty$, $m/N \rightarrow 0$) Gaussian density and for sufficiently large m (say $m \gtrsim 6$ for one-body Hs, $m \gtrsim 10$ for two-body) a good Gaussian. A fluctuation calculation, also parallel with that for GOE, but now in terms of

Hermite-polynomial excitations of the Gaussian, gives for wavelengths much larger than the level-spacing D ($\zeta \ll d$) a very strong attenuation that increases very rapidly with particle number and excitation order ζ. Specifically the amplitude A_ζ, given for GOE in Equations 24 and 25, has with EGOE a factor $\binom{m}{k}^{2-\zeta}$ with k the particle rank. Since the low-ζ excitations that survive can be incorporated into the density itself, we have the remarkable prediction that the spectrum should display locally only short-wavelength excitations, $\zeta \sim d$. That the power spectrum has this form has been verified by Monte-Carlo calculations (which also give fluctuation measures in agreement with experiment) but not yet directly for the experimental data. This demonstrates the CLT action, but, because the EGOE calculations so far successful are valid only for $\zeta \ll d$, we have no formal proof that the fluctuations escaping the CLT action should be of GOE type. The experimental results (and EGOE Monte-Carlo results of Figure 2) demonstrate that they are. Except for that gap in the theory, we have an excellent demonstration of the separation of the density and its fluctuations.

We have seen by now that even for interacting particles the EGOE-average density is asymptotically Gaussian. That this result is valid for "almost all" GOE members was first shown by Grenander (1963). It follows for GOE from the moment variances given by Equation 24 with $p = q$ and $\alpha = 0$. The fully cross-correlated term ($\zeta = p$) gives $\sum_{p,p}^2 = 2\zeta/d^2$, which is modified when we add in the $\zeta < p$ terms to give (Mon & French 1975)

$$2p\binom{p-1}{\mu}^2 \bigg/ d^2$$

with μ the integral part of $(p-1)/2$. Since the normalization ($dv^2 = 1$) preserves the scale of the density (radius $= 2$), the d^{-2} dependence of the moment variances yields the desired result and can be described as the ergodic behavior of the density. The result obviously follows for the $\alpha \neq 0$ ensembles and is valid for the embedded ensembles as well (Mon & French 1975). The close relationship between fluctuations and ergodic behavior follows, of course, from the fact that the covariances of the density moments are also the moments of the two-point correlation function.

We have now an essentially complete demonstration of the origin of the Gaussian behavior, the exceptional cases being understood (French 1972) in terms of a cluster expansion (see also Kota & Potbhare 1980). Ergodic behavior for the density by no means implies ergodicity for the fluctuations. But, using procedures that are more closely analogous to those used in classical studies of ergodic behavior and drawing on results of Dyson (1970) and Mehta (1971) for higher-order correlation functions, Pandey (1979) gave general proofs of ergodicity for the fluctuation measures of the classical ensembles.

5. CONCLUSION

There are many things that we have not mentioned. One is that, in a very wide range of circumstances, transition matrix elements T_{ij} and expectation values T_{ii} in the Hamiltonian basis behave as Gaussian random variables when they are renormalized by their locally averaged RMS values. This is the basis of the Porter-Thomas (1956) χ_1^2 distribution of the transition strengths, which has been well-verified experimentally and can also be used to set a limit on NTRI. In fact it enters also into the TRI calculation of Section 4.2, specifically in evaluating the v^2 parameter as an expectation-value variance.

Another related example of the use of spectral averaging to calculate a basic parameter of the fluctuation theory is the prediction of collectivity that may be generated by a given Hamiltonian in a given model space. When most of the strength from a given starting state goes to only a few final states, the strength sum must be strongly fluctuating in the neighborhood of the given state (Draayer et al 1977). As a final example involving cooperation between spectral averaging and fluctuations, we mention the decomposition of H into a schematic part that generates the collectivity and the remainder that damps it. The decomposition appropriate to the given model space is carried out by the methods of Section 2 and the remainder can be represented by a random ensemble.

Among the recent applications not discussed in this introductory review are spectral averaging for β-decay giant resonances (Kar 1981) and for electromagnetic transitions (Halemane et al 1981), and important technical advances in methods of trace evaluation (Chang & Wong 1979, 1980, Grimes et al 1979; B. D. Chang, J. P. Draayer, and S. S. M. Wong, private communication). There are also interesting possibilities of applications to very different classes of systems (J. C. Parikh, private communication).

We have said little about the limitations and uncertainties of the subject. Is it in principle adequate to use statistical methods in the ground-state domain, and can they really be extended to unbounded spaces? The answer to the first question is probably negative. It is therefore essential to produce a satisfactory combination of statistical methods with a restricted microscopic theory for the ground-state region. The second question is still open; if the answer is positive the subject should have much wider applications.

ACKNOWLEDGMENTS

We are indebted to O. Bohigas, R. U. Haq, and A. Pandey for making available to us details of their recent work, and to J. P. Draayer, T. R. Halemane, K. Kar, A. Pandey, J. C. Parikh, V. Potbhare, and S. S. M. Wong for many valuable discussions. Part of this work was done while the first-named author was visiting at Physical Research Laboratory, Ahmedabad, India; he is especially indebted on that account to S. P. Pandya and J. C. Parikh. We thank Mrs. Edna Hughes for preparing the manuscript.

Literature Cited

Arvieu, R., Moszkowski, S. A. 1966. *Phys. Rev.* 145:830–37
Bethe, H. A. 1936. *Phys. Rev.* 50:332–41
Brody, T. A., Flores, J., Mello, P. A. 1972. *Rev. Mex. Phys.* 21: 141–66
Chakraborty, M., Kota, V. K. B., Parikh, J. C. 1980. *Ann. Phys. NY* 127:413–35
Chang, B. D. 1978. *Nucl. Phys. A* 304: 127–40
Chang, B. D., Draayer, J. P. 1979. *Phys. Rev.* C 20:2387–90
Chang, B. D., Wong, S. S. M. 1979. *Comput. Phys. Commun.* 18:35–61
Chang, B. D., Wong, S. S. M. 1980. *Comput. Phys. Commun.* 20:191–211
Chang, F. S., Zucker, A. 1972. *Nucl. Phys. A* 198:417–29
Chang, F. S., French, J. B., Thio, T. H. 1971. *Ann. Phys. NY* 66:137–88
Draayer, J. P., French, J. B., Wong, S. S. M. 1977. *Ann Phys. NY* 106:472–524
Dyson, F. J. 1962. *J. Math. Phys.* 3:1191–98
Dyson, F. J. 1970. *Commun. Math. Phys.* 19:235–50
Dyson, F. J., Mehta, M. L. 1963. *J. Math. Phys.* 4:701–12
Favro, L. D., McDonald, J. F. 1967. *Phys. Rev. Lett.* 19: 1254–56.
French, J. B. 1972. In *Dynamic Structure of Nuclear States*, ed. D. J. Rowe, L. E. H. Trainor, S. S. M. Wong, T. W. Donnelly, pp. 154–204. Univ. Toronto Press. 585 pp.
French, J. B., Draayer, J. P. 1979. In *Group Theoretical Methods in Physics*, ed. W. Beiglbock, A. Bohm, E. Takasugi, pp. 394–407. Berlin: Springer. 540 pp.
French, J. B., Mello, P. A, Pandey, A. 1978. *Ann. Phys. NY* 113:277–93
Garg, J. B., ed. 1972. *Statistical Properties of Nuclei*. New York: Plenum. 665 pp.
Grenander, U. 1963. *Probabilities on Algebraic Structures*. New York: Wiley. 218 pp.
Grimes, S. M. 1980. In *Moment Methods in Many-Fermion Systems*. ed. B. J. Dalton, S. M. Grimes, J. P. Vary, S. A. Williams, pp. 17–32. New York: Plenum. 511 pp.
Grimes, S. M., Bloom, S. D., Hausman, R. F. Jr., Dalton, B. J. 1979. *Phys. Rev.* C 19:2378–86
Halemane, T. R., Abbas, A., Zamick, L. 1981. *J. Phys. G* 7:1639–45
Haq, R. U., Pandey, A., Bohigas, O. 1982. *Phys. Rev. Lett.* 48: 1086–89
Haq, R. U., Wong, S. S. M. 1980. *Phys. Lett.* 93B:357–62
Hecht, K. T., Draayer, J. P. 1974. *Nucl. Phys. A* 223:285–319
Kahana, S., Lee, H. C., Scott, C. K. 1969. *Phys. Rev.* 185:1378–1400
Kar, K, 1981. *Nucl. Phys. A* 368:285–318
Kota, V. K. B. 1981. *Ann. Phys. NY* 134:221–58
Kota, V. K. B., Potbhare, V. 1979. *Nucl. Phys. A* 331:93–116
Kota, V. K. B., Potbhare, V. 1980. *Phys. Rev.* C 21:2637–42
Kuo, T. T. S., Brown, G. E. 1965. *Phys. Lett.* 18:54–58
Langhoff, P. W. 1980. See Grimes 1980, p. 195
Lieb, E. H., Mattis, D. C., eds. 1966. *Mathematical Physics in One Dimension*, p. 20. New York: Academic. 565 pp.
Lougheed, G. D., Wong, S. S. M. 1975. *Nucl. Phys. A* 243:215–28
Mehta, M. L. 1967. *Random Matrices*. New York: Academic. 259 pp.
Mehta, M. L. 1971. *Commum. Math. Phys.* 20:245–50
Mon, K. K., French, J. B. 1975. *Ann. Phys. NY* 95:90–111
Nissimov, H., Arvieu, R., Bohigas, O. 1972. *Nucl. Phys. A* 190:514–32
Pandey, A. 1979. *Ann. Phys. NY* 119:170–91

Pandey, A. 1981. *Ann. Phys. NY* 134:110–27

Parikh, J. C. 1978. *Group Symmetries in Nuclear Structure*. New York: Plenum. 277 pp.

Parikh, J. C., Wong, S. S. M. 1972. *Nucl. Phys. A* 182:593–605

Porter, C. E., Thomas, R. G. 1956. *Phys. Rev.* 104:483–91

Potbhare, V. 1977. *Nucl. Phys. A* 289:373–80

Potbhare, V., Pandya, S. P. 1976. *Nucl. Phys. A* 256:253–70

Quesne, C. 1975. *J. Math. Phys.* 16:2427–31

Quesne, C. 1976. *J. Math. Phys.* 17:1452–57

Quesne, C., Spitz, S. 1974. *Ann. Phys. NY* 85:115–51

Quesne, C., Spitz, S. 1978. *Ann. Phys. NY* 112:304–27

Ratcliff, K. F. 1971. *Phys. Rev. C* 3:117–43

Vincent, C. M. 1967. *Phys. Rev.* 163:1044–50

Von Neumann, J., Wigner, E. P. 1929. *Phys. Z.* 30:467–70

Weyl, H. 1946. *The Classical Groups*, pp. 251–75. Princeton, NJ: Princeton Univ. Press. 320 pp. 2nd ed.

Wigner, E. P. 1955. *Ann. Math.* 62:548–64

Wilson, K. G. 1979. *Sci. Am.* 241(2):158–79

Ann. Rev. Nucl. Part. Sci. 1982. 32:65–115
Copyright © 1982 by Annual Reviews Inc. All rights reserved

THEORY OF GIANT RESONANCES

K. Goeke and J. Speth[1]

Institut für Kernphysik, Kernforschungsanlage Jülich, D-5170 Jülich, West Germany; and Lehrstuhl für Theoretische Kernphysik; Universität Bonn, D-5300 Bonn, West Germany

CONTENTS

1. INTRODUCTION

Nuclear giant resonances were observed more than 30 years ago with the discoveries of Baldwin & Klaiber (1947, 1948) following the very

---

[1]Supported by the US Department of Energy under Contract No. DE-AC02-76ER13001.

0163-8998/82/1201-0065$02.00

first indications of Bothe & Gentner (1937) and the theoretical prediction of Migdal (1944). The signature of the phenomena discovered
was a strong increase of the photon absorption cross section at higher
excitation energies, showing a clear giant dipole resonance (GDR)
behavior. Soon after the first pioneering work these resonances
were found to be a general feature of all nuclei, except the lightest
ones, changing their form and width smoothly with the nuclear mass
number A.

These GDR were usually explained by means of macroscopic models
by Goldhaber & Teller (1948) and Steinwedel & Jensen (1950). The
Steinwedel-Jensen hydrodynamical model, originating also from Goldhaber & Teller (1948) describes the nucleus as a classical, incompress-
ible, two-fluid liquid drop with a fixed surface. The restoring force is
assumed to be proportional to the volume symmetry energy of the
Bethe-Weizsäcker mass formula. The Goldhaber-Teller model considers the oscillations of the nondeformed neutron and proton spheres
against each other, the restoring force corresponds to the surface
symmetry energy of the extended mass formula by Myers & Swiatecki
(1969, 1974). Without further modification, neither of these models
agrees with experiment.

Soon after the experimental detection and the theoretical investigations, researchers started discussing the possibility of giant resonances
different from the GDR (Danos 1952). This was motivated by the
phenomenology of the liquid-drop model, which exhibits various oscillations with different multipolarities and isospins depending on whether
the protons and neutrons vibrate in phase (isoscalar, $\Delta T = 0$) or in
opposite phase (isovector, $\Delta T = 1$). Here the isovector mode of a given
multipolarity shows an excitation energy higher than the corresponding
isoscalar one since the separation of protons and neutrons requires
further energy in addition to the deformation. The fascinating development in this field has been indeed the actual experimental discovery of
these (electric) giant resonances with multipoles different from $L = 1$. It
started with the isoscalar giant quadrupole resonances (GQR) detected
in inelastic electron scattering (Pitthan & Walcher 1971, Fukuda &
Torizuka 1972) and in inelastic proton scattering (Lewis & Bertrand
1972). These excitations and other isoscalar resonances have been
investigated in great detail recently by means of the selective properties
of inelastically scattered α particles.

Besides the isoscalar GQR, the isoscalar giant monopole resonances
(GMR) are also now experimentally well established throughout the
periodic table (Youngblood et al 1977). This "breathing mode" is of
particular interest since its excitation energy is directly connected with

the compression modulus K of the finite nucleus and thereby also of infinite nuclear matter. In a hydrodynamical model with a compressible fluid, Walecka (1962) and Werntz & Überall (1966) obtained $E(0^+) \propto \sqrt{K}$. The K is a very fundamental property of the effective nucleon-nucleon interaction and therefore of crucial importance. Today the only reliable values of K are extracted from the GMR (Blaizot 1980).

Giant resonances (GR) can be defined by the following characteristic features: First, the GR changes its form, width, and centroid energy smoothly with mass number A. Second, the width of the GR is small compared to its excitation energy. Third, the resonance exhausts a large fraction of the energy-weighted sum rule. This last property is particularly important. Sum rules, as for example that of Thomas, Reiche, and Kuhn (Levinger 1960), relate the strength distribution of the excited states integrated over all energies to some simple, experimentally detectable, ground-state properties such as the proton and neutron number, the rms radius, a multipole moment, etc. Hence the only theories suitable for describing GR are those whose corresponding model states satisfy the sum rules in the same way as the exact states.

The simplest many-body theory to satisfy the sum rules is the random phase approximation (RPA). This fundamental feature makes the RPA the basic microscopic theory for describing giant resonances. Indeed, as reviewed in this article, it has been extensively applied to the description of electric GR throughout the periodic table and it can be considered as rather successful. These calculations reveal two basic quantities of the effective nucleon-nucleon interaction: the compression modulus and the symmetry energies. However, the hope of better understanding the details of the force has not been fulfilled. It seems to be that the newly discovered magnetic resonances in charge exchange reactions are quite different in this respect and reveal remarkable features of the spin-isospin properties of the nuclear forces.

Besides the RPA, many other microscopic approaches have been and are still being applied to the description of giant resonances. These include, on the one hand, large-amplitude collective theories such as the adiabatic time-dependent Hartree-Fock theory (ATDHF) or the generator coordinate method (GCM), and on the other hand, the explicit evaluation of various mean energies of the RPA strength distribution. It is interesting to note that sum rule techniques provide a general tool for relating these theories to each other and to the RPA in a definite and precise way. A similar feature holds for recent macroscopic approaches in the framework of a fluid dynamical formalism. They appear to be approximate solutions of Landau's kinetic equation, which itself is the semiclassical limit of the RPA in an infinite medium. Therefore, as reviewed in Section 2, basically all presently used theories approximate the RPA or evaluate by simple means certain properties of the RPA spectrum. Thus the RPA appears as the fundamental microscopic theory for the description of centroid energies and transition probabilities of giant resonances. There are only a few formalisms that allow one to evaluate more details of the strength distribution, e.g. the spreading width. These approaches include the coupling of RPA states to 2p-2h configurations in one or the other form and are reviewed in Section 3.

Concerning the detailed analysis of experimental data, there exists a serious problem because in several cases giant resonances of different multipolarity are not well separated energetically. The most famous example of this is the giant monopole resonance, which, in medium and heavy mass nuclei, is nearly degenerate with the electric dipole reso-

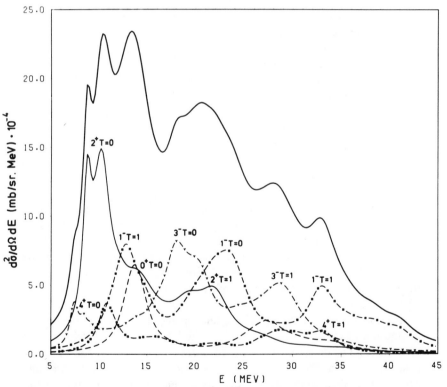

Figure 1 The electron spectrum for 90-MeV (e,e′) at 75° of ^{208}Pb calculated with microscopic RPA form factors (see Figure 4). The numbers in the figure denote the angular momentum, parity, and isospin of the corresponding cross-section contribution. The thick full line is the sum of all the different contributions (Wambach et al 1978).

nance (GDR). This fact was predicted theoretically by Ring & Speth (1973) but not until several years later was the monopole clearly established experimentally (Youngblood et al 1977). In Figure 1 the theoretical scattering cross section of 90-MeV electrons on ^{208}Pb is shown. The various types of giant resonances are the result of microscopic calculations discussed in Section 3. The figure gives an impression of the (at least theoretically expected) complexity of the experimental data. In Section 4 we show that microscopic nuclear structure models reduce to some extent the ambiguities of the conventional distorted-wave Born approximation (DWBA) analysis in those cases.

Using a highly energetic proton beam, Goodman et al (1980) found that the 1^+ Gamow-Teller resonance (GTR) is the dominant part of the (p,n) cross section at very forward angles. In the meantime, higher multipoles with $\Delta L = 1$ like 0^-, 1^- ($\Delta S = 1$), and 2^- resonances have been observed experimentally at slightly larger angles. These collective spin-isospin modes appearing energetically somewhat above the isobaric analogs of the target ground state have attracted the interest of theorists for two reasons: (*a*) They allow a selective investigation of the spin-isospin part of the particle-hole (ph) interaction, which is not well known so far. (*b*) Only ~ 30–50% of the theoretically expected GT transition strength has been found in these experiments. We show in Section 5 that this so-called quenching of the GT strength might be explained by an admixture of the $\Delta(1232)$-isobar nucleon-hole excitations into the low-lying proton-particle neutron-hole GT states. In addition, we demonstrate that these spin-isospin flip states offer the possibility of investigating the effects of the one-boson exchange potential inside the nucleus.

The gross properties of magnetic resonances and more detailed investigations of the electric resonances by means of coincidence techniques will probably be the subject of experimental and theoretical activities in this field.

2. THEORY OF GIANT RESONANCES

This section reviews the basic microscopic theories suitable for describing collective motion in finite Fermi systems. We here view the nucleus as a system of independently moving quasi-particles in a way which allows an easy transition to infinite nuclear matter.

First, we recall some basic concepts of time-dependent Hartree-Fock, Landau theory, and transport equations. Second, the resulting equations are considered in their small-amplitude limit, assuming the giant resonances to be harmonic vibrations around some static equilibrium state. This results basically in the random phase approximation and the corresponding zero-sound equation in nuclear matter. Various sum rules are formulated with respect to the RPA strength distribution associated with a given transition operator. These are used to relate other classical and semiclassical theories like the generator coordinate method or adiabatic time-dependent Hartree-Fock to RPA. In the last subsection, fluid dynamical and hydrodynamical approaches and their relationship to RPA and Landau's theory are reviewed and the concept of zero sound is discussed.

2.1   Basic Concepts

We shall assume that the complete description of collective motion is contained in the one-body density matrix $\rho = \rho(\mathbf{r}, \mathbf{r}', t) = \rho^*(\mathbf{r}', \mathbf{r}, t)$. The ρ is taken to be idempotent, $\rho^2 = \rho$. Thus we consider the system consisting of independent quasi-particles forming a many-quasi-particle state $|\phi(t)\rangle$. The equation of motion for $\rho(t)$ and $|\phi(t)\rangle$ is obtained by a variational solution of the time-dependent Schrödinger equation: $\langle \delta\phi(t)|H - i\hbar\partial/\partial t|\phi(t)\rangle = 0$. The variation corresponds precisely to considering the total energy $E[\rho] = \langle\phi(t)|H|\phi(t)\rangle$, being a functional of the density matrix $\rho(t)$, as a classical Hamiltonian with respect to the complex field $\rho(\mathbf{r}, \mathbf{r}', t)$:

$$ i\hbar \frac{\partial\rho(\mathbf{r}, \mathbf{r}', t)}{\partial t} = \frac{\delta E[\rho]}{\delta\rho^*(\mathbf{r}, \mathbf{r}', t)} = \frac{\delta E[\rho]}{\delta\rho(\mathbf{r}, \mathbf{r}', t)}. \qquad 1. $$

This is known as the time-dependent Hartree-Fock (TDHF) equation (Kerman & Koonin 1976):

$$ i\hbar \frac{\partial\rho}{\partial t} = [h[\rho], \rho], \qquad\qquad 2a. $$

which is explicitly written as

$$ i\hbar \frac{\partial\rho(\mathbf{r}, \mathbf{r}', t)}{\partial t} $$
$$ = \int d^3r[h(\mathbf{r}, \bar{\mathbf{r}}, t)\,\rho(\bar{\mathbf{r}}, \mathbf{r}', t) - \rho(\mathbf{r}, \bar{\mathbf{r}}, t)h(\bar{\mathbf{r}}, \mathbf{r}', t)] \qquad 2b. $$

with the nonlocal single-particle Hamiltonian $h[\rho(t)]$ (TDHF Hamiltonian).

It is useful for a comparison with fluid dynamical models and for a discussion of nuclear matter properties to introduce the Wigner transform of the one-body density matrix

$$n_{\mathbf{p}}(\mathbf{R}, t) = \int d^3s \exp\left(-\frac{i}{\hbar}\,\mathbf{p}\cdot\mathbf{s}\right) \rho\left(\frac{\mathbf{R}+\mathbf{s}}{2}, \frac{\mathbf{R}-\mathbf{s}}{2}, t\right) \qquad 3a.$$

and of the Hartree-Fock Hamiltonian

$$\varepsilon_{\mathbf{p}}(\mathbf{R}, t) = \int d^3s \exp\left(-\frac{i}{\hbar}\,\mathbf{p}\cdot\mathbf{s}\right) h\left(\frac{\mathbf{R}+\mathbf{s}}{2}, \frac{\mathbf{R}-\mathbf{s}}{2}, t\right). \qquad 3b.$$

In terms of $n_{\mathbf{p}}(\mathbf{R}, t)$ and $\varepsilon_{\mathbf{p}}(\mathbf{R}, t)$ the TDHF Equations 2 can be written as

$$\frac{\partial}{\partial t}\,n_{\mathbf{p}}(\mathbf{R}, t) - \frac{2}{\hbar}\,\varepsilon_{\mathbf{p}}(\mathbf{R}, t)\sin\left(\frac{\hbar}{2}\overleftrightarrow{\Lambda}\right)n_{\mathbf{p}}(\mathbf{R}, t) = 0. \qquad 4.$$

Here the operator $\overleftrightarrow{\Lambda}$ corresponds to the Poisson-bracket $\overleftrightarrow{\Lambda} = \overleftarrow{\nabla}_{\mathbf{R}}\cdot\overrightarrow{\nabla}_{\mathbf{p}} - \overleftarrow{\nabla}_{\mathbf{p}}\cdot\overrightarrow{\nabla}_{\mathbf{R}}$. Equation 4 is strictly equivalent to the TDHF Equations 2. In the context of Landau theory (Baym & Pethick 1978) it is often called the quantum kinetic equation. If one expands it in powers of $\hbar$ and retains only terms that do not vanish for $\hbar \to 0$, one obtains from Equation 4 the classical result

$$\frac{\partial n_{\mathbf{p}}(\mathbf{R}, t)}{\partial t} + \overrightarrow{\nabla}_{\mathbf{p}}\,\varepsilon_{\mathbf{p}}(\mathbf{R}, t)\cdot\overrightarrow{\nabla}_{\mathbf{R}}\,n_{\mathbf{p}}(\mathbf{R}, t)$$
$$- \overrightarrow{\nabla}_{\mathbf{R}}\,\varepsilon_{\mathbf{p}}(\mathbf{R}, t)\cdot\overrightarrow{\nabla}_{\mathbf{p}}\,n_{\mathbf{p}}(\mathbf{R}, t) = 0. \qquad 5.$$

This is exactly Landau's kinetic equation (Landau 1956, 1957, 1958) with a vanishing collision term [Vlasov equation (Vlasov 1954)], where the $\varepsilon_{\mathbf{p}}(\mathbf{R}, t)$ appear as quasi-particle energies. From this one obtains the Boltzmann transport equation (without collision term) by further limiting processes. One ignores the $\mathbf{R}$ and t dependence of the quasi-particle velocity, i.e. $\mathbf{v} = \overrightarrow{\nabla}_{\mathbf{p}}\,\varepsilon_{\mathbf{p}}(\mathbf{R}, t) = \text{const.}$ and considers from $\overrightarrow{\nabla}_{\mathbf{R}}\,\varepsilon_{\mathbf{p}}(\mathbf{R}, t)$ only the term due to the single-particle potential, i.e. $\mathbf{f} = -\overrightarrow{\nabla}_{\mathbf{R}}\,U(\mathbf{R}) = \text{const.}$ Then one gets immediately the Boltzmann equation without a collision term:

$$\left(\frac{\partial f}{\partial t} + \mathbf{v}\cdot\overrightarrow{\nabla}_{\mathbf{R}} + \mathbf{f}\cdot\overrightarrow{\nabla}_{\mathbf{p}}\right)n_{\mathbf{p}}(\mathbf{R}, t) = 0. \qquad 6.$$

Here the $n_{\mathbf{p}}(\mathbf{R}, t)$ is to be interpreted as a classical distribution function.

2.2 Small Amplitudes

Giant resonances are commonly considered to be well described as harmonic vibrations of small amplitude around some equilibrium density matrix ρ_0. Expanding the total energy $E[\rho]$ in terms of $\delta\rho = \rho - \rho_0$ to second order yields

$$E[\rho] = E[\rho_0] + \sum_{\alpha\beta} \frac{\delta E}{\delta\rho_{\alpha\beta}} \delta\rho_{\alpha\beta} + \frac{1}{2} \sum_{\alpha\beta\gamma\delta} \frac{\delta^2 E}{\delta\rho_{\alpha\beta}\delta\rho_{\gamma\delta}} \delta\rho_{\alpha\beta}\, \delta\rho_{\gamma\delta}. \qquad 7.$$

Here $\delta\rho_{\alpha\beta}$ are matrix elements of $\delta\rho$ in an arbitrary discrete or continuous representation. The functional derivatives have the following physical interpretation: The first derivative to be taken at the equilibrium density $(\delta E/\delta\rho_{\alpha\beta})_0 = (\delta E/\delta\rho^*_{\beta\alpha})_0 = h^0_{\beta\alpha}$ is the stationary single-particle Hamiltonian (stationary Hartree-Fock). The second derivative $\delta^2 E/\delta\rho_{\alpha\beta}\,\delta\rho_{\gamma\delta} = \delta h_{\beta\alpha}/\delta\rho_{\gamma\delta}$ describes how the single-particle Hamiltonian reacts to a small change in the density matrix. The equation of motion for $\delta\rho$ can immediately be obtained from the TDHF Equations 2 by linearization. Inserting $\rho = \rho_0 + \delta\rho$ yields immediately the Hartree-Fock equation for the equilibrium density $[h^0, \rho_0] = 0$ and the linear response equation $i\hbar\partial/\partial t\delta\rho = [h^0, \delta\rho] + [\delta h, \rho_0]$ with $\delta h = (\delta h/\delta\rho)\delta\rho$. For a collective motion with period ω, one can assume $\delta\rho(t) = \delta\bar\rho \exp(-i\omega t) + \delta\bar\rho^\dagger \exp(i\omega t) = \delta\rho^\dagger$ with a time-independent $\delta\bar\rho$. Inserting this into the linear response equation and separating the positive and negative frequencies yields for the $\exp(-i\omega t)$ terms

$$\hbar\omega\, \delta\bar\rho = [h^0, \delta\bar\rho] + [\delta h, \rho_0] \qquad 8.$$

and its hermitian conjugate for the $\exp(i\omega t)$ terms. This is the RPA equation. One brings it easily to the usual form by considering separately the particle-hole elements and hole-particle elements of $\delta\bar\rho$: $X_{\mathrm{ph}} = \delta\bar\rho_{\mathrm{ph}}$ and $Y_{\mathrm{ph}} = \delta\bar\rho_{\mathrm{hp}}$ respectively. They are the only non-vanishing matrix elements of $\delta\rho$. This yields immediately

$$\begin{pmatrix} A & B \\ -B^* & -A^* \end{pmatrix} \begin{pmatrix} X_{\mathrm{ph}} \\ Y_{\mathrm{ph}} \end{pmatrix} = \hbar\omega \begin{pmatrix} X_{\mathrm{ph}} \\ Y_{\mathrm{ph}} \end{pmatrix}. \qquad 9.$$

The matrices A and B are given by

$$A_{\mathrm{php'h'}} = (\varepsilon_p - \varepsilon_h)\delta_{\mathrm{pp'}}\, \delta_{\mathrm{hh'}} + \left(\frac{\delta h_{\mathrm{ph}}}{\delta\rho_{\mathrm{p'h'}}}\right)_0$$

$$B_{\mathrm{php'h'}} = \left(\frac{\delta h_{\mathrm{ph}}}{\delta\rho_{\mathrm{h'p'}}}\right)_0, \qquad 10.$$

where the ε_p and ε_h are the particle and hole energies, respectively. It is helpful for the discussion of nuclear matter properties and of fluid dynamical theories to use the Wigner transform $\delta n_{\mathbf{p}}(\mathbf{R}, t)$ of $\delta\rho(\mathbf{r}, \mathbf{r}', t)$. The expansion Equation 7 reads now

$$E[\rho] = E[\rho_0] + \int d^3R \sum_{\mathbf{p}} \varepsilon_{\mathbf{p}}(\mathbf{R}) \delta n_{\mathbf{p}}(\mathbf{R}, t)$$

$$+ \frac{1}{2} \int d^3R \, d^3R' \sum_{\mathbf{p}, \mathbf{p}'} f(\mathbf{p}, \mathbf{p}', \mathbf{R}, \mathbf{R}') \delta n_{\mathbf{p}}(\mathbf{R}, t) \delta n_{\mathbf{p}'}(\mathbf{R}', t). \quad 11.$$

Here we have set $\sum_{\mathbf{p}} = 4 \int d^3p/(2\pi\hbar)^3$ where for sake of simplicity spins and isospins are assumed to amount just to the factor 4. The undisturbed quasi-particle energies are $\varepsilon_{\mathbf{p}}^0(\mathbf{R}) = [\delta E/\delta n_{\mathbf{p}}(\mathbf{R})]_0$ and are the Wigner transforms of the static single-particle Hamiltonian $h^0(\mathbf{r}, \mathbf{r}')$. The particle-hole interaction is given by

$$f(\mathbf{p}, \mathbf{p}', \mathbf{R}, \mathbf{R}') = \left[\frac{\delta^2 E}{\delta n_{\mathbf{p}}(\mathbf{R}) \delta n_{\mathbf{p}'}(\mathbf{R}')}\right]_0 = \left[\frac{\delta\varepsilon_{\mathbf{p}}(\mathbf{R})}{\delta n_{\mathbf{p}'}(\mathbf{R}')}\right]_0. \quad 12.$$

For many commonly used interactions, e.g. of Skyrme type, this interaction turns out to be local

$$f(\mathbf{p}, \mathbf{p}', \mathbf{R}, \mathbf{R}') = \delta(\mathbf{R} - \mathbf{R}') f(\mathbf{p}, \mathbf{p}', \mathbf{R}).$$

Furthermore, in infinite systems, for excitations of a long wavelength or equivalently for particle-hole excitations with momentum q very much smaller than the Fermi momentum $|\mathbf{q}| = |\mathbf{p}_p - \mathbf{p}_h| = |\mathbf{p}'_p - \mathbf{p}'_h| \ll p_F$, the particle-hole interaction can be assumed to be a function of p_F and the angle between $\mathbf{p}$ and $\mathbf{p}'$. It is then natural to expand it in spherical harmonics

$$f(\mathbf{p}, \mathbf{p}') = \frac{1}{N_0} \sum_{lm} F_l \frac{4\pi}{2l+1} Y_{lm}^*(\hat{\mathbf{p}}') Y_{lm}(\hat{\mathbf{p}}), \quad 13.$$

with $N_0 = \frac{3}{2}\rho_0/\varepsilon_F$ being the density of single-particle states at the Fermi surface. The F_l are the well-known Landau parameters.

In this long wavelength nuclear matter limit, one can assume the $\delta n_{\mathbf{p}}(\mathbf{R}, t)$ to be small-amplitude fluctuations around $n_{\mathbf{p}}^0 = 4\theta(|\mathbf{p}| - |\mathbf{p}_F|)$, which is a simplified uniform equilibrium distribution, i.e. $n(\mathbf{p}, \mathbf{R}, t) = n_{\mathbf{p}}^0 + \delta n(\mathbf{p}, \mathbf{R}, t)$. The corresponding undisturbed quasi-particle energies are also consistently independent of $\mathbf{R}$, i.e. $\varepsilon_{\mathbf{p}}^0(\mathbf{R}, t) = \varepsilon_{\mathbf{p}}^0$ and $\varepsilon(\mathbf{p}, \mathbf{R}, t) = \varepsilon_{\mathbf{p}}^0 + \delta\varepsilon(\mathbf{p}, \mathbf{R}, t)$ with $\delta\varepsilon(\mathbf{p}, \mathbf{R}, t) = \sum_{\mathbf{p}'} f(\mathbf{p}, \mathbf{p}') \delta n(\mathbf{p}', \mathbf{R}, t)$. One thus obtains the small amplitude limit of the Vlasov Equation 5 as

$$\frac{\partial}{\partial t} \delta n_{\mathbf{p}}(\mathbf{R}, t) + \vec{\nabla}_{\mathbf{p}} \, \varepsilon_{\mathbf{p}}^0 \cdot \vec{\nabla}_{\mathbf{R}} \, \delta n_{\mathbf{p}}(\mathbf{R}, t)$$

$$- \vec{\nabla}_{\mathbf{R}} \, \delta\varepsilon_{\mathbf{p}}(\mathbf{R}, t) \cdot \nabla_{\mathbf{p}} \, n_{\mathbf{p}}^0 = 0. \quad 14.$$

For a periodic motion $\delta n_{\mathbf{p}}(\mathbf{R}, t) = \delta n_{\mathbf{p}}(\mathbf{q}, \omega) \exp[i(\mathbf{q} \cdot \mathbf{R} - \omega t)]$. Equation 14 can be transformed to

$$(\omega - \mathbf{q} \cdot \mathbf{v_p})\delta n_{\mathbf{p}}(\mathbf{q}, \omega) + \left(\frac{\partial n_{\mathbf{p}}^0}{\partial \varepsilon_{\mathbf{p}}}\right) \mathbf{q} \cdot \mathbf{v_p} \sum_{p'} f(p, p')\delta n_{\mathbf{p}}(\mathbf{q}, \omega) = 0,$$

15.

where $\mathbf{v_p} = \vec{\nabla}_{\mathbf{p}}\, \varepsilon_{\mathbf{p}}$. This is the RPA equation in the limit of an infinite medium (Landau 1956, 1957, 1958, Baym & Pethick 1978, Blaizot 1980). It corresponds precisely to Equation 9 and describes a wave with wave vector $\mathbf{q}$ and frequency ω. It is well known to describe the zero sound since the equation does not contain a collision term on its right-hand side. Hence the RPA equation can be considered as the equation describing the zero sound in a finite Fermi system.

2.3   The Random Phase Approximation

In Section 2.2 the RPA was derived in order to describe harmonic small-amplitude vibrations around the Hartree-Fock equilibrium position. It is given by Equation 9. If one denotes the various eigensolutions of the RPA equation by $|n\rangle$, where $|0\rangle$ indicates the corresponding (correlated) ground state, one can write $|n\rangle = \Omega_n^\dagger|0\rangle$ with $\Omega_n^\dagger$ being a boson creation operator $\Omega_n^\dagger = \Sigma_{ph} \{X_{ph}^{(n)}a_p^\dagger a_h - Y_{ph}^{(n)}a_h^\dagger a_p\}$. The ground state $|0\rangle$ appears now as the vacuum of the bosons $\Omega_n|0\rangle = 0$ for all n and the RPA amplitudes are given by $X_{ph}^{(n)} = \langle n|a_p^\dagger a_h|0\rangle$ and $Y_{ph}^{(n)} = \langle n|a_h^\dagger a_p|0\rangle$. Thus the matrix element of a transition operator $Q = \Sigma_{ik} Q_{ik}a_i^\dagger a_k$ between the RPA states $|n\rangle$ and the correlated ground state $|0\rangle$ is denoted as

$$|\langle n|Q|0\rangle|^2 = \sum_{ph} |Q_{ph}[X_{ph}^{(n)} + Y_{ph}^{(n)}]|^2.$$

16.

Therefore the RPA is especially suitable for the description of those excited states that differ from the ground state mainly by 1p-1h components. These are by definition the giant resonances since they exhibit large transition amplitudes between the excited states and the ground state with respect to an appropriate one-body transition operator Q. In many cases, the Q are simply the standard multipole operators $Q_L = r^L Y_{LM}$. Since the RPA is restricted to 1p-1h excitations, it is expected to be basically a good approximation for centroid energies and the total transition probabilities. The spreading of the 1p-1h transition strength (spreading width) due to a coupling to more complex configurations like 2p-2h states is not included (see Section 3.4). As far as the width of the resonances in the continuum is concerned, the RPA accounts only for the width due to the particle being in an unbound continuum state (escape width).

The basic outcome of RPA calculations and the characteristics of collective states were illustrated by Brown & Bolsterli (1959). They solved the RPA equations within a schematic model assuming a separable multipole-multipole force, $\chi Q_L \cdot Q_L$, and choosing all the ph-energies $\varepsilon_p - \varepsilon_h$ to be equal. The well-known result of that model is this: all the transition strength gets concentrated in one state, which is shifted to lower energies if the force is attractive ($\chi < 0$, usually isoscalar) and to higher energies if the force is repulsive ($\chi > 0$, usually isovector). In Figure 2 is plotted the result of a realistic RPA calculation of the $B(E2)$ strength distribution in ^{208}Pb (Rinker & Speth 1978). In the upper part the (uncorrelated) 1p-1h shell model states are shown. The major strength is spread in a range between 12 and 16 MeV ($2\hbar\omega$). The RPA result in the lower part nicely demonstrates the depletion of

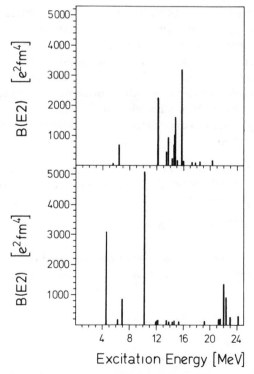

Figure 2 Comparison of the correlated and uncorrelated $B(E2)$-strength distribution in ^{208}Pb. In the upper part, the uncorrelated (no ph interaction) situation is shown; in the lower part, the RPA result is plotted (Rinker & Speth 1978).

these states and the concentration of the strength in one state at 10 MeV (isoscalar GQR) and in a couple of states beyond 20 MeV (isovector GQR).

In contrast to the schematic model, one finds in the realistic calculation a second collective isoscalar state that is much lower in energy. This is because the spin orbit force shifts the spin orbit partner with the larger spin below the Fermi surface so that it can couple with the remaining particle states to low-lying 2^+ ph configurations (upper part of Figure 2). The ph interaction mixes an appreciable part of the $2\hbar\omega$ strength into the low-lying states (lower part of the figure), which give rise to the rather collective low-lying 2^+ resonance.

2.4   Sum Rules

Sum rules have received increasing attention in the last years. Their use does not provide information as detailed as the numerical RPA calculations mentioned. Their interest lies rather in the possibility of obtaining simple expressions for some average properties of the collective excitation spectrum, which, being analytical, allow for a very transparent, sometimes even model-independent, study of systematic properties such as A dependence of excitation energies etc. Furthermore, as discussed in Section 2.5, other collective theories such as macroscopic models or the generator coordinate method can be related to RPA by sum rules in a systematic way. In addition, certain sum rules can be directly related to experimental data without making the random phase approximation.

2.4.1   EXACT SUM RULES   The response of the nucleus to the action of an arbitrary external potential Q is completely characterized by its associated strength function

$$S_Q(E) = \sum_n |\langle \Psi_n | Q | \Psi_0 \rangle|^2 \, \delta(E_n - E),$$

where $|\Psi_0\rangle$, $|\Psi_n\rangle$ are the (exact) ground and excited states of the system. The $S_Q(E)$ may have discrete and continuous parts. The strength function can be characterized by its moments (sum rules)

$$\begin{aligned} m_k(Q) &= \int_0^\infty (E - E_0)^k S_Q(E) \, \mathrm{d}E \\ &= \sum_n (E_n - E_0)^k |\langle \Psi_n | Q | \Psi_0 \rangle|^2, \end{aligned} \qquad 17.$$

with $k = 0, \pm 1, \pm 2, \ldots$. The moments are helpful to characterize how the transition strength is distributed as a function of the energy. For example, they allow one to define various excitation energies $E_k = (m_k/m_{k-2})^{1/2}$ or the average excitation energy (centroid energy) $\bar{E} = m_1/m_0$. The variance of the distribution is given by $\sigma^2 = m_2/m_0 - (m_1/m_0)^2$. Of particular interest are the odd moments, which can be written as

$$m_k(Q) = \tfrac{1}{2}(-1)^t(i)^k \langle \Psi_0 | [\bar{Q}_s, \bar{Q}_t] | \Psi_0 \rangle, \qquad 18.$$

where $\bar{Q}_n = [iH, \bar{Q}_{n-1}]$ with $\bar{Q}_0 = Q$, $\hbar = 1$, and $s + t = k$ (Bohigas et al 1979). As is apparent in Equations 17 and 18, properties of the excitation spectrum $|\Psi_n\rangle$ are related to properties of the ground state $|\Psi_0\rangle$. If Q is an isoscalar and velocity-independent local operator and if the forces used are velocity independent, then $\bar{Q}_1 = i[H, Q] = i[T, Q]$ and

$$m_1(Q) = \frac{\hbar^2}{2m} \langle \Psi_0 | (\vec{\nabla}Q)^2 | \Psi_0 \rangle, \qquad 19.$$

where T is the kinetic energy operator. For example, with the monopole operator $Q = \sum r_i^2$ one finds

$$m_1(Q) = \frac{2\hbar^2}{m} \left\langle \Psi_0 \left| \sum_i r_i^2 \right| \Psi_0 \right\rangle \qquad 20.$$

and for the multipole operator $Q = \sum_i r_i^L Y_{LM}$

$$m_1(Q) = \frac{L(2L+1)}{8\pi m} \left\langle \Psi_0 \left| \sum_i r_i^{2L-2} \right| \Psi_0 \right\rangle. \qquad 21.$$

In some cases the right-hand sides of Equations 20 and 21 are known experimentally and thus they provide a direct measure of the total strength contained in the strength function. Therefore they allow statements on how much of the total strength is exhausted by all the states in a given energy interval. This is how one classifies states as collective or noncollective.

Another important moment is $m_{-1}(Q)$, which is related to the polarizability of the nucleus. Consider the nucleus under the action of a weak external potential λQ. For small λ the changes of the expectation values $\langle Q \rangle$ and $\langle H \rangle$ are given by (dielectric theorem)

$$m_{-1}(Q) = -\frac{1}{2} \left(\frac{\partial \langle Q \rangle}{\partial \lambda} \right)_{\lambda=0} = \frac{1}{2} \left(\frac{\partial^2 \langle H \rangle}{\partial \lambda^2} \right)_{\lambda=0}. \qquad 22.$$

2.4.2 RPA SUM RULES The RPA analogues of the sum rules in Section 2.4.1 are obtained by replacing the exact states and energies by their random phase approximations. The analogue of Equation 18 (Goeke et al 1978) is a generalization of the theorem of Thouless (1961):

$$m_k^{\text{RPA}}(Q) = \tfrac{1}{2}(-1)^l(i)^k \langle\phi_0|[Q_s, Q_l]|\phi_0\rangle, \qquad k = \text{odd}, \qquad 23.$$

with $Q_n = [iH, Q_{n-1}]_{\text{ph}}$ and $Q_0 = Q_{\text{ph}}$, where ph indicates the 1p-1h components. Accordingly, in Equations 20–22 the exact $|\Psi_0\rangle$ has to be replaced by the HF state $|\phi_0\rangle$. In this limit, Goeke et al (1978) showed that all odd RPA sum rules are related to each other as $m_k(Q_q) = m_{k-2r}(Q_{q+r})$. Thus one sees that each odd RPA moment is given by two different sum rules, one being a generalization of the Thouless theorem, the other of the dielectric theorem.

Another remarkable feature of the hierarchy of Q_n is that it contains special cases that allow an interpretation of constrained Hartree-Fock (CHF) in terms of scaling (Goeke et al 1978): If a collective path is generated by means of CHF with $\langle\delta\phi_\lambda|H - \lambda Q_n|\phi_\lambda\rangle = 0$, standard CHF theory shows that for small λ one has $|\phi_\lambda\rangle = \exp\{\lambda[H, Q_{n-2}]\}|\phi_0\rangle$. In the special case of a local Q_{n-2} with $\nabla^2 Q_{n-2} = 0$, one obtains

$$\phi_\lambda(r_1, \ldots, r_A) = \exp\left[-\lambda \sum_i \mathbf{v}(\mathbf{r}_i) \cdot \vec{\nabla}_i\right] \times \phi_0(r_1, \ldots, r_A), \qquad 24.$$

with the velocity field $\mathbf{v}(\mathbf{r}) = \hbar^2/m \, \vec{\nabla} Q_{n-2}(\mathbf{r})$ associated to the velocity potential Q_{n-2}. Thus one can evaluate $m_3(Q_{n-2}) = m_{-1}(Q_n)$:

$$m_3(Q_{n-2}) = \frac{1}{2}\frac{\partial^2}{\partial\lambda^2} \langle\phi_0|\exp(-\lambda\overleftarrow{\nabla}\cdot\overleftarrow{v})H\exp(-\lambda\overrightarrow{v}\cdot\overrightarrow{\nabla})|\phi_0\rangle. \qquad 25.$$

The above three ways to calculate sum rules — direct evaluation of the commutators of the Thouless theorem, constrained Hartree-Fock and polarization, and generalized scaling procedure — have frequently been used in the literature to evaluate the moments m_k with $k = -1, +1, +3$ for a variety of giant resonances. One should note that, if a true HF state $|\phi_0\rangle$ is used, the results are strictly equivalent to the moments obtained from a self-consistent continuum RPA strength distribution. In this way, the isoscalar and isovector monopole resonances have been investigated by Bohigas et al (1976, 1979), Martorell et al (1976), Goeke & Castel (1979), Goeke et al (1980), and Stringari et al (1978). Some of their results, combined with values of Auerbach & van Giai (1978), are given for electric giant resonances in Table 1. The m_{-3} were evaluated by adiabatic time-dependent Hartree-Fock as described in Section 2.5. The m_0 have an error of $\pm 2\%$ since they are encircled by inequalities between the moments described by Bohigas et al (1979) and Goeke et al (1980). The force used is the Skyrme-III interaction. One realizes the decreasing variance σ of the RPA strength

Table 1   Sum rules for isoscalar and isovector monopole resonances[a]

	^{16}O		^{40}Ca		^{90}Zr		^{208}Pb	
	$T=0$	$T=1$	$T=0$	$T=1$	$T=0$	$T=1$	$T=0$	$T=1$
$\bar{E}_0 = (m_1/m_0)$ (MeV)		33.2		32.7		32.6		27.6
$E_{-1} = (m_{-1}/m_{-3})^{1/2}$ (MeV)	23.1	25.1	22.4	25.0	21.0	26.7	16.1	19.3
$E_1 = (m_1/m_{-1})^{1/2}$ (MeV)	28.2	32.1	25.6	31.8	22.1	32.1	16.9	26.8
$E_3 = (m_3/m_1)^{1/2}$ (MeV)	34.5	40.6	28.5	38.6	23.4	35.8	17.9	30.5
m_{-3} (fm^4 MeV^{-3})	0.0217	0.0150	0.118	0.0706	0.641	0.246	7.30	2.75
m_0 (fm^4)		294		1370		5520		27,100
σ (MeV)	9.9	12.3	6.3	10.6	3.8	7.9	2.9	7.1

[a] The values are compiled from Auerbach & van Giai (1978), Goeke & Castel (1979); Bohigas et al (1979), Goeke et al (1980).

distribution with increasing mass number. Thus only for nuclei with $A > 40$ is the energy E_3 close to the peak and can it be identified with the resonance energy. However, a comparison with Figure 6 shows that apparently the σ is very much larger than the width of the single, well-isolated collective peak, which is only a few hundred keV in heavy mass nuclei.

Using Skyrme's interactions and simple velocity fields for the isoscalar monopole ($L = 0$, $T = 0$), $\mathbf{v(r)} = \nabla(\mathbf{r}^2/2) = \mathbf{r}$, and the isoscalar quadrupole $\mathbf{v(r)} = \vec{\nabla}(2z^2 - x^2 - y^2) = (-x, -y, +2z)$, one obtains even analytical results for $m_3^{RPA}(Q)$. For the quadrupole, for example, one gets (Bohigas et al 1979)

$$E_3^{RPA}(L = 2, T = 0) = \left(\frac{4\hbar^2}{m\langle r^2\rangle_0}\right)^{1/2} [T + E(t_1, t_2) + \tfrac{1}{4}E_{LS} - \tfrac{1}{5}E_{Coul}]^{1/2}.$$

26.

Here the T is the kinetic energy, $E(t_1, t_2)$ is the finite-range energy originating from the terms $\propto t_1$, t_2 of the total energy, etc. Actually there are various approximations to Equation 26 in the literature (Golin & Zamick 1975, Goeke et al 1976, Suzuki 1981). If one ignores E_{LS} and E_{Coul} and approximates $[1 + E(t_1, t_2)/T] \simeq m/m^*$, one obtains

$$E_3^{RPA}(L = 2, T = 0) = \left(\frac{4\hbar^2}{m^*\langle r^2\rangle_0} T\right)^{1/2} \simeq \sqrt{2}\hbar\omega\sqrt{\frac{m}{m^*}}.$$

27.

For $m^*/m = 1$ one reproduces the old result of Mottelson (1960) and Suzuki (1973).

The isoscalar monopole sum rules are intimately related to the compression modulus of the finite nucleus. If one defines this as

$K_A = (\langle r^2 \rangle_0 / A) \partial^2 (E/A) / \partial \langle r^2 \rangle$, one obtains (Jennings & Jackson 1980, 1980a)

$$E_1^{\mathrm{RPA}} = \left(A \frac{\hbar^2 K_A^{\mathrm{CHF}}}{m \langle r^2 \rangle_0} \right)^{1/2} \qquad E_3^{\mathrm{RPA}} = \left(A \frac{\hbar^2 K_A^{\mathrm{SCL}}}{m \langle r^2 \rangle_0} \right)^{1/2}, \qquad 28.$$

where one distinguishes between K_A^{CHF} and K_A^{SCL} depending on whether the change of the square radius is evaluated by CHF or by scaling. From some determinantal inequalities between the moments, one can show that $K_A^{\mathrm{CHF}} \leq K_A^{\mathrm{SCL}}$. Only if one single state exhausts the full RPA strength with respect to $\sum r_i^2$ as transition operator are both compression moduli equal. This is apparently not the case since, as Jennings & Jackson (1980a) discussed, a comparison of the nuclear matter compression modulus K_{NM} with the large-A limits of K_A^{SCL} and K_A^{CHF} yields

$$\lim_{A \to \infty} K_A^{\mathrm{SCL}} = \frac{10}{7} \lim_{A \to \infty} K_A^{\mathrm{CHF}} = K_{\mathrm{NM}}. \qquad 29.$$

A recent investigation of Blaizot (1980) based on the presently available data reveals a value of $K_{\mathrm{NM}} = 210 \pm 30$ MeV.

Suzuki (1981a) applied sum rule techniques to the description of the mean energy of isobaric analogue states and Gamow-Teller states using the effective Hamiltonian suggested by Bohr & Mottelson (1975, 1981) and Migdal's force. His estimates of the force constants, based on the available experiments, are in reasonable agreement with those extracted by Bertsch (1981).

2.5 Collective Theories

Besides the previously mentioned RPA and the sum rule approaches, other collective theories such as the cranking model, the generator coordinate method (GCM), time-dependent Hartree-Fock, and adiabatic time-dependent Hartree-Fock (ATDHF) have also been used to describe giant resonance states. These theories are basically suitable for the description of large-amplitude nuclear collective motion as heavy ion scattering, fission, fusion, or nonharmonic vibrations. Consequently in their small-amplitude limit they do not account for the full spectrum. This is because these theories require an explicit perturbation of the nuclear ground state by means of an external constraint or by scaling procedures, whereas RPA corresponds to a complete eigenvalue problem. Thus, while RPA provides a detailed description of the full strength distribution with respect to any transition operator, the GCM, ATDHF, etc are only able to evaluate certain moments $m_k^{\mathrm{RPA}}(Q)$ of the distribution.

Actually Goeke et al (1978) showed (see Goeke 1979 for review) that these theories can be intimately related to RPA by means of sum rule techniques (Martorell et al 1976, Bohigas et al 1976). In the following we show this for ATDHF and GCM since all the other theories of this sort can be identified after minor modifications with either of them (Goeke 1976).

If one disturbs the nuclear ground state by some external field Q and constructs accordingly a collective path $(|\phi_\lambda\rangle)$ by means of CHF, i.e. $\langle \delta\phi_\lambda | H - \lambda Q | \phi_\lambda \rangle = 0$, then the GCM gives the total stationary wave function of the system as $|\Psi_\alpha\rangle = \int d\lambda \, f_\alpha(\lambda) |\phi_\lambda\rangle$. By varying the total energy $\langle \Psi | H | \Psi \rangle$ with respect to the weight function $f(\lambda)$, one obtains the Hill-Wheeler equation:

$$\int [\mathcal{H}(\lambda, \lambda') - E\mathcal{N}(\lambda, \lambda')] f(\lambda') \, d\lambda' = 0 \qquad\qquad 30.$$

where $\mathcal{H}(\lambda, \lambda') = \langle \phi_\lambda | H | \phi_{\lambda'} \rangle$ and $\mathcal{N}(\lambda, \lambda') = \langle \phi_\lambda | \phi_{\lambda'} \rangle$. The integral equation can be made discrete and solved numerically. However, one obtains further insight into the physical interpretation of the eigenvalue E by assuming the Gaussian overlap approximation. By well-known techniques the integral Equation 30 can now be transformed to a Schrödinger equation for the collective wave function. In the limit of small values of λ this reads (Goeke & Reinhard 1980 and references therein)

$$H_{\text{GCM}} = -\frac{\hbar^2}{2M_{\text{GCM}}} \frac{\partial^2}{\partial \lambda^2} + \frac{1}{2} \lambda^2 \left(\frac{\partial^2 V}{\partial \lambda^2} \right)_{\lambda=0} , \qquad\qquad 31.$$

with $V(\lambda) = \langle \phi_\lambda | H | \phi_\lambda \rangle$ being the classical potential energy surface. The second derivative is a spring constant and to be identified with $2m_{-1}(Q)$ in case of CHF with Q as constraint and with $2m_3(Q)$ in case of Q being the velocity potential. The mass parameter M_{GCM} can be expressed in terms of the RPA matrices A and B with the result that $M_{\text{GCM}}(\lambda) \lesssim 2m_{-3}(Q)$. Thus the excitation energy of GCM in the small-amplitude limit is

$$E_{\text{GCM}}^{\text{CHF}} \geq \left[\frac{m_{-1}^{\text{RPA}}(Q)}{m_{-3}^{\text{RPA}}(Q)} \right]^{1/2} \quad \text{and} \quad E_{\text{GCM}}^{\text{SCL}} \geq \left[\frac{m_3^{\text{RPA}}(Q)}{m_1^{\text{RPA}}(Q)} \right]^{1/2} \qquad 32.$$

for CHF and scaling, respectively. The failure of the GCM mass not to be identical to m_{-3}^{RPA} corresponds to the well-known fact that GCM in case of translations does not reproduce the total mass of the system but gives a slightly smaller value. This can be remedied (Goeke & Reinhard 1980) by using a two parameter collective path $\phi_{\lambda\dot\lambda}$. In such a case the equalities hold in Equations 32. Thus the results of the GCM are approximately embedded in the RPA and RPA sum rule formalism, at least in the case of small amplitudes and in the Gaussian overlap approach.

The generator coordinate method, with and without Gaussian overlap assumptions, has been applied by various authors to the monopole and quadrupole giant resonances (Abgrall & Caurier 1975, Caurier et al 1973, Flocard & Vautherin 1975, Krewald et al 1975, 1976, Giraud & Grammaticos 1975). In some of these papers, and this holds also for the ATDHF ones quoted below, there was a confusing variation in the theoretical position of the resonances, in particular for the monopole. This resulted partly from the use of interactions with quite different properties, as for example the compression modulus. The main reason, however, was that the resulting energy, i.e. E_{-1} for CHF and E_3 for scaling, was often falsely interpreted as "the" excitation energy of the giant resonance although for nuclei with $A \leq 40$ the strength distribution is rather broad.

The ATDHF theory (Goeke & Reinhard 1978, Baranger & Vénéroni 1978) starts from a time evolution along a given collective path $\phi_{\lambda\dot\lambda}$ with small values of the collective velocity $\dot\lambda$. An adiabatic expansion in powers of $\dot\lambda$ yields in second order the classical Hamiltonian $\langle\phi_{\lambda\dot\lambda}|H|\phi_{\lambda\dot\lambda}\rangle = \frac{1}{2}M_{\text{ATDHF}}(\lambda)\dot\lambda^2 + \langle\phi_\lambda|H|\phi_\lambda\rangle$. One obtains for the collective mass in case of CHF with constraint Q and scaling with velocity field Q (Goeke 1976, Goeke et al 1978)

$$M^{\text{CHF}}_{\text{ATDHF}}(\lambda) = 2m^{\text{RPA}}_{-3}(Q) \quad \text{and} \quad M^{\text{SCL}}_{\text{ATDHF}}(\lambda) = 2m^{\text{RPA}}_1(Q). \qquad 33.$$

Together with Equation 23 for the spring constant, one obtains therefore in the harmonic limit $E^{\text{CHF}}_{\text{ATDHF}} = E^{\text{RPA}}_{-1}$ and $E^{\text{SCL}}_{\text{ATDHF}} = E^{\text{RPA}}_3$. Thus the ATDHF results in the small-amplitude limit are directly related to the RPA sum rules. In this limit the ATDHF has been applied to monopole and quadrupole vibrations by Engel et al (1975), Blaizot & Grammaticos (1974), Vautherin (1975), Goeke (1977), Goeke & Castel (1979), and Goeke et al (1980). Apparently the ATDHF is an appropriate tool to evaluate $m^{\text{RPA}}_{-3}(Q)$ and there it adds a new feature to the pure sum rule results, which basically concentrate on m^{RPA}_{-1}, m^{RPA}_1, and m^{RPA}_3. The value of $m^{\text{RPA}}_{-3}(Q)$ for isoscalar and isovector monopole resonances can be found in Table 1.

Apparently neither GCM nor ATDHF in their small-amplitude limit are able to calculate anything that could not be evaluated by RPA or by RPA sum rules. In fact, by going through the hierarchy of Q_n constraints they just evaluate all odd RPA moments. The virtue of GCM and ATDHF, however, lies in their numerical simplicity. Furthermore, these theories can be extended to large amplitudes, a question little studied in the context of giant resonances. In ^{16}O Goeke (1977) observed a 15% reduction of the E_3^{RPA} due to large-amplitude effects of the breathing mode. For other nuclei this deserves further study.

2.6 Macroscopic Models

Macroscopic models of collective nuclear motion and of giant resonances in particular have always complemented microscopic calculations. They appeal to our intuitive understanding of collective motion in terms of a convection current $\mathbf{j}(\mathbf{r})$ and a local density $\rho(\mathbf{r})$ much more than a time-dependent nonlocal density matrix $\rho(\mathbf{r}, \mathbf{r}', t)$ as in fully microscopic approaches. Furthermore, macroscopic models are very easy to handle numerically and can in a simple way relate systematic features of collective excitations to gross properties of the energy functional and the corresponding nuclear ground state. In this respect they have very much in common with sum rule approaches and, indeed, to some extent both are identical.

Some macroscopic models are based on sum rules of Noble (1971), Deal & Fallieros (1973), and Suzuki & Rowe (1977), which read for a single state $|\omega_\lambda\rangle$ exhausting the total sum rule value

$$\langle 0|\hat{j}(\mathbf{r})|\omega_\lambda\rangle\bigg| = -\frac{i\omega_\lambda}{2mS_\lambda}\langle 0|r^\lambda Y_{\lambda\mu}|\omega_\lambda\rangle\rho_0(r)\,\vec{\nabla}r^\lambda Y_{\lambda\mu}(\hat{\mathbf{r}}) \qquad 34a.$$

$$\langle 0|\hat{\rho}(\mathbf{r})|\omega_\lambda\rangle = -\frac{\lambda}{2mS_\lambda}\langle 0|r^\lambda Y_{\lambda\mu}|\omega_\lambda\rangle r^{\lambda-1}\frac{\partial\rho_0}{\partial r}\,Y_{\lambda\mu}(\hat{\mathbf{r}}). \qquad 34b.$$

Here $\hat{j}$ and $\hat{\rho}$ are the operators associated with $\mathbf{j}(\mathbf{r})$ and $\rho(\mathbf{r})$ and $S_\lambda = \lambda(2\lambda+1)A\langle r^{2\lambda-2}\rangle/8\pi m$ of the ground state. The essential feature is that now one can fulfill both sum rules and the continuity equation simultaneously if the $\rho_0(\mathbf{r}, \mathbf{r}')$ of the ground state is deformed by means of a time-dependent scaling transformation similar to that of Equation 24:

$$\rho_+(\mathbf{r}, \mathbf{r}', t) = \exp\tfrac{1}{2}[\mathbf{S}(\mathbf{r}, t)\cdot\vec{\nabla} + \vec{\nabla}\cdot\mathbf{S}(\mathbf{r}, t)$$
$$+ \mathbf{S}(\mathbf{r}', t)\cdot\vec{\nabla}' + \vec{\nabla}'\cdot\mathbf{S}(\mathbf{r}', t)]\rho_0(\mathbf{r}, \mathbf{r}'). \qquad 35.$$

Assuming $\mathbf{S}(\mathbf{r}, t) = \alpha(t)\mathbf{v}(\mathbf{r})$, one can express the total energy in terms of a kinetic part $\propto\dot{\alpha}^2$ and a potential part $\propto\alpha^2$ from which the excitation energy can be derived (Engel et al 1975, Bertsch 1975, Sagawa & Holzwarth 1978, Suzuki 1980). Using the velocity fields of Section 2.4.2 and Skyrme's interaction, one obtains exactly Equation 26 for the excitation energy of the isoscalar quadrupole resonance. From this, one concludes that the above-described macroscopic models precisely evaluate $(m_3/m_1)^{1/2}$ of the corresponding RPA strength distribution.

One interesting new feature these calculations have revealed (Bertsch 1975, Sagawa & Holzwarth 1978, Holzwarth & Eckart 1978) concerns a hydrodynamic description of nuclear collective motion. If one evaluates the Wigner transform $n_p(\mathbf{R}, t)$ of the scaled one-body density matrix, one obtains instead of an isotropic $n_p(\mathbf{R}) = \theta[p^2 - p_F^2(\mathbf{R})]^2$ as it is in the ground state [with $p_F^3(\mathbf{R}) = \frac{3}{2}\pi^2\rho(\mathbf{R})$], a highly anisotropic local Fermi surface $n_p(\mathbf{R}, t) = \theta[p^2 - p_F^2(\mathbf{R}, \vartheta, \varphi, t)]$, where ϑ and φ are related to the directions of distortion of $\rho(\mathbf{r}, \mathbf{r}')$. In case of a quadrupole

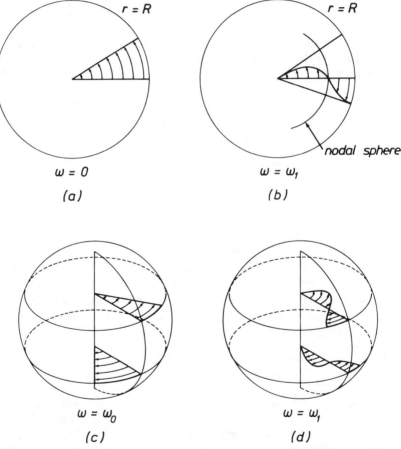

Figure 3 Velocity field for the lowest unnatural (magnetic) parity modes of a spherical Fermi-liquid droplet: (*a*) 1^+ rigid rotation at $\omega=0$; (*b*) first excited mode; (*c*) lowest excited 2^- twist mode; (*d*) second excited 2^- mode (Holzwarth & Eckart 1979).

distortion, for example, the Fermi surface at **R** is an ellipsoid in momentum space. This feature, associated to zero sound, is ignored in purely hydrodynamical models where, owing to the assumption of frequent nuclear collisions, the local Fermi surface is always assumed to be spherical. However, as we know, the mean free path of a nucleon in a nucleus is rather large and hence relaxation mechanisms to reduce the deformation of the Fermi surface should be negligible. Thus one must conclude that for low-lying excitations nuclear hydrodynamics is not justified. For the isoscalar GQR, hydrodynamical models (Eckart et al 1981) yield a wrong A dependence in disagreement to experiment, $\hbar\omega_2 \sim 63 \, A^{-1/3}$, which is accurately reproduced by $(m_3^{RPA}/m_1^{RPA})^{1/2}$.

The above approaches basically evaluated the excitation energy of imposed modes by using a properly chosen velocity field associated with $\mathbf{S}(\mathbf{r}, t)$. The first attempts to determine $\mathbf{S}(\mathbf{r}, t)$ by variational principles were made by Holzwarth & Eckart (1979). They invoke semiclassical approximations for the kinetic energy in the spirit of a generalized Thomas-Fermi ansatz, outgoing from a deformed local Fermi surface, and derive linearized equations of motion for $\mathbf{S}(\mathbf{r}, t)$. The application of this fluid dynamical formalism to non-normal parity modes in spherical nuclei suggests the existence of a twist mode, the restoring force of which originates solely from the deformation of the Fermi surface. Some velocity fields are given in Figure 3. Actually the fluid dynamical theory is a well-defined approximation to Landau's kinetic equation without a collision term (Vlasov equation) in the small-amplitude limit; see Equation 15 (Yukawa & Holzwarth 1981). This can be seen by assuming $\delta n_\mathbf{p}(\mathbf{q}, \omega) = -(\partial n_\mathbf{p}^0/\partial\varepsilon_\mathbf{p}) \, \delta a_\mathbf{p}(\mathbf{q}, \omega)$ with $|p| = p_F$ and performing a multipole expansion $\delta a_\mathbf{p}(\mathbf{q}, \omega) = \Sigma a_{lm}(\mathbf{q}, \omega)Y_{lm}(\hat{p})$. Together with Equation 15, one obtains an infinite chain of linear equations for the amplitudes a_{lm}. If one truncates this at $l \leq 1$, one obtains ordinary hydrodynamics; for $l \leq 2$ one obtains the fluid dynamics. This corresponds to the findings of Nix & Sierk (1980), who take second-order velocity moments of the linearized small-amplitude Boltzmann Equation 6. It is very satisfying that these theories can thus fully be related to RPA, TDHF, and Landau's theory.

3. APPLICATION OF THE RPA TO GIANT RESONANCES

Most of the microscopic calculations of giant resonances have been performed in the framework of the RPA. Therefore we restrict ourselves in the following discussion to those results. In order to solve the

RPA equations, one needs single-particle energies, single-particle wave functions, and the ph interaction. The calculations of the various groups differ mainly in the manner in which these input data are determined. Here one distinguishes between two different methods: shell model RPA and self-consistent RPA.

3.1 Shell Model RPA

In this approach one starts from a Saxon-Woods or harmonic oscillator potential from which one obtains the single-particle energies and the corresponding wave functions. In addition, one has to choose an appropriate ph interaction, usually making an ansatz of the following form:

$$F^{ph}(\mathbf{r}_1, \mathbf{r}_2) = (f_0 + f'_0 \boldsymbol{\tau}_1 \cdot \boldsymbol{\tau}_2 + g_0 \boldsymbol{\sigma}_1 \cdot \boldsymbol{\sigma}_2 + g'_0 \boldsymbol{\sigma}_1 \cdot \boldsymbol{\sigma}_2 \boldsymbol{\tau}_1 \cdot \boldsymbol{\tau}_2) \cdot F(\mathbf{r}_1, \mathbf{r}_2). \qquad 36.$$

Noncentral contributions like the tensor force are not included in this ansatz. The expression in the parentheses is the so-called exchange mixture and $F(\mathbf{r}_1, \mathbf{r}_2)$ gives the radial behavior of the force (which, in principle, could be different for each channel). In many calculations a "separable" force is used, which is very convenient in numerical applications.

The various groups who investigate giant resonances come to similar conclusions as far as excitation energies and transition strengths are concerned, although they differ in the choice of the ph interaction. The schematic multipole-multipole force was used in the calculation of giant multipole resonances by Soloviev and co-workers (see Soloviev 1976, 1978) and by Bes et al (1975). A more sophisticated version of a separable force, suggested by Knüpfer & Huber (1976a), is known as the "model of separable residual interaction" (MSI). In the Landau-Migdal theory (Migdal 1967, Speth, Werner & Wild 1977), which has been extensively applied to giant resonances, a density-dependent zero-range force is used in the actual calculations. The results obtained by several groups are summarized in review papers by Kamerdzhiev (1979) and Speth & van der Woude (1981).

In the following we discuss a few typical examples for the various microscopic approaches. In Figure 4 a 1p-1h RPA spectrum of ^{208}Pb is shown (Wambach 1979). Here the ph basis includes nearly all $4\hbar\omega$ excitations and the single-particle resonances are approximated as quasi-bound states. Therefore, this excitation spectrum consists of discrete states only. Similar results were obtained by Kamerdzhiev (1973), who actually did the first microscopic calculation within that

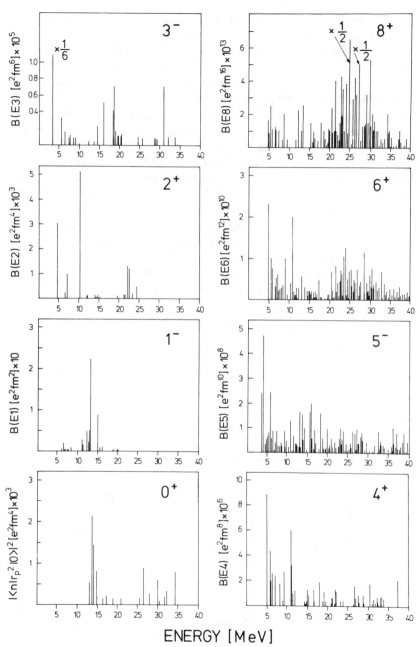

Figure 4   Theoretical spectrum of ^{208}Pb calculated within the RPA (Wambach 1979).

model (see Figure 5), Grecksch, Knüpfer & Huber (1975), Borzov & Kamerdzhiev (1975), and Ring & Speth (1973, 1974).

One important result follows immediately from Figure 4: The giant multipole resonances are not well separated energetically from each other. This fact makes the experimental identification in some cases very difficult, as described in Section 3.2.

We also notice that only for excitations with small angular momentum is the transition strength well concentrated as predicted from the schematic model. Only in this case is the basic assumption of the schematic model (degenerated ph energies) fulfilled approximately. The main reason is again the spin-orbit interaction, which shifts the state with the largest spin into the next lower major shell. With increasing angular momentum we obtain more and more coupling possibilities so that the approximate $2\hbar\omega$ spacing of the unperturbed ph energies disappears. This is nicely demonstrated in Figure 4 in the case of the 5^-, 6^+, and 8^+ strength distributions.

In all the calculations mentioned so far the theoretical resonances do not have any width because the unbound, single-particle resonances

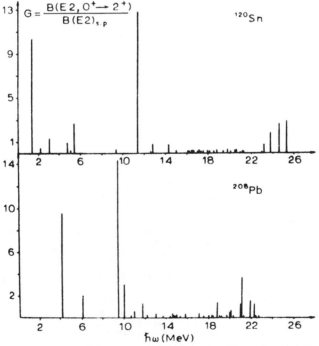

Figure 5   B(E2) distribution (in single-particle units) in (a) ^{120}Sn and (b) ^{208}Pb, calculated by Kamerdzhiev (1972).

have been described as quasi-bound states. In these cases one made the continuum discrete by diagonalizing the Saxon-Woods Hamiltonian in a finite basis of harmonic oscillators (Ring & Speth 1974) or by putting the nucleus into a box (Bertsch & Tsai 1975, Liu & Brown 1976). An extension of the RPA is the continuum RPA, which takes into account correctly the width of the single-particle resonances with the ansatz

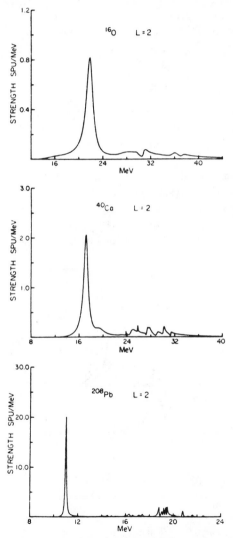

Figure 6   Continuum RPA calculations (inclusion of the escape width) of the GQR (Shlomo & Bertsch 1975).

$\psi_\lambda(E)\rangle = \Omega_\lambda^+(E)|0\rangle$. Here Ω_λ^+ is a generalization of the boson creation operator given in Section 2.3 (Krewald et al 1974). In such an approach the escape width is taken into account but it is only a part of the experimental width. Results of such calculations are shown in Figure 6 for the quadrupole resonances in ^{16}O, ^{40}Ca and, ^{208}Pb. In all cases the experimental centroid energies of the isoscalar GQR agree nicely with the theoretical ones; the experimental width, however, is much larger. This arises from the "spreading" width, discussed below. The extremely narrow width of the isoscalar GQR in ^{208}Pb is due to the Coulomb and centrifugal barrier that prevents the resonances from decaying by particle emission.

3.2 Self-consistent RPA

The self-consistent RPA is from a conceptional point of view more fundamental than the shell model RPA. It basically consists of a Hartree-Fock calculation and a successive RPA calculation using the single-particle energies and wave functions as they come out of HF. Both calculations use the same interaction, mostly taken to be phenomenological and of Skyrme type. In some cases a Brueckner G matrix has been used (Müther et al 1976). It is clear from Section 2.1 that the ph interaction to be used in RPA has to be evaluated from the energy by the functional derivatives. Calculations of this sort were performed for

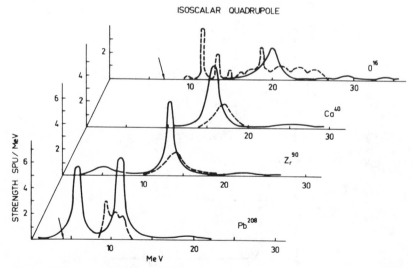

Figure 7 Self-consistent RPA calculations in a discrete basis (the width is artificially included) of the GQR in several nuclei (Liu & Brown 1976). The broken curves denote the experimental data.

doubly closed shell nuclei by Bertsch & Tsai (1975), Liu & Brown (1976), Krewald et al (1977), Blaizot et al (1976), Blaizot & Gogny (1977), and Blaizot (1980). In this approach one has fewer free parameters and takes less information from experiments (e.g. one uses the HF s.p. energies). Therefore one may expect that the general agreement between theory and experiment is not as good as it is in the previous cases.

In Figure 7 we give one example of a self-consistent RPA calculation. The widths of the resonances ·in this calculation by Liu & Brown are artificially included; the authors used a discrete basis. Self-consistent continuum RPA calculations, e.g. by Krewald et al (1977), Liu & van Giai (1976), Cavinato et al (1982), give results similar to those shown in Figure 6. Also, in these cases the theoretical widths are much smaller than the experimental ones.

3.3   Dynamical Theory of Collective States

One basic difference between the shell model RPA and the self-consistent RPA relates to the single-particle spectrum. It is well known that the single-particle spectra of medium and heavy mass nuclei calculated within the HF approach deviate appreciably from the empirical ones. The empirical low-lying spectrum is more compressed than the theoretical one. Actually the theoretical result depends on the effective mass used in a specific HF calculation. Realistic forces derived from a Brueckner G matrix give an effective mass in nuclear matter of $m^*/m = 0.6$–0.7 whereas the empirical spectrum corresponds in the average to $m^*/m \gtrsim 1$ (Brown et al 1963). Bertsch & Kuo (1968) and Hamamoto & Siemens (1976) showed that the coupling of the single-particle (hole) states to nuclear vibrations gives rise to a compression of the single-particle spectrum, which explains essentially the differences between the HF and the empirical spectrum. The nucleon effective mass and correspondingly the single-particle spectrum can be regarded as made up of two pieces (Figure 8). The first part (Figure 8a) is the HF contribution of the effective interaction, and the second part (Figure 8b) is the phonon contribution. In the framework of the self-consistent RPA only the first piece is included in the single-particle energies. In the shell model RPA, on the other hand, where one uses an empirical single-particle spectrum (which corresponds to $m^*/m \sim 1$), both contributions are included implicitly. Therefore this version of the RPA also includes (in principle) processes of the form shown in Figure 9. In the original version of the dynamical theory of collective states, Brown & Speth (1979) and Brown, Dehesa & Speth (1979) argued that the phonon

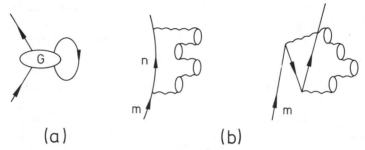

(a)   (b)

Figure 8 Interactions leading to the nucleon effective mass: (*a*) the Hartree-Fock contribution; (*b*) the contributions arising from coupling to collective modes.

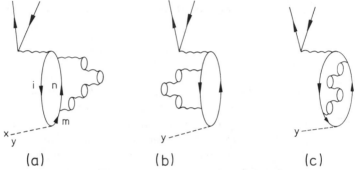

(a)   (b)   (c)

Figure 9 Coupling of collective modes to a resonance that is viewed as having been started by an external field, denoted by *y*.

contribution (Figure 8*b*) is different for high-lying and low-lying single-particle states. Therefore it is not possible to include this piece through a constant m^*/m, but one has to introduce an energy-dependent effective mass (Jeukenne et al 1976). This energy dependence, however, introduces another serious problem if such a spectrum is used within the RPA (see Figure 9). Here, in general the phonon contributions are off energy, which means that the corresponding contributions of Figures 8*b* and 9 are different from each other. This difference is small for the low-lying collective resonances and non-collective states (which are close to the unperturbed ph energies). Therefore it is necessary and meaningful to use here the empirical single-particle spectrum. In the case of giant resonances, however, the compression effect due to the phonons is much smaller than for low-lying states. Therefore one has to use a single-particle spectrum close to the HF result. In order to simulate this energy dependence the authors suggested an effective mass that depends on the energy of the vibration investigated. A more elaborated version of this approach was recently given by Brown & Rho

(1981). In the meantime the energy dependence of the effective mass has become much better understood through the work of Bernard & Mahaux (1981) and Mahaux and Ngô (1981). Even in the original version of the dynamical theory, the authors pointed out that the energy dependence introduced through the phonon coupling has to be introduced explicitly (Section 3.4). The results of such an extended theory obtained by Bortignon & Broglia (1981) and Wambach et al (1982) justify the essential features of the dynamical theory. This concept resolved some long-standing problems: (a) It explains why one should use empirical single-particle energies rather than the HF spectrum in order to describe low-lying resonances and noncollective states. (b) It explains why shell model RPA calculations of the giant electric dipole resonance (using reasonable effective nucleon-nucleon forces) always give results 2–3 MeV lower than the experimental excitation energies.

3.4 2p-2h RPA Calculations

So far we restricted the discussion to the 1p-1h RPA. This is appropriate for the description of the centroid energy and the total transition strength of giant resonances. But the width and fine structure of the resonances can be understood theoretically only if 2p-2h configurations are taken into account. The residual interaction mixes the 1p-1h components into the 2p-2h ones, which spreads the transition strength. This spreading width of the resonances is the dominant contribution to the total width, at least for the isoscalar GQR and the breathing mode. In ^{16}O and ^{208}Pb the experimental width of the GQR is of the order of 8 MeV and 2 MeV, respectively. This is much larger than the escape widths, shown in Figure 6. The inclusion of 2p-2h configurations in the framework of a shell model diagonalization gives a good qualitative description of the experimental facts in ^{16}O (Hoshino & Arima 1976, Knüpfer & Huber 1976a,b). For numerical reasons such calculations are restricted to light nuclei. Therefore methods have been developed that take into account the 2p-2h configurations in an approximate way. Extensive calculations along this line were first performed by Soloviev (1976), Malov et al (1976), Kyrchev et al (1978).

In the core-coupling RPA by Dehesa et al (1977) the effects of collective phonons are taken into account by coupling them to the single-particle and single-hole states. The various terms contributing to the spreading width in this approach are shown in Figure 9. In the actual calculation by Dehesa (1977), the effect of the graph of Figure 9c (which renormalizes the ph interaction) was included in an approximate way only: The strength of the ph force has been readjusted in order to reproduce the centroid energy of the corresponding resonances. As examples, the quadrupole and monopole strength distributions of ^{208}Pb are shown in Figure 10. It is obvious from the comparison with the escape width shown in Figure 6 that for the isoscalar GQR the spreading width is by far the most important one. As the escape width is extremely small in heavy mass nuclei in the giant quadrupole and monopole region, one should not necessarily expect these resonances to appear as one broad bump with a width of a few MeV; they may be fragmented into several narrow peaks that stick up from a much broader part. The actual form of the giant resonances therefore depends on the ratio of escape to spreading width. If the escape width is much larger than the mean distance between two fragments, the resonance will appear as one single broad peak. For the opposite situation, i.e. a small escape width, the various fragments can be resolved in high resolution experiments. The comparison of the Figures 6 and 10 suggests that in general we have a mixture of these two limiting cases. This discussion

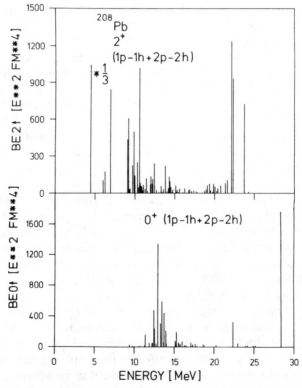

Figure 10 Monopole and quadrupole strength distribution in ^{208}Pb calculated within the core-coupling RPA (Dehesa 1977).

also explains, at least qualitatively, the high resolution electron data from Darmstadt (Meuer et al 1981). Calculations considering simultaneously the escape and spreading width have not yet been performed.

Another method for calculating giant resonance decay was proposed by Bertsch et al (1979) in the framework of the "nuclear field theory" and by Wambach et al (1982) in a Fermi-liquid formulation. These approaches are similar to the core-coupling RPA of Dehesa et al (1977) but the numerics is simpler and the models can be used with realistic level densities.

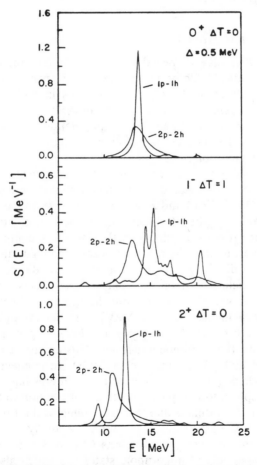

Figure 11 Strength distribution of the monopole, dipole, and quadrupole resonances in ^{208}Pb calculated within the extended RPA, which includes the 2p-2h configurations (spreading width). An arbitrary width of 0.5 MeV has been folded into the discrete spectra in order to simulate the escape width (Wambach et al 1982).

Figure 11 shows the calculated strength functions of the giant monopole, dipole, and quadrupole resonances in ^{208}Pb (Wambach et al 1982). These can be compared directly with Dehesa's calculation shown in Figure 10 because in both cases a Landau-Migdal type of force has been used. Both calculations are performed in a discrete basis, but in the calculation by Wambach et al a purely artificial width of 0.5 MeV was added to each state to simplify the mathematics. Similar results were obtained by Bortignon & Broglia (1981). This calculation also assigns an arbitrary width to each discrete level in order to simulate the decay width of the resonances.

3.5 Deformed Nuclei

Deformed nuclei play a special role in the giant dipole resonances because the statically deformed ground state gives rise to the very spectacular splitting of the dipole resonances in the rare earth and uranium region. This effect was first theoretically predicted by Danos (1958) and Okamoto (1958) and then experimentally detected by Fuller & Weiss (1958). The interesting question is whether this splitting exists also for other multipolarities.

The giant multipole resonances in deformed nuclei in the heavier mass region were investigated within the quasi-particle RPA by Zawischa & Speth (1975, 1976) and Zawischa et al (1978). One typical result of the distribution of the $B(E2)$ and $B(E0)$ strength is shown in Figure 12. First glance reveals a splitting of the various K components of the GQR. However, neither the coupling to 2p-2h configurations (Soloviev 1976, Malov et al 1976, Kyrchev et al 1978) nor the coupling to the continuum is included. If one estimates those effects by introducing artificially a line width extracted from the neighbored spherical ^{144}Sm (full width, half maximum = 3.9 MeV), one obtains only one broad bump, the width of which is roughly 1–2 MeV larger. This is in fair agreement with the experimental results (Kishimoto et al 1975, Schwierczinski et al 1975, Bertrand et al 1978). In connection with the breathing mode denoted by F in Figure 12 it is interesting to see that this resonance is split into two pieces (dotted contribution in the peak C). This unexpected splitting is due to the coupling of the breathing mode with the $K^\pi = 0^+$ component of the GQR. Actually the effect is well known from the low-lying β vibrations (A), which are not only the $K^\pi = 0^+$ components of quadrupole states but which also possess an enhanced $B(E0)$ value. Very recently some evidence for such an effect has been discovered (Buenerd et al 1980, Garg et al 1980a, b, Morsch et al 1981).

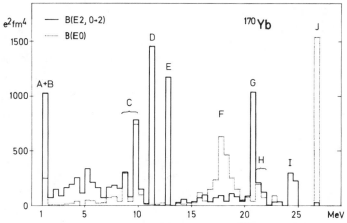

Figure 12 Distribution of $B(E2)$ and $B(E0)$ strength in ^{170}Yb. The $B(E\lambda)$ strength is summed in intervals of 0.5 MeV. The peaks are identified by capital letters. A: lowest $K^{\pi} = 0^+$ excitation. B: lowest $K^{\pi} = 2^+$ excitation. C, D, and E: isoscalar giant quadrupole resonances for $K^{\pi} = 0^+$, 1^+, and 2^+ components, respectively. F: $K^{\pi} = 0^+$ state ($\Delta T = 0$), predominantly of a breathing-mode type (the states C and F are both superpositions of β vibrations and breathing mode). G, H, and I: isovector quadrupole resonances, $K^{\pi} = 0^+$, 1^+, and 2^+ components, respectively. J: this $K^{\pi} = 0^+$ state ($\Delta T = 1$) is predominantly an isovector breathing mode (Zawischa et al 1978).

Similar investigations within the Hill-Wheeler generator coordinate method have been performed in light nuclei by Caurier et al (1973) and Abgrall & Caurier (1975). Within a simple phenomenological model Abgrall et al (1980) also investigated the situation in the rare earth region.

In light nuclei a completely microscopic description of the giant resonances was given by Schmid & Do Dang (1977, 1978) and Schmid (1980). They used angular momentum projected, deformed Slater determinants as nuclear structure wave functions and also incorporated the continuum. The theoretical results are in fair agreement with the experimental facts.

4. MICROSCOPIC ANALYSIS OF EXPERIMENTAL DATA

In order to analyze the results of hadron and electron scattering experiments one generally uses the distorted-wave Born approximation (DWBA). The major complication in the analysis of giant resonances compared to bound states is that the peaks to be analyzed (after the subtraction of the background, which is an additional and essentially unsolved problem) are generally a superposition of several resonances of different multipolarities (see Figure 1). For that reason a conventional analysis in the framework of phenomenological transition densities and transition potentials is not applicable in this context since it does not allow a unique answer. This is demonstrated in the following.

In the case of inelastic electron scattering, we restrict our discussion to the first Born approximation. The differential cross section is give as

$$\frac{d\sigma}{d\Omega} = 4\pi\sigma_M \left[\frac{q_\mu^2}{q^2} F_L^2 + \left(\frac{1}{2} \frac{q_\mu^2}{q^2} + \tan^2 \frac{\theta}{2} \right) F_T^2 \right] \qquad 37.$$

where σ_M, F_L, and F_T are the Mott cross section and the longitudinal and transverse form factors, respectively. In the analysis of electric resonances (except at very large angles) only the longitudinal form factor enters. It is given as $F_L^{(J)} \propto \int_0^\infty \rho_{if}(r) j_J(kr) r^2 \, dr$ where $j_J(kr)$ is the spherical Bessel function and $\rho_{if}(r)$ is the (radial part of the) transition density which includes all the nuclear structure information. The quantity $\rho_{if}(r)$ introduces the model dependence into the analysis. In general the phenomenological transition densities used so far deviate appreciably from the microscopic ones calculated within the RPA. One example that played an important role in connection with the discovery of the breathing mode is shown in Figure 13. In the upper part of Figure 13 we compare the RPA transition density of the electric dipole state at 12.9 MeV shown in Figure 4 with various phenomenological ones. The difference in the densities gives rise to rather different (e, e′) cross sections, as shown in the lower part of this figure. It is obvious that the magnitude of the monopole strength (which is situated in the same energy region) deduced from an (e, e′) experiment depends strongly on the model for the E1 resonance used in the analysis (see Figure 1).

In the case of hadron scattering one gets the following form of the differential cross section of a particle with the initial velocity v_i which is scattered inelastically at a nucleus:

$$\frac{d\sigma}{d\Omega} = \frac{2\pi}{hv_i} |T_{fi}|^2 \bar{\rho}_f(E), \qquad 38.$$

where $\bar{\rho}_f$ denotes the density of final states of the nucleus. The direct part of the transition amplitude T_{fi} is given by (we neglect all spin and isospin dependence):

$$T_{fi}^{\text{dir}} = \int \chi_f^{(-)}(\mathbf{k}_f, \mathbf{r}) U_{fi}(\mathbf{r}) \chi_i^{(+)}(\mathbf{k}_i, \mathbf{r}) \, d^3 r, \qquad 39.$$

where $\chi^{(\pm)}$ are the optical model wave functions and $U_{fi}(\mathbf{r})$ is the transition potential:

$$U_{fi}(\mathbf{r}) = \int \rho_{fi}(\mathbf{r}_1) V(\mathbf{r} - \mathbf{r}_1) \, d^3 r_1. \qquad 40.$$

Here $V(\mathbf{r} - \mathbf{r}_1)$ is the interaction between the projectile and the target nucleons and $\rho_{fi}(\mathbf{r})$ is again the nuclear transition density. The corresponding formula for the exchange amplitude T^{exch}, being of crucial importance in the case of nucleon-nucleus scattering, may be found in Madsen's (1975) review paper.

In the analysis of hadronic experiments there occurs an additional difficulty compared to (e, e') because the interaction $V(\mathbf{r} - \mathbf{r}_1)$ in Equation 40 between the projectile and the nucleons in the target must be known. This interaction has been more or less parametrized in earlier works and the free parameters have been adjusted to known experimental results (Halbert & Satchler 1974, Halbert et al 1975). In the more recent approaches one tries to calculate microscopically this interaction starting from the free nucleon-nucleon interaction (Brieva & Rook 1977, 1978a, b, Geramb et al 1978, Petrovich & Love 1981).

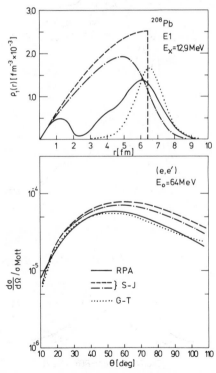

Figure 13 Comparison of various giant electric dipole transition densities for ^{208}Pb and the corresponding cross sections for inelastically scattered electrons. The S-J and G-T correspond to the Steinwedel-Jensen (1950) and the Goldhaber-Teller (1948) model, respectively.

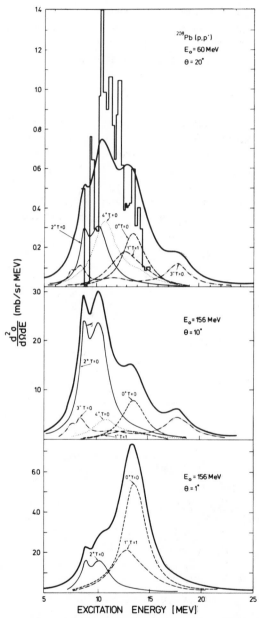

Figure 14 Spectra in ^{208}Pb (p, p′) at different energies and scattering angles (Wambach et al 1979). The theoretical results are derived from the RPA spectra shown in Figure 4. The full thick line in each figure is the sum of all theoretical results. The experimental data shown as a histogram in the upper part of the figure are taken from Bertrand (1976).

Two examples demonstrate the complications of an analysis of hadronic data:

1. Figure 14 shows spectra for ^{208}Pb(p, p'). The transition densities are based on 2p-2h RPA calculations (Dehesa 1977) and the Eikemeier & Hackenbroich (1971) potential has been used as projectile-nucleon interaction (Wambach et al 1979). The experimental data are taken from Bertrand (1976). Again the (theoretically expected) complexity of the spectra is shown. In the upper part, for example, one notices that at this specific angle the 4^+ contribution is as big as the GQR contribution. The comparison of the cross sections at three different angles shows how the angular distribution can be used to unravel different multipolarities.

2. In Figure 15 we show the angular distribution of inelastically scattered α particles into the giant resonance region of ^{208}Pb (Wambach 1979). The various theoretical cross sections were calculated like those done by Wambach et al (1979). The experimental data are taken from

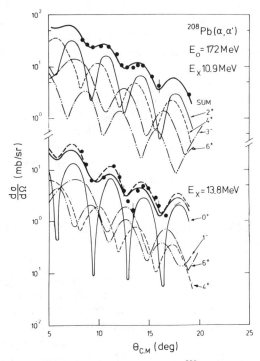

Figure 15   Angular distributions for giant resonances in ^{208}Pb at 10.9 and 13.8 MeV from the (α, α') reaction using 172-MeV α particles. The experimental data are taken from Morsch et al (1980) and the theoretical results are derived from the RPA results shown in Figure 4 (Wambach et al 1979).

Table 2 Sum-rule strength in the GQR-region of ^{208}Pb

	$L = 2$	$L = 3$	$L = 4$	$L = 6$
RPA result	65	5	26	16
Fit 1	59	5	16	16
Fit 2	74	25	5	

Morsch et al (1980). It is obvious that only the superposition of the contributions of all the various (theoretical) multipole resonances can explain such experimental cross sections. This might be considered as a nice confirmation of the RPA results that predicts these overlapping multipole resonances, but it is by no means an experimental proof. For comparison we also show in Table 2 the result of an analysis of the same data using phenomenological transition densities (Morsch et al 1980). Within the experimentals errors, Morsch et al (1980) could not distinguish between the two fits shown in Table 2. In the meantime Fit 2 can be ruled out by sum rule arguments since there exist evidence that the major part of the isoscalar E3 resonance is around 17 MeV.

The investigation of collective states with high energetic protons ($E_p \geq$ 100 MeV) is of special interest because this probe penetrates deeper inside the nucleus than alpha particles. Therefore these high energetic protons are much more sensitive to the details of the transition densities than are other hadronic probes (Osterfeld et al 1979). An analysis of such experimental data within the conventional collective model might give $B(EL)$-values that are smaller by up to a factor of two than those extracted by microscopic techniques. Experimental results (Marty et al 1975, Bertrand et al 1981, Djalali et al 1981) support this theoretical conclusion.

5. SPIN-ISOSPIN RESONANCES IN CHARGE EXCHANGE REACTIONS

The various models discussed in the previous section describe nearly equally well the gross properties of the electric resonances. But the results are rather insensitive to details of the ph interaction used. One may thus conclude that the theoretical investigation of the electric giant resonances has mainly proven that the RPA is an appropriate many-body theory to describe GR. The most extensive calculations of GR have been performed within the Landau-Migdal theory. Here one approximates the ph interaction (Equation 36) by a density-dependent, zero-range force that is without any doubt a very crude approximation to the nucleon-nucleon interaction. Nevertheless the theoretical results for isoscalar and isovector electric resonances are in good agreement with the experimental data. These quantities are basically only sensitive to the f_0 and f'_0 parameters of the ph interaction (Equation 36), which are directly connected with the compressibility and symmetry energy of the nucleus (Migdal 1967).

We see in the following that the resonances in the spin-isospin channel behave quite differently in this respect. These states are sensitive to details of the interaction and offer, therefore, for the first time the possibility of a thorough and detailed investigation of the effect of one-boson exchange potentials (OBEP) inside the nucleus.

5.1 Particle-Hole Interaction in the Spin-Isospin Channel

Migdal's original ansatz of the ph force in this channel was also a zero-range force. The investigation of magnetic moments and transition probabilities in odd-mass nuclei (Migdal 1967, Speth, Werner & Wild 1977) revealed that this force should be strongly repulsive ($g'_0 \sim 0.5$–0.6). But it is well known that the energies of unnatural parity (magnetic) states are close to the unperturbed (experimental) particle-hole energies, which indicates that the spin-dependent interaction is weak. Finally, from the (theoretical) investigations of pion condensation one even expects an attractive force in this channel.

In order to overcome this difficulty, Migdal (1972) explicitly added to the zero-range part the one-pion exchange potential (OPEP), which gives rise to a momentum-dependent interaction (if one Fourier-transforms Equation 36 into the momentum space). Baym & Brown (1975) argued that one must also consider the effect of the ρ exchange potential, because the tensor force from the ρ exchange cancels that from the π exchange and at short distances this is of crucial importance in connection with pion condensation and precritical phenomena. The simplest ansatz of the $\sigma\tau$ part of Equation 36 including all these effects can be written (in momentum space) as (Anastasio & Brown 1977):

$$F_{\sigma\tau}^{\text{ph}}(\mathbf{q}) = C_0 \cdot g'_0 \boldsymbol{\sigma} \cdot \boldsymbol{\sigma}' \boldsymbol{\tau} \cdot \boldsymbol{\tau}' - \frac{4\pi}{m_\pi^2} f_\pi^2 \frac{\boldsymbol{\sigma} \cdot \mathbf{q} \boldsymbol{\sigma}' \cdot \mathbf{q}}{q^2 + m_\rho^2} \boldsymbol{\tau} \cdot \boldsymbol{\tau}'.$$

$$-\frac{4\pi \tilde{f}_\rho^2}{m_\rho^2} \cdot \frac{(\boldsymbol{\sigma} \times \mathbf{q}) \cdot (\boldsymbol{\sigma}' \times \mathbf{q})}{q^2 + m_\rho^2} \boldsymbol{\tau} \cdot \boldsymbol{\tau}'. \qquad 41.$$

Here the tilde on $\tilde{f}_\rho^2$ indicates that the reduction of the (central part) of the ρ exchange potential due to the short-range correlations is incorporated.

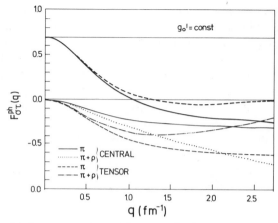

Figure 16 Graphical representation of the generalized spin-dependent interaction in momentum space (Equation 41). The full thick line is the sum of all contributions. The thick dashed line is again the sum of all contributions, but with a central part of the ρ contribution reduced by 0.4 (Speth et al 1980).

In Figure 16 the graphical representation of Equation 41 is given. Here the OPEP and ρ exchange potential are separated into a central and a tensor part. The zero-range term in Equation 41 is a constant in the q-space ($g_0' = $ const.). The thick full line is the sum of all contributions. The thick dashed line is again the sum of all contributions, however with a central part of the ρ contribution reduced by a factor of 0.4.

The effects of this generalized ansatz on magnetic properties in light and heavy mass nuclei were investigated by Speth et al (1980) using the interaction shown in Figure 16. From that figure one immediately realizes that the force is strongly repulsive if small momentum transfers are involved but it is very weak if the corresponding momentum transfer is large. Section 5.2 discusses the influence of the momentum dependence of the force on the spin-flip charge exchange resonances.

5.2   *Microscopic Description of Spin-Isospin Modes*

Only very recently were collective spin-isospin modes discovered in (p, n) and (^{3}He, t) charge exchange experiments (Doering et al 1975, Goodman et al 1980, Bainum et al 1980) in medium and heavy mass nuclei. The most prominent of these newly discovered pionic states is the 1^+, $\Delta L = 0$, $\Delta S = 1$ giant Gamow-Teller resonance (GTR), theoretically predicted many years ago by Ikeda et al (1963). In (p, n) experiments at very forward angles and with highly energetic protons ($E_p > 100$ MeV), these resonances turn out to be the dominant reaction channel. This fact is related to the energy dependence of the isospin-dependent parts of the nucleon-target interaction V in Equation 40 (Petrovich and Love 1981): The strength of the τ-dependent part of the interaction is strongly reduced if the energy of the incoming proton is increased so that the excitation of non spin-flip states is very weak for $E_p > 100$ MeV. On the other hand, the $\sigma\tau$ force is nearly independent of the proton energy; therefore beyond $E_p > 100$ MeV mainly spin-flip states are excited in (p, n) reactions. At lower energies ($E_p \lesssim 40$ MeV) the two pieces are nearly equally strong. This surprising effect can be simply understood in terms of one- and two-boson exchange contributions to the interaction (Brown, Speth & Wambach 1981).

Examples of zero-degree (p, n) spectra are shown in Figure 17. The structure of these states is very similar to the well-known isobaric analog

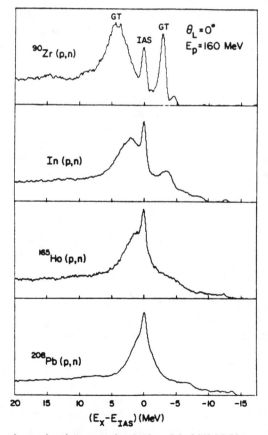

Figure 17 Zero-degree (p, n) spectra of several nuclei of 160-MeV protons. The spectra are plotted on an excitation energy scale centered on the isobaric analog state (Goodman 1982).

states (IAR). Both resonances can be described in the framework of the RPA as a superposition of proton-particle, neutron-hole states. In the case of the IAR the particle-hole pairs are coupled to 0^+, in the GTR case to 1^+. Both kinds of states are expected to be rather collective in heavy mass nuclei. Since the GTR are connected with a spin and isospin flip they allow a selective investigation of the spin-isospin part of the ph interaction. The same is true for the 0^-, 1^-, and 2^- states ($\Delta L = 1$, $\Delta S = 1$ resonances) that dominate the cross sections at slightly larger angles.

The effect of the momentum dependence of the force on a given state can qualitatively be discussed by considering the corresponding form factors, since the largest (diagonal) contributions to the direct part of the ph force are simply a double convolution of the transition density with the interaction. As examples, we show in Figure 18 the form factors of the GTR and the $\Delta L = 1$ resonances in ^{208}Pb. It is obvious from that figure that in the GTR only very small momenta are involved, which means that the GTR are sensitive to the strongly repulsive part of the $\sigma\tau$ interaction. In addition, the dominant part of the form factor is peaked in a narrow range; this indicates that the excitation energy of the GTR is insensitive to details of the ph force.

Results very similar to each other were obtained with a separable, constant force (Bohr & Mottelson 1981, Brown & Rho 1981), zero-range force (Gaarde et al 1981, Bertsch et al 1981), and the momentum-dependent force of Equation 41 where the effects of the dynamical theory were also considered (Krewald et al 1981). In all cases the major part of the GTR strength is concentrated in one single resonance that is shifted to higher energies compared to the unperturbed ph energies. On the other hand, the coupling to the $\Delta(1232)$-particle-nucleon-hole configurations that strongly modify the transition strength depends sensitively on the form of the interaction used, as discussed in the next section.

Figure 18 also shows the form factors of the 0^-, 1^-, and 2^- resonances ($\Delta L = 1$). Those of the 0^- and 1^- resonances are similar to that of GTR, but the maximum is shifted to higher momentum transfer. Therefore the repulsion of the ph force is weaker than in the GTR case (Osterfeld et al 1981). Nevertheless it is still strong enough to build up one single collective state in which the major fraction of the 0^- and 1^- strength is concentrated. The 2^- result is qualitatively different. The form factor predicts that the ph force in this specific case should be weak due to the high q-components. Actually it is so weak that there exists no longer a single collective state but five 2^- states with roughly the same strength. In addition, those states are only slightly shifted from their

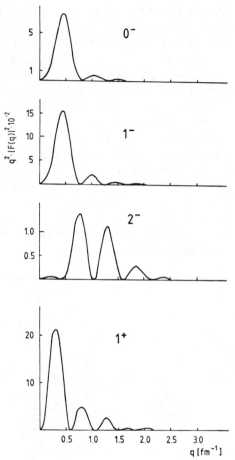

Figure 18   Fourier transform of the (p, n) form factors of $\Delta L = 0$ (GTR, 1^+) and $\Delta L = 1$ resonances (0^-, 1^-, 2^-) in ^{208}Pb (Osterfeld et al 1981).

uncorrelated particle-hole energies. In Figure 19 a microscopic analysis of (p, n) experiments in ^{208}Pb is shown (Osterfeld et al 1981). In the experiment using 160-MeV protons, the spin-flip $\Delta L = 1$ resonances are strongly excited. Here one indeed sees that the 2^- strength is much more spread out than the 0^- and 1^- ($\Delta S = 1$) strengths and that the 2^- strength is several MeV lower in energy than the other two resonances. This explains in a natural way the large experimental width of the $\Delta L = 1$ resonance.

The small contribution of the non-spin-flip 1^- ($\Delta S = 0$) resonance in the upper part of Figure 19 is due to the very weak τ part of the coupling potential (Equation 40) at this high proton energy. Actually the 1^- ($\Delta S = 0$) resonance is the most collective one and is dominant in the low-energy spectrum (lower part of Figure 19) while the contributions of the spin-flip states are comparatively small (this again results from the energy dependence of the τ part of the coupling potential). One finds, therefore, a shift in the centroid energy of the two spectra of about 2.5 MeV. This general feature is found experimentally in many medium and heavy mass nuclei (Sterrenburg et al 1980).

For a consistent description of GTR, $\Delta L = 1$, and higher magnetic multipole states (Speth et al 1980, Krewald & Speth 1980), it is crucial to consider the momentum dependence of the $\sigma\tau$ part of the interaction. With a zero-range force the excitation energies of the $\Delta L = 1$ resonances are too high compared to experiments, if one takes the strength deduced from the GTR (Bertsch et al 1981). The situation gets even worse if one considers the transition strength of the resonances, which is discussed in the next subsection.

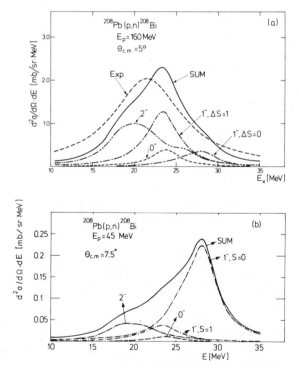

Figure 19   Charge exchange spectra for 160-MeV and 54-Mev incident proton energies scattered from ^{208}Pb. In the upper part, the theoretical spectra (Osterfeld et al 1981) are compared with the experimental data of Horen et al (1981).

5.3   Δ(1232)-Isobar Degrees of Freedom and the Strength of Spin-Isospin Resonances

The most interesting feature of the spin-isospin modes excited in charge exchange reactions is the magnitude of the transition strength. In the case of the GTR there exists a well established, model-independent sum rule (Ikeda et al 1963) that relates simply to the number of protons and neutrons: $S_{\beta -} - S_{\beta +} = 3(N - Z)$, where the left side of the equation is the difference between the β^- and β^+ GT strengths. So far only about 30% of the β^- GT strength has been detected experimentally in (p, n) reactions (e.g. Gaarde et al 1981), where the contribution due to $S_{\beta+}$ has been estimated. One example of a microscopic analysis of a (p, n) cross section in ^{208}Pb is shown in Figure 20 (Krewald et al 1981). There is an obvious discrepancy of more than a factor of 2 in the GTR case, whereas the experimental cross section of the isobaric analog state (IAR) is well reproduced theoretically. It is mainly this "quenching" of the GT-strength that has captured the interest of theorists. Missing strength has also been claimed for the M1 strength (Knüpfer et al 1978,

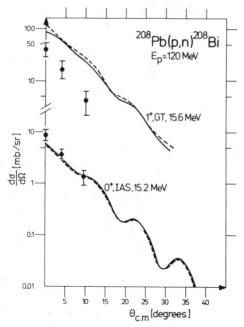

Figure 20   Theoretical cross sections of the 0^+ (IAR) and 1^+ (GTR) of the reaction ^{208}Pb (p, n) ^{208}Bi (Krewald et al 1981). The experimental results are taken from Horen et al (1980).

Richter 1979, Anantaraman et al 1981). There, however, no model-independent sum rule exists.

Conventional nuclear structure effects are unable to explain the "missing" GT strength, e.g. many-particle many-hole excitations only redistribute the strength. Several years ago nucleonic Δ-isobar degrees of freedom were suggested to play an important role in the quenching of the axial-vector coupling constant g_A, which is directly connected with the GT strength (Rho 1974, Ohta & Wakamatsu 1974, Oset & Rho 1979, Towner & Khanna 1979). [The first study of the "quenching" of g_A was performed by Ericson et al (1973) in terms of the Lorentz-Lorenz effect, without reference to the Δ-resonance.] Investigations of the GTR including the nucleonic Δ-isobar degree of freedom, consisting in fact of quark spin-isospin flips, have been performed by Bohr & Mottelson (1981), Suzuki, Krewald & Speth (1981), Brown & Rho (1981), and Osterfeld et al (1982). Similar considerations have also been performed for M1 states by Knüpfer et al (1980) and Härting et al (1981). In all these approaches the Δ resonance, coupled with a nucleon hole (Δ-h), can move part of the isovector $\sigma\tau$ strength into an energy region approximately 300 MeV above the low-lying excitations of the nucleus. The major reason why this mechanism may have a significant effect despite this enormous energy gap lies in the Pauli principle. Since there is no Pauli blocking for the Δ resonance, virtually all nucleons can share in building Δ-h states. Therefore the sheer number of possible configurations is able to bridge the energy gap if one supposes, however, that the interactions between the nucleon-particle and nucleon-hole pair (F_{NN}^{ph}) and Δ-particle nucleon-hole pair ($F_{\Delta N}^{ph}$) are roughly equally strong. This crucial assumption has been made implicitly in the work by Bohr & Mottelson (1981) and Brown & Rho (1981), who connected the two different forces (assumed to be constant in momentum space) by a scaling factor $F_{N\Delta}^{ph} = f_\pi^* / f_\pi \, F_{NN}^{ph}$, where f_π and f_π^* are then π-nucleon and π-delta coupling constants, respectively. Within this approach the low-lying GT transition probability in ^{208}Pb is reduced by a factor of 2.

There exists, however, a serious problem regarding the exchange term of the ph interaction. By definition, in the Migdal theory and in the schematic models mentioned above F^{ph} includes the direct and the exchange terms of the ph interaction. Therefore, the scaling assumptions would only be justified if the direct and the exchange contribution of nucleons and deltas had the same structure. This is, however, not the case. The cancellation between the direct and the exchange terms of, for example, the central part of the ρ exchange (which gives the dominant part to g_0' in Equation 36) is much larger for Δ-h configurations than for nucleon-hole configurations (Suzuki, Krewald & Speth 1981). Therefore it is essential to develop a microscopic model for the $\sigma\tau$ part of the ph interaction and to use it in those calculations. For such an interaction one may replace f_π with f_π^* and f_ρ with f_ρ^*, respectively. A first step in this direction was recently taken by Suzuki et al (1981) and Osterfeld et al (1982), who investigated the influence of the Δ-h configurations on spin-isospin resonances in ^{16}O and ^{48}Ca. They found that the "quenching" effect depends on the momentum transfer and the spin of a considered state. This fact might perhaps be used for experimental signature of the Δ-h "quenching" mechanism.

In the calculations by Toki & Weise (1980) and Härting et al (1981) on 1^+ states in ^{12}C and ^{48}Ca respectively, a force similar to that given in Equation 41 was used and includes the scaling assumption for g_0'. Unfortunately, these authors left out the exchange terms of the π and ρ contribution. Therefore it is likely that their conclusions concerning the Δ-h effects have to be modified.

In summary, the GT strength is definitely "quenched" by the Δ-h admixtures, but the magnitude of this effect is still uncertain.

6. SUMMARY AND OUTLOOK

One of the exciting developments in nuclear physics during the past decade has been the discovery and description of electric giant resonances different from the well-known giant dipole excitations. It is now established that they represent a general collective behavior of nuclei and that their systematic features are mainly characterized by the fact that they exhaust energy-weighted sum rules to a large extent.

The theory applied most to electric giant resonances is the random phase approximation, which is the simplest many-body theory to satisfy sum rules. This approach has proven quite suitable for the description of centroid energies, transition probabilities, and reaction cross sections for electrons and hadronic probes, and it can be directly compared with experimental data. This is of particular importance for the analysis of spectra that are often complicated mixtures of various multipoles and isospins. Concerning the properties of the effective interaction these investigations have basically fixed the values of the nuclear compression modulus and of the symmetry energy. Other properties of the force did not influence the results for electric giant resonances very much.

Besides the random phase approximation, there are other collective theories that have also frequently been applied to giant resonances. These are theories for large-amplitude collective motion, direct sum rule approaches, and fluid dynamical models. They are numerically simpler than RPA, but only applicable for not-too-detailed quantitative features and qualitative trends.Their results allow for a clear interpretation in terms of RPA sum rules. Apparently the RPA and its various derivatives cannot account for the spreading width of giant resonances. First steps toward a coupling of the 1p-1h excitations to more complex configurations were taken recently. However, for a quantitative interpretation of the width of giant resonances a 2p-2h calculation including the continuum is still required.

An interesting new feature in the field of giant resonances is the spin-isospin modes detected in charge exchange reactions. This new type of resonance gives important new information on the spin-isospin part of the particle-hole interaction. From the theoretical studies of pion condensation and its precritical phenomena, one knows that the particle-hole force in this channel is dominated by the one-pion and one-rho exchange potential. Therefore the spin-isospin modes give us the opportunity to study for the first time the effects of the corresponding exchange potentials inside the nucleus. Moreover, the "quenching" of the Gamow-Teller strength seems to offer a possibility of studying the effects of the Δ-baryon resonances on nuclear structure properties. For these reasons we feel that the spin-isospin resonances are presently the most exciting in the field of giant resonances.

ACKNOWLEDGMENTS

We wish to thank G. Holzwarth and F. Osterfeld for many helpful comments and fruitful discussions. One of us (J. S.) is grateful to Gerry Brown for numerous discussions and the hospitality he enjoyed in Stony Brook where part of the work on this paper was done.

Literature Cited

Abgrall, Y., Caurier, E. 1975. *Phys. Lett.* 56B:299–31

Abgrall, Y., et al. 1980. *Nucl. Phys. A* 346:431–48

Anantaraman, N., et al. 1981. *Phys. Rev. Lett.* 46:1318–21

Anastasio, M. R., Brown, G. E. 1977. *Nucl. Phys. A* 285:516–30

Auerbach, N., van Giai, N. 1978. *Phys. Lett.* 72B:289–92

Bainum, D. E., et al. 1980. *Phys. Rev. Lett.* 44:1751–54

Baldwin, G. C., Klaiber, G. S. 1947. *Phys. Rev.* 71:3–10

Baldwin, G. C., Klaiber, G. S. 1948. *Phys. Rev.* 73:1156–63

Baranger, M., Vénéroni, M. 1978. *Ann. Phys.* 114:123–200

Baym, G., Brown, G. E. 1975. *Nucl. Phys. A* 247:395–410

Baym, G., Pethick, C. 1978. *The Physics of Liquid and Solid Helium*, ed. K. H. Bennemann, et al 2:1–122. New York: Wiley

Bernard, V., Mahaux, C. 1981. *Phys. Rev. C* 23:888–904

Bertrand, F. E. 1976. *Ann. Rev. Nucl. Sci.* 26:457–509

Bertrand, F. E., Satchler, G. R., Horen, D. J., van der Woude, A. 1978. *Phys. Rev. C* 18:2788–91

Bertrand, F. E., Gross, E. E., Horen, D. J., Wu, J. R., Tinsley, J., McDaniels, D. K., Swenson, L. W., Liljestrand, R. 1981. *Phys. Lett.* 103B:326

Bertsch, G. F. 1975. *Nucl. Phys. A* 249:253–68

Bertsch, G. F. 1981. *Nucl. Phys. A* 354:157c–72c

Bertsch, G. F., Kuo, T. T. S. 1968. *Nucl. Phys. A* 112:204–8

Bertsch, G. F., Tsai, S. F. 1975. *Phys. Rep. C* 18:125–58

Bertsch, G. F., Bortignon, P. F., Broglia, R. A., Dasso, C. H. 1979. *Phys. Lett.* 80B:161–5

Bertsch, G. F., Cha, D., Toki, H. 1981. *Phys. Rev. C* 24:533–40

Bes, D. R., Broglia, R. A., Nilsson, B. S. 1975. *Phys. Rep.* 16C:1–56

Blaizot, J. P. 1980. *Phys Rep. C* 64:171–248

Blaizot, J. P., Gogny, D. 1977. *Nucl. Phys. A* 284:429–60

Blaizot, J. P., Grammaticos, B. 1974 *Phys. Lett.* 53B:231–3

Blaizot, J. P., Gogny, D., Grammaticos, B. 1976. *Nucl. Phys. A* 265:315–36

Bohigas, O., Martorell, J., Lane, A. M. 1976 *Phys. Lett.* 64B:1–4

Bohigas, O., Lane, A. M., Martorell, J. 1979. *Phys. Rep.* 51:267–316

Bohr, A., Mottelson, B. R. 1975. *Nuclear Structure*, Vol. 2, Ch. 6. Reading: Benjamin

Bohr, A., Mottelson, B. R. 1981. *Phys. Lett.* 100B:10–12

Bothe, W., Gentner, W. 1937. *Z. Phys.* 106:236–48

Bortignon, P. F., Broglia, R. A. 1981. *Nucl. Phys. A* 371:405–29

Borzov, I. N., Kamerdzhiev, S. 1975. *Sov. J. Nucl. Phys.* 21:15–19

Brieva, F. A., Rook, J. R. 1977. *Nucl. Phys. A* 291:299–316, 317–41

Brieva, F. A., Rook, J. R. 1978a. *Nucl. Phys. A* 297:206–30

Brieva, F. A., Rook, J. R., 1978b. *Nucl. Phys. A* 307:493–514

Brown, G. E., Bolsterli, M. 1959. *Phys. Rev. Lett.* 3:472

Brown, G. E., Gunn, J. H., Gould, P. 1963. *Nucl. Phys.* 46:598–606

Brown, G. E., Rho, M. 1981 *Nucl. Phys. A* 372:397–417

Brown, G. E., Speth, J. 1979. *Neutron Capture Gamma-Ray Spectroscopy*, ed. R. E. Chrien, W. R. Kane. New York: Plenum

Brown, G. E., Dehesa, J. S., Speth, J. 1979. *Nucl. Phys. A* 330:290–306

Brown, G. E., Speth, J., Wambach, J. 1981 *Phys. Rev. Lett.* 46:1057–61

Buenerd, M., Lebrun, D., Martin, P., de Saintignon, P., Perrin, C. 1980. *Phys. Rev. Lett.* 45:1667–70

Caurier, E., Bourotte-Bilwes, B., Abgrall, Y. 1973. *Phys. Lett.* 44B:411–15

Cavinato, M., Marangoni, M., Ottaviani, P. L., Saruis, A. M. 1982. *Nucl. Phys. A* 373:445–82

Danos, M. 1952. *Ann. Phys (Leipzig)* 10:265

Danos, M. 1958. *Nucl. Phys.* 5:23–32

Deal, T. J., Fallieros, S. 1973. *Phys. Rev. C* 7:1709–12

Dehesa, J. S. 1977. *"Microscopic Description of Giant Electric and Magnetic Multipole Resonances in Closed Shell Nuclei."* Thesis, Univ. Bonn. *Jül-Spec-1425*

Dehesa, J. S., Krewald, S., Speth, J., Faessler, A. 1977. *Phys. Rev. C* 15:1858–65

Djalali, C., Marty, N., Morlet, M., Willis, A. 1981. *IPNO-PhN 81.21*, Univ. Paris-Sud

Doering, R. R., Galonsky, A., Patterson, D. M., Bertsch, G. F. 1975. *Phys. Rev. Lett.* 35:1691–93

Eckart, G., Holzwarth, G., da Providencia, J. P. 1981. *Nucl. Phys. A* 364:1–28

Eikemeier, H., Hackenbroich, H.H. 1971. *Nucl. Phys. A* 169:407–16

Engel, Y. M., Brink, D. M., Goeke, K., Krieger, S. J., Vautherin, D. 1975. *Nucl. Phys. A* 249:215–38

Ericson, M., Figureau, A., Thévenet, C. 1973. *Phys. Lett.* 45B:19–22

Flocard, H., Vautherin, D. 1975. *Phys. Lett.* 55B:259–62

Fukuda, S., Torizuka, Y. 1972. *Phys. Rev. Lett.* 29:1109–11

Fuller, E. G., Weiss, M. S. 1958. *Phys. Rev.* 112:560–67

Gaarde, C., et al. 1981. *Nucl. Phys. A* 369:258–80

Garg, U., et al. 1980a. *Phys. Rev. Lett.* 45:1670–73

Garg, U., et al. 1980b. *Phys. Lett.* 93B:39–41

Geramb, H. V., Brieva, F. A., Rook, J. R. 1978. *Lect. Notes Phys.* 89:104–17

Giraud, B., Grammaticos, B. 1975. *Nucl. Phys. A* 255:141–56

Goeke, K. 1976. *Nucl. Phys. A* 265:301–14

Goeke, K. 1977. *Phys. Rev. Lett.* 38:212–15

Goeke, K. 1979. *Common Problems of Low and Medium Energy Nuclear Physics*, ed. B. Castel, B. Goulard, F. C. Khanna. New York/London: Plenum

Goeke, K., Castel, B. 1979. *Phys. Rev. C* 19:201–6

Goeke, K., Reinhard, P. G. 1978. *Ann. Phys.* 112:328–55

Goeke, K., Reinhard, P. G. 1980. *Ann. Phys.* 124:249–89

Goeke, K., Moszkowski, S. A., Krewald, S. 1976. *Phys. Lett.* 65B:113–15

Goeke, K., Lane, A. M., Martorell, J. 1978. *Nucl. Phys. A* 296:109–33

Goeke, K., Castel, B., Reinhard, P. G. 1980. *Nucl. Phys. A* 339:337–89

Goldhaber, M., Teller, E. 1948. *Phys. Rev.* 74:1046–49

Golin, M., Zamick, L. 1975. *Nucl. Phys. A* 249:320–28

Goodman, C. D., et al. 1980. *Phys. Rev. Lett.* 44:1755–59

Goodman, C. D., 1982. *Proc. 9th Int. Conf. High Energy Phys. Nucl. Struct. Versailles*, ed. P. Catillon, P. Radvanyi, M. Porneuf. Amsterdam: North-Holland

Grecksch, E., Knüpfer, W., Huber, M. G. 1975. *Lett. Nuovo Cimento* 14:505–10

Halbert, E. C., Satchler, G. R. 1974. *Nucl. Phys. A* 233:265–85

Halbert, E. C., McGrory, J. B., Satchler, G. R., Speth, J. 1975. *Nucl. Phys. A* 245:189–204

Hamamoto, I., Siemens, P. 1976. *Nucl. Phys. A* 269:199–209

Härting, A., Weise, W., Toki, H., Richter, A. 1981. *Phys. Lett.* 104B:261–64

Holzwarth, G., Eckart, G. 1978. *Z. Phys. A* 285:291–96

Holzwarth, G., Eckart, G. 1979. *Nucl. Phys. A* 325:1–30

Horen, D. J., et al. 1980. *Phys. Lett.* 95B:27–30

Horen, D. J., et al. 1981. *Phys. Lett.* 99B:383–86

Hoshino, T., Arima, A. 1976. *Phys. Rev. Lett.* 37:266–69

Ikeda, K. I., Fujii, S., Fujita, J. I. 1963. *Phys. Lett.* 3:271–72

Jennings, B. K., Jackson, A. D. 1980. *Nucl. Phys. A* 342:23–36

Jennings, B. K., Jackson, A. D. 1980a *Phys. Rep.* 66:141–85

Jeukenne, J. P., Lejeune, A., Mahaux, C. 1976. *Phys. Rep.* 25C:83–174

Kamerdzhiev, S. P. 1972. *Sov. J. Nucl. Phys.* 15:379–86

Kamerdzhiev, S. P. 1973, *Phys. Lett.* 47B:147–51

Kamerdzhiev, S. P. 1979. *Proc. 4th Semin. Electromagn. Interact. Nuclei at Low and Medium Energies, Moscow*. Moscow: Nauka

Kerman, A. K., Koonin, S. E. 1976. *Ann. Phys.* 100:332–58

Kishimoto, T., Moss, J. M., Youngblood, D. H., Bronson, J. D., Bacher, A. D. 1975. *Phys. Rev. Lett.* 35:552–55

Knüpfer, W., Huber, M. G. 1976a. *Phys. Rev. C* 14:2254–68

Knüpfer, W., Huber, M. G. 1976b. *Z. Phys. A* 276:99–102

Knüpfer, W., et al. 1978. *Phys. Lett.* 77B:367–70

Knüpfer, W., Dillig, M., Richter, A. 1980. *Phys. Lett.* 95B:349–54

Krewald, S., Speth, J. 1980. *Phys. Rev. Lett.* 45:417–19

Krewald, S., Birkholz, J., Faessler, A., Speth, J. 1974, *Phys. Rev. Lett.* 33:1386–88

Krewald, S., Galonska, J. E., Faessler, A. 1975. *Phys. Lett.* 55B:267–69

Krewald, S., Rosenfelder, R., Galonska, J. E., Faessler, A. 1976. *Nucl. Phys. A* 269:112–24

Krewald, S., Klemt, V., Speth, J., Faessler, A. 1977. *Nucl. Phys. A* 281:166–206

Krewald, S., Osterfeld, F., Speth, J., Brown, G. E. 1981. *Phys. Rev. Lett.* 46:103–6

Kyrchev, G., Malov, L. A., Nesterenko, V. O., Soloviev, V. G. 1978. *Sov. J. Nucl. Phys.* 25:506–10

Landau, L. D. 1956. *Zh. Eksp. Teor. Fiz.* 30:1058 (transl. *Sov. Phys. JETP* 3:920, 1957)

Landau, L. D. 1957. *Zh. Eksp. Teor. Fiz.* 32:59 (transl. *Sov. Phys. JETP* 5:101)

Landau, L. D. 1958. *Zh. Eksp. Teor. Fiz.* 35:97 (transl. *Sov. Phys. JETP* 8:70, 1959)

Levinger, J. S. 1960. *Nuclear Photo-Disintegration*. Oxford: Univ. Press

Lewis, M. B., Bertrand, F. E. 1972. *Nucl. Phys. A* 196:337–46

Liu, K. F., Brown, G. E. 1976. *Nucl. Phys. A* 265:385–415

Liu, K. F., van Giai, N. 1976. *Phys. Lett.* 65:23–26

Madsen, V. A. 1975. *Nuclear Spectroscopy and Reactions*, Pt. D. New York: Academic

Mahaux, C., Ngô, H. 1981. *Phys. Lett.* 100B:285–89

Malov, L. A., Nesterenko, V. O., Soloviev, V. G. 1976. *Phys. Lett.* 64B:247–50

Martorell, J., Bohigas, O., Fallieros, S., Lane, A. M. 1976. *Phys. Lett.* 60B:313–16

Marty, N., Morlet, M., Willis, A., Comparat, V., Frascaria, R. 1975. *Proc. Int. Symp. Highly Excited States in Nuclei*, ed. A. Faessler, C. Mayer-Böricke, P. Turek, Jül-Conf-16, Vol. 1, p. 17

Meuer, D., et al. 1981. *Phys. Lett.* 106B:289–92

Migdal, A. B. 1944. *J. Phys. USSR* 8:331

Migdal, A. B. 1967, *Theory of Finite Fermi Systems*. New York: Wiley

Migdal, A. B. 1972. *JETP (Sov. Phys.)* 34:1184–91

Morsch, H. P., Sükösd, C., Rogge, M., Turek, P., Machner, H., Mayer-Böricke, P. 1980. *Phys. Rev. C* 22:489–500

Morsch, H. P., Rogge, M., Turek, P., Mayer-Böricke, De Cewski, P. 1981. "Excitation of Isoscalar Giant Resonances, and Monopole Giant Resonance Splitting in Actinide Nuclei." *Preprint Jülich*

Mottelson, B. 1960, *Proc. Int. Conf. Nucl. Structure, Kingston.* Univ. Toronto Press; Amsterdam: North-Holland

Müther, H., Waghmare, Y. R., Faessler, A. 1976. *Phys. Lett.* 64B:125–27

Myers, W. D., Swiatecki, W. J. 1969. *Ann. Phys.* 55:395–505

Myers, W. D., Swiatecki, W. J. 1974. *Ann. Phys.* 84:186–210

Nix, J. R., Sierk, A. J. 1980. *Phys. Rev. C* 21:396–404

Noble, J. V. 1971. *Ann. Phys.* 67:98–113

Ohta, K., Wakamatsu, M. 1974. *Nucl. Phys. A* 234:445–57

Okamoto, K. 1958. *Phys. Rev.* 110:143

Oset, E., Rho, M. 1979. *Phys. Rev. Lett.* 42:47–50

Osterfeld, F., Wambach, J., Lenske, H., Speth, J. 1979. *Nucl. Phys. A* 318:45–53

Osterfeld, F., Krewald, S., Dermawan, H., Speth, J. 1981. *Phys. Lett.* 105B:257–62

Osterfeld, F., Krewald, S., Speth, J., Suzuki, T. 1982. *Phys. Rev. Lett.* In press

Petrovich, F., Love, W. G. 1981, *Nucl. Phys. A* 354:499c–534c

Pitthan, R., Walcher, T. 1971. *Phys. Lett.* 36B:563–64

Rho, M. 1974. *Nucl. Phys. A* 231:493–503

Richter, A. 1979. *Lect. Notes Phys.* 108:19–32

Ring, P., Speth, J. 1973. *Phys. Lett.* 44B:477–80

Ring, P., Speth, J. 1974. *Nucl. Phys. A* 235:315–51

Rinker, G. A., Speth, J. 1978. *Nucl. Phys. A* 360:306–96

Sagawa, H., Holzwarth, G. 1978. *Prog. Theor. Phys.* 59:1213–29

Schmid, K. W. 1980, *Anal. Fis.* 76:63–88

Schmid, K. W., Do Dang, G. 1977. *Phys. Rev. C* 15:1515–29

Schmid, K. W., Do Dang, G. 1978. *Phys. Rev. C* 18:1003–10

Schwierczinski, A., Frey, R., Spamer, E., Theissen, H., Walcher, T. 1975. *Phys. Lett.* 55B:171–74

Shlomo, S., Bertsch, G. F. 1975, *Nucl. Phys. A* 243:507–18

Soloviev, V. G. 1976. *Theory of Complex Nuclei.* Oxford: Pergamon

Soloviev, V. G. 1978. *Nukleonika* 23:1149

Speth, J., van der Woude, A. 1981. *Rep. Prog. Phys.* 44:719–86

Speth, J., Werner, E., Wild, W. 1977. *Phys. Rev. C* 33:127–208

Speth, J., Klemt, V., Wambach, J., Brown, G. E. 1980. *Nucl. Phys. A* 343:382–416

Steinwedel, H., Jensen, J. H. D. 1950. *Z. Naturforsch. Teil A* 5:413

Sterrenburg, W. A., Austin, S. M., DeVito, R. P., Galonsky, A. 1980. *Phys. Rev. Lett.* 45:1839–42

Stringari, S., Lipparini, E., Orlandini, G., Traini, M., Leonardi, R. 1978. *Nucl. Phys. A* 309:177–88, 189–205

Suzuki, T. 1973. *Nucl. Phys. A* 217:182–88

Suzuki, T. 1980. *Prog. Theor. Phys.* 64:1627–49

Suzuki, T. 1981. *Prog. Theor. Phys.* 65:910–18

Suzuki, T. 1981a. "The Giant Gamow-Teller Resonance States." *Preprint Kyoto*

Suzuki, T., Rowe, D. J. 1977. *Nucl. Phys. A* 286:307–21

Suzuki, T., Krewald, S., Speth, J. 1981. *Phys. Lett.* 107B:9–13

Thouless, D. J. 1961. *Nucl. Phys.* 22:78

Toki, H., Weise, W. 1980. *Phys. Lett.* 97B:12–16

Towner, I. S., Khanna, F. C. 1979. *Phys. Rev. Lett.* 42:51–54

Vautherin, D. 1975. *Phys. Lett.* 57B:425–28

Vlasov, A. 1945. *J. Phys. (USSR)* 9:25

Walecka, J. D. 1962. *Phys. Rev.* 126:653–62

Wambach, J. 1979. "Inelastische Streuung zu kollektiven Zuständen in doppelt magischen Kernen." Thesis, Univ. Bonn. *Jül-Spez-42*

Wambach, J., Klemt, V., Speth, J. 1978. *Phys. Lett.* 77B:245–48

Wambach, J., Osterfeld, F., Speth, J., Madsen, V. A. 1979. *Nucl. Phys. A* 324:77–98

Wambach, J., Mishra, V., Li, C. H. 1982. *Nucl. Phys. A.* In press

Werntz, C., Überall, H. 1966. *Phys. Rev.* 149:762–67

Youngblood, D.H., Rozsa, C. M., Moss, J. M., Brown, D. R., Bronson, J. D. 1977. *Phys. Rev. Lett.* 39:1188–91

Yukawa, T., Holzwarth, G. 1981. *Nucl. Phys. A* 364:29–42

Zawischa, D., Speth, J. 1975. *Phys. Lett.* 56B:225–28

Zawischa, D., Speth, J. 1976. *Phys. Rev. Lett.* 36:843–46

Zawischa, D., Pal, D., Speth, J. 1978. *Nucl. Phys. A* 311:445–76

Ann. Rev. Nucl. Part. Sci. 1982. 32:117-47
Copyright © 1982 by Annual Reviews Inc. All rights reserved

RAPID CHEMICAL METHODS FOR IDENTIFICATION AND STUDY OF SHORT-LIVED NUCLIDES

Günter Herrmann

Institut für Kernchemie, Universität Mainz, D-6500 Mainz; and Gesellschaft für Schwerionenforschung, D-6100 Darmstadt, Federal Republic of Germany

Norbert Trautmann

Institut für Kernchemie, Universität Mainz, D-6500 Mainz, Federal Republic of Germany

CONTENTS

117

0163-8998/82/1201-0117$02.00

1 INTRODUCTION

1.1 *Chemical Methods in Nuclear Studies*

For decades chemical methods provided the only high resolution technique for unravelling mixtures of radioactive species and firmly assigning atomic numbers. This approach can clearly distinguish even closely related homologous elements, such as radium and barium, or within elements with very similar chemical properties, such as lanthanides and actinides. Hence the use of chemical methods has led to major discoveries in the nuclear field: the identification of the first radioactive element in nature, polonium, by Pierre and Marie Curie; the discovery of artificial radioactivity by Irène Curie & Joliot; the identification of the first synthetic element, technetium, by Perrier & Segrè, of the first transuranium element, neptunium, by McMillan & Abelson, and of eight more transuranium elements by Seaborg et al; and the discovery of nuclear fission by Hahn & Strassmann.

In addition to determining atomic number, chemical methods can also be used for analyzing complex mixtures of radioactive species in terms of their mass number. This is because the isotopes of a given element can be distinguished by their decay properties: half-lives, energies and intensities of emitted radiations, and coincidence relationships between such radiations. With a proper choice of characteristic decay properties, this so-called radiochemical approach provides a unique resolution in atomic and mass number throughout the whole periodic table from the lightest to the heaviest nuclei. With radiochemical techniques new isotopes can be identified with a guess at their mass number, which has then to be confirmed by more direct methods such as straightforward nuclear reactions, genetic relationships with nuclides of known mass numbers, mass separation, and others.

The strength of the radiochemical approach lies in the unique chemical behavior of each element. This also means, however, that selective separation procedures for individual elements are different from each other, and that a variety of methods must be applied in the analysis of complex mixtures. In recent radiochemical procedures the number of chemical steps is reduced by taking advantage of new instrumental developments that allow the simultaneous detection of many nuclei. With high resolution γ- and x-ray spectroscopy, the production cross sections of a large number of nuclei in complex nuclear reactions can be measured, e.g. after simple separations into groups of elements. Furthermore, there is a strong trend toward purely instrumental approaches. For example, large arrays of gas discharge detectors recording position, energy loss, and time of flight of reaction products recoiling out of a target are now being used to identify individual nuclei and to measure their energy and angular distributions and mutual correlations. Methods based on the deflection in magnetic and electrical fields, a classical tool for the study of nuclear reactions, have been improved by the combination of several ion optical elements.

These instrumental developments do not make the use of radiochemical techniques in nuclear reaction studies obsolete, however. In particular, the detection of minor reaction products is still a domain of this technique. Recent examples are attempts to produce superheavy elements in the laboratory. Here the most sensitive searches were carried out with chemical methods that achieved a limit of 10^{-35} cm^2 for the production cross sections, which corresponds to a production of a few atoms per week (G. Herrmann 1979, Trautmann 1982, J. V. Kratz 1982). Another illustrative case is the detection of heavy actinides over a cross-section range of eight orders of magnitude in reactions between very heavy nuclei (Schädel et al 1978). Even for the study of major reaction channels in such reactions, radiochemistry is still an attractive technique especially after modifications that give, in addition to cross sections, data on energy and angular distributions of the reaction products (J. V. Kratz et al 1981). Another class of recent applications of radiochemical techniques concerns the calibration of instrumental methods. In order to deduce from measured γ-ray intensities the production cross sections of the emitter nuclides, the transition abundances have to be known on an absolute scale. More generally, a set of relative cross sections measured with some instrumental set-up may be put on an absolute scale by measuring the cross sections for a subset of suitable nuclei with a radiochemical method.

In the other domain of radiochemical techniques, detection of new nuclei and studies of decay properties, we note the same trend to prefer instrumental approaches. One of the major lines of research in this field is the study of nuclei at and beyond the borderline of the currently explored region. This implies short half-lives that are not accessible by the rather time-consuming radiochemical procedures in their conventional form. One instrumental approach combines rapid transport systems such as mechanical devices, gas jets, and electrostatic particle guides with high resolution radiation spectroscopy (Macfarlane & McHarris 1974). The other approach separates short-lived nuclei by deflection in magnetic and electrical fields applied either to the ionized atoms recoiling out of a target or to ions produced in an ion source into which the reaction products are introduced by evaporation from the target region (Klapisch 1974) or by a gas jet (Macfarlane & McHarris 1974). In this latter approach, use is often made of chemical selectivities in the evaporation step: under proper conditions, only a few out of many elements formed in the nuclear reaction are evaporated from the target region. An alternative way to introduce selectivity is the ionization process in the ion source itself. Without such selective steps all isobaric nuclei produced in the nuclear reaction will appear at the collector position. Chemical separation at the collector side has also been applied occasionally. For a comprehensive survey of on-line mass separation we refer the reader to the review by Ravn (1979) and of ion sources to the article by Kirchner (1981), but we discuss briefly the chemically selective steps used in these approaches.

1.2 Rapid Chemical Methods

Parallel to these instrumental developments, efforts were made to speed up chemical procedures by automation. There are arguments in favor of rapid chemical separation procedures, although on-line mass separation combined with some chemical selective step seems to be superior since it permits the isolation of individual nuclei: (a) A number of elements that can easily be separated by chemical techniques create great problems in on-line mass separation. Among these, there are several elements whose isotopes fall into interesting regions in view of their nuclear structure and decay properties. (b) In the transactinide region of the periodic table, element 104 and beyond, only short-lived nuclei are now accessible. Hence any study of their chemical behavior will require fast chemical procedures. Such studies are highly desirable in the context of the systematics of the periodic table, but not very much has been done so far because of the experimental difficulties (Zvara 1981, Hulet 1982). (c) There is an increasing use of short-lived isotopes in applied fields, in particular in nuclear medicine. Techniques worked out for rapid chemical separations may be applicable, for example, to the rapid production of labelled compounds.

In most chemical separation procedures, the principal step is a distribution between two phases, gaseous, liquid, or solid, in such a way that the element of interest occurs in the one, the bulk of the interfering elements in the other phase. By separating the two phases, a clean fraction of the desired element is obtained. In the chemical steps involved, the behavior of specific chemical species is utilized, such as that of the elemental state or of simple or complex ions or compounds. Hence one has to ensure that the elements are present in the form of the required species. If necessary, the elements have to be brought into the required form by chemically reactive steps; this may raise problems since chemical reactions are sometimes too slow in the time scale of rapid separations, whereas transfer rates between phases and their separation can be accomplished fast enough, at least in general.

There are two main approaches to rapid radiochemical separations: the batch-wise discontinuous and the continuous approach. In discontinuous procedures, a nuclide is produced, chemically separated, and measured sequentially, including its transport from the production place to the chemical processing set-up and from there to the detector system. An early example of a fully automated, programmer-controlled, discontinuous procedure is the separation of iodine from fission products (Schüssler et al 1969). The activity was produced by irradiating an aqueous solution of uranium in a reactor. After quick pneumatic transport to the chemical set-up, the solution was filtered through a reactive layer, in this case silver chloride, which takes up radioactive iodide ions by an exchange reaction. This layer was finally projected to the detector. By combining several chemical steps, rather complicated procedures can be performed within a few seconds (Trautmann & G. Herrmann 1976).

In continuous procedures the target is permanently irradiated and extraction from the target, chemical separation, and counting are continuously performed. The chemistry can be carried out in the target region itself or in a separate unit located at some distance from the target. This approach was pioneered by Zvara et al (1966) in experiments designed to prove chemically the assignment of a (then) 0.3-s spontaneously fissioning activity to element 104. Products of a heavy ion reaction recoiling out of the target were slowed down in nitrogen gas and mixed with niobium pentachloride and zirconium tetrachloride vapor in order to produce a volatile chloride of element 104. The gases passed a filter, which trapped the nonvolatile components, and a detector chamber with mica fission fragment detectors. The whole gas tract was kept at 300–350°C. In a first experiment, Zvara et al (1966) operated this procedure for 90 h and recorded eight fission events in the volatile fraction.

Experience shows that discontinuous procedures are relatively easy to accomplish. They are very convenient if the decay of the isolated nuclides has to be followed for some time. Their main disadvantage is the short effective counting time, which may lead to inadequate counting statistics even after frequent repetition of the whole sequence. Lien et al (1981) demonstrated, however, that by automation of all auxiliary functions (rabbit supply, cleaning operations, replenishing of reagents, etc) repetition rates can be much improved: about 1000 separations per working day were performed in a study of the 5-s [84]As separated from fission products by volatilization as arsenic hydride. The alternative is, of course, to use a continuous procedure that can be operated over sufficiently long times in order to allow the detection of weak decay modes and measurements with low counting efficiencies. A recent example is the operation of a continuous solvent extraction procedure for technetium (Brodén et al 1981) over about 100 h to study γ-γ coincidence and angular correlations in the β^--decay of 36-s ^{106}Tc and 5-s ^{108}Tc (Stachel et al 1981).

In designing a rapid chemical separation, one must consider the scope of the experiment and the production process of the nuclide to be studied. If weak γ-ray transitions have to be identified, the requirements on the purity of the final fraction may be more stringent than in a study of exotic decay modes like spontaneous fission, where a background of nuclides not undergoing this decay is acceptable. When thin targets can or must be used, a convenient first step is the application of a rapid gas transport system in which reaction products are carried on aerosol clusters after recoil from the target and thermalization. Such gas transport systems can be combined with a variety of separation techniques, e.g. thermochromatography (Silva et al 1977) and solvent extraction from aqueous solutions (Trautmann et al 1975). For thick targets in solid or molten form, diffusion out and evaporation from the target at high temperatures is the typical initial step (Ravn 1979). Requirements on the counting samples must also be considered. Thin samples as needed for high resolution spectroscopy of fission fragments, protons, α particles, and electrons are more easily obtained by deposition from gases than from aqueous solutions.

In the following sections we outline recent trends in the design of rapid chemical separations in more detail, including some perspectives for future developments. References to earlier work can be found in the preceding review article in this series (G. Herrmann & Denschlag 1969). We emphasize techniques that allow the isolation of individual elements from complex mixtures on a time scale of a few seconds with direct detection of their isotopes. We also include somewhat slower techniques in cases where very similar elements have to be separated. A comprehensive survey is not intended here. Part of the field, the separation of fission products, was recently reviewed by Meyer & Henry (1980). We also do not refer extensively to applications of rapid chemical methods in studies of short-lived nuclei with the exception of a few examples that illustrate the accessible half-life ranges. Since we want to show how rapid separations can be performed technically, the article is organized according to the principal chemical step, not according to elements. We discuss first separations from and in aqueous solutions and then separations via gaseous phases. In solution chemistry one takes advantage of the broad experience accumulated over long times in radio-chemical separations; gas phase chemistry is less developed but very attractive for rapid work because of the short separation times involved.

2. SOLUTION CHEMISTRY

2.1 *Precipitation*

Precipitation of a solid compound in an aqueous solution followed by its filtration is one of the traditional separation techniques in radiochemical analysis. However, precipitation and filtration may proceed too slowly for rapid separation. Therefore such a step is often replaced by an exchange reaction with a preformed precipitate: the solid is prepared in advance, and the solution with the radioactive species to be separated is quickly filtered through a thin layer of the reactive solid. The exchange of iodide ions with preformed silver chloride yielding the less soluble silver iodide has already been mentioned in Section 1.2.

The same reaction is also part of a rapid separation procedure for niobium from fission products (Ahrens et al 1976), which we describe in some detail in order to illustrate a typical approach to rapid separations from aqueous solutions, the "smashed-capsule" technique. The proce-dure consists of the following steps: (*a*) A solution of a fissile nuclide is sealed in a breakable capsule and irradiated in a pneumatic tube system of a reactor. (*b*) The capsule is projected into the separation apparatus and smashed by impact on its walls. (*c*) The solution is filtered through silver chloride to remove bromine and iodine isotopes. (*d*) Niobium is separated from a strong nitric acid solution by filtration through fiberglass filters. Under similar conditions, weighable quantities of niobium precipitate as an oxide hydrate but for trace amounts sorption on the glass surface in an ion exchange reaction is the more likely mechanism. (*e*) The glass filter with the niobium fraction is projected to the detector.

Apparatus and time schedule for this procedure are shown in Fig-ure 1. The different steps are initiated by opening and closing several stopcocks operated pneumatically at times determined by an electronic programmer. A few details should be mentioned. The initial solution contains sulfur dioxide to reduce the halogen elements to halide ions, tartaric acid for complexing antimony, and small quantities of several elements to prevent their co-adsorption on the glass filters. Projection into the apparatus takes 0.2 s. Filtration through two silver chloride layers is initiated by applying vacuum no. 1 for 0.9 s. During the last part of this time interval, stopcock no. 1 is opened to apply washing

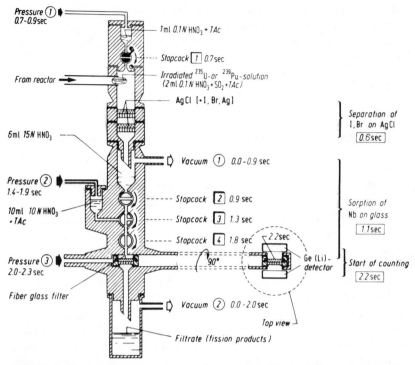

Figure 1 Rapid separation procedure for niobium by sorption on fiberglass filters after removal of halogens by exchange with preformed silver chloride: apparatus and time schedule (Ahrens et al 1976).

solution. The original and the washing solution are collected in concentrated nitric acid. By opening stopcock no. 2 0.9 s after the end of irradiation, the resulting strong nitric acid solution is filtered through two fiberglass filters; washing solution is applied from a second reservoir by turning stopcock no. 3. Then 2.0 s after the end of irradiation, the final niobium fraction is projected to the detector through pressure no. 3 with a 90° rotation to position it between two Ge(Li) detectors. Counting is started 2.2 s after the end of irradiation. With this procedure, niobium isotopes with half-lives down to 0.8 s, ^{104}Nb, were studied (Ahrens et al 1976). In a modified version taking 4.4 s, the adsorbed niobium activity was redissolved and resorbed on a second set of fiberglass filters to improve decontamination from zirconium.

Other examples of rapid ion exchange reactions with preformed precipitates are the exchange of rubidium and cesium ions with ammonium phosphormolybdate (Schüssler & G. Herrmann 1972) and of pertechnetate ions with tetraphenylarsonium perchlorate (Vine & Wahl

1981). If such a precipitate is too fine grained to permit rapid passage of an aqueous solution, fixation on a surface active carrier may help. This was demonstrated for the rapid exchange of 16 elements with silver-, cadmium-, and arsenic sulfides deposited on cellulose fibers (Schüssler & G. Herrmann 1970).

By means of isotopic exchange reactions between the same ions in solutions and in solids, very selective procedures are obtained. Exchange of iodide with silver iodide was used for a rapid separation from fission products with identification of 0.8-s ^{140}I (Schüssler & G. Herrmann 1972), and the exchange of silver ions with silver chloride in a study of silver isotopes formed in fission including 4-s ^{118}Ag (Brüchle & G. Herrmann 1982).

2.2   Electromigration

Separation by electromigration is based on differences in the ion mobilities in aqueous solutions under the action of an electric field. A paper strip is moistened with a suitable electrolyte. After the irradiated sample is placed on the paper in a small spot, a potential gradient of about 500 V cm^{-1} is applied. Depending on their mobility, different ions migrate in different directions and distances. With dilute perchloric acid as electrolyte, a number of fission products (Se, Te, Br, I, Tc, Rb + Cs, Sr + Ba, and the lanthanides) can be separated within 20 s in well-focussed spots distributed over a total distance of about 8 cm; the 16-s ^{86}Se could be observed in these experiments (Tamai et al 1973a). With the complexing electrolyte nitrilotriacetic acid, barium and strontium could be separated from each other and the bulk of fission products within 20 s, which permitted detection of 12-s ^{144}Ba (Tamai et al 1973b). In the same electrolyte, a separation of lanthanides is possible at lower potential gradients of about 100 V cm^{-1} within 1.5 to 3 min (Ohyoshi et al 1972).

2.3   Volatilization

Rapid separations based on the production of volatile species and their rapid escape offer the advantage that gases can quickly pass through additional purification steps such as chemical traps. This approach has mainly been used for elements forming volatile hydrides: arsenic, selenium, tin, antimony, and tellurium.

For illustration we discuss a separation procedure for antimony from fission products (Rudolph et al 1977). The main steps are: (a) Antimony hydride is volatilized together with the hydrides of arsenic, selenium, and tellurium by a burst of nascent hydrogen. (b) Selenium and tellurium hydrides are removed in a chemical trap. (c) Antimony and arsenic are decomposed in strong hydrochloric acid. (*d*) This solution is contacted with an organic solvent that takes up antimony but not arsenic. (*e*) The sample is projected to the detector, and counting is started 2.7 s after the end of irradiation. This procedure allowed the detection of nuclides with half-lives down to 0.75 s, ^{136}Sb.

Figure 2 shows the set-up and gives the time schedule for this technique. In the unit at the left, the capsule is smashed and the irradiated solution is mixed with concentrated hydrochloric acid. A hydrogen burst is generated by adding zinc powder to the solution. The hydrides carried by hydrogen pass first through a trap with a dilute sodium hydroxide solution, which maintains selenium and tellurium, and then through a strong hydrochloric acid solution containing potassium chlorate as an oxidizing agent for decomposition of antimony and arsenic hydride, as shown at the top of the middle part of the figure. Antimony and arsenic are separated from each other by solvent extraction in the unit below with a technique discussed in Section 2.5: the hydrochloric acid solution is filtered through a layer of plastic powder coated with di(2-ethylhexyl)orthophosphoric acid (HDEHP), which extracts antimony. After washing, the separation procedure is completed 2.7 s after the end of irradiation. About 24% of the antimony present was recovered in the final sample; the main loss occurred in the decomposition step.

As mentioned in Section 1.2, Lien et al (1981) worked out a technique that allows a high repetition rate. For antimony the main difference from the above procedure is the use of a sodium borohydride solution for generating the antimony hydride; although not as efficient as zinc powder, the use of a solute reducing agent allows rapid cleaning of the reaction vessel. This vessel is a specially designed cyclone separator by which the degassing time can be decreased to 0.1–0.2 s. Other differences to the procedure shown in Figure 2 are: the rabbit with the irradiated solution is emptied by puncture with a needle system; calcium sulfate is used for absorption of selenium and tellurium hydride, and potassium hydroxide dissolved in ethanol for the final fixation of antimony at the detector station. Counting can be started 2.7 s after the end of irradiation.

The rapid generation of hydrides by zinc powder added to a strong hydrochloric acid solution has also been used for the separation of arsenic and selenium by modifying the purification steps. In the arsenic procedure (J. V. Kratz et al 1973) the cleaning trap is filled with potassium hydroxide in ethanol in order to maintain selenium, tellurium, and antimony whereas arsenic passes through. For further purification the same decomposition and solvent extraction step is applied as

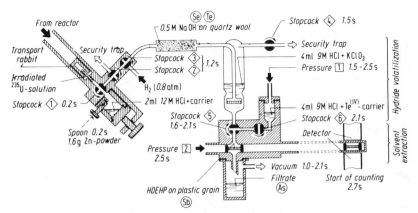

Figure 2 Rapid separation procedure for antimony by volatilization of its hydride in a burst of nascent hydrogen followed by purification through passing an absorption trap and with a final solvent extraction step: apparatus and time schedule (Rudolph et al 1977).

shown in Figure 2; arsenic is counted in the effluent from the solvent-coated filter, beginning 5.0 s after irradiation. In an even faster version with start of counting 2.5 s after irradiation, the arsenic hydride escaping from the potassium hydroxide trap is decomposed in a small column filled with silver nitrate on firebrick. With these procedures short-lived arsenic isotopes up to 0.9-s [86]As have been investigated (J. V. Kratz et al 1973, 1975). Lien et al (1981) studied the 5-s [84]As in a version of their technique in which sodium hydroxide was added to the calcium sulfate cleaning trap.

In the selenium procedure (J. V. Kratz & G. Herrmann 1970), the gases pass through a dilute sodium hydroxide solution that absorbs selenium and tellurium. After conversion into a strong hydrochloric acid solution, this fraction is filtered through a layer of plastic grains coated with tri-*n*-butyl phosphate (TBP), which takes up tellurium by solvent extraction; selenium is counted in the filtrate. The procedure takes 5.0 s and was used to study neutron-rich selenium isotopes, including 1.4-s [88]Se.

Tin hydride can be volatilized from a dilute hydrochloric acid solution by sodium borohydride. After separating antimony in a calcium sulfate trap, tin hydride can be caught in activated charcoal. This procedure was used by Izak & Amiel (1972) to study fission tin such as 41-s [132]Sn.

A different way of volatilizing the hydrides of selenium and tellurium from aqueous solutions was worked out by Tomlinson & Hurdus (1971, 1972): the isotopic exchange with preformed selenium hydride or tellurium hydride gas brought in intense contact with a fission product solution. Tomlinson & Hurdus (1971) were able to perform the selenium separation within 1.0 s and to identify 0.4-s ^{89}Se.

Volatilization has also been applied to the isolation of some other elements. Del Marmol & van Tigchelt (1972) describe a procedure for germanium based on the volatilization of germanium tetrachloride from a strong hydrochloric acid solution by applying a burst of chlorine gas. The germanium is caught in finely divided zinc probably by reduction to the less volatile dichloride. The procedure was used to identify germanium isotopes in fission with half-lives down to 1.2 s, ^{84}Ge (del Marmol & Fettweis 1972). Ruthenium has been separated within 5 s from fission products by volatilization of ruthenium tetroxide for direct detection of isotopes up to 16-s ^{110}Ru (Fettweis & del Marmol 1975). A burst of chlorine is passed through an alkaline solution producing and carrying the tetroxide, which is then caught on polyethylene pellets.

2.4   *Ion Exchange Chromatography*

Ion exchange chromatography is a chemical method for separations within groups of very similar elements such as the lanthanides and the trivalent actinides. It is based on the small differences in the distribution of these elements between a solid phase consisting of a polymer with chemically reactive groups and a solution in which the elements to be separated are present in the form of a suitable complex ion. To resolve such a mixture into individual elements, the distribution step has to be repeated very often. This is achieved by chromatography, i.e. the mixture is placed on top of a column filled with the solid phase through which the liquid phase is slowly passed. In case of the lanthanides and trivalent actinides, the elements are eluted with a complexing agent from the column in inverse order of their atomic number, e.g. lutetium (element 71) first and lanthanum (57) last.

Recent developments of high performance solid phases, together with the operation of such columns at high pressures, make it possible to decrease the time required for such a chromatographic separation. Schädel et al (1977) and Elchuk & Cassidy (1979) showed that a complete separation of the whole series of lanthanide elements can be achieved within about 20 min.

An automated separation procedure for lanthanides from fission products was first reported by Baker et al (1981). Before the chromatographic separation, the whole group of lanthanide elements is isolated by solvent extraction from a nitric acid solution into dihexyldiethyl-carbamylmethylene phosphate fixed on a resin in a column. After elution the lanthanide fraction is applied to a cation exchange column, 25 cm long by 3.2 mm inside diameter, filled with a cation exchange resin with sulfonic acid as the reactive group (Aminex A-9). Elution is carried out with an aqueous solution of α-hydroxyisobutyric acid (α-HIBA) at 90°C and 80 MPa pressure with smoothly increasing concentration and pH value of the complexing agent. Figure 3 shows a chromatogram obtained with the lanthanide fraction from spontaneous fission of ^{252}Cf under the optimal conditions for the heavier lanthanides terbium and dysprosium (Greenwood et al 1981). The time scale counts from the end of the fission product collection. By changing eluent concentration and pH value, one can optimize such chromatographic separations for a given element. Under appropriate conditions, 48-s ^{155}Pm has been identified.

A much faster separation has been worked out for yttrium, the lighter homologous element of the lanthanides that appears around dysprosium in ion exchange separations with α-HIBA. This procedure takes advantage of the fact that in low energy fission of ^{235}U, yttrium isotopes are much more abundant than fission products in the dysprosium-terbium region. Hence layers instead of columns of a cation exchange resin are sufficient to obtain within 10 s and with 7% chemical yield an yttrium fraction in which the 9.6-s ^{96}Y could be detected (Klein et al 1975). With a similar approach heavy isotopes of actinium were

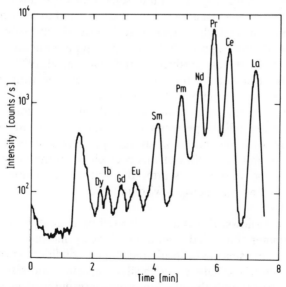

Figure 3 Chromatographic separation of lanthanides by elution from a cation exchange column with aqueous solutions of α-hydroxyisobutyric acid, showing the appearance with time of the different lanthanide elements in the effluent from the column (Greenwood et al 1981).

identified with half-lives down to 35 s, ^{232}Ac (Chayawattanangkur et al 1973).

Applications of fast ion exchange separations to the trivalent actinides have not yet been described, but an application to the first transactinide element, 104, was reported by Silva et al (1970a). This experiment was based on the expectation that, like its homologs zirconium and hafnium, the first transactinide element should pass through a cation exchange column in a weakly acid solution of α-hydroxyisobutyric acid, in contrast to the heavy actinides, which are strongly absorbed. This expectation was confirmed with the 65-s ^{261}104 produced in a heavy ion reaction. To obtain samples for α-particle spectroscopy within about one minute, a large part of the chemical system was automated. In order to produce 100 atoms of ^{261}104, hundreds of irradiations had to be carried out, each followed by the chemical separation. Only about one tenth of the produced atoms were observed as α-decay events after chemistry due to decay, counting geometry, and chemical losses.

2.5   Solvent Extraction: Discontinuous Techniques

In the distribution of chemical species between an aqueous solution and a liquid organic phase, the separation of the two phases is in general the time-determining step. For rapid separations it is advantageous to fix the organic solvent on a fine-grained carrier and to filter quickly the aqueous solution through a layer of such a quasi-solid solvent.

We illustrate this approach by a rapid separation procedure for technetium from fission products (Trautmann et al 1972). The main steps are: (a) Removal of the halogens by exchange with silver chloride. (b) Solvent extraction of the pertechnetate ion into a solution of tetraphenylarsonium chloride adsorbed on a fine-grained carrier. (c) Reextraction of the pertechnetate from this solvent and its co-precipitation with the homologous tetraphenylarsonium perrhenate. (d) Filtration and projection of this precipitate to the detector where counting can be started 7.5 s after the end of irradiation. With this procedure, technetium isotopes with half-lives down to 5 s, ^{108}Tc, have been identified.

As Figure 4 shows, this is a complicated procedure since it involves the programmer-controlled operation of many mechanical functions. The initial steps, smashed-capsule technique, conditioning of the solution, exchange of the halogens with silver chloride, and washing of this solid are identical with the niobium procedure outlined in Section 2.1 and Figure 1. This underlines the fact that such procedures can be designed on a building-block principle. The procedure is specific for technetium from the middle part of the set-up, the collection of the

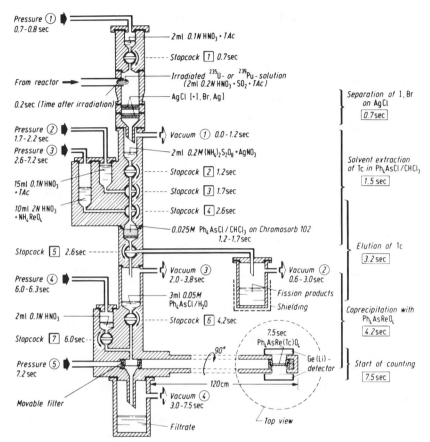

Figure 4 Rapid separation procedure for technetium by solvent extraction of the pertechnetate ion into tetraphenylarsonium chloride and its coprecipitation with tetraphenylarsonium perrhenate after removal of halogens by exchange with preformed silver chloride: apparatus and time schedule (Trautmann et al 1972).

original and washing solution passing the silver chloride layers in an ammonium persulfate solution containing traces of silver ions. In this solution, technetium is oxidized to the heptavalent state, the pertechnetate ion. By opening stopcock no. 2, this solution is filtered through a tetraphenylarsonium chloride solution in chloroform adsorbed on a fine-grained carrier; the bulk of fission products passes through and is collected in a container. Washing solution is applied from a reservoir by pressure no. 2 and turning stopcock no. 3. Then the pertechnetate ion is backextracted from the tetraphenylarsonium chloride phase with nitric acid supplied via stopcock no. 4; simultaneously, stopcock no. 5 is turned to collect the resulting eluate in a solution of tetraphenylarsonium chloride, where the presence of perrhenate ions in the eluting nitric acid causes tetraphenylarsonium perrhenate to precipitate and to carry the technetium. By operating stopcock no. 6 this precipitate is filtered on a movable filter; this final sample is washed by opening stopcock no. 7. Then, by applying pressure at position no. 5, the sample is projected to the detector where it arrives 7.5 s after the end of irradiation.

In a modified version of the technetium procedure designed for fission reactions with high yields for technetium isotopes, the precipitation step is omitted so that counting can be started after only 2.5 s. Here the fissile nuclide, ^{249}Cf, is not irradiated in solution but as an electroplated source that stays fixed in the reactor. The recoiling fission fragments are caught in a hollow cylindrical capsule coated on its inside with solid ammonium nitrate. At the end of irradiation, the capsule is transferred into the separation apparatus where the catcher material together with the fission products are dissolved in dilute nitric acid. With this modification, technetium isotopes with half-lives down to 1.0 s, ^{110}Tc, could be observed (Trautmann et al 1976).

This technique of extraction into solvents fixed on fine-grained carriers has also been used for other elements in the fission product region. Zirconium can be separated within 4 s by extraction from strong nitric acid into tri-n-butyl phosphate (Trautmann et al 1972). For molybdenum the extraction of its thiocyanate complex into a mixture of isoamylalcohol and n-butylacetate was used in the identification of nuclides with half-lives down to 3.5 s, ^{107}Mo (Tittel et al 1977). Using solvent extraction of ruthenium tetroxide into petrolether, neutron-rich ruthenium isotopes up to 3.5-s ^{113}Ru have been investigated (G. Franz & G. Herrmann 1978).

Solvent extraction has been used to study the chemical properties of very heavy elements. A solvent extraction procedure with methylisobutyl ketone containing the chelating agent thenoyltrifluoroacetone as the organic phase and buffered acetate solutions as the aqueous phase was used to demonstrate the trivalency of lawrencium in aqueous solutions. These experiments were performed manually with the α-emitting isotope 35-s ^{256}Lr (Silva et al 1970b). An automated procedure was applied by Hulet et al (1980) to show with the 65-s α-emitting isotope ^{261}104 that element 104 forms stronger chloride complexes than the trivalent actinides, and thus resembles its homolog hafnium. This was concluded from solvent extraction experiments with a column of trioctylmethylammonium chloride (Aliquat-336) on an inert carrier. Like hafnium, element 104 is extracted from concentrated hydrochloric acid and eluted at lower concentrations whereas the actinides pass through in concentrated hydrochloric acid. Six decay events of element 104 were observed in 44 runs.

A separation of the lanthanide elements and yttrium on a time scale of several minutes can be carried out by solvent extraction chromatography with di(2-ethylhexyl)orthophosphoric acid adsorbed on silicagel as the stationary and nitric acid as the mobile phase (Schädel et al 1978). In contrast to ion exchange chromatography, the individual elements appear in order of increasing atomic number.

A different approach to discontinuous solvent extraction was followed by Meikrantz et al (1981), who used annular centrifugal contactors for fast phase separation in an automated system. This system was applied to a study of palladium isotopes from ^{252}Cf fission via solvent extraction of the palladium dimethylglyoxim complex; half-lives down to 47 s, ^{115}Pd, were covered.

2.6 Solvent Extraction: Continuous Techniques

Continuous operation of fast solvent extraction procedures became possible with the introduction of high speed centrifuges for rapid phase separation (Aronsson et al 1974a). An early application to short-lived nuclei is the identification of 4.0-s ^{150}Ce (Aronsson et al 1974b). In the most recent version, a gas jet (Section 3.1) is used for rapid transport of fission products from the target to the centrifuge system (Trautmann et al 1975) to simplify its operation at a reactor. With this modification, several lanthanide nuclides were investigated, e.g. 6.2-s ^{150}Pr (Skarnemark et al 1976). Furthermore, for even faster performance much smaller centrifuges were developed and installed; in these the mean hold-up time of each of the phases is decreased to about 0.25 s (Skarnemark et al 1980). In a typical two-step procedure, the first activity reaches the detector after about 2.8 s, and the maximum count rate is observed after 5.2 s. Of this delay time, 0.7 and 1.1 s, respectively, are due to the gas transport system. Loop systems can be used to introduce additional delays in order to optimize a procedure for a certain half-life region.

The improvements achieved with this system were demonstrated by the separation of several lanthanide elements using di(2-ethylhexyl) orthophosphoric acid as the solvent. Lanthanum and praseodymium were isolated by extraction from dilute nitric acid and backextraction with a stronger nitric acid solution after oxidation of cerium to the tetravalent state. Cerium was separated by extraction in this state and stripping from the organic phase by a reducing agent. In this experiments, 2.2-s ^{147}La, 1-s ^{148}La, and 4.8-s ^{150}Ce could be observed (Skarnemark et al 1980).

To illustrate this approach we show in Figure 5 a continuous procedure for the simultaneous separation of zirconium and niobium from fission products (Brodén et al 1981). The main steps are: (*a*) Extraction of zirconium and niobium from dilute sulfuric acid into a commercially available, long-chain ternary amine (Alamine-336) in the first centrifuge C1. (*b*) Stripping of zirconium and niobium from the organic phase by dilute nitric acid in C2. (*c*) Extraction of zirconium into di(2-ethylhexyl) orthophosphoric acid in C3 from nitric acid containing hydrogen peroxide as a complexing agent for niobium. Zirconium is detected in the organic phase obtained in this step, niobium in the aqueous phase after a further purification step in a fourth centrifuge C4. The organic phases circulate within one unit or are recycled, e.g. between C2 and C1. Maximum count rates at the detectors are observed at 7 and 9 s for zirconium and niobium, respectively. This allows the investigation of nuclides with half-lives down to 1.5 s, as demonstrated by the detection of 1.5-s [103]Nb.

A few comments should be added. The fission products carried by a gas jet are dissolved in sulfuric acid in a static mixer. The gas-liquid mixture is then fed into a degassing unit in which the jet carrier gas, nitrogen, and fission krypton and xenon are separated from the aqueous solution. In step C1 zirconium and niobium are extracted with at least 80% yield with some contamination mainly by technetium. In step C2, zirconium and niobium are almost quantitatively stripped from the organic phase while technetium and other contaminants remain there. In C3 a clean zirconium fraction is obtained, but about 5% of the zirconium remain in the aqueous phase so that the zirconium extraction is repeated in C4.

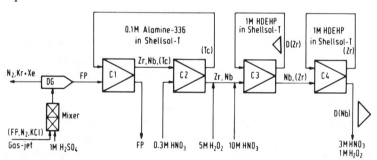

Figure 5 Flow diagram for the continuous separation of zirconium and niobium from fission products by solvent extraction with a series of high speed centrifuges (Brodén et al 1981). DG: degassing unit; D(Zr) and D(Nb): detector positions; C1, C2, C3, C4: centrifuges; FP: fission products. Alamine-336 is a commercially available, long-chain amine, Shellsol-T is an aliphatic kerosene, and HDEHP is di(2-ethylhexyl) orthophosphoric acid.

The same technique is used in fast separation procedures for technetium, bromine, and iodine from fission products, each requiring only three centrifuges (Brodén et al 1981). The technetium procedure is based on the extraction of the pertechnetate ion from dilute nitric acid using the same amine as in the preceding case. This procedure was applied in long-time measurements of decay properties of 5-s ^{108}Tc as already mentioned in Section 1.2 (Stachel et al 1981). In the bromine and iodine procedures the extraction of these elements in carbon tetrachloride is used under.conditions that allow their mutual separation. In demonstration runs, nuclides such as 16-s ^{88}Br and 6.3-s ^{138}I were seen (Brodén et al 1981).

3. GAS PHASE CHEMISTRY

3.1 *Gas Jet Transport Systems*

The gas jet transport technique was originally developed with helium as the carrier gas for on-line studies of short-lived α- and delayed proton emitters. Recoil atoms ejected from a target are thermalized in helium and then pumped through a capillary tube into an evacuated chamber where the reaction products are collected on a surface. Under optimum conditions, nuclei with half-lives less than 1 ms can be investigated if the distance to be bridged is very short (Macfarlane et al 1969), but transport over distances up to 100 m has been reported, too. A general survey can be found in the article by Macfarlane & McHarris (1974). We refer only to recent developments that may have implications for rapid chemical separations.

In the jet the recoil atoms are attached to large clusters with masses of 10^6 to 10^8 mass units produced from impurities in the helium under the action of intense radiation (Jungclas et al 1971). In this form the translational velocity becomes very small, and a mass transport due to the pressure gradient along the capillary overcomes Brownian motion even over long distances. Difficulties arise, however, under less ionizing conditions, e.g. at a reactor where clusters are not produced automatically. A large number of additives have been recommended to ensure efficient and stable operation of jet systems over long times. At present, inorganic clusters like sodium chloride aerosols (Ghiorso et al 1974, Äystö et al 1976) are preferred by most groups but organic additives such as ethylene glycol (Moltz et al 1980a) or isopentyl alcohol together with water (Lister et al 1981) are still in use. In applications to recoil atoms with high kinetic energies emerging from fission or heavy ion reactions, the long ranges require rather large stopping chambers (Georg et al 1978). In such cases one can use either a multicapillary system to transport the thermalized atoms from different regions of the chamber (Moltz et al 1980b) or a heavier stopping gas to decrease the ranges. Very stable and efficient operation of a nitrogen jet for fission fragments has been achieved by loading it with alkali halide clusters evaporating from a ceramic boat heated to temperatures 100–150°C below their melting point (Stender et al 1980). Choice of the cluster material is essential in connecting a jet system with separation techniques discussed in this article. For example, problems in the dissolution of the activity from organic clusters in aqueous solutions were noted (Silva et al 1977). This is not the case with inorganic clusters in combinations with solution (Brodén et al 1981) and gas phase chemistry (Hickmann et al 1980) nor with the ion sources of a mass separator (Mazumdar et al 1981). Connection of a helium jet system with organic clusters to a mass separator has also been successful (Moltz et al 1980a) In this context, developments of ion sources for surface ionization of actinide elements are of interest (Beyer et al 1972) and may open the way to a fast on-line mass separation of actinide elements.

It may be possible to feed thermalized recoil atoms directly into a mass separator if they can be kept in a charged state (Ärje & Valli 1981). Another recent development that deserves attention is a combination of a fast gas flow transport system with electrostatic collection of reaction products, which was shown to work for half-lives down to about 20 ms (Dufour et al 1981).

3.2   Condensation and Trapping

A rather straightforward case is the rapid separation of a gaseous element. Esterl et al (1971) described an automated system in which short-lived argon, oxygen, and nitrogen isotopes were studied by sweeping them into a counting chamber. In the case of the 0.17-s ^{33}Ar produced by bombarding carbon disulfide vapor, the target substance was removed from the gas stream by condensation at low temperatures. The same system has also been used for the transport of nongaseous activities such as 3.9-s ^{22}Mg obtained in the bombardment of neon gas (Hardy et al 1975).

When a reactive element recoils into a suitable gas, a volatile species may be produced, which can then be separated by trapping techniques. Methyl halides were formed in reactions between recoiling fission halogens and methane and were separated out of the gas by absorption of the iodine fraction on silicagel coated with silver nitrate and of the bromine fraction on silicagel coated with silver chloride. Counting can be started 0.6 s after irradiation. Both 0.25-s ^{92}Br and 0.45-s ^{141}I were identified (K.-L. Kratz & G. Herrmann 1973). Slowing down the fission products in nitrogen plus gaseous hydrochloric acid forms volatile chlorides of germanium and arsenic that can be separated by selective absorption. In the germanium fraction, 4.5-s ^{82}Ge could be observed (Zendel et al 1981). There is evidence that 2.2-s ^{26}Si recoiling out of a stack of magnesium targets into hydrogen is thermalized as silicium hydride (Hardy et al 1975).

An alternative approach is the thermal synthesis of volatile compounds. If an ethylene jet loaded with fission products is passed through a reaction zone heated to about 900°C, selenium and tellurium form volatile species that can be separated from each other by utilizing differences in their transport behavior (Zendel et al 1978): selenium, as well as the halogens and noble gases, is not attached to clusters in the jet, in contrast to tellurium and the bulk of fission products. Figure 6 illustrates a continuous separation procedure for tellurium based on these differences. The ethylene-nitrogen jet carrying the fission products enters a charcoal column that absorbs selenium, bromine, and iodine; the fission products fixed on clusters pass through, accompanied by krypton and xenon. The escaping gas is then heated to 860°C. At this temperature, the clusters are decomposed and volatile compounds of

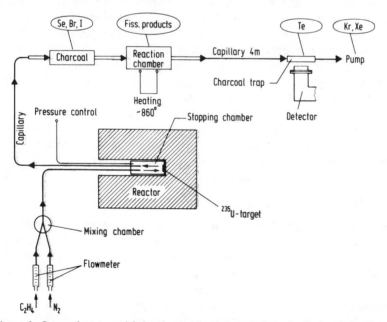

Figure 6 Set-up (not to scale) for the continuous separation of tellurium from fission products by the formation of volatile tellurium species in a hot reaction zone and their collection in a charcoal trap (Zendel et al 1978).

tellurium are formed, transported to, and absorbed in a charcoal trap at the detector. With this technique, neutron-rich tellurium isotopes up to 3.6-s [137]Te were detected. In the selenium procedure, the activities attached to clusters are first retained in paper filters whereas selenium, the halogens, and noble gases pass through and are heated in the reaction chamber. The volatile species of selenium are then adsorbed on quartz powder coated with silver nitrate. With this procedure 1.6-s [88]Se could be observed.

Selective sticking to surfaces may also be utilized in rapid separations. Grover & Lilenfeld (1972) reported large differences in the sticking of astatine atoms and astatine hydride impinging as molecular beams on various metallic surfaces.

Trapping with revolatilization was used by Matschoss & Bächmann (1979) in a rapid separation of ruthenium from fission products covering a half-life range down to a few seconds. The fission products were thermalized in a mixture of nitrogen, oxygen, and water vapor. Technetium, ruthenium, and iodine were volatilized in a tube kept at 900°C and deposited in a trap at 150°C. After flushing the deposit with hydrogen chloride, only ruthenium remains trapped there.

Gas chemical methods have been combined with mass separation in order to purify elements from isobaric contaminations (Grapengiesser & Rudstam 1973, Rudstam et al 1981). The mass-separated beam impinges on a hot collector consisting of graphite, molybdenum, or tungsten. The mixture of gaseous species liberated from the collector passes through a long evacuated quartz tube kept at medium temperature where the less volatile constituents are deposited. From the escaping gas the element of interest is removed by deposition on a catcher foil held at lower temperatures whereas the most volatile components pass this foil. Successful operation has been demonstrated by the study of short-lived fission products such as 0.47-s [141]I, 1.5-s [78]Zn, and 0.9-s [128]Cd (Rudstam et al 1981).

3.3 Chromatography

In gas chromatography differences in the distribution of volatile elements or compounds between a mobile gaseous phase and a stationary solid phase are used to separate elements in a number of different approaches. Of those, thermochromatography is the only one that has been applied so far to short-lived nuclei. In this technique the mixture of gaseous species passes through a column kept at temperatures decreasing with distance from the entrance. In such a temperature gradient the less volatile species are deposited in the high temperature region and the more volatile species in the low temperature region. For trace amounts the adsorption enthalpy of the species on the surface material is the essential parameter determining the deposition but relative volatilities indicated by boiling points may serve for a first orientation.

To illustrate this approach, we show schematically in Figure 7 an on-line method used to identify element 104 on the basis of its chemical behavior (Zvara et al 1972). As mentioned in Section 1.2, this first transactinide element should form, by analogy with its homologs zirconium and hafnium, a volatile tetrachloride whereas the chlorides of the adjacent heavy actinide elements are much less volatile. Figure 7 gives the gas chromatographic apparatus along with a histogram of the results for 3-s $^{259}104$. This nuclide was produced by bombarding ^{242}Pu with ^{22}Ne. The target also contained a small amount of samarium in order to produce the homolog hafnium; in addition scandium, a stand-in for the heavier actinides, was formed from the target backing aluminum. The recoil atoms are thermalized in nitrogen and swept to the chromatographic column, an empty glass tube of 195 cm length and 4 mm inside diameter. At the entrance to the column, vapors of the chlorinating agents titanium tetrachloride and thionyl chloride are added. In the first 30 cm, turbulent flow is generated by numerous protuberances in the

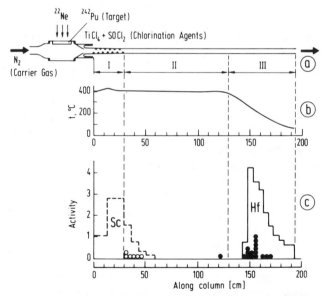

Figure 7 (a) Set-up (schematically) for on-line thermochromatography of volatile chlorides used to study chemical properties of element 104; (b) temperature gradient in the column; (c) deposition position of $^{259}104$ (*dots*) compared to the deposition position of hafnium, scandium (*lines*) and of heavy actinides (*circles*). After Zvara et al (1972).

wall in order to enhance the deposition of less volatile chlorides. This and the following section (100 cm) are kept at 400°C; in the third section the temperature decreases approximately linearly to 50° over 65 cm length. For detection of spontaneous fission decay, mica plates are inserted along the axis. As can be seen in Figure 7, scandium appears at the beginning of the column together with some fission events that can be attributed to spontaneously fissioning heavy actinide isotopes. At the end of the column where the temperature gradient is maintained, element 104 is deposited at the same position as hafnium. A total of 16 fission events were seen there. Hence, element 104 forms a volatile chloride species in the manner similar to hafnium, and is different in its chemistry from the heavy actinides.

With the same technique, the similarity of element 105 to niobium was demonstrated in an experiment using volatile bromides (Zvara et al 1976) after the bromide system had been found suitable for such procedures (Belov et al 1975, Zvara et al 1975). For a general discussion of this approach, we refer the reader to the review articles of Keller & Seaborg (1977) and Zvara (1981). The latter article also gives a procedure for the identification of element 107 by thermochromatography in a stream of humid air. This procedure is based on the formation of volatile oxides, such as a trioxide, that are known for the homologous element rhenium.

A convenient modification is the use of a gas transport system between target region and chromatographic column, as shown by Silva et al (1977). They transported fission products in a mixture of ethylene and nitrogen and produced bromides by a reaction with bromine gas; in order to destroy the clusters totally, this reaction has to be carried out at temperatures greater than 600°C. During a thermochromatographic separation, 1.7-s ^{135}Sb could be measured on-line by placing detectors at the deposition zone of antimony in the column. The on-line separation of elements forming volatile chlorides in several reactive gases such as chlorine, hydrochloric acid, thionyl chloride was studied by Hickmann et al (1980), who used potassium chloride clusters in a nitrogen jet for transportation. Conversion into volatile species was achieved by stopping the clusters in a hot quartz wool plug swept by the reactive gases. With a similar approach von Dincklage et al (1980) produced niobium pentachloride and hafnium tetrachloride from a helium jet loaded with carbon tetrachloride and measured on-line the adsorption enthalpies of these chlorides on quartz surfaces using the isotopes 18-s ^{90m}Nb, 12-s ^{160}Hf, and 17-s ^{161}Hf.

A number of thermochromatographic radiochemical separations have been performed on a somewhat slower time scale than considered here;

they may be useful as a starting point for further developments. We refer to work with oxides, hydroxides, chlorides, and oxychlorides (Adilbish et al 1979, Novgorodov et al 1980), and with fluorides (Weber et al 1973). Rapid formation of fluorides may be achieved with carbon tetrafluoride, as evidenced by the observation of molecular ions for a number of elements, including alkaline earths, lanthanides, and zirconium, in gas discharge ion sources in presence of carbon tetrafluoride (Hoff et al 1980).

Coating of the column surface with alkali and alkaline earth chlorides changes the deposition behavior of volatile species by the formation of compounds (Tsalas & Bächmann 1978). Thermochromatography on titanium columns was used to demonstrate the divalency of californium, einsteinium, fermium, and mendelevium in the metallic state (Hübener 1980). Thermochromatography in vacuum under molecular flow conditions may be an approach to separations on a very short time scale (Eichler 1977).

The separation of lanthanides and actinides by high resolution gas chromatography has been investigated using complexes formed with aluminum chloride (Zvara et al 1974). In this technique, the mixture is fed into a long glass column filled with glass balls, and the different species appear as well-separated peaks at the end of the column, which is operated at temperatures between 200 and 300°C, in most cases with an increase of temperature with time. The procedures are too slow at present; typical separation times are tens of minutes. Similar complexes with aluminum bromide (Zvarova 1979) offer some advantages. Chelate complexes with trifluoroacetones (E. Herrmann et al 1974) are other alternatives.

3.4   Evaporation from Molten Phases and Solids

A large body of data on the rapid release of volatile species from molten phases and solids has been accumulated in connection with on-line mass separation either by direct studies of release kinetics or indirectly by the observation of short-lived nuclides in such experiments. The rate-determining step can be diffusion to or evaporation from the surface as discussed in detail in the recent review article by Ravn (1979). Although rapid release from solid and molten targets has not yet found much application in rapid chemical separations, this approach may receive attention in future developments. The reason why such techniques have not been incorporated thus far in rapid chemical separation procedures may be due to the fact that rapid release requires, in general, high temperatures. Under such conditions the selectivity of the initial separation step disappears more and more so that the following steps have to be more selective. In on-line mass separation, this problem is overcome by the selectivity of mass separation plus additional selective steps such as surface ionization in the ion source.

To illustrate this further, consider the very fast and selective emanation of noble gases and halogens such as 35-ms ^{218}Rn and 32-ms ^{217}At from uranyl and lanthanum stearate at room temperature (Grover et al 1969). At moderate temperatures, radon escapes rapidly from solid thorium hydroxide hydrate (Hornshøj et al 1971) as illustrated with 1.0-s ^{200}Rn, and certain fission products are released from solid uranium tetrafluoride (Weber et al 1973). Release from molten metals and alloys can be chemically selective, too (Ravn et al 1975), as in the cases of mercury from lead, cesium from lanthanum metal, and francium from a thorium-lanthanum alloy. For example, the release of mercury from lead at 700°C is fast enough to identify 0.17-s ^{177}Hg (Hagberg et al 1979). In addition, eutectics of molten salts can be utilized as demonstrated by the release of antimony from a tellurium-lithium-potassium chloride melt (Glomset et al 1981).

Widely in use is the rapid diffusion of elements through hot graphite (Cowan & Orth 1958) in which recoil atoms from a target are stopped. A rather large number of elements rapidly diffuse through the graphite and evaporate at medium temperatures, 1300–1600°C (Borg et al 1971). Among the fission products, nuclides with half-lives around one second were identified in this way: isotopes of zinc, gallium, bromine, krypton, rubidium (Rudstam & Lund 1976), silver, cadmium, indium, tin, antimony, iodine, and cesium (Lund & Rudstam 1976). A recent example is the identification of 60-ms ^{183}Tl (Schrewe et al 1980). By combining selective surface ionization with diffusion (Klapisch 1974), very short-lived neutron-rich alkali isotopes were found, such as 4.6-ms ^{34}Na (Thibault et al 1981). Carbides of titanium, vanadium, lanthanum, and uranium were successfully used for rapid release of short-lived nuclides including 48-ms ^{201}Fr (Carraz et al 1978, 1979).

The other approach is rapid release from very hot metals (Andersen et al 1965) such as tantalum in which the recoil atoms are implanted (Bogdanov et al 1976, Kabachenko et al 1977). Release kinetics of lanthanides, alkaline earth, and alkali metals from hot tantalum foils and other materials was studied by several authors (Latuszyński et al 1974, Latuszyński 1975, Beyer et al 1978, 1981). The identification of the proton emitter 0.42-s ^{147}Tm by on-line mass separation is a recent example of the application of such techniques (Klepper et al 1982). Instead of metal foils, metal powders can be used, as shown by release measurements for, say, ytterbium from hafnium and tantalum, and for selenium from niobium and molybdenum (Carraz et al 1978). As

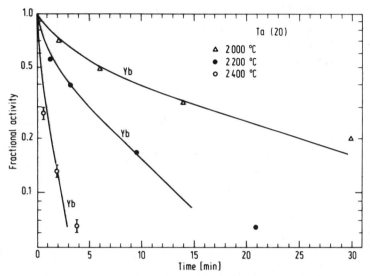

Figure 8 Release of ytterbium from tantalum powder of 20 μm average particle size as a function of time and temperature (Carraz et al 1978).

Figure 8 shows, the release is very temperature dependent. To mention only one application of such a powder target, we refer the reader to the study of 65-ms [74]Rb (D'Auria et al 1977).

4. CONCLUSIONS AND OUTLOOK

Without attempting to be comprehensive we have referred in this article to rapid chemical separation procedures for some thirty elements. This fact demonstrates that such techniques can be widely applied. So far they have mainly been used for studies on fission products, actinide and transactinide elements, i.e. in regions where radiochemical techniques traditionally play a role. Not much has been done, on the other hand, with such techniques on neutron-deficient nuclei. Half-life regions down to a few tenths of a second can be covered with rapid chemical separations. This time scale is close to the limit put by diffusion of chemical species through boundary layers between two phases and by the velocity of phase separations. A further exploration of yet unknown regions of the chart of nuclides, and studies of the chemistry of the heaviest elements whose half-lives get shorter and shorter with increasing atomic number, require chemical procedures on a time scale so far inaccessible. It should be possible to overcome the present limits by a combination of specific nuclear effects with chemical selective steps. Nuclear recoil from a solid into a gas is one example for an almost prompt transfer between two phases; electrostatic collection of ionic species is another one. The combination of such rapid phase transport techniques with chemically selective steps has still to be explored, however.

Literature Cited

Adilbish, M., Bayar, B., Votsilka, I., Zaitseva, N. G., Kovalew, A. S., Kovach, Z., Novgorodov, A. F., Fominykh, M. I., 1979. *Radiokhimiya* 21:296–307; *Sov. Radiochem.* 21:255–65
Ahrens, H., Kaffrell, N., Trautmann, N., Herrmann, G. 1976. *Phys. Rev. C* 14:211–17
Andersen, M. L., Nielsen, O. B., Scharff, B. 1965. *Nucl. Instrum. Methods* 38:303–5
Ärje, J., Valli, K. 1981. *Nucl. Instrum. Methods* 179:533–39
Aronsson, P. O., Johansson, B. E., Rydberg, J., Skarnemark, G., Alstad, J., Bergersen, B., Kvåle, E., Skarestad, M. 1974a. *J. Inorg. Nucl. Chem.* 36:2397–2403
Aronsson, P. O., Skarnemark, G., Skarestad, M. 1974b. *Inorg. Nucl. Chem. Lett.* 10:499–504
Äystö, J., Rantala, V., Valli, K., Hillebrand, S., Kortelahti, M., Eskola, K., Raunemaa, T. 1976. *Nucl. Instrum. Methods* 139:325–29
Baker, J. D., Gehrke, R. J., Greenwood, R. C., Meikrantz, D. H. 1981. *Radiochim. Acta* 28:51–54
Belov, V. Z., Zvarova, T. S., Shalaevskii, M. R. 1975. *Radiokhimiya* 17:86–93; *Sov. Radiochem.* 17:87–92
Beyer, G. J., Herrmann, E., Molnar, F., Raiko, V. I., Tyrroff, H. 1972. *Radiochem. Radioanalyt. Lett.* 12:259–69
Beyer, G. J., Knotek, O., Jachim, M., Yushkevich, Yu. V., Novgorodov, A. F. 1978. *Nucl. Instrum. Methods* 148:543–51
Beyer, G. J., Novgorodov, A. F., Kovalew, A., Prazak, F., Khalkin, V. A., Yushkevich, Yu. V. 1981. *Nucl. Instrum. Methods* 186:401–7
Bogdanov, D. D., Voboril, J., Demyanov, A. V., Karnaukhov, V. A., Petrov, L. A. 1976. *Nucl. Instrum. Methods* 136:433–35
Borg, S., Bergström, I., Holm, G. B., Rydberg, B., De Geer, L.-E., Rudstam, G., Grapengiesser, B., Lund, E., Westgaard, L. 1971. *Nucl. Instrum. Methods* 91:109–16
Brodén, K., Skarnemark, G., Bjørnstad,

T., Eriksen, D., Haldorsen, I., Kaffrell, N., Stender, E., Trautmann, N. 1981. *J. Inorg. Nucl. Chem.* 43:765–71
Brüchle, W., Herrmann, G. 1982. *Radiochim. Acta.* 30:1–10
Carraz, L. C., Haldorsen, I. R., Ravn, H. L., Skarestad, M., Westgaard, L. 1978. *Nucl. Instrum. Methods* 148:217–30
Carraz, L. C., Sundell, S., Ravn, H. L., Skarestad, M., Westgaard, L. 1979. *Nucl. Instrum. Methods* 158:69–80
Chayawattanangkur, K., Herrmann, G., Trautmann, N. 1973. *J. Inorg. Nucl. Chem.* 35:3061–73
Cowan, G. A., Orth, C. J. 1958. *Proc. 2nd United Nations Int. Conf. Peaceful Uses At. Energy, Geneva* 7:328–34
D'Auria, J. M., Carraz, L. C., Hansen, P. G., Jonson, B., Mattsson, S., Ravn, H. L., Skarestad, M., Westgaard, L. 1977. *Phys. Lett. B* 66:233–35
del Marmol, P., Fettweis, P. 1972. *Nucl. Phys. A* 194:140–60
del Marmol, P., van Tigchelt, H. 1972. *Radiochim. Acta* 17:52–54
Dufour, J. P., Del Moral, R., Fleury, A., Hubert, F., Llabador, Y., Mauhourat, M. B., Bimbot, R., Gardès, D., Rivet, M. F. 1981. *Proc. 4th Int. Conf. Nuclei far from Stability, Helsingør 1981*, pp. 711–16, CERN Rep. 81–09. Geneva: CERN
Eichler, B. 1977. *Report ZfK–346* Rossendorf bei Dresden: Zentralinst. Kernforsch.
Elchuk, S., Cassidy, R. M. 1979. *Analyt. Chem.* 51:1434–38
Esterl, J. E., Sextro, R. G., Hardy, J. C., Ehrhardt, G. J., Cerny, J. 1971. *Nucl. Instrum. Methods* 97:229–33
Fettweis, P., del Marmol, P. 1975. *Z. Phys. A* 275:359–67
Franz, G., Herrmann, G. 1978. *J. Inorg. Nucl. Chem.* 40:945–55
Georg, E., Fass, R., Lemmertz, P., Wilhelm, H. G., Wollnik, H., Hirdes, D., Jungclas, H., Brandt, R., Schardt, D., Wien, K. 1978. *Nucl. Instrum. Methods* 157:9–18
Ghiorso, A., Nitschke, J. M., Alonso, J. R., Alonso, C. T., Nurmia, M., Seaborg, G. T., Hulet, E. K., Lougheed,

R. W. 1974. *Phys. Rev. Lett.* 33:1490–93

Glomset, O., Bjørnstad, T., Hagebø, E., Haldorsen, I. R., Hjaltadottir, V., Sundell, S. 1981. See Dufour et al 1981, pp. 732–39

Grapengiesser, B., Rudstam, G. 1973. *Radiochim. Acta* 20:85–90

Greenwood, R. C., Gehrke, R. J., Baker, J. D., Meikrantz, D. H. 1981. See Dufour et al 1981, pp. 602–7

Grover, R. J., Lebowitz, E., Baker, E. 1969. *J. Inorg. Nucl. Chem.* 31:3705–20

Grover, J. R., Lilenfeld, H. V. 1972. *Nucl. Instrum. Methods* 105:189–96

Hagberg, E., Hansen, P. G., Hornshøj, P., Jonson, B., Mattsson, S., Tidemand-Petersson, P. 1979. *Nucl. Phys. A* 318:29–44

Hardy, J. C., Schmeing, H., Geiger, J. S., Graham, R. L. 1975. *Nucl. Phys. A* 246:61–75

Herrmann, E., Beyer, G. J., John, M. 1974. *Isotopenpraxis* 10:411–13

Herrmann, G. 1979. *Nature* 280:543–49

Herrmann, G., Denschlag, H. O. 1969. *Ann. Rev. Nucl. Sci.* 19:1–32

Hickmann, U., Greulich, N., Trautmann, N., Gäggeler, H., Gäggeler-Koch, H., Eichler, B., Herrmann, G. 1980. *Nucl. Instrum. Methods* 174:507–13

Hoff, P., Jacobsson, J., Johansson, B., Aagaard, P., Rudstam, G., Zwicky, H. U. 1980. *Nucl. Instum. Methods* 172:413–18

Hornshøj, P., Wilsky, K., Hansen, P. G., Lindahl, A., Nielsen, O. B. 1971. *Nucl. Phys. A* 163:277–88

Hübener, S. 1980 *Radiochem. Radioanalyt. Lett.* 44:79–86

Hulet, E. K. 1982 *Actinides '81 Proc. Int. Conf., Asilomar, 1981*. London: Pergamon. In press

Hulet, E. K., Lougheed, R. W., Wild, J. F., Landrum, J. H., Nitschke, J. M., Ghiorso, A. 1980 *J. Inorg. Nucl. Chem.* 42:79–82

Izak, T., Amiel, S. 1972 *J. Inorg. Nucl. Chem.* 34:1469–77

Jungclas, H., Macfarlane, R. D., Fares, Y. 1971. *Radiochim. Acta* 16:141–47

Kabachenko, A. P., Kuznetsov, I. V., Soo. L. G., Tarantin, N. I. 1977. *Nucl. Instrum. Methods* 147:337–40

Keller, O. L., Seaborg, G. T. 1977. *Ann. Rev. Nucl. Sci.* 27:139–66

Kirchner, R. 1981. *Nucl. Instrum. Methods* 186:275–93

Klapisch, R. 1974. In *Nuclear Spectroscopy and Reactions*, ed. J. Cerny, Vol. A, pp. 213–42. New York/London: Academic

Klein, G., Kaffrell, N., Trautmann, N.,

Herrmann, G. 1975. *Inorg. Nucl. Chem. Lett.* 11:511–18

Klepper, O., Batsch, T., Hofmann, S., Kirchner, R., Kurcewicz, W., Reisdorf, W., Roeckl, E., Schardt, D., Nyman, G. 1982. *Z. Phys. A* 305:125–30

Kratz, J. V. 1982. *Radiochim. Acta.* In press

Kratz, J. V., Franz, H., Herrmann, G. 1973. *J. Inorg. Nucl. Chem.* 35:1407–17

Kratz, J. V., Franz, H., Kaffrell, N., Herrmann, G. 1975. *Nucl. Phys. A* 250:13–37

Kratz, J. V., Herrmann, G. 1970. *J. Inorg. Nucl. Chem.* 32:3713–23

Kratz, J. V., Poitou, J., Brüchle, W., Gäggeler, H., Schädel, M., Wirth, G., Lucas, R. 1981. *Nucl. Phys. A* 357:437–70

Kratz, K.-L., Herrmann, G. 1973. *Radiochem. Radioanalyt. Lett.* 13:385–90

Latuszyński, A. 1975. *Nucl. Instrum. Methods* 123:489–94

Latuszyński, A., Zuber, K., Zuber, J., Potempa, A., Zuk, W. 1974. *Nucl. Instrum. Methods* 120:321–28

Lien, O. G., Stevenson, P. C., Henry, E. A., Yaffe, R. P., Meyer, R. A. 1981 *Nucl. Instrum. Methods* 185:351–58

Lister, C. J., Haustein, P. E., Alburger, D. E., Olness, J. W. 1981. *Phys. Rev. C* 24:260–78

Lund, E., Rudstam, G. 1976. *Phys. Rev. C* 13:1544–51

Macfarlane, R. D., Gough, R. A., Oakey, N. S., Torgerson, D. F. 1969. *Nucl. Instrum. Methods* 73:285–91

Macfarlane, R. D., McHarris, W. C. 1974. See Klapisch 1974, pp. 244–86

Matschoss, V., Bächmann, K. 1979. *J. Inorg. Nucl. Chem.* 41:141–47

Mazumdar, A. K., Wagner, H., Walcher, W., Brügger, M., Trautmann, N. 1981. *Nucl. Instrum. Methods* 186:131–34

Meikrantz, D. H., Gehrke, R. J., McIsaac, L. D., Baker, J. D., Greenwood, R. C. 1981. *Radiochim. Acta* 29:93–101

Meyer, R. A., Henry, E. A. 1980. *Proc. Workshop Nucl. Spectrosc. Fission Products, Grenoble, 1979*, pp. 59–103. Bristol/London: Inst. Phys. Conf. Ser. No. 51

Moltz, D. M., Gough, R. A., Zisman, M. S., Vieira, D. J., Evans, H. C., Cerny, J. 1980a. *Nucl. Instrum. Methods* 172:507–18

Moltz, D. M., Wouters, J. M., Äystö, J., Cable, M. D., Parry, R. F., von Dincklage, R. D., Cerny, J. 1980b. *Nucl. Instrum. Methods* 172:519–25

Novgorodov, A. F., Adilbish, M., Zaitseva, N. G., Kowalew, A., Kovacs, Z. 1980. *J. Radioanalyt. Chem.* 56:37–51

Ohyoshi, A., Ohyoshi, E., Tamai, T., Shinagawa, M. 1972. *J. Inorg. Nucl. Chem.* 34:3293–3302

Ravn, H. L. 1979. *Phys. Rep.* 54:201–59

Ravn, H. L., Sundell, S., Westgaard, L. 1975. *Nucl. Instrum. Methods* 123:131–44

Rudolph, W., Kratz, K.-L., Herrmann, G. 1977. *J. Inorg. Nucl. Chem.* 39:753–58

Rudstam, G., Aagaard, P., Hoff, P., Johansson, B., Zwicky, H. U. 1981. *Nucl. Instrum. Methods* 186:365–79

Rudstam, G., Lund, E. 1976. *Phys. Rev. C* 13:321–30

Schädel, M., Kratz, J. V., Ahrens, H., Brüchle, W., Franz, G., Gäggeler, H., Warnecke, I., Wirth, G., Herrmann, G., Trautmann, N., Weis, M. 1978. *Phys. Rev. Lett.* 41:469–72

Schädel, M., Trautmann, N., Herrmann, G. 1977. *Radiochim. Acta* 24:27–31

Schrewe, U. J., Tidemand-Petersson, P., Gowdy, G. M., Kirchner, R., Klepper, O., Plochocki, A., Reisdorf, W., Roeckl, E., Wood, J. L., Zylicz, J., Fass, R., Schardt, D. 1980. *Phys. Lett. B* 91:46–50

Schüssler, H. D., Grimm, W., Weber, M., Tharun, U., Denschlag, H. O., Herrmann, G. 1969. *Nucl. Instrum. Methods* 73:125–31

Schüssler, H. D., Herrmann, G. 1970. *Radiochim. Acta* 13:65–69

Schüssler, H. D., Herrmann, G. 1972. *Radiochim. Acta* 18:123–33

Silva, R. J., Harris, J., Nurmia, M., Eskola, K., Ghiorso, A. 1970a. *Inorg. Nucl. Chem. Lett.* 6:871–77

Silva, R. J., Sikkeland, T., Nurmia, M., Ghiorso, A. 1970b. *Inorg. Nucl. Chem. Lett.* 6:733–39

Silva, R. J., Trautmann, N., Zendel, M., Dittner, P. F., Stender, E., Ahrens, H. 1977. *Nucl. Instrum. Methods* 147:371–78

Skarnemark, G., Aronsson, P. O., Brodén, K., Rydberg, J., Bjørnstad, T., Kaffrell, N., Stender, E., Trautmann, N. 1980. *Nucl. Instrum. Methods* 171:323–28

Skarnemark, G., Stender, E., Trautmann, N., Aronsson, P. O., Bjørnstad, T., Kaffrell, N., Kvåle, E., Skarestad, M. 1976. *Radiochim. Acta* 23:98–103

Stachel, J., Kaffrell, N., Trautmann, N., Emling, H., Folger, H., Grosse, E., Kulessa, R., Schwalm, D., Brodén, K., Skarnemark, G., Eriksen, D. 1981. See Dufour et al 1981, pp. 436–42

Stender, E., Trautmann, N., Herrmann, G. 1980. *Radiochem. Radioanalyt. Lett.* 42:291–96

Tamai, T., Matsushita, R., Takada, J., Kiso, Y. 1973a. *Inorg. Nucl. Chem. Lett.* 9:1145–52

Tamai, T., Takada, J., Matsushita, R., Kiso, Y. 1973b. *Inorg. Nucl. Chem. Lett.* 9:973–79

Thibault, C., Epherre, M., Audi, G., Huber, G., Klapisch, R., Touchard, F., Guillemaud, D., Naulin, F. 1981. See Dufour et al 1981, pp. 365–67

Tittel, G., Kaffrell, N., Trautmann, N., Herrmann, G. 1977. *J. Inorg. Nucl. Chem.* 39:2115–19

Tomlinson, L., Hurdus, M. H. 1971. *J. Inorg. Nucl. Chem.* 33:3609–20

Tomlinson, L., Hurdus, M. H. 1972. *Radiochim. Acta* 17:199–202

Trautmann, N. 1982, See Hulet 1982

Trautmann, N., Aronsson, P. O., Bjørnstad, T., Kaffrell, N., Kvåle, E., Skarestad, M., Skarnemark, G., Stender, E. 1975. *Inorg. Nucl. Chem. Lett.* 11:729–35

Trautmann, N., Herrmann, G. 1976. *J. Radioanalyt. Chem.* 32:533–48

Trautmann, N., Kaffrell, N., Ahrens, H., Dittner, P. F. 1976. *Phys. Rev. C* 13:872–74

Trautmann, N., Kaffrell, N., Behlich, H. W., Folger, H., Herrmann, G., Hübscher, D., Ahrens, H. 1972. *Radiochim. Acta* 18:86–101

Tsalas, S., Bächmann, K. 1978. *Analyt. Chim. Acta* 98:17–24

Vine, E. N., Wahl, A. C. 1981. *J. Inorg. Nucl. Chem.* 43:877–83

von Dincklage, R. D., Schrewe, U. J., Schmidt-Ott, W.-D., Fehse, H., Bächmann, K. 1980. *Nucl. Instrum. Methods* 176:529–35

Weber, M., Trautmann, N., Menke, H., Herrmann, G., Kaffrell, N. 1973. *Radiochim. Acta* 19:106–13

Zendel, M., Stender, E., Trautmann, N., Herrmann, G. 1978. *Nucl. Instrum. Methods* 153:149–56

Zendel, M., Trautmann, N., Herrmann, G. 1981. *Radiochim. Acta* 29:17–20

Zvara, I. 1981. *Pure Appl. Chem.* 53:979–95

Zvara, I., Belov, V. Z., Domanov, V. P., Korotkin, Yu. S., Chelnokov, L. P., Shalaevskii, M. R., Shchegolev, V. A., Hussonois, M. 1972. *Radiokhimiya* 14:119–22; *Sov. Radiochem.* 14:115–18

Zvara, I., Belov, V. Z., Domanov, V. P., Shalaevskii, M. R. 1976. *Radiokhimiya* 18:371–77; *Sov. Radiochem.* 18:328–34

Zvara, I., Chuburkov, Yu. T., Tsaletka, R., Zvarova, T. S., Shalaevskii, M. R., Shilov, B. V. 1966. *At. Energy* 21:83–84; *Sov. At. Energy* 21:709–10

Zvara, I., Eichler, B., Belov, V. Z., Zvarova, T. S., Korotkin, Yu. S., Shalaevskii, M. R., Shchegolev, V. A.,

Hussonois, M. 1974. *Radiokhimiya* 16:720–27; *Sov. Radiochem.* 16:709–15

Zvara, I., Keller, O. L., Silva, R. J., Tarrant, J. R. 1975. *J. Chromatogr.* 103:77–83

Zvarova, T. S. 1979. *Radiokhimiya* 21:727–30; *Sov. Radiochem.* 21:627–29

Ann. Rev. Part. Sci. 1982. 32:149–75
Copyright © 1982 by Annual Reviews Inc. All rights reserved

WHAT CAN WE COUNT ON?

A Discussion of Constituent-Counting Rules for Interactions Involving Composite Systems

Dennis Sivers

Argonne National Laboratory, Argonne, Illinois 60439

CONTENTS

1. INTRODUCTION: BOUND STATES IN A FIELD THEORY

High energy physics now has a candidate theory to describe the strong interactions. This theory, quantum chromodynamics or QCD, has several attractive features to commend it — including the fact that it puts the strong, hadronic force on a similar footing with the electroweak forces. At the level that we can deal with it quantitatively, QCD is formulated as a perturbation expansion in the basis of free quarks and gluons. The states observed in the laboratory, baryons and mesons, are not explicitly present but are to be interpreted as composite systems of quarks and gluons.

149

0163-8998/82/1201-0149$02.00

Many of the problems in doing quantitative calculations in QCD perturbation theory arise from the fact that the general theory of bound states in a relativistic field theory is not understood. An extra complication occurs because we do not really know that the specific formulation of QCD we use allows bound states such as mesons and baryons to appear as the physical spectrum. In place of rigorous proofs about the important property of confinement, we currently have only plausibility arguments. Even though specific simplified models and numerical calculations on finite lattices suggest that such bound states do occur, these approaches do not as yet give much insight into the detailed properties of composite systems and our understanding of the theory remains incomplete.

The fact that we can proceed at all with quantitative calculations is highly nontrivial. Using various mathematical techniques we can isolate the part of the theory we do not understand (the structure of composite systems attributable to the "soft" or long-distance part of the forces) from the portions that we can calculate perturbatively (the "hard" or short-distance behavior where a description in terms of free quarks and gluons can be valid). Even with this separation, the predictive power of the theory is limited by the lack of understanding of bound states. For example, QCD perturbation theory cannot be used to calculate the shape of the "structure function," which gives the spectrum for the deep inelastic scattering of a lepton from a hadronic target. Only when the structure function is measured experimentally at some large value for the momentum transfer of the lepton probe, does it become possible to calculate perturbatively how the shape of the structure function should change as measurements are made at still higher values of momentum transfer. Even the presentation of the prediction is influenced by our lack of understanding of composite systems, since the evolution equation involves the theoretically unknown starting point.

The overall usefulness of the theory can be greatly extended if, within the context of perturbation theory, it can be shown possible to deal with bound states even when they cannot rigorously be shown to exist. It is with this attitude that we should approach the empirical relations which are the subject of this review. These relations, sometimes called dimensional scaling laws but usually referred to as constituent-counting rules, characterize the asymptotic behavior of scattering cross sections of bound states in terms of the number of elementary point-like constituents involved. They were first formulated by Brodsky & Farrar (1973, 1975) and, independently, by Matveev, Murddyan & Tavkheldize (1973). They have since been used so frequently in the course of discussing experimental data that they have become part of the "folklore" of particle physics phenomenology. (See, for example, the discussion in Sivers, Brodsky & Blankenbecler, 1976.) Although many processes can be treated by these methods, our discussion concentrates on three types of measurement that amply demonstrate their use. For these processes, simple versions of the rules take the following three forms.

1. Hadronic form factors at large-momentum transfer. The spin-averaged form factor for the elastic scattering of a hadron consisting of a minimum number, n_{H}, of elementary constituents can be approximated:

$$\lim_{Q^2 \to \infty} F_{\mathrm{H}}(Q^2) = C(Q^2)^{1-n_{\mathrm{H}}}. \qquad 1.$$

2. The large-x behavior of hadronic structure functions. As the Bjorken scaling variable $x = Q^2/(2p \cdot Q)$ for a scale-invariant hadronic structure function approaches one, there is a limiting behavior

$$\lim_{Q^2 \to \infty} \lim_{x \to 1} G(x, Q^2) = C(1 - x)^{2n_s - 1}, \qquad 2.$$

where n_s is the minimum number of spectator constituents in the hadronic system.

3. Fixed-angle exclusive scattering of hadrons. At high energies the fixed-angle scattering of the hadrons AB $\to$ CD achieves the asymptotic form

$$\lim_{s \to \infty} \frac{d\sigma}{dt}(\mathrm{AB} \to \mathrm{CD}) = f(\theta)s^{2-(n_{\mathrm{A}} + n_{\mathrm{B}} + n_{\mathrm{C}} + n_{\mathrm{D}})}, \qquad 3.$$

where $s = (p_{\mathrm{A}} + p_{\mathrm{B}})^2$, $t = (p_{\mathrm{A}} - p_{\mathrm{B}})^2$, $-t/s \cong (1 - \cos \theta)/2$; and the n_{H} are the minimum number of fields in each hadron.

Each of these simple empirical rules has some support from available experimental data. Their common feature is that they count the minimum number of elementary constituents in a bound system. Their derivation assumes that in high energy collisions hadrons can be conveniently described as collinear systems of elementary constituents and that the forces scattering these constituents through large-momentum transfers are scale invariant. Neither of these assumptions is exactly true in quantum chromodynamics and the rules above (Equations 1–3) are therefore subject to modification. Calculations for the form of these modifications involve details of the complicated machinery of perturbative QCD and some assumptions about nonperturbative effects. After considerable theoretical speculation and an educated peek at the data, current estimates suggest that the corrections to these formulas remain small. The apparent validity of these simple rules is an interesting, perhaps surprising, result. Their success may indicate that all the complicated, bizarre types of behavior that could, in principle, be associated with the formation of bound states in a field theory do not, in fact, appear in QCD.

At any rate, the study of these rules and their corrections in QCD perturbation theory provides an interesting exercise. We consider each of them here in more detail, along with closely related results concerning the helicity structure of hadronic amplitudes. The structure of the remainder of this article is as follows: In Section 2 we discuss QCD predictions for the hadronic form factors. The pion form factor is considered in some detail but we also look at baryon form factors. Section 3 is devoted to counting-rule predictions for the shape of deep inelastic structure functions. In Section 4 we consider QCD-based approaches to hadronic fixed-angle scattering. In Section 5 we present the QCD-modified versions of Equations 1–3 and try to draw some conclusions from the data. The intent of this review is to be as self-contained as possible without presenting the details of complicated calculations. The discussion duplicates material available elsewhere. A reader interested in pursuing the subject further can find several excellent alternative starting points in the reviews of Mueller (1981) and Brodsky (1979).

2. HADRONIC FORM FACTORS

The large-momentum behavior of hadronic form factors is widely considered to contain basic information about the composite structure of hadrons and the nature of the strong interactions at short distances. For nonrelativistic systems, there is ample basis for this belief. The large-$|q^2|$ behavior of a nonrelativistic form factor for a two-body bound state,

$$
\begin{aligned}
F(|q^2|) &= \int \exp(i\mathbf{q} \cdot \mathbf{r}) |\Psi(\mathbf{r})|^2 \, d^3\mathbf{r} \\
&= \int \Psi^*(\mathbf{k} + \mathbf{q}) \Psi(\mathbf{k}) \, d^3\mathbf{k},
\end{aligned}
\qquad 4.
$$

depends on the behavior of $\Psi(\mathbf{r})$ at small $|\mathbf{r}|$. This, in turn, is governed by the strength of the singularity in the nonrelativistic potential (Alabiso & Schierholz 1974).

For a relativistic field theory, the relation between form factors and forces is potentially more complicated. (See, for example, Artru 1982 and Sivers, Brodsky & Blankenbecler 1976.) The counting-rule expression (Equation 1) follows from the assumption that all the constituents in the hadron scatter with a large-momentum transfer and that the forces involved are asymptotically scale invariant. This assumption turns out to be "almost" correct in QCD; thus it is highly plausible that Equation 1 is subject only to logarithmic modification. In addition, because the forces in QCD are associated with spin-1 gluons and because, to leading order in m_q/Q, the qqG vertex conserves quark helicity, there are extra selection rules indicating that spin-flip form factors are suppressed.

2.1 Meson Form Factors

Let us look at the large-Q^2 behavior of the electromagnetic form factor of a $\bar{q}q$ meson such as the pion. The treatment of this problem in perturbative QCD is largely due to Brodsky & Lepage (1979a,b). These authors first developed the formalism to show that we can write the pion form factor as

$$F_\pi(Q^2) = \int_0^1 [dx_1][dx_2]\phi_0^*(x_1, \mu^2)T_H(x_1, x_2; Q^2, \mu^2)\phi_0(x_2, \mu^2) \quad 5.$$

where $[dx_i] = dx_i\, dy_i\, \delta(1 - x_i - y_i)$ and terms of order $(\text{mass})^2/Q^2$ have been neglected. The valence "wave function," $\phi_0(x, \mu^2)$ gives the amplitude for finding in the pion a quark carrying fractional momentum x and antiquark carrying fractional momentum $(1 - x)$. The connected, hard-scattering amplitude, $T_H(x_1, x_2; Q^2, \mu^2)$ is an entity that can be calculated perturbatively,

$$T_H(x_1, x_2, Q^2, \mu^2) = \alpha_s^i(\mu^2)T_B(x_1, x_2; Q^2, \mu^2)$$
$$\times [1+ \alpha_s^i(\mu^2)T_1^i(x_1, x_2; Q^2, \mu^2) + \ldots]. \quad 6.$$

The structure of this calculation is indicated schematically in Figure 1. The exact definition of the expansion parameter, $\alpha_s^i(\mu^2)$, depends on the renormalization prescription chosen. In the expression for the nth-order perturbative correction to T_H there are collinear mass singularities of the form $[\alpha_s^i \log (Q^2/\mu^2)]^n$. The form of Equation 5 displays the fact that these singularities can be factorized and absorbed into the definition of $\phi_0(x, \mu^2)$ providing that $\phi_0(x, \mu^2)$ evolves with the factorization scale,

$$\frac{\partial\phi_0(x, \mu^2)}{\partial \ln \mu^2} = \frac{\alpha_s^i(\mu^2)}{2\pi} \int_0^1 dy\, V_{qq\to qq}^{(i)}(y, x; \alpha)\phi_0(y, \mu^2). \quad 7.$$

The leading-order expression for the evolution kernel, $V_{qq\to qq}$, was given by Brodsky & Lepage (1979a,b), who also discussed solutions to Equation 7. The superscripts (i) in Equations 6 and 7 indicate that the next-to-leading order corrections to these equations can depend on how the renormalization and factorization are carried out. (For a discussion on prescription dependence, see Celmaster & Silvers 1981.)

154   SIVERS

For the pion, there is an extra constraint on the wave function. For large μ^2, it is possible to normalize $\phi_0(x, \mu^2)$ in terms of the weak decay constant for $\pi \to \mu\nu$ (Farrar & Jackson 1979),

$$f_\pi = \frac{2}{\sqrt{3}} \lim_{\mu^2 \to \infty} \int_0^1 dx \, \phi_0(x, \mu^2). \qquad 8.$$

Combining these equations and using the calculation of the next-to-leading corrections of Field, Gupta, Otto and Chang (1981), we can write the pion form factor

$$F_\pi(Q^2) = 16\pi \frac{f_\pi^2}{Q^2} \alpha_s^{\text{mom}} \left(\frac{Q^2}{2.8}\right) [1 + O(\alpha_s^2)] \qquad 9.$$

where, with the momentum space subtraction renormalization and the specific ansatz for factorization used, the identification of a scale $\mu^2 = Q^2/2.8$ absorbs the $O(\alpha_s)$ correction. Except for the logarithmic variation of $\alpha_s^{\text{mom}}(Q^2/2.8)$, the expression validates the specific form of the counting rule (Equation 1). We return to the question of comparing Equation 9 with experimental data in Section 5.

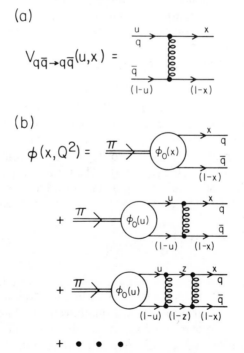

Figure 1 Elements in the calculation of the pion form factor.

This same approach can be applied to other mesonic form factors, and without too much difficulty it is possible to read off some of the consequences. In addition to giving the $1/Q^2$ power behavior, the hadron scattering amplitude, T_H, conserves quark helicities in the limit that we neglect quark masses. For regions of the x_i away from the endpoints ($x_i = 0, 1$), the wave functions represent states of vanishing orbital angular momentum so the meson helicities are the sum of the quark helicities. Looking at time-like form factors, this gives several selection rules for allowed and forbidden two-meson systems in e^+e^- annihilation. For example, the process $e^+e^- \rightarrow \pi^+\rho^-$ should fall off asymptotically with an extra power of $1/Q^2$ compared to $e^+e^- \rightarrow \pi^+\pi^-$. A more complete discussion of these helicity selection rules can be found in Brodsky & Lepage (1981).

2.2 Baryon Form Factors

The treatment of baryon form factors in QCD perturbation theory closely parallels that of meson form factors discussed above. As might be expected, however, this is potentially complicated by the fact that the minimum number of quarks involved in the process is three, instead of two. In analogy to Equation 6, the hard-scattering contribution to a baryon electromagnetic form factor can be written

$$F_B(Q^2) = \int [dx_1][dx_2]\phi^*(x_1, \mu^2)T_H(x_1, x_2; Q^2, \mu^2)\phi(x_2, \mu^2), \quad 10.$$

where $[dx_i] = dx_i\, dy_i\, dz_i\, \delta(1 - x_i - y_i - z_i)$ and T_H is the minimally connected amplitude for $\gamma^*3q \rightarrow 3q$. An explicit form for the lowest-order expression can be found in the work of Brodsky & Lepage (1980). The wave function of the proton is not normalized by any (presently measured) decay rate so the leading-order calculation contains the overall normalization as a free parameter. The perturbative calculation follows the same lines as that for the meson case and yields the leading-order expression for the nonflip proton form factor

$$G_M(Q^2) = C_i \frac{[\alpha_s^i(Q^2)]^2}{[Q^2]^2} \sum_{n,m} a_{nm}[\alpha_s^i(Q^2)]^{\gamma_n + \gamma_m}[1 + O(\alpha_s)]. \quad 11.$$

The γ_n are the same anomolous dimensions found in deep inelastic scattering and arise from the solution to the evolution equation for $\phi(x, \mu^2)$. Neither the a_{nm} nor the next-to-leading-order corrections to the hard-scattering amplitude have yet been calculated. Except for the logarithmic variation of $\alpha_s(Q^2)$, this expression conforms to the expectation of the counting rules. Because the hard-scattering amplitude in Equation 10 conserves quark helicity and because the transverse momentum integrations projecting out of the $\phi(x_i, \mu^2)$ isolate $L_z = 0$ baryon spin-flip form factors should be suppressed and there should exist connections between the normalization of nonflip form factors based on SU_6. A collection of the SU_6 constraints for baryon form factors can be found in the article by Brodsky, Lepage & Zaidi (1981).

The endpoints, $x_i \rightarrow 1$, in the integration of both Equations 5 and 10 must be analyzed carefully since the hard-scattering amplitudes are singular in that limit. As first pointed out by Drell & Yan (1970) and by West (1970), the contribution to hadronic form factor from the endpoint region requires that large momentum be transferred from the virtual probe to only one "active" constituent. Partons that are slow in the rest frames of both the initial and final state can have longitudinal momenta of approximately the same magnitude as transverse momenta in both frames and need experience only soft scattering. The complications involved in estimating the contributions of this region are more trouble-some for baryon form factors since with two soft spectator quarks there are infrared logarithms in perturbation theory as well as possible nonperturbative interactions between them not adequately described by Equation 10.

The assumption that the endpoint region dominates the elastic form factor leads to the Drell-Yan-West relation.

$$\lim_{x \rightarrow 1} G(x, Q^2) \sim C(1 - x)^{\delta+1} \quad \longleftrightarrow \quad \lim_{Q^2 \rightarrow \infty} F(Q^2) \sim C\left(\frac{K^2}{Q^2}\right)^{\delta}, \qquad 12.$$

where one parameter, δ, determines the asymptotic power behavior of the form factor and of the inelastic structure function. Because experimentally this relationship seems valid for the proton (with $\delta \cong 2$), the possibility of endpoint dominance of form factors has been considered seriously by many theorists.

The application of perturbation theory as sketched above, when combined with reasonable assumptions of the shape of ϕ, suggests that the endpoints of the integral in Equation 10 do not dominate. However, this type of argument does not give a final answer to the question of what the Drell-Yan-West mechanism contributes to the form factor. The issue remains controversial because of the possibility of real non-perturbative effects, but arguments based on the selected resummation of certain higher-order diagrams suggest that the endpoint regions are suppressed. This is discussed further by Duncan & Mueller (1980) and by Mueller (1981). This may be a question best answered experimentally. One way to look for possible endpoint contributions is to examine form factors involving a net change of hadron helicities. When "soft" quarks are involved, the arguments given above equating the hadron's helicity to the sum of the helicities of its valence quarks are not valid. The quark helicity frame and hadronic helicity frame cannot be equated when the quark has transverse momentum comparable to its longitudinal momentum. Therefore the relative normalization of helicity flip and helicity nonflip form factors can provide a clue to the possible contribution of endpoint regions. If helicity flip form factors are found experimentally to vanish with an extra power of $1/Q^2$ as predicted from the naive application of expressions like Equation 10, then the endpoints are probably harmless.

3. THE SHAPE OF DEEP INELASTIC STRUCTURE FUNCTIONS

The large-momentum scattering of a lepton from a hadronic target is one of the classic testing grounds for QCD perturbation theory. The usual treatments of this subject emphasize that, while the evolution with Q^2 of deep inelastic scattering can be calculated perturbatively, the overall shape of the structure function can depend on details of the forces at long distances. However, it is possible to argue that near the kinematic limit $x = 1$ the behavior of the structure function depends on short-distance effects. This is because as $x \rightarrow 1$ the struck quark is driven off shell

$$k^2 \simeq -\frac{(k_T^2 + m_q^2)}{(1 - x)},$$
13.

and the forces driving the particle far off shell can be calculated perturbatively by diagrams such as those shown in Figure 2.

When the leading $(1 - x)$ power is estimated from such diagrams, the probability for finding a quark at large x in a proton is

$$G_{q(+)/p(+)}(x, Q^2) \sim a(Q^2)(1 - x)^3$$
$$G_{q(-)/p(+)}(x, Q^2) \sim a'(Q^2)(1 - x)^5$$
14.

where the $+$, $-$ refer to quark and proton helicities. This correlation between the spin of the proton and the spin of the leading quark is a new

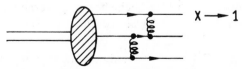

Figure 2   Connected diagram leading to an off-shell quark with $x \rightarrow 1$.

feature of QCD not found in the original version of the counting rules (Equation 2) but is a necessary consequence of the spin structure of QCD couplings (Farrar & Jackson 1975, Vainstein & Zakharov 1978). As shown in Section 5, the correlation seems to be verified by the trend of the data on spin-spin asymmetries in deep inelastic ep scattering.

Diagram such as that shown in Figure 2 can be drawn in which the proton contains more than its minimal set of valence quarks; we can thereby generalize the counting rules (Equation 14) to gluons and antiquarks. Since the gluon distribution in a proton is not measured by any convenient experiment, it is particularly desirable to have a theoretical estimate of its shape to use in phenomenological studies.

However, there is a better way to obtain the large-x behavior of these nonvalence distributions, one which takes into account the forces in QCD that lead to scaling violations. For example, a simple estimate of the shape of the gluon distribution at large x can be obtained from convoluting Equation 14 with a bremsstrahlung spectrum for the emission of a gluon from a quark. This gives (Close & Sivers 1979)

$$G_{g(+)/p(+)}(x, Q^2) \simeq b(Q^2)(1 - x)^4$$
$$G_{g(-)/p(+)}(x, Q^2) \simeq b'(Q^2)(1 - x)^6. \qquad 15.$$

Continuing down to the level of antiquarks in the "sea," one can convolute the gluon distributions (Equation 15) with the probability for a gluon to produce a $q\bar{q}$ pair to find the large-x behavior:

$$G_{\bar{q}(+)/p(+)}(x, Q^2) \simeq c(Q^2)(1 - x)^5$$
$$G_{\bar{q}(+)/p(+)}(x, Q^2) \simeq c'(Q^2)(1 - x)^7. \qquad 16.$$

Again, the $+$ and $-$ refer to helicities. The appearance of helicity factors in these expressions constitutes a major improvement on the original counting rules. This factor, $(1 - x)^{2|\Delta s_z|}$, with $\Delta s_z = s_z^{(\text{target})} - s_z^{(\text{constituent})}$ reflects the helicity conservations of QCD vertices. One thing, however, missing in Equations 14–16 is the depletion effect of scaling violations on the $x \rightarrow 1$ behavior. The shapes of the quark and gluon distributions necessarily evolve with Q^2 because a probe with increasingly high momentum can resolve more of the structure of the internal constituents. This Q^2 behavior is governed by an integro-differential equation from Altarelli & Parisi (1977):

$$\frac{\partial}{\partial \ln Q^2} G_{q/A}(x, Q^2) = \frac{\alpha_s(Q^2)}{2\pi} \int dy \, dz \, \delta(x - yz)P(y)G_{q/A}(z, Q^2), \qquad 17.$$

where $P(y)$ can be calculated from the fundamental three-point functions in the theory. For large x, this evolution equation has an approximate solution (Buras & Gaemers, 1978) that, when combined with Equation 14, yields for the nonsinglet valence distribution an expression

$$G_{q/A}(x, Q^2) \simeq c(\alpha_s)(1 - x)^{2n_s - 1 + 2|\Delta s| + \Delta \zeta},$$ 18.

where Δs is the helicity suppression factor given in Equations 14–16 and

$$\Delta \zeta = \frac{4}{3} \frac{1}{\pi} \int_{\mu_0^2}^{Q^2(1-x)} \frac{dt}{t} \, \alpha_s(t)$$ 19.

controls the evolution due to perturbative radiation. Equation 18 represents a significant modification of the original counting-rule expression in Equation 2. Even more interesting, perhaps, is the fact that the expressions for the gluon distribution (Equation 15) and the anti-quark distributions (Equation 16) contain pieces that dominate those estimated from spin-corrected counting rules using diagrams such as are deduced from Figure 2 with extra constituents. The treatment of scaling violations in perturbation theory includes a $(1 - x)^4$ piece of the gluon distribution, whereas if we drew a diagram with a large-x gluon and three spectator quarks we would find $2n_s - 1 + |\Delta s_z| = 6$. A reason for this discrepancy is that the quark emitting a hard gluon cannot be treated as a "spectator" in calculating the large-x behavior of the gluon distribution. It has been suggested (see Brodsky 1979) that it is sometimes convenient to separate "intrinsic" and "extrinsic" components of the wave function, where the intrinsic components obey the original type of counting rules and the extrinsic are generated dynamically. Any such separation would be highly Q^2 dependent and, for most applications, Equations 15 and 16 seem preferable.

The treatment of the counting rules discussed above gives abstract quark and gluon probabilities and does not directly give the spectrum for a deep inelastic scattering. To obtain that piece of information, it is necessary to consider the effects of possible "higher-twist" contributions. These are contributions to the deep inelastic scattering cross section that are damped by extra powers of $1/Q^2$. For example, for the pion, Equation 18 gives a form

$$G_{q/\pi}(x, Q^2) \simeq c(1 - x)^{2 + \Delta \zeta}.$$ 20.

However, an analysis of the diagrams leading to this result suggest that the deep inelastic scattering from a pion has a longitudinal component where the structure function has a large-x behavior:

$$F_\pi^L(x, Q^2) \simeq \frac{c_L x}{Q^2}.$$ 21.

The reason this type of subasymptotic term is called a higher-twist contribution involves history not relevant here but the nomenclature is now well established. One type of dynamic mechanism that can lead to extra powers of $1/Q^2$ in the cross section is an effect that is coherent over some finite length scale. The coefficients of the subasymptotic corrections associated with these coherence phenomena often have fewer powers of $(1 - x)$, and it therefore may be difficult in the large-x region to separate their contributions to the cross section from those of the point-like mechanisms. In particular, Blankenbecler & Brodsky (1974) point out that the original form of the counting rules (Equation 3) can be applied to the probability distributions $G_{a/A}(x, \mu^2)$ where a is itself composite and that the rules can provide guidance in estimating the importance of higher-twist effects. A discussion of higher-twist effects in different processes can be found in work by Berger (1980) together with a more thorough set of references concerning efforts to understand them.

The appearance of higher-twist effects in QCD complicates the application of what has been called the "correspondence principle" by Bjorken & Kogut (1973). This principle hypothesizes a smooth connection between the form of exclusive processes for $M^2/s \to 0$ and a corresponding exclusive cross section at fixed angle

$$\int_{\Delta M^2} dM^2 \, \frac{d\sigma}{dt \, dM^2} (\text{AB} \to \text{CX}) \cong N_D \frac{d\sigma}{dt} (\text{AB} \to \text{C}''\text{D}'').$$ 22.

N_D is a normalization constant factor that takes into account the number of "allowed states." This is a generalization of a "duality" originally proposed for deep inelastic scattering by Bloom & Gilman (1970). There are many problems with using an equation such as Equation 22 uncritically. (See, for example, the discussion of Berger & Jones 1981). However, sum rules of this type can be useful in estimating the size of coherent effects in various processes in a way similar to the approach of Shifman, Vainstein & Zakharov (1979) for e^+e^- annihilation.

4.   FIXED-ANGLE HADRONIC SCATTERING

The treatment of hadronic form factors presented above suggests the possibility that exclusive fixed-angle hadronic scattering can be systematically studied in perturbative QCD. Using similar assumptions, we can express the fixed-angle scattering amplitude for the process AB → CD as

$$\lim_{\substack{s\to\infty \\ t/s=\text{const}}} \langle \lambda_C \lambda_D | M | \lambda_A \lambda_B \rangle$$

$$= \sum_{\lambda i} \int \prod_i [dx_i] \phi^*_{C,\lambda_C}(x_{Ci}, \lambda_i) \phi^*_{D,\lambda_D}(x_{Di}, \lambda_{Ci})$$

$$\times T_H(x_i, \lambda_i; s, t, \alpha_s) \phi_{A,\lambda_A}(x_{Ai}, \lambda_i) \phi_{B,\lambda_B}(x_{Bi}, \lambda_i), \qquad 23.$$

a form that mimics Equations 5 and 10 but where the helicity structure is explicitly displayed. The approach, pioneered by Brodsky & Lepage (1980), argues that, to leading contributions, T_H can be calculated in perturbation theory from connected tree diagrams in which each hadron is replaced by a set of collinear valence quarks. This conjecture means that the overall amplitude in Equation 23 would contain a factor

$$\alpha^i_s(\mu^2) \left[\frac{\alpha^i_s(\mu^2)}{s} \right]^{(n/2)-2} \qquad 24.$$

(where n is the total number of quarks) and, except for the logarithmic variation associated with the expansion parameter, be consistent with the counting rule given in Equation 3. As in the case of form factors there are important spin constraints because of the conservation of quark helicity in T_H. For finite x_{Hi}, the transverse momentum integrations in the definitions of the $\phi_{H,\lambda}$ project out of $L_z = 0$, so the form of Equation 23 suggests there should be a selection rule for hadronic helicities (Brodsky & Lepage 1980):

$$\lambda_A + \lambda_B = \lambda_C + \lambda_D. \qquad 25.$$

In order to examine the validity of Equation 23 in more detail, we must confront some potentially intimidating theoretical complexities. Among them we note three separate problems:

1. the vast number of distinct Feynman diagrams,
2. possible contributions from disconnected diagrams, and
3. the contribution from the endpoints of the integrations in Equation 23.

The first of these problems stands in the way of explicitly using Equation 23 in a straightforward way. For a distinct routing of N quarks in a high-energy scattering, there are more than $(N)^{N-1}$ distinct ways of connecting them with gluons. When next-to-leading-order diagrams are considered, the explosion in numbers is even more dramatic. The large numbers make traditional approaches to calculating diagrams extremely inefficient. It may be necessary, in order to make progress in calculating hadronic scattering amplitudes, to develop Monte Carlo techniques to approximate the amplitude by keeping the contributions from some set of "dominant" diagrams. Given the delicate cancellations known to occur in gauge theory calculations of simpler processes, it would probably require sophisticated new approaches to make such approximations work.

The second problem introduces a new feature not found in the discussion of form factors. There are possible contributions to the hadron-hadron scattering amplitude coming from diagrams that, at the quark level, are disconnected. The importance of these diagrams was first pointed out by Landshoff (1974) and they are frequently referred to as Landshoff diagrams. These diagrams have, in general, fewer powers of $\alpha_s(\mu^2)$ and fewer powers of $1/s$ than the connected diagrams and, hence, do not obey the counting rules. The possibility that they dominate over the other configurations must be considered carefully.

Let us look at this explicitly for a simple case of meson-meson elastic scattering $m_{(A\bar{B})}m_{(C\bar{D})} \to m_{(A\bar{B})}m_{(D\bar{D})}$ where A, B, C, and D are distinct quark flavors so that we can neglect quark interchange and quark-antiquark annihilation diagrams.

In a direct perturbation theory approach to this process, the leading-order diagrams are disconnected at the quark level and give contributions that do *not* obey the counting rules of Equation 3. The contribution of the two diagrams in Figure 3 gives a factor

$$M_4 \propto \frac{8\pi i}{(stu)^{1/2}} \int [dx] \phi_C^* \phi_D^* \{T_{qq}^2\} \phi_A \phi_B, \qquad 26.$$

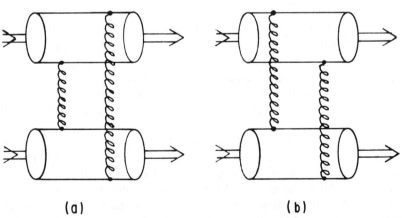

(a)   **(b)**

Figure 3 Order g^4 contributions to the amplitude $m_{(A\bar{B})}m_{(C\bar{D})} \to m_{(A\bar{B})}m_{(C\bar{D})}$ discussed in the text.

where T_{qq} is the quark-quark elastic scattering amplitude. Connected diagrams of the form assumed in Equation 23 do not appear until $O(g^6)$ in the amplitude. Therefore, for diagrams yielding the counting rules to dominate at high energy, there must be some cancellation mechanism involving the lowest-order diagrams in the perturbative expansion

$$|M|^2 = g^8|M_4|^2 + g^{10}2\text{Re}(M_4 M_6^*) + g^{12}\{2\text{Re}(M_4 M_8^*)$$
$$+ |M_6|^2\} + \dots \dots \qquad \qquad 27.$$

Arguments suggesting that this type of cancellation might occur have been based on a selected summation of certain higher-order diagrams. For example, diagrams that appear at $O(g^6)$ can be conveniently grouped into five categories:

(*i*)   Disconnected when hadronic vertices are removed;
(*ii*) Minimally connected with no 3G vertex, one gluon connects $q\bar{q}$ in the same hadron;
(*iii*) Minimally connected with no 3G vertex, all gluons connect quarks in different hadrons;
(*iv*) Minimally connected with one 3G vertex; and
(*v*)   Minimally connected with two 3G vertices or one 4G vertex.

Examples of diagrams in each of the different categories are shown in Figure 4.

If we are looking for a mechanism to suppress the Landshoff mechanism, we can find one in those diagrams in category (*i*) above. When the wave functions of the mesons are defined to be damped when constituents are highly virtual, those diagrams that remain disconnected at higher order sum to produce a "Sudhakhov factor" (Cornwall & Tiktopoulos 1976)

$$S_{qq}(t, t_0; \alpha_s) \cong \exp\left[-\frac{\alpha_s}{3\pi} \log^2\left(\frac{t}{t_0}\right) \right], \qquad \qquad 28.$$

which can be absorbed into the on-shell quark-quark scattering amplitude in Equation 26. If these factors are inserted into Equation 26, the result then vanishes asymptotically faster than any power of the energy. This exercise does not, by itself, guarantee that we can ignore disconnected diagrams. For example, when we examine diagrams of next-to-leading order in category (*ii*) we see that there are kinematic regions when the denominator for the gluon line connecting a $q\bar{q}$ pair in the same meson can vanish. The resulting singularity is sometimes called a "pinch" singularity since it occurs in a way related to pinches that appear when contours in momentum space integrations are distorted. In

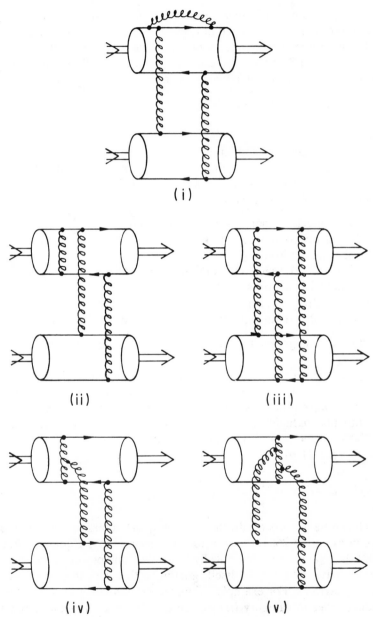

Figure 4   Examples of $O(g^6)$ Feynman diagrams in categories (i) through (v).

order to calculate the observed physical amplitude, this singularity must be removed by absorbing it into the hadronic wave function. This done, a diagram with a pinch singularity gives a specific contribution identical in structure to a disconnected diagram. Since diagrams in other categories have similar singularities, the complete treatment of the high energy limit in perturbation theory involves detailed assumptions about the behavior of wave function as well as the hard-scattering amplitude. Considerably more study on the nature of the possible cancellation mechanisms is necessary before we can be confident of a calculation that assumes disconnected diagrams can be neglected. Mueller (1981) looked at this question assuming that a "leading-log" expansion is valid for both the scattering amplitude and the evolution of the wave functions. He finds a balance between the Sudhakov suppression of processes involving "on-shell" constituents and the extra powers of momentum that appear in "off-shell" hard-scattering amplitudes. After complicated manipulations, the result is an amplitude that displays a power behavior intermediate between that of the Landshoff double-scattering mechanism and the constituent-counting rules.

As the artificial meson-meson scattering example considered above makes clear, the possible role of the Landshoff mechanism in a physically observable cross section such as pp $\rightarrow$ pp is a subject open for conjecture. For example, Donnachie & Landshoff (1979) suggest that the triple-scattering amplitude (which for pp $\rightarrow$ pp is analogous to Equation 26) yields an s-independent cross section

$$\frac{d\sigma}{dt} \cong C_{\text{TS}} \frac{1}{t^8},$$
29.

which is relevant to fixed large-t data, $|t| \geq m_{\text{p}}^2$ but $|t/s| \ll 1$.

Even if the question of disconnected diagrams is understood after further study, the interpretation of Equation 23 is sensitive to the contribution of the endpoints in the region of integration. The problem for hadronic scattering amplitudes is quite similar to the situation for form factors except that there are many more configurations to be considered. In analogy with the discussion in Section 2, it turns out to be the helicity structure of hadronic amplitudes that gives the most information about the possible contribution of these special kinematic regions. In particular, the helicity selection rule (Equation 25) can be modified by them (Szwed 1981).

For elastic pp scattering, the fact that several spin observables can be measured gives us a handle on this question. Five independent helicity amplitudes appear for this process. They are $\langle++|++\rangle$, $\langle+-|+-\rangle$, $\langle+-|-+\rangle$, $\langle++|+-\rangle$, and $\langle++|--\rangle$. The helicity selection rule in the context of the hard-scattering model for exclusive scattering suggests that, at high energies and fixed angles, the $\langle++|+-\rangle$ and $\langle++|--\rangle$ amplitudes should be suppressed by extra powers of momentum. The polarization or analyzing power for pp $\rightarrow$ pp can be written

$$\sigma P = -\text{Im}[(\langle++|++\rangle + \langle++|--\rangle + \langle+-|+-\rangle \\ - \langle+-|-+\rangle)\langle++|+-\rangle^*]$$ 30.

and the two-spin asymmetry A_{sl} is given by

$$\sigma A_{sl} = \text{Re}[(\langle++|++\rangle + \langle++|--\rangle - \langle+-|+-\rangle \\ + \langle+-|-+\rangle)\langle++|+-\rangle^*].$$ 31.

The helicity selection rule therefore implies that these two observables should vanish at high energies and fixed angles (Farrar et al 1979, Brodsky et al 1979):

$$P = 0$$ 32.

$$A_{sl} = 0.$$

There are, in addition, three other independent two-spin asymmetries

$$\sigma A_{nn} = \text{Re}(\langle++|++\rangle\langle++|--\rangle^* - \langle+-|+-\rangle\langle+-|-+\rangle^*)$$

$$\sigma A_{ll} = \tfrac{1}{2}(|\langle+-|+-\rangle|^2 + |\langle+-|-+\rangle|^2 - |\langle++|++\rangle|^2 - |\langle++|--\rangle|^2)$$ 33.

$$\sigma A_{ss} = \text{Re}(\langle++|++\rangle\langle++|--\rangle^* + \langle+-|+-\rangle\langle+-|-+\rangle^*).$$

The quark helicity conservation hypothesis then gives the relationship

$$A_{nn} = -A_{ss}$$ 34.

and, for pp $\rightarrow$ pp at 90° where $\langle+-|+-\rangle = -\langle+-|-+\rangle$, because of identical particle effects

$$2A_{nn}(\pi/2) - A_{ll}(\pi/2) = 1.$$ 35.

Spin-dependent measurements can therefore help decide the validity of ignoring endpoints in the hard-scattering approach to exclusive processes. It is interesting to note that, in general, the disconnected diagrams discussed above conserve quark helicity even though they violate the counting rules. Therefore, high energy measurements in disagreement with Equations 32, 34, or 35 signal either the presence of new, nonperturbative forces or significant contributions from kinematic regimes where the connection between quark spin and hadron spin is broken. We return to a discussion of the experimental situation in the next section.

This brief discussion illustrates the great amount of theoretical work still to be done on exclusive hadronic scattering processes. The fact that there are a large number of different processes, each with a distinct experimental signature, should make it possible to eventually test theoretical ideas rather cleanly. The general approach embodied in the hard-scattering model of Equation 23 seems promising but, as yet, there are no complete calculations that deal in a quantitative way with all the possible problems. Work on this subject is just beginning.

5. LESSONS FROM EXPERIMENT

In the discussion above, we frequently referred to the fact that the counting rules appear to have experimental support. A part of the interest in the subject can be attributed to the fact that for several years there existed a large amount of experimental data consistent with the simple form of the rules expressed as Equations 1–3, while the weight of theoretical opinion leaned toward the view that the scattering of strongly interacting bound systems could not be so simple. The theoretical situation has changed in recent years: a more thorough analysis of QCD perturbation theory seems to endorse the rules' underlying validity as approximations to the behavior of composite systems, even while suggesting specific modifications and corrections to the forms of Equations 1–3.

It is now appropriate to look in more detail at experiment and determine to what extent can data be used to test the more explicit predictions that have emerged from recent theoretical studies. This comparison may also suggest areas where further work will be fruitful.

5.1 *Hadronic Form Factors*

The discussion in Section 2 suggests that Equation 1 be modified to read

$$\lim_{Q^2 \to \infty} F_H(Q^2) = C_H \left[\frac{\alpha_s^i(Q^2)}{Q^2} \right]^{1-n_H} f[\alpha_s^i(Q^2)], \qquad 36.$$

where f can be calculated perturbatively.

The first experimental data to consider involve the pion form factor. Equation 9 is compared with experimental data in Figure 5, where the value of the QCD expansion parameter,

$$\alpha_s^{\text{mom}}(\mu^2)\big|_{\mu^2 = 30\,\text{GeV}^2} = 0.200, \qquad 37.$$

is chosen to agree with that found in other phenomenological analyses (Buras 1981). [For those QCD experts who prefer to deal with the Λ's sometimes chosen to parametrize the different definitions of $\alpha_s(\mu^2)$, this

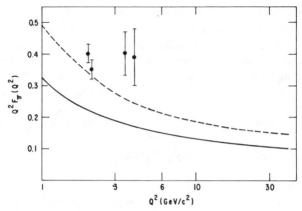

Figure 5 Data on the pion form factor compared with two curves representing theoretical calculations. The predictions are normalized using Equation 37. Except for this choice, the calculations follow Field et al (1981).

choice corresponds to $\Lambda_{\text{mom}} \cong 300$ MeV.] Also shown in Figure 5 is a form with a factorization hypothesis different from that used by Field et al (1981). The difference between the two curves provides an estimate of the contribution of higher-order terms associated with the evolution of $\phi_\pi(x, \mu^2)$. This is one case where the usual plea of theorists for data at higher values of Q^2 stands up to closer scrutiny. Existing data are in regions where there are large theoretical ambiguities. However, it is possible to continue the QCD prediction to the time-like region using the known analytic structure of $F_\pi(s)$, and therefore measurements of $e^+e^- \rightarrow \pi^+\pi^-$ for $Q^2 \gtrsim 12$ GeV^2 could prove to be important. The discussion given by Machet (1981) provides a good starting point for the formalism involved in making the comparison.

As might be expected, data on the proton magnetic form factor is significantly better than that on the pion. However, since neither the normalization nor the higher-order corrections to the theoretical expression are known, it is necessary to do some guesswork to achieve the theoretical curve for $Q^4 G_M(Q^2)$ shown in Figure 6. One of the assumptions that goes into this curve is the hypothesis that the higher-order corrections are of similar magnitude to those in the pion form factor. Another, closely related, assumption involves ignoring the possible complications from the endpoint of the integration region and using a form for the hadronic wave function which vanishes in that limit. Because of these additional assumptions, the comparison between theory and experiment cannot be considered decisive. Interestingly enough, it can be seen from the figure that the simple counting rule

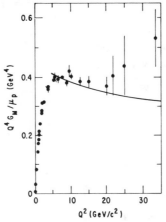

Figure 6 Data on the proton form factor compared to a theoretical curve based on Equation 11. The curve uses Equation 37 and the assumptions discussed in the text.

result

$$Q^4 G_M(Q^2) \rightarrow \text{constant} \qquad 38.$$

does rather well and that the logarithmic variations found in the explicit calculations do not appear to be necessary to describe the data. The theory suggests that this situation is unlikely to remain at higher Q^2. For more information on the experimental situation concerning hadronic form factors, the reader should examine the article by Brodsky (1979).

5.2 *Large-x Behavior of Distribution Functions*

From the discussion in Section 3, we found that the counting rule for the large-x behavior of distribution functions must be modified by a helicity factor and by scaling violations to read

$$G_{q/A}(x, Q^2) = C(\alpha_s)(1-x)^{2n_s - 1 + 2|\Delta s| + \Delta \zeta} \qquad 39.$$

where $\Delta \zeta$ is given by Equation 19.

An indication of the experimental results on the large-x behavior of the proton's deep inelastic structure function can be found in Figure 7 where $(1-x)^3 \nu W_2(Q^2, x)$ is plotted against x for three different values of Q^2. The errors on the experimental points include estimates of errors due to the uncertainty in σ_L/σ_T in this region. There seems to be some evidence here for an underlying $(1-x)^3$ behavior with additional Q^2-dependent structure in agreement with Equation 39. The helicity dependence of the $G_{q/p}(x, Q^2)$ is supported by data from Baum et al (1980) indicated in Figure 8. As a matter of fact, these data can be

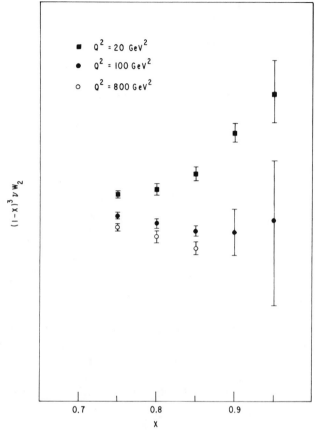

Figure 7 The value $(1 - x)^3 \nu W_2(x, Q^2)$ for different values of Q^2.

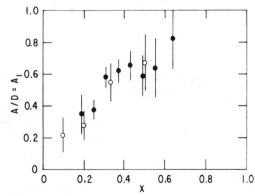

Figure 8 Data from Baum et al (1980) on quark polarization distribution in a proton.

considered among the strongest direct evidence for the belief that the underlying hadronic forces are associated with a spin-1 vector gluon. As indicated in Section 3, there are theoretical problems in using the counting rules for nonvalence constituents.

As yet there are no data that give significant information about the shape of the gluon distribution at large x. It is not possible to distinguish between the $(1 - x)^{6 + \zeta(G)}$ behavior of a spin-corrected constituent-counting ansatz and the $(1 - x)^{4 + \zeta(G)}$ obtained by convoluting a valence quark distribution that obeys Equation 39 with the probability for a quark to produce a hard gluon. In fact, it is not clear exactly what process could be chosen to measure the gluon distribution. In the simple parton model, a process such as the photoproduction of heavy mass pairs could be identified as coming from the hard subprocess $\gamma G \rightarrow h\bar{h}$ (where G is a gluon and h represents a heavy quark) and chosen to "measure" the gluon distribution. In a more general treatment of interacting system it becomes a matter of convention to separate contributions from the subprocesses $\gamma q \rightarrow h\bar{h}q$ and $\gamma G \rightarrow h\bar{h}$. Since the Altarelli-Parisi equations couple the different constituent distributions, the question of the large-x power behavior of the nonvalence gluon distribution has so far remained largely academic.

For antiquarks, it has been argued (Close & Sivers 1979) that the $G_{\bar{q}(+)/\text{p}(+)}(x, Q^2) \sim (1 - x)^5$ structure mentioned in the text is associated with observable asymmetry in $pp \rightarrow \ell^+ \ell^- X$. Again, the shape of the "sea" distribution at large x has not been subjected to an unambiguous test. A closely related question involves the shape and normalization of the "near-mass-shell" heavy quarks content of the proton wave function associated with an intrinsic $|qqqh\bar{h} >$ state. (See, for example Brodsky, Peterson & Sakai 1981.)

5.3   Fixed-Angle Exclusive Scattering

The discussion in Section 4 demonstrates that the constituent approach to exclusive hadron-hadron processes at fixed angle is still subject to considerable theoretical uncertainty. In view of this uncertainty, it is heartening that the simple constituent-counting rules (Equation 3) give an accurate guide to the systematics of fixed-angle cross sections involving different quantum numbers. This indicates that it may be possible to work through the theoretical difficulties and obtain some real predictions. The form of the counting rules extracted from Equation 23 is

$$\lim_{t \to \infty} \frac{d\sigma}{dt} (AB \rightarrow CD) = \left[\frac{\alpha_s^i(r_i t)}{t} \right]^{(\Sigma n) - 2} f^i(\theta, \alpha_s^i) \qquad 40.$$

where r_i is a reduction factor $(r_i \cong 1/\Sigma n)$ that picks out a typical momentum transfer.

The exclusive process that has been given the most thorough experimental study is pp → pp. The naive counting-rule prediction for the power fall-off of fixed-angle cross section is s^{-10}. A fit to experimental data gives a behavior $s^{-9.7 \pm 0.4}$, which supports the simple counting arguments. This does not mean that the data rule out any of the difficulties discussed above. In fact, Donnachie & Landshoff (1979) pointed out that data at large, but fixed, t and $s \to \infty$ are consistent with the s-independent form of the triple-scattering cross section in Equation 29. A comparison between theory and experiment is shown in Figure 9, where the normalization of the theoretical curve is calculated using some reasonable assumptions about the hadronic wave function.

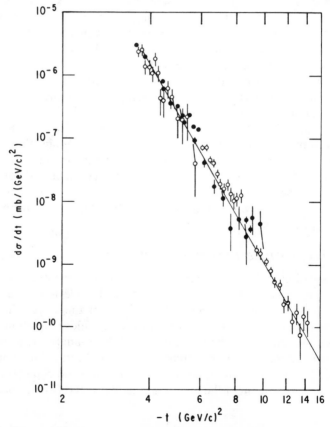

Figure 9 Cross section for pp → pp compared to calculation of Landshoff mechanism. Figure from Donnachie & Landshoff (1979).

To illustrate the theoretical uncertainty Brodsky & Lepage (1980) advocated a "hybrid" approach to pp → pp. They point out that data for $|t| > 5$ Gev^2 can be fit by an empirical form

$$\frac{d\alpha}{dt} = \frac{1}{16\pi s^2}\left[(4\pi\alpha_s)^3\left(\frac{m_p}{t}\right)^4\frac{s}{m_p}K_L + (4\pi\alpha_s)G_m^2(t)K_{HS}\right]^2, \qquad 41.$$

where $K_L \cong 0.05$ normalizes the Landshoff contribution and $K_{HS} \cong 500$ approximates the effect of the thousands of hard-scattering diagrams. It can be considered a preliminary success of the hard-scattering approach that the normalization of the pp → pp fixed-angle cross section is large,

$$\frac{\left(\dfrac{d\sigma}{dt}\right)(pp \to pp)}{\dfrac{\pi\alpha_s^2}{s^2}G_M^4(t)} \gg 1, \qquad 42.$$

which implies a large number of possible diagrams.

Data on pp spin-observables give another indication that theoretical approaches based on hard scattering are on the right track but that we have not yet found the final answer. As summarized by Yokasawa (1980), large-angle data give indications that Equations 32 are valid and that polarization and $A_{s\ell}$ vanish asymptotically. However, as discussed by Brodsky, Carlson & Lipkin (1979) and by Farrar et al (1979), it is difficult to understand the structure in A_{nn} observed by Crabb et al (1978) in terms of the hard-scattering model with quark interchange.

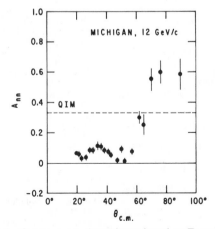

Figure 10 Data on A_{nn} for pp → pp. An estimate based on Equation 23 and SU$_6$ wave functions gives $A_{nn} \cong 1/3$ with little structure.

Nevertheless, recent data on $A_{\ell\ell}$ (Yokasawa 1981) suggests that the helicity conservation sum rule may be valid for $P_{LAB} = 12$ GeV/c.

6. CONCLUSIONS

This is not a subject amenable to firm conclusions. We have exposed enough of the theoretical complications to see that the counting rules can, with appropriate modifications, approximate the high-energy behavior of cross sections involving composite systems. However, because of all the possible complications, it is not trivial that they do so. The interplay between theory and experiment will continue to provide important information. At present, experimental results support the basic approach of using perturbation theory and point-like quarks to obtain high-energy scattering amplitudes, but experimental results do not yet provide quantitative tests of specific calculations.

It should be noted that, for the purposes of our discussion, the terms "constituent" and "quark" were almost interchangeable. Although most physicists believe there should exist "glueballs," hadrons consisting of vector gluon constituents, there is no reason to believe that a description of such states in terms of a small number of constituent gluons will be simple.

Literature Cited

Alabiso, C., Schierholz, G. 1974. *Phys. Rev. D* 10:960
Altarelli, G., Parisi, G. 1977. *Nucl. Phys. B* 126:298
Artru, X. 1982. *Phys. Rev. D* 24:1662
Baum, G., et al 1980. *20th Int. Conf. High Energy Phys.* Univ. Wis., Madison
Berger, E. L. 1980. *Z. Phys. C* 4:289
Berger, E. L., Jones, D. 1981. *Phys. Rev. D* 23:1521
Bjorken, J. D., Kogut, J. 1973. *Phys. Rev. D* 8:1371
Blankenbecler, R., Brodsky, S. J. 1974. *Phys. Rev. D* 10:2973
Bloom, E. D., Gilman, F. J. 1970. *Phys. Rev. Lett.* 25:1140
Brodsky, S. J. 1979. *Proc. 1979 SLAC Inst. Part. Phys.* SLAC-224, p. 133. Stanford, Calif: SLAC
Brodsky, S. J., Farrar, G. R. 1973. *Phys. Rev. Lett.* 31:1153
Brodsky, S. J., Farrar, G. R. 1975. *Phys. Rev. D* 11:1309
Brodsky, S. J., Lepage, G. P. 1979a. *Phys. Rev. Lett.* 43:545
Brodsky, S. J., Lepage, G. P. 1979b. *Phys. Lett.* 87B:359
Brodsky, S. J., Lepage, G. P. 1980. *Phys. Rev. D* 22:2157

Brodsky, S. J., Lepage, G. P. 1981. *Phys. Rev. D* 24:2848
Brodsky, S. J., Carlson, C., Lipkin, H. J. 1979. *Phys. Rev. D* 20:2278
Brodsky, S. J., Lepage, G. P., Zaidi, S. A. A. 1981. *Phys. Rev. D* 23:1152
Brodsky, S. J., Peterson, C., Sakai, N. 1981. *Phys. Rev. D* 23:2724
Buras, A. J. 1981. *Fermilab-Conf-81/69-THY*. Batavia, Ill: Fermi Natl. Lab.
Buras, A. J., Gaemers, K. 1978. *Nucl. Phys. B* 132:249
Celmaster, W., Sivers, D. 1981. *Phys. Rev. D* 23:227
Close, F., Sivers, D. 1979. *Phys. Rev. Lett.* 39:1116
Cornwall, J., Tiktopoulos, G. 1976. *Phys. Rev. D* 13:3370
Crabb, D. G., et al 1978. *Phys. Rev. Lett.* 41:1257
Donnachie, A., Landshoff, P. V. 1979. *Z. Phys. C* 2:55
Drell, S. P., Yan, T. M. 1970. *Phys. Rev. Lett.* 24:181
Duncan, A., Mueller, A. 1980. *Phys. Rev. D* 21:1636
Farrar, G. R., Gottlieb, S., Sivers, D., Thomas, G. H. 1979. *Phys. Rev. D* 20:202

Farrar, G. R., Jackson, D. R. 1975. *Phys. Rev. Lett.* 35:1416

Farrar, G. R., Jackson, D. R. 1979 *Phys. Rev. Lett.* 43:246

Field, R. D., Gupta, R., Otto, S., Chang, L. 1981. *Nucl. Phys. B* 186:429

Landschoff, P. V. 1974. *Phys. Rev. D* 10:1024

Machet, B. 1981. *Z. Phys. C* 8:215

Matveev, V. A., Murddyan, R. M., Tavkheldize, A. V. 1973. *Lett Nuovo Cimento* 7:719

Mueller, A. H . 1981. *Phys. Rep.* 73:238

Shifman, M. A., Vainstein, A. I., Zakharov, V. I. 1979. *Nucl. Phys. B* 47:385

Sivers, D., Brodsky, S. J., Blankenbecler, R. 1976. *Phys. Rep.* 23C:1

Szwed, J. 1981. Jagellonian Univ. preprint

Vainstein, A. I., Zakharov, V. I. 1978. *Phys. Lett.* 72B:368

West, G. B. 1970. *Phys. Rev. Lett.* 24:1206

Yokasawa, A. 1980. *Phys. Rep.* 64:50

Yokasawa, A. 1981. *Proc. Int. Symp. High Energy Phys. with Polarized Beams & Polarized Targets*, ed. C. Joseph, J. Soffer, p. 261. Basel: Birkhäuser Verlag

Ann. Rev. Nucl. Part. Sci. 1982. 32:177–209
Copyright © 1982 by Annual Reviews Inc. All rights reserved

QUANTUM CHROMODYNAMICS:
The Modern Theory of
the Strong Interaction

Frank Wilczek

Institute for Theoretical Physics, University of California,
Santa Barbara, California 93106

CONTENTS

0163-8998/82/1201-0177$02.00

1. INTRODUCTION

"You know I admire classical artists like Rembrandt and Bonestell, and don't care for abstractions or chromodynamics." (Poul Anderson 1970)[1]

The nature of the strong interaction responsible for holding nuclei together and for a wealth of phenomena observed in high energy collisions has been a main concern of physics since the 1930s. It now appears that we have a precise microscopic theory that in principle solves the problem, in the same sense that quantum electrodynamics solves atomic spectroscopy and chemistry. In this article I review the nature of the theory, the evidence for it, and its implications for other parts of physics.

The modern theory of the strong interaction, quantum chromodynamics or QCD, is a "radically conservative" theory in Wheeler's sense. That is, we extrapolate a few fundamental principles as far as we can, accepting "paradoxes" that fall short of actual contradictions! It resulted from taking such general principles as locality, causality, and renormalizability very seriously and reconciling them with a few outstanding experimental facts. We have learned that certain relativistic quantum theories contain unexpected richness — such phenomena as asymptotic freedom, confinement, and anomalies — but no revision of the basic principles of relativity and quantum mechanics has proved necessary for understanding the strong interaction.

The plan of this review is as follows. In Section 2 I schematically review the relevant phenomenology of strong interactions and the semi-phenomenological models used to organize these facts. In Section 3 the fundamental field theory of quarks and gluons, QCD, is presented and discussed qualitatively. Section 4 contains a general discussion of scaling in quantum field theory, with both heuristic and formal arguments for asymptotic freedom in QCD. In Section 5 quantitative results of QCD are compared with relevant experiments. The last section describes some implications of the modern theory of the strong interaction for other parts of physics.

The bulk of this review is ahistorical and little effort has been given to tracing references. For this reason, and since it helps to provide perspective on the detailed presentation, I now present a brief history of some of the main developments leading to our present understanding.

The modern theory of the strong interaction began in 1963 with the independent introduction of the concept of quarks by Gell-Mann (1964) and Zweig (1964). Originally the quarks were introduced as a rational-

---

[1] This is, so far as I know, the earliest published reference to chromodynamics.

ization of hadron spectroscopy; the observed spectrum of mesons and baryons could be readily understood as bound states of quark-antiquark (qq̄) and three quarks (qqq). The existence of three different species or "flavors" (u, d, s) of spin-$\frac{1}{2}$ quarks with different quantum numbers (electric charge, isospin, strangeness) but approximately the same strong interaction rationalized the immensely successful SU(3) "eight-fold way" symmetry introduced by Gell-Mann earlier. Although isolated quarks were not, and have not to this day been, observed, a very successful phenomenology was developed by Dalitz (1969) and others based on simple models of hadrons as bound states of localized but essentially noninteracting and structureless spin-$\frac{1}{2}$ quarks.

The idea that structureless quarks could form a *fundamental* basis for the description of hadrons became plausible when relations abstracted from the quantum field theory of quarks, namely the algebra of quark currents and their divergences, were successfully applied. For an account of these ideas including many of the original papers see Adler & Dashen (1968). For the use of current divergences see especially Gell-Mann et al (1968).

A second major idea, that each flavor of quark should come in three colors, lies somewhat deeper. The basic idea that quarks carry a new internal quantum number was introduced by Han & Nambu (1965) but not quite in the now-accepted form. Hints also came from work (Greenberg 1964, Nambu 1966, Greenberg & Zwanzinger 1966) on the saturation problem — why do qq̄ and qqq states have low masses but not q, qq, ...? This fact can be understood if there is an SU(3) symmetry among the different colors of quarks and for some reason color singlets are much lighter than nonsinglets.

Quark models indicated that the wave functions of baryons qqq should be symmetrical in the interchange of the spatial, spin, and flavor quantum numbers of the quarks. Bose symmetry for spin-$\frac{1}{2}$ particles contradicts the principles of relativistic field theory (e.g. Streater & Wightman 1964). Antisymmetry in color quantum numbers can restore the expected Fermi statistics. Later, calculations of the decay rate for $\pi^0 \rightarrow \gamma\gamma$ (Adler 1969, Bell & Jackiw 1969) and of the cross section for $e^+e^- \rightarrow$ hadrons (Bjorken 1969) further strengthened the case for the color degree of freedom.

A third major line of ideas leading to QCD was the parton model introduced by Feynman (1969, 1972). These ideas, based on profound intuitions gleaned from observations on high energy hadron-hadron collisions (in particular, the extreme smallness of transverse momentum in typical collisions), suggested that hadrons contained point-like constituents with simple properties. Some of these constituents were soon identified as quarks, and strikingly successful predictions for high energy electron-hadron and neutrino-hadron scattering were made from the quark-parton model. Especial mention should be made of the SLAC electroproduction experiments of Miller et al (1972), which were crucial in establishing the ideas.

The parton model, though strikingly successful as a description of data, was an intuitive model and did not have an honest microscopic basis. A microscopic basis for analyzing some of the processes described by the parton model in quantum field theory had been developed over the years by Gell-Mann & Low (1954), Wilson (1971), Callan (1970), and Symanzik (1970) among others. In 1973 Gross & Wilczek (1973a) and Politzer (1973; see also Caswell 1974, Jones 1974) found that nonabelian gauge theories possess the property of asymptotic freedom, which could begin to justify the parton ideas. [Coleman & Gross (1973) showed that nonabelian gauge theories are essentially unique in this respect.] The work of 't Hooft (1973) on renormalization of gauge theories, was essential for these developments. At this stage, a synthesis of the previous ideas concerning quarks, color, and partons was at hand. The color degree of freedom of the quarks could be gauged to yield an asymptotically free theory — using the degrees of freedom determined spectroscopically to justify the dynamical hypotheses of the parton model (Gross & Wilczek 1973b, Fritzsch et al 1973, Weinberg 1973).

At this stage we possessed a mathematically well-defined Lagrangian field theory that had to be regarded as a serious candidate for the microscopic basis of the strong interaction. Progress in understanding and testing the implications of the theory over the past eight years has been striking, as the following, and the article by Söding & Wolf (1981), document.

2. SEMI-PHENOMENOLOGICAL MODELS AND THEIR RELATIONSHIP TO QCD

In this section some of the most important semi-phenomenological models in strong interaction physics — the bag model, potential models of quarkonium, the string model, and the parton model — are described schematically. The intent is to relate the concepts of these models to the underlying microscopic theory where possible. The basic ideas of potential models and many parton model ideas are indeed now known to be mathematical consequences of QCD. The microscopic basis of bag and string models is not yet completely clear, although there are attractive qualitative ideas. (Logically, this section should come after Sections 3 and 4.)

2.1 Bag Model

The bag model is an improved form of the quark model, in which the quarks are treated relativistically, (Chodos et al 1974). A "bag" degree of freedom is explicitly introduced, such that the physics inside the bag is different from that outside. Inside the bag quarks are massless, outside infinitely massive. There is also a finite difference between energy density inside and outside; "bag vacuum" has higher energy than normal vacuum. Hadronic bags are then formed around quarks. Their size is determined by a balance between the kinetic energy it requires to localize the quarks inside (according to the uncertainty principle) and the volume energy associated with the bag vacuum.

Formally, one considers the action

$$\mathcal{L} = [\bar{q}(x)(i\overleftrightarrow{\partial})q(x) - B]\theta_R(x), \qquad\qquad 1.$$

where $\overleftrightarrow{\partial} = \frac{1}{2}\gamma_\mu(\overrightarrow{\partial}_\mu - \overleftarrow{\partial}_\mu)$ is the symmetrized derivative, $q(x)$ is the quark field, B the bag constant, and θ_R the step function, which is one inside and zero outside the bag region R. Notice that, as discussed above, quarks propagate only inside the bag region and this region is associated with a finite volume energy. The equations of motion follow by demanding that the action be stationary to local variations in $q(x)$ and R. In the applications to light hadrons one employs a solution where the bag region R is spherical. The quarks propagate freely inside, do not go outside, and satisfy the condition that the flux of energy going out of the bag boundary vanishes.

In this framework, one may compute masses of mesons and baryons as well as their magnetic moments and other static quantities. The results are generally satisfactory, except that the pseudoscalar meson masses come out too large [the reasons for this are partly understood (Donoghue & Johnson 1979)].

The ideas of QCD contribute to the bag model in two essential ways:

First, of course, Equation 1 should in QCD be extended to include gluon degrees of freedom. The gluon fields then also vanish outside the bag region. According to Gauss's Law, this is only possible if the contents of the bag form an overall color singlet. This argument, given three colors of quarks, shows why hadronic bags should be q$\bar{\text{q}}$ or qqq states.

Second, one may attempt to include the exchange of gluons as a correction to the free propagation of quarks in the bag. If this is done, the description of the details of the meson and baryon spectrum improves; notably the $\Sigma^0 - \Lambda$ mass difference is explained (DeRujula et al 1975, DeGrand et al 1975).

In a more general way, asymptotic freedom suggests the almost free propagation of quarks at short distances assumed in the bag model. There have been several attempts to derive something like the bag model on a more microscopic basis (Callan et al 1976; Friedberg & Lee 1978; H. Hansson, K., Johnson, C. Peterson, private communication), but this important problem cannot be regarded as being completely solved.

2.2 Quarkonium and Strings

It was early hoped that the rich spectroscopy of the J/ψ system and related heavy $q\bar{q}$ bound states would provide the "hydrogen atom" of QCD. This may yet come to pass. The qualitative resemblance between the ψ family spectrum and postronium is certainly striking. Theoretically, QCD does indeed predict that for sufficiently heavy quarks the $q\bar{q}$ bound states will be spatially small (of order $1/\alpha_s m$, where α_s is the effective strong coupling and m the quark mass) and may self-consistently be described as bound by their lowest order, Coulomb-like interaction (Appelquist & Politzer 1975a,b). Quantitatively, however, the quarks hitherto observed (i.e. as hadron constituents) are not nearly heavy enough and the simple Coulomb picture needs amendment.

Phenomenological analyses (Eichten et al 1976) have been carried out in great detail assuming a Coulomb plus linear potential between the quark and antiquark:

$$U(R) = -\frac{4}{3} \alpha_s \frac{1}{R} + \sigma R. \qquad\qquad 2.$$

This is motivated for small R by asymptotic freedom and for large R by string ideas (see below).

Hasenfratz & Kuti (1978) analyzed the problem as follows. The quarks, being heavy, move slowly. In the spirit of the Born-Oppenheimer approximation, one regards the quarks as fixed a distance R apart and by calculating the energy of this configuration derives a potential $U(R)$. The color electrostatic field surrounding the fixed quarks exerts a stress that can stabilize a bag region, as for light hadrons. The heavy quark potential in this scheme can be calculated using as input only the constant B determined from light hadron spectroscopy. The results are very good.

When the heavy quarks are not too far apart, the bag surrounding them is nearly spherical. As they are taken very far apart, the way to minimize the volume energy associated with the bag is to make the bag tubular — a flux tube of color electric field. The geometry explains the linear nature of the potential at large distances (in fact the linear behavior sets in precociously, even for nearly spherical bags!).

When light quark-antiquark pairs are in states of high angular momentum we can expect the bag enclosing them to stretch out into a similar flux tube or string. The energy associated with a relativistic string (see Rebbi 1974) with tension K and angular momentum J goes as

$$E^2 = 2\pi KJ + \text{const.} \qquad\qquad 3.$$

We have long known that in fact high angular momentum excitations of mesons and baryons do obey Equation 3 (see Frautschi 1963); it is one of the most striking features of hadron spectroscopy.

From the above sketch it should be clear that the problems of calculating the "long-distance" behavior of the heavy quark potential and of proving Equation 3 and calculating K from the microscopic theory are closely related and both are closely related to justifying the bag model. Some progress has been made by numerical methods (see Section 5.5). (In the literature σ and K are commonly identified. I do not know of any rigorous basis for this, although it works well numerically.)

2.3   Parton Model

The parton model has been reviewed by its inventor and others (Feynman 1972 and Close 1979). One leading idea, that quarks behave essentially as free particles in large momentum regimes, is fully justified within QCD as a first approximation in many circumstances (compare Section 5). Some other parton ideas (regarding Regge behavior and quark quantum numbers in jets, for example) have not to my knowledge been derived within QCD.

3. FORMULATION OF QCD AND ITS GENERAL CONSEQUENCES

3.1   Local Gauge Invariance

Quantum electrodynamics (QED) and quantum chromodynamics can both be derived from such general principles as relativistic invariance and renormalizability plus the principle of local gauge invariance. Looked at this way, QCD is a very simple generalization of QED. We derive these theories in parallel below.

In quantum mechanics the conservation of charge is expressed as the commutation of the charge operator and the time-development operator (Hamiltonian):

$$[H, Q] = 0. \qquad\qquad 4.$$

The unitary transformations $\exp(iQ\theta)$ therefore leave the equations of motion unchanged. Of course, this is an example of the general connection between symmetries and conservation laws in quantum mechanics. The symmetry associated with conservation of charge is called a gauge symmetry.

Applied to a field $\psi(x)$, which creates quanta of charge q (and destroys quanta of charge $-q$), the gauge symmetry acts as a phase:

$$\psi'(x) = \exp(iQ\theta)\psi(x)\exp(-iQ\theta) = \exp(iq\theta)\psi(x) \qquad 5.$$

or

$$[Q, \psi(x)] = q\psi(x). \qquad\qquad 6.$$

At this stage the gauge symmetry is simply equivalent to conservation of charge; in order that an interaction

$$\Delta\mathcal{L} = \psi_1(x)\psi_2(x)\ldots\psi_n(x) \qquad\qquad 7.$$

be invariant under the gauge symmetry, it is both necessary and sufficient that

$$\sum_{i=1}^{n} q_i = 0 \qquad\qquad 8.$$

so that charge is neither created nor destroyed.

(This discussion has been phrased in the language of second quantization, where the fields create and destroy particles. A parallel discussion could be carried through in a first-quantized formulation; then the link between symmetries and conservation laws would be forged through Noether's theorem.)

To implement color symmetry we let $\psi(x)$ be a three-component vector and generalize Equation 5 so that arbitrary unitary transformations in this internal space are allowed. For every unitary transformation Ω in color space we have then a corresponding hermitian operator ω so that

$$\psi'(x) \equiv \exp(i\omega)\psi(x)\exp(-i\omega) = \Omega\psi(x). \qquad 9.$$

The space SU(3) of unitary transformations in color space is generated by eight infinitesimal transformations — a conventional choice of basis is the set $\lambda^a/2$, $a = 1, \ldots, 8$ introduced by Gell-Mann (see

Gell-Mann & Ne'eman 1964). In terms of these, we can write

$$\Omega = \exp\left(ig \frac{\lambda^a}{2} \theta^a\right)$$

and $\omega = \exp(iT^a\theta^a)$; then the analog of Equation 6 is

$$[T^a, \psi(x)] = g\frac{\lambda^a}{2}\psi(x). \qquad 10.$$

The passage to a *local* symmetry is made by postulating that θ in Equation 5 or ω and Ω in Equation 9 can depend on the space-time position x. This requires some adjustment, because derivatives transform inhomogeneously:

QED: $\partial_\mu\psi'(x) = \exp[iq\theta(x)][\partial_\mu\psi(x) + iq\partial_\mu\theta(x)\psi]$ 11.

QCD: $\partial_\mu\psi'(x) = \Omega(x)[\partial_\mu\psi(x) + \Omega^{-1}(x)\partial_\mu\Omega(x)\psi(x)].$ 12.

We need derivatives in order to construct a reasonable Lagrangian. Otherwise the equations of motion will give only constraints, not interesting dynamics! The inhomogeneous transformation law, however, makes it very awkward to construct Lagrangians invariant under the local symmetry. To remedy this, the derivative is modified by a correction term so that $D'\psi'$ and ψ' have the same homogeneous transformation law. In equations:

QED: $D'_\mu\psi'(x) = \exp[iq\,\theta(x)]D_\mu\psi(x)$ 13.

$$D_\mu \equiv \partial_\mu + iqA_\mu(x) \qquad 14.$$

$$A'_\mu(x) = A_\mu(x) + \partial_\mu\theta(x); \qquad 15.$$

QCD: $D'_\mu\psi'(x) = \Omega(x)D_\mu\psi(x)$ 16.

$$D_\mu = \partial_\mu + igB_\mu(x) \qquad 17.$$

$$B'_\mu(x) = \Omega(x)B_\mu(x)\Omega^{-1}(x) + \frac{1}{ig}[\partial_\mu\Omega(x)]\Omega(x)^{-1}. \qquad 18.$$

D_μ is called the covariant derivative. In the case of QED, $A_\mu(x)$ is a vector field of real numbers — the usual four-vector potential; in the case of QCD, $B_\mu(x)$ is a vector field of 3×3, traceless, Hermitean matrices. The trace part of $B_\mu(x)$ corresponds to an overall phase transformation of ψ. This transformation is totally independent of (commutes with) the other unitary transformations in color space. The corresponding "charge" is therefore in principle independent of the coupling g. There are very good reasons (Lee & Yang 1961) to take the corresponding coupling, which physically would mediate a long-range force coupled to baryon number, equal to zero. Equation 15 is of course the familiar "gauge transformation" of electromagnetism.

We can now easily construct the invariant kinetic energy for the matter field ψ; if it is a Dirac spinor field, for instance

$$L_{\text{kin}} = \bar{\psi}\overleftrightarrow{D}_\mu \gamma_\mu \psi. \tag{19}$$

We are still left with the problem of constructing a kinetic energy term for A_μ, which transforms inhomogeneously. In electromagnetism it is familiar that in the field strength

$$F_{\mu\nu} = \partial_\mu A_\nu - \partial_\nu A_\mu \tag{20}$$

the inhomogeneous term in Equation 15 cancels, so that

$$\mathscr{L}_{\text{field}} = -\tfrac{1}{4}F_{\mu\nu}F_{\mu\nu} \tag{21}$$

is a suitable invariant kinetic energy. In QCD a similar but better-hidden construction works. The commutator of two covariant derivatives contains derivatives of B_μ, and is guaranteed to transform homogeneously. In equations we have

$$[D_\mu, D_\nu]\psi = ig[B_\mu, B_\nu] \tag{22}$$

$$G_{\mu\nu} = \partial_\mu B_\nu - \partial B_\mu + ig[B_\mu, B_\nu] \tag{23}$$

and combining Equations 22 and 16, we find the transformation law for $G'_{\mu\nu}$:

$$G'_{\mu\nu}(x) = \Omega(x)G_{\mu\nu}(x)\Omega(x)^{-1}. \tag{24}$$

Therefore

$$\mathscr{L}_{\text{field}} = -\tfrac{1}{4}\,\text{tr}\, G_{\mu\nu}G_{\mu\nu} \tag{25}$$

is the invariant kinetic energy for B_μ analogous to Equation 21.

3.2 Lagrangian of QCD: Renormalizability and Canonical Form

We now can construct Lagrangians for the interactions of colored quarks with local color symmetry. Indeed, given any Lagrangian with a global color symmetry we need merely change ordinary into covariant derivatives to obtain a locally symmetric Lagrangian.

The possibilities are much more limited, however, if we insist that our theory be *renormalizable*. Heuristically, this requirement may be stated as follows. In the "natural" units for action and velocity $\hbar = c = 1$ (which we adopt throughout), the action

$$S = \int d^4x \, \mathscr{L}(x) \tag{26}$$

is dimensionless so $\mathscr{L}(x)$ has units of $(\text{mass})^4$. The form of the kinetic energy in Equation 19 therefore indicates that the fermion field ψ has units $(\text{mass})^{3/2}$; the kinetic energies expressed in Equations 21 and 25 indicate that the potentials A_μ, B_μ have units $(\text{mass})^1$; finally the couplings e and g are dimensionless. A term $\mu\bar{\psi}\psi$ in $\mathscr{L}$ appears with a coefficient μ with units $(\text{mass})^1$ (indeed, this term simply represents a mass for the fermion!), a term $K\,\text{tr}\,(G_{\mu\nu}G_{\mu\nu})^2$ would require the units of K to be $(\text{mass})^{-4}$, and so forth. Now if we compute any physical process in a power series in K, we find the answer in a power series of the form

$$a_0 + a_1 K\Lambda^4 + a_2 K^2\Lambda^8 + \cdots, \qquad\qquad 27.$$

where $a_0, a_1, a_2, \ldots$ have a common dimension and Λ, with units of $(\text{mass})^1$, represents a large momentum or energy scale at which we cut off the integrals (also, logarithms of Λ will appear — see Section 4 below). Each successive term in the series diverges worse than the last in the absence of a cutoff, owing to large contributions from virtual particles of large energy and momentum. Couplings with dimensions of mass to a negative power — or, equivalently, operators in the Lagrangian with mass dimension greater than four — are called *unrenormaliz-able*. The usual renormalization procedures for dealing with infinities (for which see Section 4) do not suffice to make sense of such theories and they may not really exist.

If we suppose that unrenormalizable couplings do not appear in the Lagrangian, the possible gauge-invariant terms are really very limited. Let us suppose we have n species (flavors), labelled by $j = 1, \ldots, n$ of color triplet spin-$\frac{1}{2}$ quark fields ψ_j in addition to the gauge field B_μ. Then the allowed terms are as follows:

$$Z(-\tfrac{1}{4}\,\text{tr}\,G_{\mu\nu}G_{\mu\nu}) \qquad\qquad 28.$$

$$Z^{\text{L}}_{jk}\bar{\psi}_j(i\overleftrightarrow{D}_\mu\gamma_\mu)\left(\frac{1-\gamma_5}{2}\right)\psi_k = Z^{\text{L}}_{jk}\bar{\psi}^{\text{L}}_j(i\overleftrightarrow{D}_\mu\gamma_\mu)\psi^{\text{L}}_k \qquad\qquad 29.$$

$$Z^{\text{R}}_{jk}\bar{\psi}_j(i\overleftrightarrow{D}_\mu\gamma_\mu)\left(\frac{1+\gamma_5}{2}\right)\psi_k = Z^{\text{R}}_{jk}\bar{\psi}^{\text{R}}_j(i\overleftrightarrow{D}_\mu\gamma_\mu)\psi^{\text{R}}_k \qquad\qquad 30.$$

$$M_{jk}\bar{\psi}_j\psi_k = M_{jk}(\bar{\psi}^{\text{L}}_j\psi^{\text{R}}_k + \bar{\psi}^{\text{R}}_j\psi^{\text{L}}_k) \qquad\qquad 31.$$

$$M^5_{jk}\bar{\psi}_j(i\gamma_5)\psi_k = M^5_{jk}i(\bar{\psi}^{\text{L}}_j\psi^{\text{R}}_k - \bar{\psi}^{\text{R}}_j\psi^{\text{L}}_k) \qquad\qquad 32.$$

$$\frac{\theta g^2}{32\pi^2}\,\varepsilon_{\mu\nu\rho\sigma}\,\text{tr}\,G_{\mu\nu}G_{\rho\sigma} \qquad\qquad 33.$$

where the Zs and Ms are hermitian matrices, θ is a real number, and $\overleftrightarrow{D}_\mu = \frac{1}{2}(\overrightarrow{D}_\mu - \overleftarrow{D}_\mu)$ is the symmetrized derivative, and $\psi_j^{L,R} \equiv \frac{1}{2}(1 \mp \gamma_5)\psi_j$ are the left- and right-handed components of ψ_j. Equation 28 represents the kinetic energy for the gluons; Equations 29 and 30 represent kinetic energies for the left- and right-handed quarks, respectively; Equations 31 and 32 are generalized mass terms for the quarks, and Equation 33 is the θ term to be discussed separately in Section 3.4.

The Lagrangian containing terms in Equations 28–33 can be brought into a simple canonical form by the following steps:

1. The gauge field B_μ^{old} is replaced by $B_\mu^{\text{new}} \equiv Z^{1/2}B_\mu^{\text{old}}$, and similarly $g^{\text{new}} = Z^{-1/2}g^{\text{old}}$. In terms of the new variables, the kinetic energy (Equation 28) appears with coefficient $Z = 1$.

2. Similarly, new left- and right-handed quark fields can be introduced so that Z_{ij}^L and Z_{ij}^R become unit matrices. (This assumes that the original Z_{ij}^L and Z_{ij}^R and nonsingular, i.e. that there really are n quark degrees of freedom for both helicities.)

3. Without undoing the simplification introduced by step 2, we are still permitted to make redefinitions $\psi_j^{L\,\text{new}} = U_{jk}\psi_k^{L\,\text{old}}$, $\psi_j^{R\,\text{new}} = V_{jk}\psi_k^{R\,\text{old}}$ where U and V are unitary matrices. An exercise in linear algebra teaches us that with a judicious choice of U and V we can eliminate M_{jk}^5 and convert M_{jk} into a diagonal matrix.

With these redefinitions, the Lagrangian assumes the canonical form

$$\mathscr{L}_{\text{QCD}} = -\tfrac{1}{4}\,\text{tr}\ G_{\mu\nu}G_{\mu\nu} + \sum_j \bar{\psi}_j(i\overleftrightarrow{D}_\mu\gamma_\mu - M_j)\psi_j + (\theta\ \text{term}). \qquad 34.$$

Before praising the many virtues of $\mathscr{L}_{\text{QCD}}$, it may be appropriate to add a word on the criterion of renormalizability, which is crucial in forbidding more complicated terms. A first reaction is that the apparent bad behavior (Equation 27) of unrenormalizable theories may simply reflect the mathematical inappropriateness of a perturbative expansion in K. However, experience in rigorous constructive field theory has so far been that dangerous high-energy behaviors indicated in perturbative analyses of model field theories persist in the complete theory. Numerical studies of discretized (lattice) field theories will shed more light on this question. By its very nature a lattice formulation of field theory provides a nonperturbative cutoff of dangerous high energy or short-distance behavior, and the question becomes whether or not one can extract a sensible, Lorentz-invariant limiting theory as the lattice grid is made finer. Such studies have barely begun. The most complete is the beautiful work by Shenker & Tobochnik (1980).

There is already impressive *experimental* evidence as to the relevance of the criterion of renormalizability. For example, it is only the criterion of renormalizability that tells us that a term, $\bar{\mu}\sigma_{\mu\nu}F_{\mu\nu}\mu$, representing an extra contribution to the magnetic moment of the muon, is forbidden. The extraordinary agreement with experiment of the prediction of QED without this extra term (Farley & Picasso 1979) is strong evidence for the renormalizability criterion. Also, the development of the modern (and very successful) gauge theories of the weak interaction was largely motivated by the desire to have a renormalizable theory to replace the old unrenormalizable current-current theory.

3.3 *Symmetry Properties and Their Stability*

The Lagrangian $\mathscr{L}_{QCD}$ of Equation 34 has many important symmetry properties:

1. Discrete symmetries P, C, and T: In the absence of the θ term, $\mathscr{L}_{QCD}$ is manifestly invariant under the discrete space-time symmetries of parity and time reversal, and under charge conjugation. (The θ term destroys P and T symmetry; it is discussed separately in Section 3.4.)

2. Flavor conservation: $\mathscr{L}_{QCD}$ is invariant under the phase rotations $\psi_j \rightarrow \exp(i\theta_j)\psi_j$, which as we discussed is equivalent to conservation of the additive quantum number that counts the number of quarks of flavor type j. These conservation laws correspond to the conservation of strangeness, baryon number, I_3 = the third component of isospin, electric charge, charm, etc by the strong interactions.

3. Approximate flavor symmetry: Insofar as quark masses m_i can be ignored, $\mathscr{L}_{QCD}$ is invariant under transformations of the type $\psi_j = U_{jk}\psi_k$ where U_{jk} is a unitary matrix. The existence of two very light quarks, u and d, and a third fairly light one, s, will then account for the accurate isospin and approximate SU(3) invariance of the strong interactions.

It is especially to be noticed that the coupling constant g is the same for every kind of quark. This is because the transformation law (Equation 18) for the gauge potential itself involves g. It would be mathematically inconsistent, therefore, to have different couplings for different quarks. (This is to be contrasted with QED, where in principle A_μ may couple to fields of arbitrary charge. QED in itself neither requires nor explains the quantization of charge.)

4. Approximate chiral symmetry: Again ignoring the masses of the u, d, and s quarks, $\mathscr{L}_{QCD}$ is invariant under separate unitary rotations

$$\psi_j^{L'} = U_{jk}\psi_k^L, \quad \psi_j^{R'} = V_{jk}\psi_k^R$$

among left- and right-handed light quarks, generating the group $U(3)_L \times U(3)_R$ of chiral symmetries. The diagonal subgroup $(V = U)$ of SU(3) and baryon number symmetries has been mentioned above. The remaining chiral SU(3) symmetries $(V = U^{-1}, \det U = 1)$ are not directly manifested in the hadron spectrum. The hypothesis that these symmetries are spontaneously broken can explain why the pseudoscalar mesons π and K are so light (approximate Nambu-Goldstone bosons) and various successful soft-pion theorems (see Adler & Dashen 1968, Gell-Mann et al 1968, Weinberg 1970).

Whether these chiral symmetries are predicted to break spontaneously according to $\mathscr{L}_{QCD}$ is a difficult dynamical question that is not yet completed settled — although all indications are that the breakdown is very likely (Coleman & Witten 1980). In any case two important consequences regarding chiral symmetry can be read off from $\mathscr{L}_{QCD}$. Firstly, the chiral symmetries should become exact in high-energy, high-momentum regimes where the quark masses are negligible. In particular, equal-time commutators between axial and vector currents (which involve spatial δ functions, i.e. infinitely small distances or large momenta) satisfy the $U(3) \times U(3)$ algebra exactly, with deep and successfully verified consequences (Adler & Dashen 1968, Gell-Mann et al 1968), including the Adler-Weisberger sum rule for g_A. Secondly, the quark mass terms dictate a very simple pattern for explicit SU(3) $\times$ SU(3) breaking. Treating the mass terms as a perturbation, one arrives at the very successful Gell-Mann–Okubo formula (Gell-Mann 1962, Okubo 1962) relating the masses of pseudoscalar mesons

$$3m_\eta^2 + 2m_{\pi^+}^2 = 2m_{K^0}^2 + 2m_{K^+}^2 + m_{\pi^0}^2 \qquad\qquad 35.$$

among other things.

To summarize, there is an almost perfect match between the known symmetry properties of the strong interactions and the symmetry properties of the candidate Lagrangian $\mathscr{L}_{QCD}$ (Equation 34). This "almost" is addressed in Section 3.4.

An important theoretical advantage of QCD is that all these symmetries are *stable* against radiative corrections (Weinberg 1973). After all, we know the parity invariance and strangeness conservation, for instance, are broken by the weak interactions. The symmetry breaking induced by these interactions is guaranteed to be small, however, for the following reason. They will generate various interactions as corrections to $\mathscr{L}_{QCD}$. If these interaction terms are operators of dimension ≤ 4, then, as we have seen, suitable redefinition of the gauge and quark fields will restore the canonical form (Equation 34) with all its symmetries. Terms involving operators of dimension $D > 4$, e.g. the usual current-

current interaction $\bar{\psi}\gamma_\mu(1 - \gamma_j)\psi\bar{\psi}\gamma_\mu(1 - \gamma_j)\psi$ of dimension 6, will by dimensional analysis appear with a coefficient of dimension $(\text{mass})^{-(D-4)}$. The relevant mass is the large mass m_W of the W boson, so such corrections are always small. If our theory of the strong interaction allowed breaking of P or strangeness by operators of dimension $D \leq 4$, we would expect order α violation of these symmetries; there would be no extra suppression of weak effects due to the heaviness of the W boson.

3.4 U(1) Problem, Anomalies, θ Term

One apparent approximate symmetry of $\mathscr{L}_{QCD}$ is *not* observed in the strong interactions. This is the chiral U(1) symmetry under which the light left-handed quark fields u, d, and s are multiplied by a common phase $\exp(i\rho)$ and the corresponding right-handed quark field by the opposite phase $\exp(-i\rho)$. This symmetry is associated with approximate conservation of the axial "light baryon number" current

$$j_\mu^5 = \bar{u}\gamma_\mu\gamma_5 u + \bar{d}\gamma_\mu\gamma_5 d + \bar{s}\gamma_\mu\gamma_5 s \qquad\qquad 36.$$

$$\partial_\mu j_\mu^5 \propto \text{light quark masses}\quad (\text{wrong!}). \qquad\qquad 37.$$

This symmetry appears superficially to be of the same nature as the other chiral symmetries. If this were so, however, we would predict, in addition to the eight observed light pseudoscalars, a ninth light one, an SU(3) singlet associated with the spontaneous breakdown of axial baryon number symmetry. There is a particle, the η' meson, with the correct quantum numbers but it is much too heavy. The absence of the ninth light pseudoscalar is called the U(1) problem. For a review, see Weinberg (1975).

The reason for the failure of Equation 37 lies quite deep. Equation 37 is correct at the classical level, but fails quantum mechanically. Perhaps the most direct way to see the reason for the failure is the following. The amplitude for the axial current j_μ^5 to produce two gluons by the triangle graph is linearly divergent at large internal momentum by power counting. To define the amplitude properly, this infinity must be regulated (compare Section 4). A convenient regularization procedure is that introduced by Pauli & Villars (1949). One formally adds heavy spin-$\frac{1}{2}$ quarks obeying Bose statistics into the theory. Because these quarks have opposite statistics, they contribute to the loop with the opposite sign. For very large internal momenta the contribution of the two kinds of quarks cancel as the mass difference becomes unimportant. This removes the infinity in an unambiguous way; at the end, the mass of the fictitious Bose quark is sent to infinity.

This regularization procedure has the great virtue of preserving P, C, T, Lorentz, gauge, and flavor invariances, but the regularized axial current including the Bose quarks U, D, and S

$$j_\mu^5(\text{reg}) = \bar{u}\gamma_\mu\gamma_5 u + \bar{U}\gamma_\mu\gamma_5 U + \ldots \qquad 38.$$

is far from conserved; its divergence includes both light and heavy quark pieces:

$$\partial_\mu j_\mu^5(\text{reg}) = 2(m_u\bar{u}\gamma_5 u + m_U \bar{U}\gamma_5 U) + \ldots \qquad 39.$$

Now the amplitude for $\bar{U}\gamma_5 U$ to go into two gluons by the triangle graph converges at large momentum and is proportional to M^{-1}. Combining this with Equation 39 we find an "anomalous" correction to approximate axial baryon symmetry. The correct form of Equation 37, involving the regulated current, is

$$\partial_\mu j_\mu^5 = \frac{g^2}{32\pi^2}\, \varepsilon_{\mu\nu\rho\sigma}\, \text{tr}\, G_{\mu\nu}G_{\rho\sigma} + (\text{quark mass terms}). \qquad 40.$$

It can be argued that the anomaly equation (Equation 40) holds independently of the details of the regulator procedure. It shows that the unwanted extra U(1) symmetry is not present in QCD (Actually, there is an additional subtlety here, as will appear shortly.)

A similar anomaly is crucial to the analysis of $\pi^0 \rightarrow \gamma\gamma$ decay (Adler 1969, Bell & Jackiw 1969). Soft-pion theorems allow one to express amplitudes for processes with pions at small momentum in terms of amplitudes involving divergences of the axial currents with the same quantum numbers. The amplitude for $\pi^0 \rightarrow \gamma\gamma$ is controlled by the anomalous contribution to the divergence of axial I_3:

$$\partial_\mu(\bar{u}\gamma_\mu\gamma_5 u - \bar{d}\gamma_\mu\gamma_5 d) = \frac{\alpha}{4\pi}\, \varepsilon_{\mu\nu\rho\sigma}F_{\mu\nu}F_{\rho\sigma}, \qquad 41.$$

where $F_{\mu\nu}$ is the electromagnetic field strength.

The agreement of the calculated rate for $\pi^0 \rightarrow \gamma\gamma$ with experiment is a remarkable triumph of quantum field theory. It should be remarked that the color degrees of freedom are crucial here — one gets a factor of three in the amplitude because of three different colors circulating in the anomalous triangle graph; without color the calculated rate would be low by a factor of nine!

The other apparent mismatch between the observed symmetries of the strong interaction and the symmetries of QCD regards the θ term, which breaks P and T invariance. At first one might be tempted to ignore the θ term, since it is formally a total divergence:

$$\varepsilon_{\mu\nu\rho\sigma} \text{ tr } G_{\mu\nu}G_{\rho\sigma} = 4\varepsilon_{\mu\nu\rho\sigma} \text{ tr } \partial_\mu \left(A_\nu \partial_\rho A_\sigma + \frac{2g}{3i} A_\nu A_\rho A_\sigma \right) \equiv \partial_\mu R_\mu .$$
42.

However, further (and very deep) analysis shows that R_μ is a singular operator, not really properly defined in the full quantum field theory (Belavin et al 1975, 't Hooft 1976a, b, Callan et al 1976, Jackiw & Rebbi 1976). This is fortunate, since the anomaly term in Equation 40 is of the same form. If the anomaly term could be expressed as a total divergence, we could easily absorb it into j_μ^5 and produce a new approximate conservation law, landing us back in the U(1) problem.

The θ term cannot be neglected, and the observed P and T invariance of the strong interaction (in particular, the smallness of the neutron electric dipole moment) forces us to conclude that $\theta = 0$ to high accuracy ($\theta \lesssim 10^{-8}$). It is mildly disappointing to be forced in this way to add P or T invariance as a separate assumption. What is worse, the θ term, being of operator dimension $D = 4$, lacks stability against radiative corrections in the sense discussed above. We would in general expect it to be induced, at a level much larger than $\theta \lesssim 10^{-8}$, by weak interactions.

QCD by itself cannot explain the smallness of the θ term. Larger unified theories may do this (Peccei & Quinn 1977, Weinberg 1978, Wilczek 1978). In any case, for the description of the strong interaction proper it is both necessary and sufficient to take $\theta = 0$.

3.5 Perturbative Content of $\mathscr{L}_{QCD}$

With $g = 0$, $\mathscr{L}_{QCD}$ represents a field theory of free spin-$\frac{1}{2}$ fermions — quarks — and vector bosons. On expanding the traceless hermitian matrix field $B_\mu(x)$ in terms of the λ matrices

$$B_\mu(x) = B_\mu^a(x) \frac{\lambda^a}{\sqrt{2}}, \quad a = 1, \dots, 8$$
43.

$$\tfrac{1}{2} \text{ tr } \lambda^a \lambda^b = \delta^{ab},$$

we find that the gluon kinetic energy decomposes, so there are eight independent vector gluon degrees of freedom. To first order in g, we find gluon couplings changing each color of quark to the other ones. Six gluons mediate this sort of interaction; the other two, like the photon in QED, couple to but do not change colors. In addition there are, because of the commutator term in $G_{\mu\nu}$, trilinear couplings of the gluons to each other in order g. This self-coupling of the gauge fields, which of course is absent for QED, leads to profound dynamical consequences, as is discussed in Section 4.

A perturbative treatment of QCD is not adequate for calculating the physical spectrum. The quarks and gluons are not manifested as free physical particles, but do manifest themselves strikingly in jets and other high-energy phenomena (Sections 4 and 5) as well as in their symmetries as outlined above.

4.   SCALING IN QUANTUM FIELD THEORY

"By now you are probably convinced that all the laws of physics are symmetrical under any kind of change whatsoever, so now I will give a few that do not work. The first one is change of scale...."

(Feynman 1967)

4.1   *Quantum Mechanics and Scaling Violation*

It has been recognized at least since the time of Galileo that physical laws are not invariant under a uniform change of the size of all objects. Galileo discusses this in the second day of his masterwork *Dialogues Concerning Two New Sciences*, where you will find a picture of the distorted thighbone necessary to support a scaled-up dog.

An obvious reason for the lack of scale invariance is the appearance of explicit scales — e.g. the masses of elementary particles — in the fundamental laws of physics. Gravitational interactions, being governed by Newton's constant G_N with dimensions of (length)^2 or (mass)$^{-2}$ (in units where $\hbar = c = 1$), also lead to macroscopic scaling violations.

In QCD there are two sources of scale symmetry breaking. One is the obvious one; the appearance of quark masses. (The gluon, like the very analogous photon in QED, is required by gauge invariance to be massless.) The other is a more subtle, quantum mechanical effect. The gauge coupling g is formally dimensionless, but, if we think about defining it in terms of "measurable" quantities a scale inevitably appears. For example, since $\mathcal{L}_{QCD}$ contains a quartic coupling among the gluon fields A_μ with coefficient g^2, one might hope to extract g^2 from the gluon-gluon scattering amplitude. However, the scattering amplitude for real, on-shell gluons (with invariant mass squared, $p^2 = 0$) is plagued with severe infrared divergences. We can identify g^2 as the amplitude for scattering off-shell gluons with invariant mass squared $p^2 = -\mu^2$; but then we must expect that g^2 will depend on μ. By the way, we cannot avoid introducing a scale by taking $\mu^2 \to \infty$, because of ultraviolet divergences. In this way, even a dimensionless coupling introduces a hidden scale.

A more picturesque explanation of the hidden scale is as follows. We can calculate the interaction energy between two static charges (infinitely heavy quarks); to lowest order it is just the Coulomb-like energy $-\frac{4}{3}g^2/r$. However, in quantum mechanics the physical vacuum must be regarded as a polarizable medium. Charged pairs are always virtually present as quantum mechanical fluctuations, and can partially screen — or, as we shall see, antiscreen — the charges. Therefore in reality the energy will depend on r in a nontrivial way. If we define the interaction energy as

$$E(r) \equiv -\frac{4}{3}\frac{g^2(r)}{r},$$

44.

then once again it appears that the definition of g necessarily requires specification of a scale of length.

If we can isolate processes that involve only very large momenta and energies, or equivalently very small distances and times — we discuss how this is done in Sections 4.4 and 5 — then we expect that quark masses become unimportant. Under these circumstances, scale invariance is violated only by the quantum mechanical effect just described.

4.2   Formulation of the Renormalization Group

The definition of g involves picking some scale μ, but this scale is at our disposal. The mathematical equation showing how the same theory may be expressed with different μ is called the renormalization group equation.

Let us first, for simplicity, suppose the quark masses to be zero. This should be adequate for large momenta. In order to define the terms appearing in $\mathcal{L}_{\text{QCD}}$ properly we must normalize the fields B_μ and ψ and the coupling g. As discussed above, this must be done by considering amplitudes involving virtual particles with squared invariant mass equal to $-\mu^2$. If we change μ, then to have the same theory we must change the normalization of the fields — say by the factors $Z_B^{1/2}(\mu)$, $Z_\psi^{1/2}(\mu)$ — and let $g = g(\mu)$ be a function of μ. Then for any amplitude (Green's function) of the theory involving virtual gluons and quarks with four-momenta $q_1, \ldots, q_m; p_1, \ldots, p_n$ we have the statement that μ is redundant:

$$Z_B^{-m/2}(\mu)\, Z_\psi^{-n/2}(\mu) G[q_1, \ldots, q_m;\, p_1, \ldots, p_n;\, g(\mu);\, \mu]$$

45.

$$= \text{independent of } \mu,$$

where the last two slots in G indicate the coupling and normalization point dependence. The Z factors also depend on $g(\mu)$.

We also have a relation following from ordinary "engineering" dimensional analysis:

$$G[\lambda q_1, \ldots, \lambda q_m; \lambda p_1, \ldots, \lambda p_n; g(\mu); \mu]$$
$$= \lambda^{m+(3/2)n-4} G[q_1, \ldots, q_m; p_1, \ldots, p_n; g(\mu); \lambda^{-1}\mu]. \qquad 46.$$

To understand the exponent of λ, recall that gluon and quark fields have mass dimensions 1 and 3/2, respectively. The -4 in the exponent reflects that an energy-momentum-conserving $\delta^4(\sum q_i + \sum p_i)$ is extracted in the definition of G. Combining Equations 45 and 46, we arrive at the most useful form of the equation:

$$G(\lambda q_i; \lambda p_j; g(\mu); \mu)$$
$$= \left(\frac{Z_B(\lambda\mu)}{Z_B(\mu)}\right)^{-m/2} \left(\frac{Z_\psi(\lambda\mu)}{Z_\psi(\mu)}\right)^{-n/2} \lambda^{m+(3/2)n-4} G[q_i; p_j; g(\lambda\mu); \mu].$$
$$47.$$

This formidable-looking equation has a simple meaning. It tells us that we can calculate the effect of scaling all the four-momenta in the amplitude by supplying some normalization factors and using a new coupling constant $g(\lambda\mu)$. This will be useful if $g(\lambda\mu)$ becomes small; then we can calculate the λ dependence perturbatively. In QCD, as we shall see, $g(\lambda\mu)$ becomes small for large λ so that the large-momentum behavior of amplitudes is calculable; in fact it approximates that of a free ($g = 0$) theory. This is the phenomena called asymptotic freedom.

4.3   The Effective Coupling of QCD

The differential form of Equation 47 is very transparent. We obtain it by differentiating with respect to λ and then setting $\lambda = 1$:

$$\mathcal{D}G = \left[-4 + m(1 + \gamma_B) + n\left(\frac{3}{2} + \gamma_\psi\right) + \beta \frac{\partial}{\partial g}\right] G, \qquad 48.$$

where

$$\mathcal{D} \equiv p_i \frac{\partial}{\partial p_i} + q_j \frac{\partial}{\partial q_j} \qquad 49.$$

is the dilation operator, which generates infinitesimal scale transformations, and

$$\gamma_B = \frac{\partial}{\partial \ln \mu} \ln Z_B(\mu)^{-1/2} \qquad\qquad 50.$$

$$\gamma_\psi = \frac{\partial}{\partial \ln \mu} \ln Z_\psi(\mu)^{-1/2} \qquad\qquad 51.$$

$$\beta = \frac{\partial}{\partial \ln \mu} g(\mu). \qquad\qquad 52.$$

The γs are known as anomalous dimensions. This name is very appropriate to describe Equation 48, because when $\gamma \neq 0$ the fields do not scale according to naive dimensional analysis. The anomalous dimensions and β are functions of $g(\mu)$.

The β function may be calculated perturbatively. The principle of the calculation is very simple. We can calculate any two Green's functions — say with $m = 2$, 3 and $n = 0$ — and demand consistency with Equation 48. This allows us to extract both γ_B and β

The sources of μ dependence in the calculation by Feynman graphs are the divergences found in perturbation theory. For example, the higher order corrections to the propagator look like

$$G(p, -p)_{\rho\sigma} = -\frac{ig_{\rho\sigma}}{p^2} \left[1 + g^2 \left(c_1 \ln \frac{\Lambda^2}{p^2} + c_2 \right) + \cdots \right], \qquad 53.$$

where Λ is an ultraviolet cutoff and ρ and σ are the polarizations of the vector gluons. The renormalization philosophy is to recognize that, while this perturbation series for G is ill-defined, the series for $\partial/\partial p \, G(p, -p)$ is perfectly finite (and in fact is all we encounter in Equation 48). $G(p, -p)$ is normalized equal to its naive value $ig_{\rho\sigma}/\mu^2$ at one reference scale $p^2 = -\mu^2$; it is then calculable for any other value of p. The divergences in $G(p, -p)$ leave their mark on finite quantities like $\partial/\partial p \, G(p, -p)$ in the form of "anomalous," nonscaling, p dependence.

The actual calculation of β is now reasonably straightforward. The result is (Gross & Wilczek 1973a, Politzer 1973, Caswell 1974, Jones 1974)

$$\beta(g) = -\frac{1}{16\pi^2}\left(11 - \frac{2}{3}n_F\right)g^3 - \left(\frac{1}{16\pi^2}\right)^2 \left(102 - \frac{46}{3}n_F\right)g^5$$

$$\equiv -b_0 g^3 - b_1 g^5 + \cdots, \qquad\qquad 54.$$

where n_F is the number of quark flavors.

It is of the utmost significance that for small g and $n_F < 17$ the β function is negative. This means, according to Equation 52, that the effective coupling decreases as we increase the momentum scale. It is therefore self-consistent to assume that $g(\mu)$ is small for large μ; then for larger μ it gets still smaller.

With the form of Equation 54, Equation 52 for the coupling can be solved at large μ to yield

$$\frac{1}{g^2(\mu)} \xrightarrow[\mu \to \infty]{} \frac{1}{2b_0} \ln \mu + \frac{b_1}{b_0} \ln \ln \mu \qquad\qquad 55.$$

with corrections that are finite as $\mu \to \infty$. The effective coupling approaches zero essentially logarithmically with the scale for large momenta. Naive scale invariance is restored asymptotically for extremely large momenta.

4.4   Heuristic Discussion of Asymptotic Freedom

The decrease of coupling with increase of momentum scale is at first sight very surprising. Consider a point charge q. In general, we would expect the quantum fluctuations discussed in Section 2 to shield the charge. After all, for a virtual pair the one charged oppositely to q will be attracted to q. Shielding means that the effective charge seen at large distances (small momenta) becomes small — the opposite behavior to asymptotic freedom. In other words, we might expect the polarization cloud surrounding a charged particle to be densest near the particle. In an asymptotically free theory, the particle is instead structureless at short distances.

Asymptotic freedom is indeed a rare phenomenon in quantum field theories. Zee (1973) and Coleman & Gross (1973) showed that it is basically unique to nonabelian gauge theories. For several years the physical basis of asymptotic freedom seemed extremely obscure. However, I think the following considerations (which I learned from K. Johnson) remove much of the mystery (see Hughes 1981 and references therein).

Asymptotic freedom means that the vacuum antiscreens charge; it acts like a dielectric medium with $\varepsilon < 1$. For a vacuum we must have the product $\varepsilon \mu$ of electric and magnetic susceptibilities equal to one (so that the speed of light $= c$). We need then to understand why $\mu > 1$. It is somewhat easier to visualize the magnetic than the electric response because a constant electric field can create pairs.

The response of a charged particle to a magnetic field has two pieces. First, the magnetic field leads to convective currents that induce magnetic fields opposing the original field (Landau diamagnetism).

Second, if the particles have a magnetic moment, the moment tends to align with the field and increase it (Pauli paramagnetism). The magnitudes of these effects are proportional to

$$\delta\mu \propto -\tfrac{1}{3}q^2 \qquad \text{(diamagnetism)} \qquad\qquad 56.$$

$$\delta\mu \propto +\gamma^2 q^2 s(s+1) \quad \text{(paramagnetism)} \qquad 57.$$

where q is the charge, s the spin, γ the gyromagnetic ratio. The relative factor $\tfrac{1}{3}$ is related to the three-dimensionality of space. The gyromagnetic ratio $\gamma = 2$ for both the spin-$\tfrac{1}{2}$ and spin-1 fields is implicit in $\mathscr{L}_{\text{QCD}}$. For the spin-$\tfrac{1}{2}$ quarks we then seem to have

$$\delta\mu_{1/2} \propto q^2(-\tfrac{1}{3} + 3) = q^2(+\tfrac{8}{3}) \quad \text{(wrong!)} \qquad 58.$$

and for the gluons

$$\delta\mu_1 \propto q^2(-\tfrac{1}{3} + 8) = q^2(\tfrac{23}{3}) \quad \text{(wrong!)} \qquad 59.$$

while a spin-0 particle would contribute

$$\delta\mu_0 \propto q^2(-\tfrac{1}{3}). \qquad\qquad\qquad 60.$$

Equation 58 is not correct because the Fermi statistics of the virtual quarks reverses the sign (in Feynman graphs, the fermion loops are accompanied by a negative sign). Thus for quarks

$$\delta\mu_{1/2} \propto q^2(-\tfrac{8}{3}). \qquad\qquad\qquad 61.$$

For the massless spin-1 particles we must subtract off the contribution from unphysical longitudinal modes, which contribute like a spin-0 particle (with an additional minus sign, because these modes have negative probability). Thus

$$\delta\mu_1 \propto q^2(+\tfrac{22}{3}). \qquad\qquad\qquad 62.$$

The results expressed in Equations 61 and 62 for spin-0 and spin-$\tfrac{1}{2}$ particles show us that ordinary electromagnetism coupled to such particles is not asymptotically free, $\delta\mu < 0$. It is the interactions of the vector particles, in particular their magnetic moment, that leads to asymptotic freedom.

The above remarks can without too much difficulty be made into an actual derivation of the first term in the β function (Equation 54). [The ratio of color (charges)^2 of gluons and quarks

$$\frac{\displaystyle\sum_{\text{glue}} q^2}{\displaystyle\sum_{\text{quarks}} q^2} = 6$$

can be read off from the ratio of ordinary electric (charges)[2] of vector mesons $\rho^{\pm}$, $K^{*\pm}$ versus quarks u, d, and s since the same mathematical group SU(3) is relevant!]

4.5   Application to Experiment

Asymptotic freedom allows us to calculate the large momentum behavior of Green's functions using Equation 47, or equivalently by integrating Equation 48. This is adequate for only a small number of applications. Unfortunately scale transformations involve uniform multiplication of all components of the four-momenta, so they are totally inadequate to describe scattering experiments involving projectiles and targets of definite mass.

One type of process to which scaling may be applied reasonably directly is the "antenna pattern" of energy flow in e^+e^- annihilation into hadrons (Sterman & Weinberg 1977, Brown & Ellis 1981). The electron-positron pair may be made to annihilate at different center-of-mass energies $\sqrt{s}$; this is a mass parameter we can scale in the laboratory. (Of course the purely electromagnetic interaction of electrons and positrons is accurately known; we are not interested in them except as sources of a virtual photon of mass $\sqrt{s}$. We need not worry about their fixed mass.) If we ask for the cross section that energy fractions $f_1\sqrt{s}, \ldots, f_n\sqrt{s}$ flow into fixed solid angles $\Omega_1, \ldots, \Omega_n$, then still the only dimensional parameter is $\sqrt{s}$. So the $\sqrt{s}$ dependence of these cross sections, for large $\sqrt{s}$, should be calculable perturbatively — an essentially direct application of scaling.

Very many other processes can be addressed perturbatively by factorizing them into a product of momentum-independent structure functions and kernels to which scaling and the renormalization group apply (for this philosophy, see especially Politzer 1977). We shall not now go into these ingenious techniques, which have already generated not only an enormous literature but even an enormous review literature (e.g. Reya 1981, Mueller 1981 and references therein). Several representative results along with relevant experimental data are quoted in Section 5.

5.   CONFRONTATION WITH EXPERIMENT

The confrontation of QCD with experiment is a rapidly developing and rich subject. This section gives only a schematic summary of a few of the most recent results; for more details see the indicated references, reviews, and the article of Söding & Wolf (1981) that appeared in the *Annual Review of Nuclear and Particle Science* last year.

5.1 e^+e^- Annihilation

As mentioned at the close of Section 4, the energy distribution of hadron products in e^+e^- annihilation is predicted very directly from scaling and asymptotic freedom. A beautiful analysis of the experimental data has been carried out by Brown & Ellis (1981), who find good agreement with QCD predictions.

A small effective coupling $\bar{g}$ at large $\sqrt{s}$ means that the processes may be approximated by perturbative calculations. In particular, to first approximation the energy flow in e^+e^- annihilation is described by the "two-jet" quark production. The next order in $\bar{g}$ allows "three-jet" processes, where a gluon is emitted.

Three-jet events have been observed (Brandelik et al 1980a). Their frequency of course gives a direct measure of $\bar{g}^2$. The angular correlations among the three jets depend on the spin of the gluons and may be regarded as a test of their vector nature (Brandelik et al 1980b).

5.2 Deeply Inelastic Scattering

Historically the scattering of electrons and neutrinos by nuclei at large momentum transfer gave the first strong experimental indication of approximate scaling (Miller et al 1972) and were the first to be treated in detail using QCD (Gross & Wilczek 1974, Georgi & Politzer 1974).

Roughly speaking, the scattering process is factorized into an amplitude for finding a quark with given momentum fraction inside the nucleon, and the amplitude for scattering at large momentum transfer by the quark. The latter process can be treated perturbatively. The first term gives essentially the parton model result; in the next order, corrections arise because of the possibility of emitting a gluon in the scattering or scattering from a gluon.

There are very good experimental indications for the pattern of corrections to scaling predicted by QCD, as discussed in detail by Söding & Wolf (1981). Recent work of Jaffe & Soldate (1981) sharpens the analysis. Direct studies of the predicted jets in $\mu + p \rightarrow \mu + X$ have also given impressive results in agreement with QCD (Aubert et al 1981).

QCD allows one to derive parton model sum rules as a first approximation and also to compute corrections. For example, the spin-$\frac{1}{2}$ nature of quarks gives a characteristic angular distribution for the final lepton in deeply inelastic scattering [the Callan & Gross relation (1969)]. In QCD one finds this true to order $\bar{g}^2$, with calculable corrections (Zee et al 1974). There is also a sum rule for the energy-momentum carried by the charged constituents of the proton; of course, the energy-momentum carried by neutral particles is not detected directly in deeply inelastic scattering. Comparison of this sum rule long ago indicated the need for neutral (nonquark!) constituents in the proton. In QCD these constituents are identified as the vector gluons and one can describe quantitatively how they modify the sum rule (Gross & Wilczek 1974, Georgi & Politzer 1974).

5.3 Two-Photon Processes

In high energy e^+e^- scattering, an important role is played by two-photon processes. The collision of these virtual photons often produces hadrons. At high energies this process can be described perturbatively — it is much like deeply inelastic scattering, but here the nucleon target is replaced by a photon (in principle, with adjustable mass!). These processes give particularly clean tests of QCD, since the internal structure of a photon is simpler than that of a nucleon.

The basic jet process was studied recently, with good evidence for the angular and momentum distribution expected from QCD at high transverse momentum (Bartel et al 1981). The photon structure function is observed in the 2γ process when one photon is highly virtual and the other near the mass shell. This is a particularly striking QCD test since so much of it is calculable and because the modification of ordinary parton model predictions is particularly large (Witten 1977). Crude data do agree with QCD predictions (Bergen et al 1981); we may look foward to tighter comparisons soon.

5.4 Quarkonium

Bound states of very heavy quark-antiquark pairs annihilate when these particles approach within a Compton wavelength of one another. This occurs, then, at short distances and should be calculable perturbatively. (Appelquist & Politzer 1975a,b).

The leading process for a charge conjugation odd state is annihilation into three gluons. The ratio of this hadronic annihilation to the electromagnetic annihilation into $\mu^+\mu^-$ allows a measure of $\bar{g}^2$. This sort of measurement is particularly important because the result is directly proportional to $\bar{g}^6$ (i.e. not an order $\bar{g}^2$ correction to the parton model as in several other processes mentioned above), and because both theory and experiment can be done accurately.

In a very impressive analysis, MacKenzie & LePage (1981) calculated the QCD prediction, including corrections, and compared it to experiment. They find consistency between the result for ψ, ψ', and Υ decays, and determine a value for $\bar{g}$ consistent with other determinations (see Section 5.6).

It will be most fruitful to extend this analysis to other heavy quark states (besides the 1^{--} family) and to gather further experimental results. Also, the related annihilation into two gluons and a photon has been studied (Brodsky et al 1978).

5.5   Lattice Field Theory Calculations

The results discussed in Sections 5.1 to 5.4 involve the behavior of hadrons at short distances or large momenta and were based on perturbation theory. A host of important questions — notably whether or not color is confined in QCD and what the masses of hadrons are — cannot be addressed by these methods.

Wilson (1975) showed how QCD may be formulated on a lattice, in such a way that gauge invariance is preserved. The lattice of course provides a (non-Lorentz invariant) cutoff of the short-distance divergences of the theory. Quantum field theory on a finite lattice is then a completely finite system; all calculations are reduced to well-defined integrals, albeit in a space of enormous dimensionality. Creutz (1980) and others demonstrated that Monte Carlo techniques can be used to evaluate the relevant integrals on high speed computers.

The main calculation that has been done so far concerns the energy between a colored source and its antisource (one may think of infinitely heavy quark and antiquark) as a function of their separation. It is found that for large separations the potential rises linearly with separation:

$$U(R) \to \sigma R + \cdots . \qquad\qquad 63.$$

The important qualitative result here is that colored sources cannot be isolated with finite energy — color confinement.

In a string picture, σ should be identified with the string tension. It can be compared to the Regge slope (Equation 3) or to the potential derived in heavy quark spectroscopy.

On a sufficiently large lattice one may calculate the corrections to the Coulomb potential at short distances as well as σ. This permits one quantitatively to relate α_s to σ.

To date all practical Monte Carlo calculations have been done in the pure gluon theory, with no quarks. Newly available techniques promise to make inclusion of quarks computationally feasible. It seems probable to the present author that honest QCD calculations of light hadron masses with tolerable (10%) accuracy will be available within the decade, possibly much sooner.

5.6   Summary of Experimental Status of QCD

The various quantitative tests of QCD (e^+e^- annihilation jets, deep inelastic scattering, quarkonium decay) may all be fitted consistently with a common value of the coupling

$$0.14 \lesssim \alpha_s(10 \text{ GeV}) \lesssim 0.18 \qquad 64.$$

or of the conventional Λ parameter (roughly, the scale at which α_s becomes large)

$$100 \text{ MeV} \lesssim \Lambda_{\overline{\text{MS}}} \lesssim 250 \text{ MeV}. \qquad 65.$$

The lattice calculations, fitted to the string tension σ, are also consistent with these values. (The inclusion of quarks is conjectured to be a small effect numerically; clearly this needs real testing.) Rapid progress can be expected in the measurement and analysis of heavy quarkonium decays, in measurement e^+e^- annihilation processes at higher energies (in particular, under the Z resonance) and in numerical work on lattice field theory.

There are other processes for which QCD makes important semi-quantitative predictions, although I do not expect them to compete with those above as quantitative tests. Most notable is the prediction of asymptotic scaling

$$\frac{1}{E}\frac{d\sigma}{dp_\perp} \propto \frac{1}{p_\perp^4} \qquad 66.$$

for hadron production at large transverse momentum in proton-proton (or $p\bar{p}$) collisions. This cross section is much larger than expected from old extrapolations of low $p_\perp$ data, although recently an approach to Equation 66 has been found experimentally. Similarly, high mass $\mu^+\mu^-$ pairs and Z bosons produced in hadronic collisions are predicted to have large, scaling cross sections at high $p_\perp$ (Fritzsch & Minkowski 1977).

6. IMPLICATIONS FOR OTHER BRANCHES OF PHYSICS

6.1 Weak Interactions

Most of our experimental information on weak interactions comes from observation of processes involving hadrons. For this reason increased understanding of the strong interaction has often stimulated or accompanied progress in weak interactions. Examples include the development of the Cabibbo theory based on SU(3) symmetry and many applications of current algebra to description of weak decays.

The more recent developments, particularly asymptotic freedom, have brought some new applications:

1. Control of large virtual momentum exchanges permits a more

accurate and rigorous treatment of radiative corrections to universality in semileptonic weak decays (Sirlin 1978).

2. Contribution of heavy quarks to K_L-K_S mixing necessarily involves large momentum transfers and can therefore be analyzed using asymptotic freedom. Historically, estimates of such contributions were important in obtaining upper bounds on the charmed-quark mass (Gaillard et al 1975). More recently, interesting upper bounds on the top-quark mass have been claimed (Hagelin 1979, 1981).

3. Similarly, a heavy quark contribution to the isoscalar axial neutral current was derived by Collins et al (1978).

4. Of course, the simplicity of strong interactions at large momentum permits us to extract cleanly weak interaction parameters in this regime, as for instance in deep inelastic neutrino scattering, in parity-violating electroproduction, and eventually in Z-meson decay.

6.2   Unification of Interactions

The mathematical kinship of QCD and the SU(2) × U(1) theory of weak interactions, both based on local gauge symmetry, suggests that both might derive from a common ancestor. Moreover, asymptotic freedom, the decrease of the effective strong coupling at large momenta, suggests that in this regime the strong interaction will eventually cease to be any "stronger" than the weak and electromagnetic interactions, again suggesting the possibility of unification.

A remarkable calculation based on these ideas was done by Georgi, Quinn & Weinberg (1974). They followed the evolution of the effective couplings $g_3(M)$, $g_2(M)$, $g_1(M)$ of the SU(3) × SU(2) × U(1) theories as a function of momentum, according to Equation 54. If we demand the unification condition that all three couplings become equal at a single momentum scale M_0, a constraint among g_1, g_2, and g_3 at low energy follows: the two parameters $g_3(M_0)[=g_2(M_0)=g_1(M_0)]$ and M_0 determine the three low energy parameters. The resulting prediction, which can be quoted as a prediction for $\sin^2 \theta_W$, is remarkably successful. Because the renormalization effects are only logarithmic in M, a very large scale ($M_0 \simeq 10^{15}$ GeV) is required for unification.

A very economical model embodying the unification idea was created by Georgi & Glashow (1974), who linked SU(3) × SU(2) × U(1) in the larger group SU(5). This model predicts proton decay at an experimentally interesting rate as a result of exchange of superheavy gauge particles in SU(5) [but not in SU(3) × SU(2) × (1)], which have mass $\sim M_0$.

Besides providing the general framework and motivation as described above, QCD and asymptotic freedom are crucially important in the more detailed theory of proton decay. The QCD scale parameter Λ sets the scale for the unification mass M_0, so that the proton decay rate (due to exchange of particle of mass $\sim M_0$) goes roughly as Λ^{-4}. There are additional calculable renormalization effects that influence both the overall scale of proton decay and predictions for branching ratios (Buras et al 1978, Wilczek & Zee 1979). Should proton decay be observed, we will have the opportunity to test quantum field theory at hitherto undreamed-of scales!

Unified theories also link quark and lepton masses, and again QCD is crucial in evaluating the predictions (Buras et al 1978). In the simplest version of SU(5), the charge-$\frac{1}{3}$ quark and charged-lepton effective masses are equal at M_0. Calculable strong interaction renormalization effects increase the effective quark mass by about a factor of three as we go from M_0 down to laboratory scales, which is suspiciously close to the observed ratio $m_b/m_\tau \cong 3$ of bottom-quark to tau-lepton masses!

6.3 Hadronic Matter in Extreme Conditions

Collision between heavy ions at high energies may produce a fireball of hadronic matter with an effective temperature $T \gtrsim 200$ MeV. This may be very significant because there are theoretical indications that at such temperatures the equation of state of hadronic matter may change drastically. Roughly speaking, we expect to pass from a dense liquid of identifiable protons and neutrons to a plasma of (locally) free quarks.

Important numerical evidence for a phase transition in the pure gluon theory — QCD with no quarks — has appeared from lattice calculations (McLerran & Svetitsky 1981, Kuti et al 1981, Kajantie et al 1981). It is remarkable that the phase transition occurs at a temperature considerably less than the mass of the lightest glueball in the spectrum. At first sight this seems almost paradoxical, since just below the transition temperature we have a very dilute gas of glueballs. It is quite plausible, however, that $\rho(m)$, the density of state as a function of mass, goes like $\rho(m) \propto m^d \exp(m/T_0)$ for large m where d is a small number. This is suggested by Hagedorn's fireball picture (1965, 1968), by dual string models (Huang & Weinberg 1972), and by bag models. This spectral density would lead to a sharp transition at $T = T_0$ as many very heavy glueballs suddenly appear! (This picture was developed in conversations with K. Johnson and R. Pisarski.)

Numerical results for the theory including quarks should soon be available. It is an interesting theoretical challenge to predict how including quarks will modify T_0. It is a major experimental challenge to produce and detect these extreme states of matter in the laboratory.

Quarks become locally free at high densities as well as at high temperatures. This may be relevant to describing the deep interior of neutron stars. (Collins & Perry 1975).

6.4 *The Very Early Universe*

In big-bang cosmology arbitrarily high temperatures and densities are in principle reached at very early times. Prior to the modern developments in QCD, it was not known how to extrapolate the equation of state to this regime. Asymptotic freedom tells us that the behavior of matter at ultrahigh temperatures is extremely simple — we arrive essentially at a plasma of weakly interacting quarks, gluons, leptons, ... With the dynamics under control, it has become possible to calculate the cosmological consequences of particle interactions at ultrahigh energies. A remarkable consequence is a possible explanation for the cosmic asymmetry between matter and antimatter. (For a semi-popular account, see Wilczek 1980).

6.5 *Mathematical Methods*

The mathematical challenge of computing the hadronic mass spectrum and other low energy parameters directly from $\mathscr{L}_{QCD}$ has exercised the ingenuity of theorists and led to the development of methods that will be very powerful tools for solving other quantum field theory problems. The Monte Carlo approach to numerical field theory, and its extension to fermions (see especially Kuti 1981 and references therein), are outstanding examples. Here work motivated by QCD had led to enormous progress in treating problems of condensed matter physics.

ACKNOWLEDGEMENT

I wish to thank C. Callan, S. Treiman, A. Zee, and especially D. Gross for instruction over the years. Conversations with T. DeGrand, K. Johnson, and J. Kuti were very useful in preparing the manuscript.

The research was supported in part through the National Science Foundation, Grant PHY77-27084.

Literature Cited

Adler, S. L. 1969. *Phys. Rev.* 177:2426
Adler, S. L., Dashen, R. 1968. *Current Algebras*. New York: Benjamin
Anderson, P. 1970. *Tau Zero*. New York: Doubleday
Appelquist, T., Politzer, H. D. 1975a. *Phys. Rev. Lett.* 34:419
Appelquist, T., Politzer, H. D. 1975b. *Phys. Rev. D* 12:1404

Aubert, J. J., et al. 1981. *Phys. Lett.* 100B:433
Bartel, W., et al. 1981. *Phys. Lett.* 107B:163
Belavin, A. A., Polyakov, A. M., Schwartz, A. S., Tyupkin, Y. S. 1975. *Phys. Lett. B* 59:85
Bell, J. S., Jackiw, R. 1969. *Nuovo Cimento* 60A:47

Bergen, C., et al. 1981. *Phys. Lett.* 107B:168

Bjorken, J. D. 1969. *Phys. Rev.* 179:1547

Brandelik, R., et al. 1980a. *Phys. Lett.* 94B:437

Brandelik, R., et al. 1980b. *Phys. Lett.* 97B:453

Brodsky, S. J., DeGrand, T. A., Hogan, R. R., Coyne, D. G. 1978. *Phys. Lett.* 73B:203

Brown, L. S., Ellis, S. D. 1981. *Phys. Rev. D* 24:2383

Buras, A. J., Ellis, J., Gaillard, M. K., Nanopoulos, D. V. 1978. *Nucl. Phys. B* 135:66

Callan, C. G. 1970. *Phys. Rev. D* 2:1541

Callan, C. G., Dashen, R. F., Gross, D. J. 1976. *Phys. Lett. B* 63:334

Callan, C. G., Gross, D. J. 1969. *Phys. Rev. Lett.* 22:156

Caswell, W. E. 1974. *Phys. Rev. Lett.* 33:244

Chodos, A., Jaffe, R. L., Johnson, K., Thorn, C. B., Weisskopf, V. F. 1974. *Phys. Rev. D* 9:3471

Close, F. 1979. *An Introduction to Quarks and Partons*. New York: Academic

Coleman, S., Gross, D. J. 1973. *Phys. Rev. Lett.* 31:1343

Coleman, S., Witten, E. 1980. *Phys. Rev. Lett.* 45:100

Collins, J. C., Perry, M. J. 1975. *Phys. Rev. Lett.* 34:1353

Collins, J., Wilczek, F., Zee, A. 1978. *Phys. Rev. D* 18:242

Creutz, M. 1980. *Phys. Rev. D* 21:2308

Dalitz, R. 1969. In *The Quark Model*, ed. J. J. J. Kokkedee. New York: Benjamin

DeGrand, T., Jaffe, R. L., Johnson, K., Kiskis, J. 1975. *Phys. Rev. D* 12:2060

DeRujula, A., Georgi, H., Glashow, S. L. 1975. *Phys. Rev. D* 12:147

Donoghue, J. F., Johnson, K. 1979. *Phys. Rev. D* 21:1974

Eichten, E., et al. 1976. *Phys. Rev. Lett.* 36:500

Farley, F. J. M., Picasso, E. 1979. *Ann. Rev. Nucl. Part. Sci.* 29:243–82

Feynman, R. P. 1967. *The Character of Physical Law*. Cambridge, Mass: M.I.T. Press

Feynman, R. P. 1969. *Phys. Rev. Lett.* 23:1415

Feynman, R. P. 1972. *Photon-Hadron Interactions*. New York: Benjamin

Frautschi, S. 1963. *Regge Poles and S-Matrix Theory*. New York: Benjamin

Friedberg, R., Lee, T.D. 1978. *Phys. Rev. D* 18:2623

Fritzsch, H., Gell-Mann, M., Leutweyler, H. 1973. *Phys. Lett.* 47B:365

Fritzsch, H., Minkowski, P. 1977. *Phys. Lett.* 69B:316

Gaillard, M. J., Lee, B. W., Rosner, J. L. 1975. *Rev. Mod. Phys.* 47:277

Gell-Mann, M. 1962. *Phys. Rev.* 125:1067

Gell-Mann, M. 1964. *Phys. Lett.* 8:214

Gell-Mann, M., Low, F. 1954. *Phys. Rev.* 95:1300

Gell-Mann, M., Ne'eman, Y. 1964. *The Eightfold Way*. New York: Benjamin

Gell-Mann, M., Oakes, R. J., Renner, B. 1968. *Phys. Rev.* 175:2195

Georgi, H., Glashow, S. L. 1974. *Phys. Rev. Lett.* 34:438

Georgi, H., Politzer, H. D. 1974. *Phys. Rev. D* 9:416

Georgi, H., Quinn, H. R., Weinberg, S. 1974. *Phys. Rev. Lett.* 33:451

Greenberg, O. W. 1964. *Phys. Rev. Lett.* 13:598

Greenberg, O. W., Zwanzinger, D. 1966. *Phys. Rev.* 150:1177

Gross, D. J., Wilczek, F. 1973a. *Phys. Rev. Lett.* 30:1343

Gross, D. J., Wilczek, F. 1973b. *Phys. Rev. D* 8:3633

Gross, D. J., Wilczek, F. 1974. *Phys. Rev. D* 9:980

Hagedorn, R. 1965. *Nuovo Cimento Suppl.* 3:147

Hagedorn, R. 1968. *Nuovo Cimento Suppl.* 6:311

Hagelin, J. S. 1979. *Phys. Rev. D* 20:2893

Hagelin, J. S. 1981. Harvard preprint HUTP-81/A020

Han, M. Y., Nambu, Y. 1965. *Phys. Rev.* 139B:1006

Hasenfratz, P., Kuti, J. 1978. *Phys. Rep.* 40C:75

Huang, K., Weinberg, S. 1972. *Phys. Rev. Lett.* 25:895

Hughes, R. J. 1981. *Nucl. Phys. B* 186:376

Jackiw, R., Rebbi, C. 1976. *Phys. Rev. Lett.* 37:172

Jaffe, R., Soldate, M. 1981. *M.I.T. preprint*

Jones, D. R. T. 1974. *Nucl. Phys. B* 75:531

Kajantie, K., Montonen, C., Pietarinen, E. 1981. *Z. Phys. C* 9:253

Kuti, J. 1981. *ITP preprint* 81:151

Kuti, J., Polonyi, J., Szlachanyi, K. 1981. *Phys. Lett.* 98B:199

Lee, T. D., Yang, C. N. 1961. *Phys. Rev.* 22:1954

MacKenzie, P. B., LePage, G. P. 1981. *Phys. Rev. Lett.* 47:1244

McLerran, L. D., Svetitsky, B. 1981. *Phys. Lett.* 98B:195

Miller, G., et al. 1972. *Phys. Rev. D* 6:3011

Mueller, A. 1981. *Phys. Rep.* 73C:238

Nambu, Y. 1966. In *Preludes in Theoretical Physics*, ed. A. DeShalit. Amsterdam: North-Holland

Okubu, S. 1962. *Prog. Theor. Phys. (Kyoto)* 27:949

Pauli, W., Villars, F. 1949. *Rev. Mod. Phys.* 21:434

Peccei, R. D., Quinn, H. R. 1977. *Phys. Rev. D* 16:1791

Politzer, H. D. 1973. *Phys. Rev. Lett.* 30:1346

Politzer, H. D. 1977. *Nucl. Phys. B* 129:301

Rebbi, C. 1974. *Phys. Rep.* 12C:1

Reya, E. 1981. *Phys. Rep.* 69C:195

Shenker, S. H., Tobochnik, J. 1980. *Phys. Rev. B* 22:4462

Sirlin, A. 1978. *Rev. Mod. Phys.* 50:573

Söding, P., Wolf, G. 1981. *Ann. Rev. Nucl. Part. Sci.* 31:231–93

Sterman, G., Weinberg, S. 1977. *Phys. Rev. Lett.* 39:1436

Streater, R., Wightman, A. 1964. *PCT, Spin and Statistics, and All That*. New York: Benjamin

Symanzik, K. 1970. *Commun. Math. Phys.* 18:277

't Hooft, G. 1973. *Nucl. Phys. B* 33:73

't Hooft, G. 1976a. *Phys. Rev. Lett.* 37:8

't Hooft, G. 1976b. *Phys. Rev. D* 14:3432

Weinberg, S. 1970. *Brandeis Summer School Lectures*, ed. S. Deser, M. Grisaru, H. Pendelton. Cambridge, Mass: M.I.T. Press

Weinberg, S. 1973. *Phys. Rev. Lett.* 31:494

Weinberg, S. 1975. *Phys. Rev. D* 11:3583

Weinberg, S. 1978. *Phys. Rev. Lett.* 40:223

Wilczek, F. 1978. *Phys. Rev. Lett.* 40:279

Wilczek, F. 1980. *Scientific American* 243(6):82–90

Wilczek, F., Zee, A. 1979. *Phys. Rev. Lett.* 43:1571

Wilson, K. G. 1971. *Phys. Rev. D* 3:1818

Wilson, K. G. 1975. *Phys. Rev. D* 10:2445

Witten, E. 1977. *Nucl. Phys. B* 120:189

Zee, A. 1973. *Phys. Rev. D* 7:3630

Zee, A., Wilczek, F., Treiman, S. 1974. *Phys. Rev. D* 10:2881

Zweig, G. 1964. CERN Rep. Th 401:412. Unpublished

Ann. Rev. Nucl. Part. Sci. 1982. 32:211–33
Copyright © 1982 by Annual Reviews Inc. All rights reserved

ELECTRIC-DIPOLE MOMENTS OF PARTICLES

Norman F. Ramsey

Mount Holyoke College, South Hadley, Massachusetts 01075; and Harvard University, Cambridge, Massachusetts 02138

CONTENTS

INTRODUCTION

The definition of the electric-dipole moment of a particle is as follows. Let the coordinate z be measured from the center of mass of a particle and let ρ_{JJ} be the electric charge density inside the particle, whose angular momentum is $\mathbf{J}$ with quantum number J and whose orientation state is given by $m = J$ relative to the z axis. The dipole moment eD is then defined by

$$eD = \int \rho_{JJ} z \, d\tau, \qquad\qquad 1.$$

where $d\tau$ is a differential volume element and e is conventionally taken to be the protonic charge. If the particle is charged, this definition implies that the center of charge of the particle is displaced from the center of mass if $D \neq 0$. If, on the other hand, the particle has no net

0163-8998/82/1201-0211$02.00

charge, the definition implies a greater positive charge in one hemisphere and a correspondingly greater negative charge in the other.

It is easy to see (Purcell & Ramsey 1950, Ramsey 1953, 1956, Golub & Pendlebury 1972) from the following argument that the electric-dipole moment must vanish if there is symmetry under the parity transformation (P) for which $\mathbf{r} \rightarrow -\mathbf{r}$ or the time-reversal transformation (T) for which $t \rightarrow -t$. Since the orientation of the particle can only be specified by the orientation of its angular momentum, the dipole moment D and the angular momentum $\mathbf{J}$ must transform their signs the same way under P and T if D is to have a nonzero value and if there is to be P and T symmetry. But D changes sign under P whereas $\mathbf{J}$ does not, so D must vanish if there is P symmetry. Likewise D does not change sign under T but $\mathbf{J}$ does, so D must vanish if there is T symmetry. More rigorous proofs of the above can be given (Ramsey 1953, 1956, Golub & Pendlebury 1972). If one makes the usual assumption of CPT symmetry (where C is the charge conjugation transformation, $Ze \rightarrow -Ze$), the existence of an electric-dipole moment would also imply a failure of CP symmetry. In molecules, electric-dipole moments do exist and are attributed to degenerate states, but degeneracy for the neutron would contradict the well-established fact that neutrons obey the Pauli exclusion principle.

For many years it was believed that the above P-symmetry argument precluded the existence of an electric-dipole moment for elementary particles, until Purcell & Ramsey (1950) pointed out that the assumption of P symmetry must be based on experiment and that few experiments on particles were sensitive to P. They pointed out that a sensitive test of P symmetry would be a search for an electric-dipole moment. This analysis led to the first experiment on the electric-dipole moment of the neutron (Smith et al 1957), which lowered the experimental limit on $|D|$ from 10^{-14} cm to 5×10^{-20} cm. Subsequently, the work in 1957 of Lee & Yang and of Wu et al showed that there was a failure of P symmetry in the decay of the kaon and in other weak interactions including the beta decay of nuclei. Despite this failure of P symmetry, which removed one argument against the existence of an electric-dipole moment, Landau (1957a,b) and others (Lee & Yang 1957, Wu et al 1957) showed that the parity argument against an electric-dipole moment could be replaced by the above argument based on time-reversal invariance. However, Jackson et al (1957) and Ramsey (1958) emphasized that time-reversal invariance, like parity at an earlier period, was merely an assumed symmetry that must rest on an experimental basis; there was little direct experimental evidence at that time in the strong and weak interactions. In 1964 Christenson et al discovered the *CP*-violating decay of the K_L^0 meson into two charged pions; if one assumes *CPT* symmetry as discussed above, the result implies a violation of *T* symmetry. A direct indication of a violation of *T* symmetry has also been found in the K_L^0 decay by Schubert et al (1970; see also Casella 1968, 1969) but so far all experimentally observed manifestations of either *CP* or *T* symmetry violations have been limited to the $K^0 - \bar{K}^0$ system.

Since the discovery of *CP* violations in the K_L^0 decay, a number of theoretical predictions have been made for particle electric-dipole moments on the basis of theories developed to account for the K_L^0 decay. The different predictions cover a wide range of values; some were as large as 10^{-19} cm for the *D* of the neutron and most were 10^{-22} cm or larger. As discussed below, the lower experimental limits on *D* have been followed by lower theoretical calculations, with 10^{-23} to 10^{-28} cm being typical predictions of current theories. This is discussed in greater detail in the final section of this report.

Experimental measurements and theoretical predictions of electric-dipole moments are possible for many different particles, including neutrons, protons, electrons, muons, and hyperons, but the most sensitive tests of theories have so far come from the neutron. The reason for this is the neutron's zero electric charge. As first pointed out in this connection by Purcell & Ramsey (1950) and further discussed by Schiff (1963), an electrically charged particle in equilibrium under the action of only electrostatic forces must be subject to zero electric field since otherwise the charged particle would be accelerated. Since the electric-dipole interaction energy is proportional to the applied electric field, the electric-dipole moments of charged particles should produce no observable interactions. As discussed by Schiff, exceptions to this theorem can occur (*a*) if the particle is significantly acted on by forces other than electric, (*b*) if the particle or nucleus involved has a finite size and structure, or (*c*) if relativistic spin-dependent effects are included. Although these exceptions permit observations of electric-dipole moments of electrically charged particles, the sensitivities of experiments with charged particles are ordinarily much less than those with neutrons. An exception is the electron, for which the experimental dipole limit is almost as low as the limit of the neutron. Most theories, however, predict an electric-dipole moment of the electron at least m/M or 10^{-3} times smaller than that for the neutron. Consequently, the most sensitive tests of electric-dipole moment theories are those with the neutron. For this reason the neutron experiments are discussed first and in greatest detail; measurements on the electron, proton, muon, and hyperons are mentioned only briefly.

NEUTRON MIRRORS

All of the recent measurements that set a limit to the neutron electric-dipole moment utilize neutron mirrors. Therefore, before describing these experiments it is best to review briefly the operation of neutron mirrors.

The passage of slow neutrons through matter can be described in terms of a wave with an index of refraction n. The index of refraction is given (Fermi et al 1950) by

$$n = \left(1 - \frac{\lambda^2 N a_{\text{coh}}}{\pi} \pm \frac{\mu_{\text{M}} B}{\frac{1}{2} M v^2}\right)^{1/2}, \qquad\qquad 2.$$

where λ is the neutron wavelength, N is the number of nuclei per cm^3, a_{coh} is the neutron coherent forward scattering length, μ_{M} is the neutron magnetic moment, B is the magnetic induction, $\frac{1}{2} M v^2$ is the neutron kinetic energy, and the $\pm$ depends on whether the neutron spin and $\mathbf{B}$ are parallel or antiparallel. As in the case of fiber optics, total reflection occurs for a glancing angle θ less than the critical glancing angle θ_c. As for light,

$$\cos \theta_c = n. \qquad\qquad 3.$$

Typically, polarized neutrons at 80 m s^{-1} may be totally reflected from a suitable wall material at a 5° glancing angle and below 6 m s^{-1} they may be totally reflected at all angles of incidence. Since a_{coh} is positive with some surface materials and negative with others, the index of refraction may be less than unity, making possible total external as well as total internal reflection. Copper, quartz, beryllium, and beryllia, for example, give total external reflection of neutrons. It is of interest to note that neutrons at 6 m s^{-1} correspond to an energy of 2×10^{-7} eV, a temperature of 0.002 K, and a wave length of 670 Å. Since the neutron in a surface reflection interacts coherently with many nuclei in the reflecting material, there is little exchange of energy between the neutrons and the atoms of the containing walls so the neutron retains its low effective temperature even when the reflecting surface is at room temperature.

Neutron mirrors are used for several purposes in the various experiments on the neutron electric-dipole moment. The existence of total reflection makes possible low-loss neutron-conducting pipes, which can be hollow if the surface material is chosen to give total external reflection. The use of such neutron-conducting pipes overcomes the usual fall of intensity with distance and thereby compensates in part the loss of intensity by the selection of only extremely slow neutrons.

Since the departure of n from unity in Equation 2 is proportional to λ^2, at very low velocities (less than $6\,\mathrm{m\,s^{-1}}$) and with suitable wall materials, the neutrons are totally reflected at all angles of incidence. As a result, totally reflecting neutron storage bottles may be made with storage times exceeding $100\,\mathrm{s}$. As discussed below, such storage bottles are now used in the most sensitive experiments on the neutron electric-dipole moment.

In Equation 2 the index of refraction depends on the neutron spin orientation through the $\pm$ term in that equation. Consequently, with suitable materials and an appropriate value of **B**, neutrons of one spin orientation may be totally reflected while those of opposite orientation are not. This property is used in current experiments both to polarize the neutrons and to analyze the extent of their polarization. One way this has been done is by transmitting the neutrons through a hollow pipe, one portion of which has magnetized ferromagnetic walls, making the transmission of the pipe much greater for one spin orientation than another. With neutrons of less than $6\,\mathrm{m\,s^{-1}}$ a particularly effective means of polarization is the insertion of a thin magnetized foil in the beam path. One orientation is reflected and the other transmitted.

MEASUREMENTS WITH NEUTRON BEAMS

The earliest measurements of the neutron electric-dipole moment were by neutron beam experiments. The first was that of Smith et al (1957) with later experiments by Miller et al (1967, 1968), Shull & Nathan (1967), Baird et al (1969), Cohen et al (1969), K. Smith and M. Pendlebury (1968, private communication), Apostolescu et al (1970), and Dress et al (1973, 1977, 1978a,b). The greatest sensitivity at each of the stages was attained in the experiments of Dress, Miller, Ramsey, and their associates. Since the latest neutron beam experiment, that of Dress et al (1977, 1978a,b) provides the greatest sensitivity, only this beam experiment is described here.

The apparatus used in this experiment measures with high precision the precessional frequency of the neutron spin in a weak magnetic field by means of a neutron beam magnetic resonance apparatus similar to that used for measuring the magnetic moment of the neutron. A strong electrostatic field is then applied successively parallel and antiparallel to the magnetic field. If the neutron had an electric-dipole moment, the torque of this dipole moment in the electric field would make the precessional frequency of the neutron spin somewhat greater with the electric field in one direction and somewhat less in the opposite. From the point of view of quantum mechanics, the electric field E produces

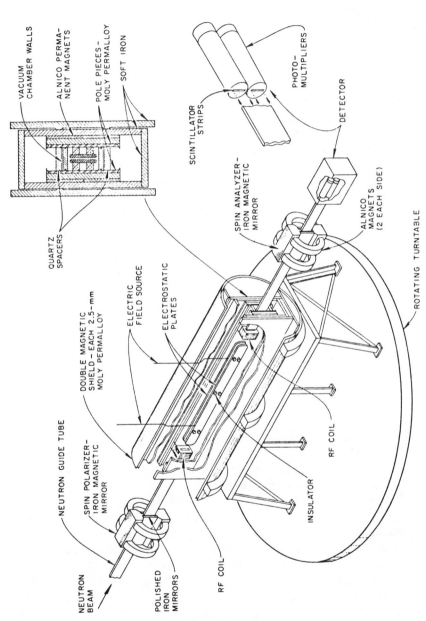

Figure 1 Pictorial representation of the neutron beam spectrometer. The insert (*upper right*) is a cross-sectional view through the midpoint of the apparatus; it indicates various materials used in construction. The gap between the magnetic poles is 9 cm.

an energy separation $2eDE$ between the $m = +\frac{1}{2}$ and $-\frac{1}{2}$ states of the neutron so the frequency shift $\Delta\nu$ when the field is applied is given by

$$\Delta\nu = 2eDE/h. \qquad\qquad\qquad 4.$$

The frequency should shift by double this amount when the field is reversed. By setting an experimental limit on $\Delta\nu$, a limit is thereby set on the electric-dipole moment of the neutron. The main requirements in the experiment are to achieve a very high sensitivity and to eliminate spurious effects that might lead to a false apparent electric-dipole moment or obscure an actual moment.

A schematic view of the apparatus is shown in Figure 1. The neutron beam comes from the cryogenic moderator at the Institute Laue-Langevin (ILL) reactor. The neutrons are conducted from the moderator through a neutron-conducting tube of rectangular cross section on whose surface they are totally reflected at glancing angles of 2° or less.

As shown in Figure 1, the neutron beam goes through a portion of the pipe in which the walls consist of magnetized iron. The neutron beam is polarized by transmission through this portion of the pipe as discussed in the previous section of this report. The analyzing device to determine if there has been a change in the neutron spin orientation is a second magnetized portion of pipe. If the neutron spin remains unaltered between the first and second of these regions with magnetized walls, most of the neutrons will be transmitted by the second region. If, on the other hand, the neutrons have been reoriented by approximately 180° between the two iron mirror sections, the neutrons whose orientations have changed will not be totally reflected in the second magnetic mirror, with a consequent reduction in beam intensity. A weak uniform static magnetic field is provided in the intermediate region so the neutron precesses in that region with its characteristic Larmor precession frequency, $\nu_L = 2\mu_M B/h$. When a suitable oscillatory magnetic field is also applied at this frequency, the neutron undergoes a resonant reorientation that, according to the above analysis, is accompanied by an observable reduction of the detected beam intensity. The oscillatory field is applied in two separate phase-coherent segments since, as shown by Ramsey (1953, 1956, 1980a) such successive oscillatory fields provide a narrower resonance, especially if the magnetic field is not perfectly uniform. If the two separated segments of the oscillatory field have a 90° phase shift between them, the shape of the resonance is that of a dispersion curve with the steepest portion of the slope at the spin precession frequency, as shown in Figure 2. If the frequency of the oscillator is set so that the detected neutron intensity is at the position of the steepest slope, the presence of a neutron electric-dipole moment can

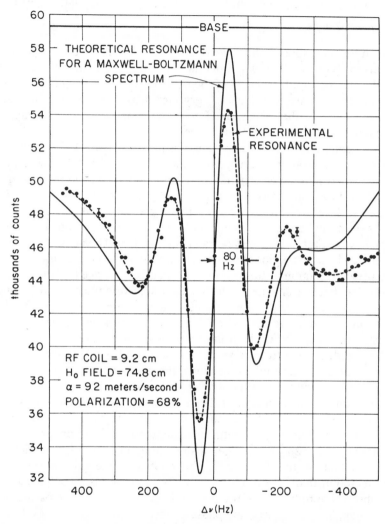

Figure 2 Typical magnetic resonance with the neutron beam apparatus for a phase shift of $\pi/2$ between the two oscillatory fields. The calculated transition probability for a Maxwell-Boltzmann distribution characterized by a temperature of 1 K is shown by the full curve. The departure of the experimental curve (*dashed*) from theory when far from resonance is to be expected from the known departure of the beam velocity from a Maxwell-Boltzmann distribution. Length of each of the two RF coils = 9.2 cm, length of uniform magnetic field = 74.8 cm, most probable neutron velocity $\alpha = 92 \, \text{m s}^{-1}$, polarization = 68%.

be detected by successively reversing a strong electrostatic field. If there is an electric-dipole moment, the torque due to the electric field will increase the precessional frequency of the neutron for one orientation of the field and decrease it for the opposite. At a fixed frequency of the oscillator, this change, $4eDE/h$, in the precessional frequency of the neutron spin will then be detectable with high sensitivity as a change in the neutron beam intensity.

The electric field is applied over a length of 196 cm and typically has a value of about $100 \, kV \, cm^{-1}$. The static magnetic field was about 17 G and the neutron beam was 89% polarized.

Great care in the experiment must be taken to avoid spurious effects that could simulate a nonexistent electric-dipole moment or mask an existing one. Fortunately, a number of things can be done to eliminate or minimize such effects. Since these are discussed in detail in the original article (Dress et al 1977) and several review articles (Ramsey 1981a, 1982), they are listed but not discussed in detail here. They include successive reversals in the relative phase of the two oscillatory fields, reversals of the leads to the source of the electrostatic potential, measurements with successive reversals of the reversing switches when no electric field is present, symmetrical elimination of data when a spark occurs to avoid effects from changed magnetic fields caused by the spark currents, and care to make $\mathbf{E}$ and $\mathbf{B}$ accurately parallel to reduce the effect of the apparent magnetic field $\mathbf{E} \times \mathbf{v}/c$ when the neutron moves with velocity $\mathbf{v}$ through the electric field $\mathbf{E}$. (If $\mathbf{E}$ and $\mathbf{B}$ are accurately parallel, the $\mathbf{E} \times \mathbf{v}/c$ field will be perpendicular to $\mathbf{B}$ and will not lead to a change in the magnetic precession frequency when $\mathbf{E}$ is reversed but kept at the same magnitude.) As further checks on a possible $\mathbf{E} \times \mathbf{v}/c$ effect the measurements were made at different neutron velocities and with the apparatus at intervals rotated 180° to reverse the neutron velocity relative to the apparatus and hence to reverse the sign of the $\mathbf{E} \times \mathbf{v}/c$ effect.

The results of the neutron beam experiment (Dress et al 1977, 1978a,b) are

$$D = (+0.4 \pm 1.5) \times 10^{-24} \, cm \qquad\qquad 5.$$

or

$$|D| < 3 \times 10^{-24} \, cm. \qquad\qquad 6.$$

In other words, the neutron electric-dipole moment, if it exists at all, is less in magnitude than 3×10^{-24} cm. To emphasize the smallness of this result, it should be noted that for this value the corresponding bulge of one unit of positive charge in one hemisphere of the neutron would correspond to only 0.01 cm if the neutron were expanded to the size of the Earth.

MEASUREMENTS WITH BOTTLED NEUTRONS

The neutron electric-dipole moment experiments, as so far described, depend upon the fact that neutrons at a velocity of, say, $80 \, \mathrm{m \, s^{-1}}$ will be totally reflected by many materials at glancing angles of 5°. As discussed in the section on neutron mirrors, when the velocity of the neutrons diminish, the glancing angle for total reflection increases until finally at a velocity less than $6 \, \mathrm{m \, s^{-1}}$ total reflection can be obtained even at normal incidence on many surfaces and it is possible to store neutrons in an enclosed bottle. For many years it was apparent (Ramsey 1957, 1969, Kleppner et al 1958, 1960, Shapiro 1968a) that electric-dipole moment experiments with bottled neutrons would be particularly sensitive, but for a long time suitable sources for adequate quantities of ultra-cold neutrons were not available.

Zel'dovich (1959), Vladimirskii (1961), Doroshkevich (1963), Luschikov et al (1969), Shapiro (1968b), and Luschikov (1977) discussed bottled ultra-cold neutrons and Shapiro and his associates stored neutrons in bottles for up to 20 s. Techniques of ultra-cold neutrons and storage bottles were improved by Golub & Pendlebury (1972), A. Steyerl and W. B. Trustedt (1973, private communication), Taran (1973), Lobashov et al (1973), Miller (1974), Harvard-Sussex-Rutherford-ILL (1974, 1975), Golub et al (1979), and Ageron et al (1980). Specific experiments to use neutron bottles to measure the electric-dipole moment were discussed by Ramsey (1957, 1969), Shapiro (1968a), Golub & Pendlebury (1972), Taran (1973), Lobashov et al (1973), Miller (1974), Harvard-Sussex-Rutherford-ILL (1974, 1975), Altarev et al (1978a,b, 1980, 1981a,b), Dombeck et al (1979), and by others.

The first experiment with bottled neutrons to provide a useful limit to the neutron electric-dipole moment was that of Altarev and his associates in Leningrad (Altarev et al 1978a,b, 1980, 1981a,b). They used the apparatus shown in Figure 3 with the neutrons stored in a double-neutron bottle. The electric fields in the two halves of the double bottle are in opposite directions to provide a first-order cancellation of magnetic-field fluctuations. The authors used a new form of the separated oscillatory field method called the adiabatic method (Ezhov et al 1976) of separated oscillatory fields. In this method the static magnetic fields in the oscillating field regions are inhomogeneous in such

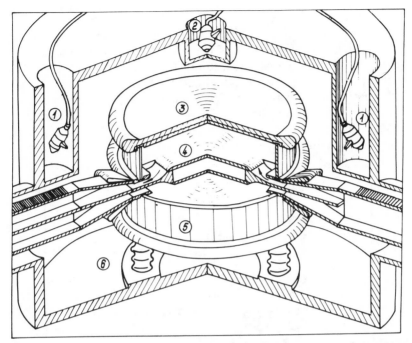

Figure 3 General view of the Leningrad double-bottle neutron magnetic resonance spectrometer: 1 indicates a magnetometer for control of magnetic field, 2 a magnetometer for field stabilization, 3 the top electrode, 4 the central electrode, 5 the chamber wall made of beryllium-oxide-coated fused quartz, and 6 the bottom electrode. The initial and final oscillatory field coils are shown on the neutron pipes at the left- and right-hand sides of the figure. The spectrometer is surrounded by three layers of magnetic shielding.

a way that the neutrons adiabatically reach the resonance condition in the first field and depart from it adiabatically in the second region.

Typical resonances obtained in the Leningrad experiment are shown in Figure 4. The results of this experiment (Altarev et al 1978a,b, 1980, 1981a,b) are

$$D = (2.3 \pm 2.3) \times 10^{-25} \, \text{cm} \qquad\qquad 7.$$

or

$$|D| < 6 \times 10^{-25} \, \text{cm}. \qquad\qquad 8.$$

At the ILL, a Harvard-Sussex-Rutherford-ILL collaboration (1974, 1975) is also using bottled ultra-cold neutrons to measure the neutron electric-dipole moment. The method depends on using ultra-cold neutrons of approximately $6 \, \text{m s}^{-1}$. These neutrons are led by a neutron-conducting pipe into the apparatus shown in Figure 5. The neutrons are stored in a cylinder approximately 25 cm in diameter and 10 cm high with the top plates being metallic—beryllium or copper—and the sides of the cylinder being insulators of beryllia or quartz. After the neutron bottle is filled, a shutter is closed, storing the neutrons for 30–100 s. The oscillatory field is applied as initial and final coherent pulses, usually with a $\pi/2$ relative phase shift. The resonance is observed in a fashion similar to the neutron beam experiment: the neutrons are polarized on passage through the indicated polarizing foil, analyzed during their return passage through the foil, and counted at the indicated ultra-cold

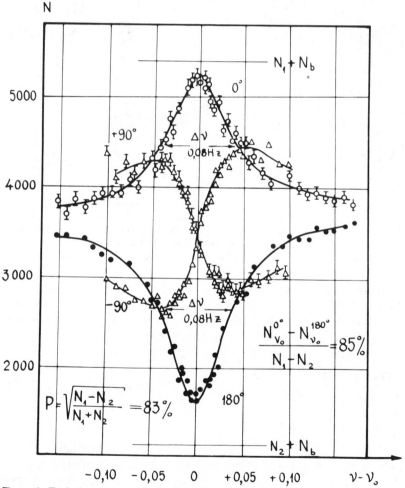

Figure 4 Typical resonance with Leningrad spectrometer.

$(N_{\nu_0}^{0°} - N_{\nu_0}^{180°})\,(N_1 - N_2)^{-1} = 85\%,\ P = [(N_1 - N_2)/(N_1 + N_2)]^{1/2} = 83\%.$

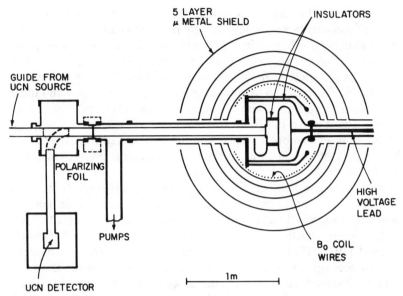

Figure 5 Schematic diagram of apparatus for measuring the neutron electric-dipole moment with bottled neutrons at the ILL.

neutron (UCN) detector. The observations are at the steepest point of the resonance curve. The change in beam intensity correlated with the application of an electric field is then examined to set a limit to the neutron electric-dipole moment. A typical resonance curve obtained with the new ILL apparatus is shown in Figure 6.

The use of stored ultra-cold neutrons offers two particularly important advantages. The resonance curve with stored neutrons in Figure 6 is 3500 times narrower than that of Figure 2. Furthermore, as mentioned above, a large fraction of running time in the beam experiment must be devoted to eliminating the $E \times v/c$ effect. Since it is the average value of v that is important, this effect is drastically diminished when the neutrons enter and leave by the same exit hole with an 80-s storage time instead of passing through the apparatus once at a velocity of $90 \, \mathrm{m \, s^{-1}}$. As a result of the reduced effective magnetic field from $E \times v/c$, it is also possible to use a much weaker static magnetic field with an accompanying reduction in the stability requirement for the current that provides the static magnetic field.

Although the new experiments with bottled neutrons will have the above marked advantages, it must be recognized that they will still have many difficulties. The limit has by now been pushed to such a low value that care must be taken to avoid all possible systematic effects. Although some of these are intrinsically reduced in an experiment with bottled neutrons, other serious problems remain. For example, problems due to stray magnetic field (especially when associated with reversals of the electric field) and to magnetic-field changes resulting from electrical sparks can be just as serious in absolute terms and more serious in relative terms with bottled neutrons. These problems have already caused much difficulty in the beam version of the experiment and should be even more formidable in the bottled-neutron experiment, which seeks to lower the limit for the neutron electric-dipole moment by a factor of 100–1000.

With an electric field of $30 \, \mathrm{kV \, cm^{-1}}$ and a multilayer Mumetal or Moly-Permalloy magnetic shield, it should be possible initially to achieve a sensitivity limit on the electric-dipole moment of $10^{-25} \, \mathrm{cm}$. Two members of the collaboration (R. Golub & J. M. Pendlebury) proposed (1975, 1977, 1979) the use of cold liquid $^4\mathrm{He}$ in a neutron bottle to accumulate ultra-cold neutrons in one bottle while the resonance is observed in another. This procedure should provide much greater neutron densities in the bottle where the resonances are studied. When these increased densities are available, a sensitivity limit of $10^{-26} \, \mathrm{cm}$ or lower should be obtained for the neutron electric-dipole moment. However, as the sensitivity limit is pushed to successively lower values the problem of magnetic field fluctuations, especially any that are coupled jointly to the electric and magnetic field reversals, will probably determine the limit to D that can be reliably established. How low this limit can be will be determined by experience during the next few years. A proposal by Ramsey (1980b) to use optically pumped $^3\mathrm{He}$

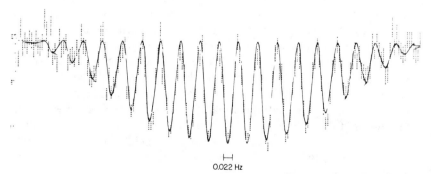

$\vdash\!\dashv$
0.022 Hz

Figure 6 Neutron magnetic resonance obtained experimentally with bottled neutrons. Two phase-coherent but successive oscillatory fields were applied and the change in the neutron intensity was observed as the oscillatory frequency was varied. The observed resonance width of 0.022 Hz is to be contrasted to the 80 Hz of Figure 2.

or a similar gas as a magnetometer should contribute markedly to lowering the limit on D. ^{3}He will be optically pumped to align the nuclear spins and then introduced into the neutron bottle at low pressure along with the neutrons. Then oscillatory fields will be applied simultaneously at the neutron and ^{3}He resonance frequencies. The ^{3}He will be subsequently pumped to a high magnetic field region where its polarization will be measured in an NMR spectrometer. In this fashion the ^{3}He will measure the magnetic field in the same region of space and at the same time as the neutron measurements, except for a small correction for the effect of gravity on the ultra-cold neutrons.

ELECTRIC-DIPOLE·MOMENTS OF OTHER PARTICLES

Electric-dipole moment measurements have been made for particles other than the neutron, including electrons, muons, protons, and hyperons. However, as discussed in the introduction, these measurements so far provide less sensitive tests of theories than do the experiments with neutrons. Furthermore, many of these experiments were described in an excellent review article by Sandars (1975). For these reasons, experiments on particles other than the neutron are discussed only briefly here and references are given to the original articles in which detailed descriptions can be found.

Electron

The earliest indications of a limit to the electric-dipole moment for the electron were based on an analysis by Salpeter (1958) and Feinberg (1969) of existing data from the Lamb-Retherford experiment. These set a limit of 2×10^{-13} cm for $|D|$. This limit was lowered to 4×10^{-16} cm by the analysis of Crane's experiment (Garwin & Lederman 1959, Nelson et al 1959) on the magnetic moment or $g - 2$ value of the electron.

Sandars & Lipworth (1964) then did an atomic beam resonance experiment with cesium that was similar in principle to the above neutron beam experiment and thereby set a limit to the electric-dipole moment of the Cs atom. They then showed (Sandars & Lipworth 1964, Sandars 1968) that, because of relativity and the admixture of a nearby p state into the s state, their observed limit for the atomic electric-dipole moment corresponded to a much lower limit for the electric-dipole moment of the electron. In this way, they set a lower limit of 2×10^{-21} cm for the electric-dipole moment of the electron. Subsequently, the sensitivity of the Cs experiments was improved by Weisskopf et al (1967, 1968a,b), leading to a limit of 3×10^{-24} cm. Finally, with a metastable xenon beam, Player & Sandars (1970) set a limit to $|D|$ for the electron of 2×10^{-24} cm. The most recent electron measurement is much less sensitive but is of interest because it was made with a totally different technique. Vasil'ev & Kolicheva (1978) obtained the limit of 3×10^{-22} cm by seeking to detect a small change in magnetic flux caused by the flip of the electric-dipole moment when an electric field was supplied to the sample.

Muon

Several experiments have set limits on the electric-dipole moment for both positive and negative muons. These experiments utilized the muon storage ring at CERN, which has primarily been used to measure the magnetic moment or $g - 2$ for the muon. One of these determinations, by Bailey et al (1977), is based on assigning the entire difference between the experimental value of $g - 2$ and the quantum electrodynamic theoretical value to an electric-dipole moment for the muon. This led to a limit for D_μ of 0.74×10^{-18} cm. The same group with a modified procedure also set a direct experimental upper limit (Farley et al 1978) to D_μ of 1.05×10^{-18} cm for both the positive and negative muon.

Proton

The first indication of a limit to the electric-dipole moment of the proton was the value 1×10^{-13} cm set by Purcell & Ramsey (1950) and Sternheimer (1959) from an analysis of the spacing of the known atomic hydrogen energy levels. This limit was lowered to 3×10^{-15} cm by a study of the free precession of the proton, as described by Sandars (1975). Khriplovich (1976), from the limit set on the electric-dipole moment of the Cs atom, concluded that $|D|$ for the proton must be less than 5.5×10^{-19} cm.

Harrison et al (1969) devised an ingenious, but somewhat complicated, molecular beam resonance method for measuring D for the proton. Sandars (1967) and Hinds & Sandars (1980) increased the sensitivity by lengthening the apparatus to 20 m. They chose to study the molecule ^{205}TlF because it is extremely sensitive to a nuclear electric-dipole moment; the large size and structure of the Tl nucleus permits an exception to the general theorem that a charged point particle in equilibrium under electrostatic forces has no contribution to its energy that is linear in its electric-dipole moment. The experiment also depends on the nonlinear variation of the molecular electric field over the nucleus.

A schematic diagram of the apparatus used in this experiment is shown in Figure 7. The proton studied is the odd $S_{1/2}$ nucleon in the Tl nucleus which is bound in a TlF molecule. The primary transition used is the nuclear resonance $m_{\text{Tl}} = -\frac{1}{2} \rightarrow \frac{1}{2}$ for the $J = 1$, $m_J = -1$ rotational state. Since m_J does not change, there would ordinarily be no difference in electrostatic deflection properties with and without a transition and the resonance would not be observed in a normal molecular beam deflection and refocusing experiment. However, the resonance can be observed if four separated oscillatory fields are used, with the first inducing a transition from an $m_J = 0$, $m_{\text{Tl}} = -\frac{1}{2}$ state to the $m_J = -1$, $m_{\text{Tl}} = -\frac{1}{2}$ state whereas the last oscillatory field induces the same transition in the opposite direction. The middle two oscillatory fields (Ramsey 1980a) then serve as coherent separated oscillatory fields to induce the primary transition, which becomes observable since the molecule after such a transition is left in the $m_J = -1$ state by the final oscillatory field and will therefore not be refocused in the final inhomogeneous electric field.

As in other electric-dipole moment experiments, the direction of the electric field is successively reversed and the corresponding resonance frequency shift is observed. When both groups of experiments are combined the value of D is $(-1.4 \pm 6) \times 10^{-21}$ cm or, with the upper limit taken as twice the experimental error, the limit on $|D|$ for the proton is 12×10^{-21} cm. A more sensitive version of this experiment by

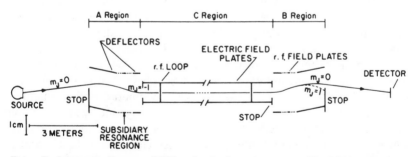

Figure 7 Schematic diagram of TlF molecular beam apparatus. The TlF in state $(1, 0, -\frac{1}{2}, -\frac{1}{2})$ is focused to an approximately parallel beam in the first inhomogeneous magnets. While the molecules are still in A region, the transition $(1, 0, -\frac{1}{2}, -\frac{1}{2}) \rightarrow (1, -1, +\frac{1}{2}, -\frac{1}{2})$ is induced by an oscillatory electric field. The opposite transition is induced in the B region after which the molecules in state $(1, 0, -\frac{1}{2}, -\frac{1}{2})$ are then focused in the final quadrupole. If the transition $(1, -1, +\frac{1}{2}, -\frac{1}{2}) \rightarrow (1, -1, -\frac{1}{2}, -\frac{1}{2})$ is induced in the C region, the final radiofrequency oscillatory will not induce a transition to the states $(1, 0, \pm\frac{1}{2}, -\frac{1}{2})$ so the molecules will not be focused.

Wilkening, Larson & Ramsey (Wilkening 1981, Ramsey 1981) has lowered this limit by a factor of 3, giving $(1.3 \pm 2.0) \times 10^{-21}$ cm for D or a limit of 4×10^{-21} cm.

Hyperon

The neutral hyperon Λ^0 is ordinarily produced with partial polarization and its polarization can be measured from the asymmetries in the decay $\Lambda^0 \to p\pi^-$. Even though no external electric field is supplied, there is in the rest frame of the hyperon an electric field proportional to $(\mathbf{v}/c) \times \mathbf{B}$. If there is a Λ^0 electric-dipole moment, it will interact with the above electric field. This interaction then determines an experimental limit for the Λ^0 electric-dipole moment as was first done by Gibson & Green (1966). From such an analysis Baroni et al (1971) later obtained a limit for D of 8×10^{-15} cm. Subsequently, this experiment was repeated using a new apparatus by Pondrom et al (1981), who found the upper limit on D to be 1.5×10^{-16} cm.

THEORIES OF ELECTRIC-DIPOLE MOMENTS

Almost all theories before 1964 predicted zero for the electric-dipole moment of all particles since theories were usually required to be symmetric under P prior to 1957 and under T prior to 1964. As discussed in the introduction, a theory that is symmetric under either P or T must necessarily lead to a zero value for the electric-dipole moment of any particle. Subsequent to the experiments performed in 1957 (Lee & Yang 1957, Wu et al 1957), showing the existence of P symmetry violations in the weak interactions, and in 1964 (Christenson et al 1964), showing violations of CP symmetry in the decay of K_L^0, almost all theories have predicted nonzero values for particle electric-dipole moments. Although the experiments of Christenson et al (1964) imply a violation of CP symmetry and the experiment of Schubert et al (1970, see also Casella 1968, 1969) implies a violation of T symmetry, both are limited to the neutral kaon system. Failures of CP or T symmetry are yet to be found in any other system. As a result, it is not yet established as to which of the fundamental interactions is associated with the CP and T violations. It is even possible that these violations are not in any of the established interactions but are instead in a new superweak interaction. For this reason it is not surprising that the different theories predict widely different values for electric-dipole moments, depending on the model assumed.

Kobayashi & Maskawa (1973) showed that a quantum chromodynamics gauge theory based on only four quarks and a single Higgs scalar

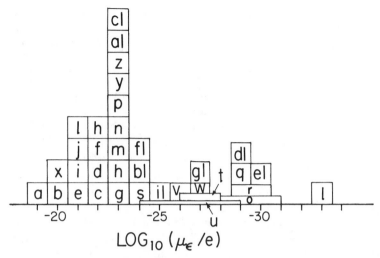

$$LOG_{10} \left(\mu_\epsilon / e \right)$$

Figure 8 Theoretical predictions of the neutron electric-dipole moment. Each lettered block corresponds to a different theory with the references to the different theories given by the corresponding letters listed below. The bases of the different theories are indicated in brackets after the references with the following interpretation: EM indicates a theory that attributes the *CP* and *T* violations to electromagnetic interactions, W attributes them to the weak forces, MW to a milliweak force, SW to a new superweak force, S to the strong force, and MS to millistrong forces. G indicates a gauge theory, H indicates a gauge theory in which the *CP* violation is associated with multiple Higgs bosons, and Q indicates that the *CP* violation is attributed to more than four quarks. The ΔS value indicates the change in strangeness considered in each theory.

a. Feinberg (1965) (EM)
b. Salzman & Salzman (1965) (EM)
c. Barton & White (1969) (EM)
d. Broadhurst (1970) (EM)
e. Babu & Suzuki (1967) (MW,$\Delta S = 0$)
f. Meister & Rhada (1964) (MW, ΔS=0)
g. Gourishankar (1968) (MW, ΔS=1)
h. McNamee & Pati (1969) (MW, ΔS=0, 1)
i. Nishijima & Swank (1967) (MW, ΔS=0)
j. Nishijima (1969) (MW, ΔS=0)
k. Boulware (1965) (MW, $\Delta S = 0$)
l. Wolfenstein (1964a,b) (SW, $\Delta S = 2$)
m. Pais & Primack (1973a,b) (MW)
n. Lee (1973, 1974) (MW)
o. Okun (1969) (SW)
p. Mohapatra (1972) (MW)
q. Frenkel & Ebel (1974a) (MW)
r. Wolfenstein (1974) (SW)
s. Weinberg (1976) (MW)
t. Pakvasa & Tuan (1975) (MW)

u. Mohapatra & Pati (1975) (MW)
v. Clark & Randa (1975) (MS)
w. Chodos & Lane (1972) (MW)
x. Feinberg & Mani (1965) (W, $\Delta S = 1$)
y. Gourishankar (1968) (MW, $\Delta S = 1$)
z. Filipov et al (1968) (EM)
a1. McNamee & Pati (1969) (MW, $\Delta S = 0, 1$)
b1. Barton & White (1969) (EM, MW, $\Delta S = 0, 1$)
c1. McCliment & Teeters (1970) (MW)
d1. Frenkel & Ebel (1974a,b)
e1. Nanopoulos & Yildiz (1979) (Q)
f1. Eichten et al (1980) (MW, H)
g1. Ellis et al (1980, 1981) (this paper has the interesting characteristic that it establishes an order-of-magnitude lower limit to D of 3×10^{-28} cm)
h1. Crewther et al (1979)
i1. Shizuya & Tye (1980) (MW, H)
j1. Epstein (1980)

multiplet cannot account for even the known CP and T violations in the K_L^0. Therefore, most current theories attribute the T violation either to a six (or more) quark theory or to the exchange of multiple Higgs bosons.

Most theories predict smaller electric-dipole moments for muons and electrons than for neutrons or protons and the latter are predicted to be comparable to neutrons. For example, a model studied by Lee (1973) in which CP violation is associated with the Higgs bosons predicts D for the neutron, muon, and electron to be 10^{-23}, 10^{-25}, and 10^{-32} cm, respectively. On the other hand, Pais & Primack (1973a,b) in a theory that incorporates CP violation in the leptonic weak current predict D for the neutron, muon, and electron to be 10^{-24}, 10^{-20}, and 10^{-24} cm, respectively. Even in this case, however, the most critical comparison between theory and experiment is with the neutron. For this reason, in the following presentation of theoretical predictions of electric-dipole moments, numerical values are given only for the neutron.

The predictions of various theories are presented graphically in Figure 8. Each theoretical prediction is represented by a block of equal area so the height of the block is correspondingly diminished if the prediction spans several decades. From Figure 8 it is apparent that it is highly desirable to lower the experimental limit on the neutron electric-dipole moment to distinguish between the different theories. In particular, of the two categories of theories most favored at present, the ones that attribute the CP violation to the exchange of Higgs bosons predict values of $D \sim 10^{-24}$ cm, whereas those that attribute it to additional quarks give $D \sim 10^{-28}$ cm. The experiments now in progress should soon distinguish between these two categories of theories.

Literature Cited

Ageron, P., Astrus, J. M., Verdier, J. 1980. *Preliminary Project of an Ultra-Cold Neutron Source at the High Flux Reaction.* Grenoble: Inst. Laue–Langevin

Altarev, I. S., et al. 1978a. *Fundamental Physics of Reactor Neutrons and Neutrinos*, ed. T. von Egidy, p. 79. Bristol: Inst. Phys.

Altarev, I. S., et al. 1978b. *Leningrad Nuclear Physics Institute, Preprint No. 430*

Altarev, I. S., et al. 1980. *Nucl. Phys. A* 341:269

Altarev, I. S., et al. 1981a. *Leningrad Nuclear Physics Institute, Preprint No. 636*

Altarev, I. S., et al. 1981b. *Phys. Lett.* 102B:13

Apostolescu, S., Ionescu, D. R., Ionescu-Bujur, M., Meiterts, S., Petroscu, M. 1970. *Rev. Roumaine Phys.* 15:343

Babu, P., Suzuki, M. 1967. *Phys. Rev.* 162:1359

Bailey, J., et al. 1977. *Phys. Lett.* 68B:191

Baird, J. K., Miller, P.D., Dress, W. B., Ramsey, N. F. 1969. *Phys. Rev.* 179:1285

Baroni, G., et al. 1971. *Lett. Nuovo Cim.* 2:1256

Barton, G., White, E. D. 1969. *Phys. Rev* 184:1660

Boulware, D. G. 1965. *Nuovo Cimento* 40:1041

Broadhurst, D. J. 1970. *Nucl. Phys. B* 20:603

Casella, R. C. 1968. *Phys. Rev. Lett.* 21:1128

Casella, R. C. 1969. *Phys. Rev. Lett.* 22:554

Chodos, A., Lane, K. 1972. *Phys. Rev. D* 6:596

Christenson, J. H., Cronin, J.W., Fitch, V. L., Turlay, R. 1964. *Phys. Rev. Lett.* 13:138

Clark, R. B., Randa, J. 1975. *Phys. Rev. D* 12:3564

Cohen, V. W., Lipworth, E., Nathan, R., Ramsey, N. F., Silsbee, H. B. 1969. *Phys. Rev.* 177:1942

Crewther, R. J., Di Vecchia, P., Veneziano, G., Witten, E. 1979. *Phys. Lett.* 88B:123

Dombeck, T. W., Lynn, J. W., Werner, S. A., Brun, T., Carpenter, J., Krohn, V., Ringo, R. 1979. *Rep. PP79–153, TR79–085.* Univ. Md: Dept. Phys.

Doroshkevich, A. 1963. *Sov. Phys. JETP* 16:56; 1962. *J. Exp. Theor. Phys. USSR* 43:79

Dress, W. B., Miller, P. D., Pendlebury, J. M., Perrin, P., Ramsey, N.F. 1977. *Phys. Rev. D* 15:9

Dress, W. B., Miller, P.D., Pendlebury, J. M., Perrin, P., Ramsey, N. F. 1978a. See Altarev et al 1978a, p. 11

Dress, W. B., Miller, P. D., Pendlebury, J. M., Perrin, P., Ramsey, N. F. 1978b. *Phys. Rep.* 43:410

Dress, W. B., Miller, P. D., Ramsey, N. F. 1973. *Phys. Rev. D* 7:3147

Eichten, E., Lane, K., Preskill, J. 1980. *Phys. Rev. Lett.* 45:225

Ellis, J., Gaillard, M. K., Nanopoulos, D. V., Rudaz, S. 1980. *Phys. Lett.* 99B:101

Ellis, J., Gaillard, M. K., Nanopoulos, D. V., Rudaz, S. 1981. *Nature* 293:41

Epstein, G. N. 1980. *Phys. Rev. Lett.* 44:905

Ezhov, V. F., Ivanov, S. N., Lobashov, V. M., Nazarenko, V. A., Porsev, G. D., Serdyuk, O. V., Serebrov, A. P., Taldaev, R. R. 1976. *Zh. Eskp. Teor. Fiz. Pisma Red.* 24:39

Farley, F. J. M., Field, J. H., Flegel, W., Hattersley, P. M., Krienen, F., Langen, F., Picasso, E., von Ruden, W. 1978. *J. Phys. G* 4:345

Feinberg, G. 1965. *Phys. Rev.* 140:B1402

Feinberg, G. 1969. *Atomic Physics*, Vol. 1. New York: Plenum

Feinberg, G, Mani, H. S. 1965, *Phys. Rev.* 137:B636

Fermi, E., Orear, J., Rosenfeld, A. H., Schluter, R. 1950. *Nuclear Physics*, pp. 201. Univ. Chicago Press

Filipov, A. T., Oziewcz, Z., Pikulski, A.

1968. *Joint Inst. Nucl. Res. Dubna Rep., USSR*

Frenkel, J., Ebel, M. E. 1974a. *Univ. Wis. Rep.* Madison, Wis.

Frenkel, J., Ebel, M. E. 1974b. *Nucl. Phys. B* 83:177

Garwin, R., Lederman, L. 1959. *Nuovo Cimento* 11:776

Gibson, W. K., and Green, K. 1966. *Nuovo Cimento A* 45:882

Golub, R., Mampe, W., Pendlebury, J. M., Ageron, P. 1979. *Sci. Am.* 240: June, p. 106

Golub, R., Pendlebury, J. M. 1972. *Contemp. Phys.* 13:519

Golub, R., Pendlebury, J.M. 1975. *Phys. Lett.* 53A:133

Golub, R., Pendlebury, J.M. 1977. *Phys. Lett.* 62A:337

Golub, R., Pendlebury, J. M. 1979. *Rep. Prog. Phys.* 42:439

Gourishankar, G. R. 1968. *Can. J. Phys.* 46:1843

Harrison, G. S., Sandars, P. G. H., Wright, S. J. 1969. *Phys. Rev. Lett.* 22:1263

Harvard-Sussex-Rutherford-ILL. 1974. This collaboration includes C. Baker, J. Byrne, R. Golub, K. Green, B. Heckel, A. Kilvington, W. Mampe, J. Morse, J. M. Pendlebury, N. Ramsey, K. Smith, and T. Sumner. The experiment is described in *A Proposal to Search for the Electric Dipole Moment of the Neutron Using Bottled Neutrons.* Grenoble, France: Institute Laue-Langevin (1974) and Oxford: Rutherford Lab. (1975)

Harvard-Sussex-Rutherfold-ILL 1975 *Rutherford Lab. Rep.*

Hinds, E. A., Sandars, P. G. H. 1980. *Phys. Rev. A* 21:471, 480

Jackson, J. D., Treiman, S. B., Wyld, H.W. Jr. 1957. *Phys. Rev.* 106:517

Khriplovich, I. B., 1976. *Zh. Eksp. Teor. Fiz.* 71:51 (transl: *Sov. Phys. JETP* 44:25)

Kleppner, D., Ramsey, N. F., Fijelstadt, P., Goldenberg, M. 1958. *Phys. Rev. Lett.* 1:232

Kleppner, D., Ramsey, N. F., Fijelstadt, P., Goldenberg, M. 1960. *Phys. Rev. Lett.* 8:361

Kobayashi, M., Maskawa, K. 1973. *Prog. Theor. Phys.* 49:652

Landau, L. 1957a. *Nucl. Phys.* 3:127

Landau, L. 1957b *Zh. Eksp. Teor. Fiz.* 32:405 (transl: *Sov. Phys. JETP* 5:336)

Lee, T. D. 1973. *Phys. Rev. D* 8:1226

Lee, T. D. 1974. *Phys. Rep.* 9:143

Lee, T. D., Yang, C. N. 1957. *Phys. Rev.* 105:1671

Lobashov, V. M., Porsey, G. D., Sereb-rov, A. P. 1973. *Konstantinov Institute of Nuclear Physics of the Academy of Sciences of USSR, Preprint No. 37*; 1974. *Yad. Fiz. USSR* 19:300

Luschikov, V. I. 1977. *Phys. Today* 30 (6):42

Luschikov, V. I., Pokotilosky, Yu. N., Strelkov, A. V., Shapiro, F. L. 1969. *Zh. Eksp. Teor. Fiz. Pisma Red.* 9:40 (transl: *Sov. Phys. JETP Lett.* 9:23)

McCliment, E. R., Teeters, W. D. 1970. *Nuovo Cimento A* 68:657

McNamee, P. J., Pati, J. C. 1969. *Phys. Rev.* 178:2273

Meister, N. T., Rhada, T. K. 1964. *Phys. Rev.* 135:B769

Miller, P. D. 1974. *Proc. 2nd Int. Sch. Neutron Phys, Alushta, Crimea, USSR*. Unpublished

Miller, P. D., Dress, W. B., Baird, J. K., Ramsey, N. F. 1967. *Phys. Rev. Lett.* 19:381

Miller, P. D., Dress, W. B., Baird, J. K., Ramsey, N. F. 1968. *Phys. Rev.* 170:1200

Mohapatra, R. N. 1972. *Phys. Rev. D* 6:2026

Mohapatra, R. N., Pati, J. C. 1975. *Phys. Rev. D* 11:659

Nanopoulos, D. V., Yildiz, A. 1979. *Phys. Lett.* 87B:53; 1980. *Ann. Phys.* 127:126

Nelson, D. F., Schapp, A. A., Pidd, R. W., Crane, H. W. 1959. *Phys. Rev. Lett.* 2:492

Nishijima, K. 1969. *Prog. Theor. Phys.* 41:739

Nishijima, K., Swank, L. J. 1967. *Nucl. Phys. B* 3:565

Okun, L. B., 1969. *Comments Nucl. Part Phys.* 3:135

Pais, A., Primack, J. 1973a. *Phys. Rev. D* 8:625

Pais, A., Primack, J. 1973b. *Nucl. Phys. B* 77:3036

Pakvasa, S., Tuan, S. F. 1975. *Phys. Rev. Lett.* 34:553

Player, M. A., Sandars, P. G. H. 1970. *J. Phys. B* 3:1620

Pondrom, L., Handler, R., Sheaff, M., Cox, P. T., Dworkin, J., Overseth, O. E., Devlin T., Schachinger, L., Heller, K. 1981. *Phys. Rev. D* 23:814

Purcell, E. M., Ramsey, N. F. 1950. *Phys. Rev.* 78:807

Ramsey, N. F. 1953. *Nuclear Moments*, p. 23. New York: Wiley

Ramsey, N. F. 1956. *Molecular Beams*. Oxford Univ. Press

Ramsey, N. F. 1957. *Rev. Sci. Instrum.* 28:57

Ramsey, N. F. 1958. *Phys. Rev.* 109:222

Ramsey, N. F. 1969. *Proposal to ILL for Neutron Electric Dipole Moment Experiment*. Grenoble, France: Inst. Laue-Langevin. Unpublished

Ramsey, N. F. 1980a. *Phys. Today* July, p. 25

Ramsey, N. F. 1980b. *Bull. Am. Phys. Soc.* 25:9

Ramsey, N. F. 1981. *Atomic Physics VII*, ed. D. Kleppner, F. Pipkin, p. 65. New York: Plenum

Ramsey, N. F. 1981a. *Comments Nucl. Part. Phys.* 10:227-42

Ramsey, N. F. 1982. *Rep. Prog. Phys.* 45: In press

Salpeter, E. E. 1958. *Phys. Rev.* 112:1642 112:1642

Salzman, F., Salzman, G. 1965. *Phys. Rev. Lett.* 15:91

Sandars, P. G. H. 1967. *Phys. Rev. Lett.* 19;1396

Sandars, P. G. H. 1968. *J. Phys. B* 1:499, 511

Sandars, P. G. H. 1975. *Atomic Physics IV*, p. 71. New York: Plenum

Sandars, P. G. H., Lipworth, E. 1964. *Phys. Rev. Lett.* 13:529, 14:718

Schiff, L. I. 1963. *Phys. Rev.* 132:2194

Schubert, K. R., Wolff, B., Chollet, J. C., Gaillard, J. M., Jane, M. R., Ratcliffe, T. J., Repellin, J. P. 1970. *Phys. Lett.* 31B:662

Shapiro, F. L. 1968a. *Usp. Fiz. Nauk* 95:145 (transl *Sov. Phys. Usp.* 11:345)

Shapiro, F. L. 1968b. *Proc. 3rd Winter Sch. Leningrad Physico-Tech. Inst. Acad. Sci. USSR* p. 35

Shizuya, K., Tye, S. H. H. 1980. *Cornell Lab. Nucl. Sci. Rep. CLNS 80/458*

Shull, C., Nathan, R. 1967. *Phys. Rev.* 19:384

Smith, J. H., Purcell, E. M., Ramsey, N. F. 1957. *Phys. Rev.* 108:120

Sternheimer, R. 1959. *Phys. Rev.* 113:828

Taran, Yu. V. 1973. *JINR, Dubna, Preprint P3–7149*. Dubna, USSR

Vasil'ev, B. V., Kolicheva, E. V. 1978. *Zh. Eksp. Teor. Fiz.* 74:466 (transl: *Sov. Phys. JETP* 47:243)

Vladimirskii, V. 1961. *Zh. Eksp. Teor. Fiz.* 39:1062 (transl: *Sov. Phys. JETP* 12:740)

Weinberg, S. 1976. *Phys. Rev. Lett.* 37:657

Weisskopf, M. C., Carrico, J. P., Gould, H., Lipworth, E., Stein, T. S. 1967. *Phys. Rev. Lett.* 19:741

Weisskopf, M. C., Carrico, J. P., Gould, H., Lipworth, E., Stein, T. S. 1968a. *Phys. Rev. Lett.* 21:1645

Weisskopf, M. C., Carrico, J. P., Gould, H., Lipworth, E. Stein, T. S. 1968b. *Phys. Rev.* 174:125

Wilkening, D. 1981. "Molecular Beam Test of Time Reversal Symmetry in the TlF Molecule." Ph. D. thesis. Harvard Univ. Unpublished

Wolfenstein, L. 1964a. *Phys. Rev. Lett.* 13:562

Wolfenstein, L. 1964b. *CERN Rep.* TH 1837

Wolfenstein, L. 1974. *Nucl. Phys. B* 77:375

Wu, C. S., Ambler, E., Hayward, R. W., Hoppes, D. D., Hudson, R. P. 1957. *Phys. Rev.* 105:1413

Zel'dovich, Y. A. B. 1959. *Zh. Eksp. Teor. Fiz.* 36:1952 (transl: *Sov. Phys. JETP* 9:1389)

Ann. Rev. Nucl. Part. Sci. 1982. 32. 235–69
Copyright © 1982 by Annual Reviews Inc. All rights reserved

GAMMA-RAY ASTRONOMY

R. Ramaty

Laboratory for High-Energy Astrophysics, NASA/Goddard Space Flight Center, Greenbelt, Maryland 20771

R. E. Lingenfelter

Center for Astrophysics and Space Sciences, University of California at San Diego, La Jolla, California 92093

CONTENTS

1. INTRODUCTION

The gamma-ray region is one of the last energy bands of the electromagnetic spectrum to be opened to astronomical observations. Early attempts to detect gamma rays were often frustrated by difficulties in

235

0163–8998/82/1201–0235$02.00

distinguishing fluxes of cosmic origin from those produced in the atmosphere and the detectors. But more recent observations made from balloons, satellites, and space probes have detected gamma rays from many astronomical objects, including the Sun, the Moon, neutron stars, interstellar clouds, the center of our Galaxy and the nuclei of active galaxies.

Cosmic gamma rays are produced in a variety of physical processes. Nuclear deexcitation, neutron capture, and positron annihilation produce the lines first observed from solar flares by Chupp et al (1973) and from the lunar surface by Metzger et al (1974). Bremsstrahlung and π^0 meson decay can lead to observed (Clark, Garmire & Kraushaar 1968) gamma-ray emission from the interstellar medium. Bremsstrahlung, Compton scattering, and gyrosynchrotron radiation in multibillion degree and trillion gauss plasmas are probably responsible for the continuum emission seen (Klebesadel, Strong & Olson 1973) from gamma-ray bursts. Electron-positron pair production by photon-photon collisions, expected in similar plasmas, could be the source of the annihilation radiation detected (Leventhal, MacCallum & Stang 1978) from the Galactic Center as well as from gamma-ray bursts (Mazets et al 1981a,b). Radiation produced by particles accelerated along curved magnetic field lines at neutron star polar caps may be responsible for gamma-ray emission observed (Kniffen et al 1974) from pulsars. Nonthermal synchrotron and Compton emissions from relativistic electrons could produce the broad spectrum of gamma rays seen (e.g. Grindlay et al 1975, Swanenburg et al 1978) from active galaxies. And the superposition of emission from similar objects at cosmological distances, together with possible matter-antimatter annihilation radiation, could lead to the observed (e.g. Fichtel, Kniffen & Hartman 1973) diffuse gamma-ray background.

Gamma-ray observations provide important information on a variety of astronomical objects and sites. In solar flares, gamma-ray line observations provide information on particle acceleration mechanisms by giving a measure of the flare energy that resides in energetic nucleons and of the acceleration time of these nucleons. In the interstellar medium, gamma-ray continuum observations map the product of the densities of cosmic rays and interstellar gas and demonstrate that the cosmic rays are of galactic origin. For gamma-ray bursts, line obervations, by showing evidence for redshifts in strong gravitational fields and absorptions in intense magnetic fields, suggest that neutron stars are the sources of these bursts. For active galactic nuclei, hard x-ray and gamma-ray continuum observations, by indicating that a major fraction of the observed luminosity is in the gamma-ray band, suggest that powerful nonthermal sources, perhaps massive black holes, power these objects. For the nucleus of our Galaxy, the observed electron-positron annihilation line, also seems to require such a hole.

These achievements notwithstanding, several of the promises of gamma-ray astronomy have not yet been fulfilled. Chief among these is the observation of gamma-ray lines from processes of nucleosynthesis in supernovae and novae.

Balloon-borne detectors have made pioneering observations in gamma-ray astronomy, but much of the recent progress has been the result of space missions carrying instruments above the atmosphere. The space vehicles that have contributed most significantly to gamma-ray astronomy are two Orbiting Solar Observatories (OSO-3 and OSO-7), the Apollo Spacecraft, the second Small Astronomical Satellite (SAS–2), the European satellite COS-B, two High-Energy Astrophysical Observatories (HEAO-1 and HEAO-3), the Solar Maximum Mission (SMM), and a variety of space probes with gamma-ray burst experiments on board. Future progress in gamma-ray astronomy, however, will only be possible if additional flight opportunities are made available to instruments that are both more sensitive and have more resolving power than those flown so far.

In the present article we describe gamma-ray observations from the solar system, from rapid nonsolar transients, from quasi-steady galactic sources, and from extragalactic sites. We consider the most reliable and statistically significant observations; gamma-ray observations, in some cases, are still limited by counting statistics that can lead to questionable results. We particularly emphasize the physical processes responsible for astrophysical gamma-ray production. These involve processes of atomic physics (positronium formation and annihilation), of low-energy nuclear physics (deexcitation lines), of medium-energy particle physics (π^0 meson production), of electromagnetism and magnetohydrodynamics (electron-positron pair production, confinement, and annihilation), as well as the more exotic processes in and around neutron stars, close to black holes, and possibly even in the early universe.

2. SOLAR SYSTEM GAMMA RAYS

Gamma rays have been observed from both the Sun and the Moon. Solar gamma-ray emission is produced by particles accelerated in flares, while lunar gamma rays result from galactic cosmic-ray interactions with the lunar surface and the decay of long-lived natural radioisotopes.

2.1   Solar Flares

Particles accelerated by solar flares interact with the ambient solar atmosphere and create gamma rays, both lines and continuum. Line emission results from the interactions of protons and nuclei, while the continuum is from relativistic electron bremsstrahlung. The first detailed calculation of the expected energetic particle interaction rates in flares, carried out by Lingenfelter & Ramaty (1967), predicted observable gamma-ray line fluxes at the Earth.

Gamma-ray lines from solar flares were first observed by Chupp et al (1973) with a NaI spectrometer flown on board the OSO-7 satellite. The lines were observed at 0.511 MeV from positron annihilation, at 2.223 MeV from neutron capture on ^{1}H, and at 4.438 and 6.129 MeV from deexcitations of nuclear levels in ^{12}C and ^{16}O, respectively. These lines, as well as other nuclear deexcitation lines, were observed from a number of subsequent flares by detectors on the HEAO-1 (Hudson et al 1980), HEAO-3 (Prince et al 1982), and SMM (Chupp et al 1981, Chupp & Forrest 1981, Chupp 1982) satellites.

Gamma-ray continuum from solar flares was first observed by Peterson & Winckler (1959) with a balloon-borne detector. This continuum below about one MeV is electron bremsstrahlung, now routinely observed from many solar flares (e.g. Kane et al 1980). But at higher energies, Doppler-broadened unresolved nuclear lines make a significant contribution to the continuum, and in the energy range from 4 to 7 MeV nuclear radiation from C, N, and O constitutes the dominant radiation mechanism (Ramaty, Kozlovsky & Suri 1977, Ibragimov & Kocharov 1977). Continuum emission at higher energies is only rarely observed (Chupp & Forrest 1981). This emission could be a combination of electron bremsstrahlung and π^0 meson decay (Lingenfelter & Ramaty 1967, Crannell, Crannell & Ramaty 1979).

The strongest predicted and observed line from solar flares is that at 2.223 MeV from neutron capture on hydrogen, ^{1}H(n, γ)^{2}H. Studies of neutron production in flares (Lingenfelter et al 1965, Ramaty, Kozlovsky & Lingenfelter 1975) indicate that most of the neutrons responsible for this line result from the breakup of helium by protons at energies greater than about 20 MeV per nucleon, ^{4}He(p, pn)^{3}He and ^{4}He(p, 2pn)^{2}H, with lesser contributions from spallation of heavier nuclei and from π^+ production, ^{1}H(p, nπ^+)^{1}H. The neutron production may take place above the photosphere, but the 2.223-MeV line emission comes from captures in the photosphere where the density is high enough ($>10^{16}$ H cm^{-3}) for the bulk of neutrons to be slowed down and captured before they decay. Calculations (Wang & Ramaty 1974) of neutron slowing down and capture in the solar atmosphere show that the principal capture reactions are $^1\mathrm{H}(n, \gamma)^2\mathrm{H}$ and $^3\mathrm{He}(n, p)^3\mathrm{H}$. Even though $^3\mathrm{He}$ is only a minor constituent of the solar atmosphere, $^3\mathrm{He}/^1\mathrm{H} \sim 5 \times 10^{-5}$ (Geiss & Reeves 1972, Hall 1975), its thermal capture cross section is 1.6×10^4 times that of hydrogen.

Comparisons of the observed (Chupp et al 1981, Chupp 1982, Prince et al 1982) time dependence of the intensity of prompt nuclear deexcitation lines to that of the 2.223-MeV line show delays of $\sim 10^2$ s due to the mean thermal neutron capture time. The time required for the neutrons to slow down is much less than that required for their capture. A capture time of $\sim 10^2$ s implies (Wang & Ramaty 1974) that the mean density of the gas where the neutrons are captured is $\sim 10^{17}\,\mathrm{H\,cm^{-3}}$, a density corresponding to a depth of ~ 300 km into the photosphere. Independent evidence for neutron capture in the photosphere comes from the relative attenuation, or limb darkening, of the neutron capture line from solar flares occurring close to the visible limb of the Sun. Comparisons (Chupp 1982) of the neutron capture line fluence to that of nuclear deexcitation lines show that the capture line for limb flares is attenuated by a factor of 10 or more over that for disk flares. This attenuation results from Compton scattering in the photosphere (Wang & Ramaty 1974) and implies (Ramaty, Lingenfelter & Kozlovsky 1982) a column density of $\sim 10^{25}\,\mathrm{H\,cm^{-2}}$ for the limb flares, consistent with a density of $\sim 10^{17}\,\mathrm{H\,cm^{-3}}$. The width of the 2.223-MeV line, determined by the photospheric temperature, is expected to be very narrow ($\sim 100\,\mathrm{eV}$), a result consistent with the high-resolution HEAO-3 observations (Prince et al 1982), which set an upper limit of several keV on the width of this line.

A significant fraction of the fastest ($\gtrsim 100\,\mathrm{MeV}$) neutrons can travel as far as the Earth before they decay, resulting (Lingenfelter et al 1965) in detectable neutron fluxes at the Earth following large flares. High-energy solar neutrons were observed from a large flare in 1980 (Chupp & Forrest 1981).

The next most intense solar flare line is that at 0.511 MeV from the annihilation of positrons. There are many astrophysically important positron production mechanisms, but in solar flares the 0.511-MeV line results (Ramaty, Kozlovsky & Lingenfelter 1975) from nuclear interactions producing short-lived radionuclei (e.g. $^{11}\mathrm{C}$, $^{13}\mathrm{N}$, $^{15}\mathrm{O}$, $^{17}\mathrm{F}$) and π^+ mesons, which decay by positron emission, as well as excited $^{16}\mathrm{O}$ in the 6.052-MeV level, which decays by electron-positron pair emission. The initial energies of the positrons range from several hundred keV to tens of MeV, but only a few annihilate at these high energies. The bulk of the positrons slow down to energies comparable with those of the ambient electrons, where annihilation takes place either directly or via positronium (Stecker 1969). For a recent review of the physics of positronium see Berko & Pendleton (1980).

Positronium in astrophysical sites is formed by radiative combination with free electrons and by charge exchange with neutral hydrogen (Ramaty & Lingenfelter 1973, Crannell et al 1976); 25% of the positronium atoms decay from the singlet state and 75% from the triplet state. Singlet positronium annihilation and direct annihilation produce a line at 0.511 MeV, while triplet positronium annihilates into three photons that form a continuum below 0.511 MeV. But if the ambient density is $\gtrsim 10^{15}\,\mathrm{H\,cm^{-3}}$, as may be the case for solar flare positrons, then most of the positronium will be broken up by collisions before it can decay (Crannell et al 1976). The width of the 0.511-MeV line from solar flares depends on the temperature of the annihilation region, and can range from a few keV to tens of keV, depending on whether the annihilation takes place predominantly in the cool photosphere or the hot flare plasma. Measurements of the positronium continuum and the width of the 0.511-MeV line could thus provide important information on the positron annihilation site, but such observations are not yet available.

A variety of gamma-ray lines are produced by the deexcitation of nuclear levels. In solar flares these levels are populated by inelastic collisions [e.g. $^{12}\mathrm{C}(\mathrm{p}, \mathrm{p}')^{12}\mathrm{C}^{*4.44}$], spallation reactions [e.g. $^{20}\mathrm{Ne}\,(\mathrm{p}, \mathrm{p}\alpha)^{16}\mathrm{O}^{*6.13}$], nonthermal fusion reactions [e.g. $^{4}\mathrm{He}(\alpha, \mathrm{p})^{7}\,\mathrm{Li}^{*0.478}$], and the decay of radionuclei produced by spallation reactions [e.g. $^{16}\mathrm{O}(\mathrm{p}, \mathrm{p}2\mathrm{n})^{14}\mathrm{O}(\mathrm{e}^{+})^{14}\mathrm{N}^{*2.31}$]. Using laboratory measurements (e.g. Dyer et al 1981) of the excitation functions of a great number of such reactions, Ramaty, Kozlovsky & Lingenfelter (1979) calculated theoretical gamma-ray spectra produced by the interaction of energetic particles in cooler ambient matter, assuming a variety of energetic particle spectra.

In the solar atmosphere two line components are produced: a narrow component resulting from the deexcitation of ambient nuclei excited by interactions with energetic protons and α particles, and a broad component from the deexcitation of energetic heavy nuclei excited by interactions with ambient hydrogen and helium. The widths of the narrow lines, broadened by the recoil velocities of the heavy target nuclei, are on the order of 1–2% of the line energy, while those of the broad lines, reflecting the velocities of the fast nuclei themselves, are about an order of magnitude larger. If the elemental and isotopic compositions of both the energetic particles and the ambient medium resemble those of the solar photosphere, the strongest narrow lines are at 6.129 MeV from ^{16}O, 4.438 MeV from ^{12}C, 2.313 MeV from ^{14}N, 1.779 MeV from ^{28}Si, 1.634 MeV from ^{20}Ne, 1.369 MeV from ^{24}Mg, 1.238 MeV and 0.847 MeV from ^{56}Fe, all produced primarily by direct excitation of these nuclei, and at two lines, 0.478 MeV from ^{7}Li and 0.431 MeV from ^{7}Be, which result from the reactions, ^{4}He(α, p)^{7}Li* and ^{4}He(α, n)^{7}Be*. The role of these reactions for producing gamma-ray lines in astrophysics was first pointed out by Kozlovsky & Ramaty (1974). As already mentioned, the broad lines, together with many un-resolved narrow lines, contribute significantly to the gamma-ray con-tinuum, especially in the 4–7-MeV range.

The implications of the gamma-ray observations of solar flares concern the timing of the acceleration, the confinement of particles at the Sun, the fraction of the total flare energy that resides in energetic nucleons, the chemical and isotopic abundances, and the possible beaming of the energetic particles. In particular the gamma-ray observa-tions show (Von Rosenvinge, Ramaty & Reames 1981, Chupp 1982, Ramaty, Lingenfelter & Kozlovsky 1982) that a few percent of the total flare energy resides in protons and nuclei, accelerated to tens of MeV per nucleon on time scales of a few seconds in closed magnetic loops with little escape into the interplanetary medium. Further analysis of data should provide important and potentially unique information on abundances and on geometric effects such as beaming. The latter would follow from shifts in the peak line energies (Ramaty & Crannell 1976) and modifications in the line widths (Kozlovsky & Ramaty 1977).

2.2 Lunar and Planetary Surfaces

The most intense gamma-ray line and continuum emission from the Moon results from interactions of galactic cosmic rays with the lunar surface material. The strongest lines are from excitations and captures of secondary neutrons generated by relativistic primary cosmic-ray particles in nuclear cascades of spallation interactions. The secondary electrons and positrons of these cascades produce the bulk of the continuum emission by bremsstrahlung. Decay of the natural radionu-clei ^{40}K, ^{232}Th, and ^{238}U, remnants of nucleosynthesis prior to the for-mation of the solar system, also produce several intense gamma-ray lines.

Detailed studies (Reedy, Arnold & Trombka 1973, Reedy 1978) of cosmic-ray secondary particle interactions showed that the two most intense lines from the lunar surface are at 1.779 and 6.129 MeV from deexcitation of the two most abundant nuclei, ^{28}Si and ^{16}O, excited by inelastic neutron scattering. But these calculations also predicted many other detectable lines from less abundant elements, the nuclear deexcitation lines at 0.847, 1.369, and 2.210 MeV from inelastic excitation of ^{56}Fe, ^{24}Mg, and ^{27}Al and the neutron capture lines on ^{48}Ti at 6.762 MeV and on ^{56}Fe at 7.631 and 7.646 MeV. These studies also predicted line intensities from the decay of natural radionuclei comparable to those produced by cosmic-ray interactions. The three strongest lines are those at 1.461 MeV from decay of ^{40}K (half-life $\sim 1.28 \times 10^9$ yr), at 2.615 MeV from decay of ^{208}Tl in the decay chain of ^{232}Th (half-life$\sim 1.41 \times 10^{10}$yr), and at 0.609 MeV from the decay of ^{214}Bi in the ^{238}U (half-life $\sim 4.47 \times 10^9$ yr) decay chain.

All of these lines have been observed (Metzger et al 1974, Bielefeld et al 1976) with NaI gamma-ray spectrometers on the lunar-orbiting Apollo 15 and 16 spacecraft. These detectors mapped a sizeable portion of the lunar surface and the relative line strengths revealed significant regional variations. The lunar mare regions showed enrichments of a factor of three or more in Fe, Ti, K, Th, and U, and depletions of as much as 50% in Al and Ca abundances with respect to the lunar highland regions. These observations provide important constraints on the differentiation and thermal evolution of the Moon.

Reedy (1978) pointed out that similar observations by orbiting gamma-ray spectrometers could also provide maps of the surface compositions of Mercury and Mars, which have atmospheres thin enough for the surface gamma-ray emission to be observed.

3.   RAPID GAMMA-RAY TRANSIENTS

Temporal variability is a common property of a large fraction of the astronomical sources of high-energy radiation. In fact, many gamma-ray sources have so far been observed only by their intense transient emission. The most common class of these transients, known as gamma-ray bursts, appear suddenly and persist for times ranging from a fraction of a second to a few minutes. We first consider these bursts, including the possibly unique March 5, 1979, burst. We then briefly review the properties of two very unusual transients that last for tens of minutes and have only been seen in line emission.

3.1 *Gamma-Ray Bursts*

Gamma-ray bursts were discovered accidentally in 1967 (Klebesadel, Strong & Olson 1973) by detectors on board the Vela satellites whose primary purpose was to monitor artificial nuclear detonations in space. Detailed reviews of the observational properties of gamma-ray bursts can be found in Hurley (1980), Vedrenne (1981), and Cline (1981) and a catalogue of recent bursts is given by Mazets et al (1981a). The early theories of gamma-ray bursts were reviewed by Ruderman (1975), while more recent reviews of current theoretical ideas were given by Woosley (1982), Lamb (1982), and Ramaty, Lingenfelter & Kozlovsky (1982). A variety of other recent observational and theoretical papers can be found in the book by Lingenfelter, Hudson & Worrall (1982). Here we give a relatively brief discussion of both the observations and theories.

Gamma-ray bursts are generally observed in the photon energy range from a few tens of keV to several MeV with typical event durations ranging from about 0.1 to 30 s. The observed burst energy fluences (>30 keV) range from about 10^{-7} to 10^{-3} erg cm^{-2}, and the frequency of occurrence of bursts ranges from a few per year with fluences $>10^{-4}$ erg cm^{-2} to a few thousand per year with fluences $>10^{-7}$ erg cm^{-2}. At fluences less than 10^{-5} erg cm^{-2}, the frequency of bursts falls below that expected from an unbounded, isotropic, and homogeneous distribution of sources. Therefore, the observed frequency distribution requires a source distribution of finite extent, implying a galactic origin (Fishman et al 1978). The average distances of the observed sources, however, are still uncertain. Therefore, galactic source distributions have been constructed, which can reproduce the observed sizes and frequencies with typical burst energies of 10^{37} ergs (Higdon 1982) to 10^{40} ergs (Jennings 1982).

Because the bursts originate from unpredictable celestial directions at unpredictable times, they have generally been observed with detectors that have large fields of view. Such instruments, however, have poor angular resolution and can determine source positions to an accuracy of only a few degrees (Mazets & Golenetskii 1981). But much more accurate burst positions can be determined from arrival-time differences using a network of instruments placed on widely separated interplanetary space probes (e.g. Cline 1981). The presence of sharp temporal features in the burst time profiles allows the measurement of differences in arrival times of wavefronts of the order of a few milliseconds over baselines separated by hundreds of light seconds. For the strongest and most rapidly varying bursts, such measurements yield angular resolutions on the order of arc seconds.

The first precise source position determined by arrival-time differences is that of the March 5, 1979, burst (Evans et al 1980). This burst, the most intense observed so far, was detected by instruments on nine different spacecraft (see Cline 1980). The resultant positional error box, of size 0.1 arc min^2 (Cline 1982), lies within the supernova remnant N49 in the Large Magellanic Cloud (LMC), a neighboring galaxy at a distance of 55kpc. If the burst source is at this distance, the total radiated energy is ~10^{44} ergs, which is at least four orders of magnitude larger than that of a typical galactic gamma-ray burst. But since the March 5 burst also exhibited a number of remarkable and possibly unique observational properties, which we discuss in more detail below, it appears (Cline 1980, Klebesadel et al 1982) to belong to a separate class of transients, which are less frequent but more energetic than the typical galactic bursts.

Small positional error boxes have also been determined for a few other gamma-ray bursts (Cline et al 1981, Laros et al 1981). But searches at longer wavelengths (e.g. Hjellming & Ewald 1981) have not produced unambiguous associations of the burst sources with identifiable astronomical objects. Nevertheless, an exciting development has occurred recently with the discovery (Schaefer 1981) of an optical flash in one of these error boxes, that of the November 19, 1978, burst (Cline et al 1981). This flash was found on an archival plate of the Harvard College Observatory exposed on November 17, 1928. Since good arguments exist (Schaefer 1981) that the 1928 flash and 1978 burst were from the same source, this discovery provides the strongest evidence to date for repetitions of galactic gamma-ray bursts and demonstrates that the bursts can also be monitored optically.

A new dimension in the study of gamma-ray bursts has been added by the discovery of emission lines and absorption features in their energy spectra (see Teegarden 1982 for review). The absorption features, observed (Mazets et al 1981b, Dennis et al 1982) at energies below about 100 keV, are probably due to cyclotron absorption in the intense (~10^{12} gauss) magnetic fields that are expected around neutron stars (Baym & Pethick 1975).

The most commonly observed emission line falls in the energy range from 0.40 to 0.46 MeV, as seen by low resolution NaI detectors in the spectra of a third of the most intense gamma-ray bursts (Mazets et al 1981b). In the spectrum of the November 19, 1978, burst, a Ge detector has resolved two emission lines at ~0.42 MeV and ~0.74 MeV, which the NaI detectors saw as one broad feature from 0.3 to 0.8 MeV (Teegarden & Cline 1980). Line emission in the range of 0.4 to 0.46 MeV is probably optically thin e^+e^- annihilation radiation redshifted by the strong gravitational field of a neutron star. In an optically thick region, however, stimulated annihilation radiation could produce a line at ~0.43 MeV without a gravitational redshift (Ramaty, McKinley & Jones 1982). The line at 0.74 MeV could be either collisionally excited and gravitationally redshifted 0.847 MeV emission from ^{56}Fe (Teegarden & Cline 1981), or gravitationally redshifted single-photon e^+e^- annihilation (Daugherty & Bussard 1980, Katz 1982) radiation at 1.022 MeV in a very strong ($\gtrsim 10^{13}$ gauss) magnetic field. In all cases, the implied redshifts of 0.1 to 0.3 are consistent with those expected from neutron stars.

The $\sim$0.43 MeV e^+e^- annihilation line was also seen (Mazets et al 1979) from the March 5, 1979, burst suggesting that the source of this burst was also a neutron star. But as mentioned above, other characteristics of this burst seem to place it in a class different from that of the typical galactic bursts. These characteristics include (Cline 1980, 1982) the extremely rapid rise time ($<2 \times 10^{-4}$ s) of the impulsive emission spike, the relatively short duration ($\sim$0.15 s) and high luminosity of this spike, the 8-s pulsed emission following the impulsive spike, the subsequent outbursts of lower intensity from apparently the same source direction on March 6, April 4, and April 24, 1979, and, as already mentioned, the coincidence of the positional error box with an extragalactic supernova remnant.

Current theoretical ideas on gamma-ray bursts generally involve strongly magnetized neutron stars. These ideas have developed, in part, as a result of the detailed March 5 observations, even though it is quite likely that the underlying energy source of this burst is not typical of all gamma-ray bursts.

The most probable energy source of gamma-ray bursts is either gravitational or nuclear. Magnetic field annihilation, responsible for energy generation in solar flares (e.g. Sturrock 1980), can be shown on the basis of total energetics to be inadequate for gamma-ray burst production. The absence of evidence for bulk antimatter in our Galaxy or in neighboring galaxies (Steigman 1976) also makes matter-antimatter annihilation an unlikely processes for energy generation in gamma-ray burst sources.

Gravitational energy can be released impulsively from a neutron star when a large amount of solid matter such as an asteroid or comet is accreted onto its surface (Harwit & Salpeter 1973, Colgate & Petschek 1981). Such accretion releases about 100 MeV per nucleon, the potential energy at the neutron star surface. Gravitational energy could also be released in a corequake of a neutron star (Tsygan 1975, Ramaty et al 1980). Such quakes can set up neutron star vibrations, which dissipate mainly by gravitational radiation. A fraction of the vibrational energy, however, can be converted into magnetoacoustic waves, which dissipate by accelerating particles in the magnetosphere. Radiation from these particles is then responsible for the observed gamma-ray emission.

Alternatively, impulsive energy release from neutron stars could result from a nuclear detonation of degenerate matter accumulated over a relatively long period of time by accretion of gas (Woosley & Taam 1976, Woosley 1982). Such detonations release several MeV per nucleon from the burning of helium to the iron peak nuclei.

All three of these processes, solid body accretion, a corequake, or a nuclear detonation, appear to be quite capable of providing the 10^{37} to 10^{40} ergs required for typical galactic gamma-ray bursts. But to account for the $\sim 10^{44}$ ergs of the March 5, 1979, burst, very large amounts of accreted matter must be involved and this probably rules out solid body accretion and nuclear detonation for this burst. Corequakes, which could in principle release energies up to a fraction of the gravitational binding energy of a neutron star ($\sim 10^{53}$ erg, Borner & Cohen 1973), appear to be adequate for the March 5 burst (Ramaty et al 1980). But no detailed calculations on these possibilities have yet been published.

An issue comparable in importance to the energy source is the radiation mechanism and the nature of the emitting region. Electron-positron annihilation, as already mentioned, is probably responsible for the observed emission line between 0.40 and 0.46 MeV. Since these lines have relatively narrow widths requiring a narrow and well-defined range of gravitational redshift, the emitting material must be confined to a thin region close to the neutron star surface. It was first proposed by Ramaty et al (1980) that this confinement is achieved by the strong magnetic field ($\sim 10^{12}$ gauss) of a neutron star. Magnetic confinement is necessary especially for the March 5 burst where the inferred radiation pressure greatly exceeds the gravitational pull of the neutron star. Magnetic fields similarly play an important role in nuclear detonation models of galactic bursts (Woosley 1982) where magnetic confinement of the nuclear burning products, or lack of it, may constitute the difference between a gamma-ray burst and an x-ray burster. For a recent review of x-ray bursters, see Lewin & Joss (1981). Lastly, if the absorption features, observed below 100 keV in gamma-ray bursts, are due to cylotron absorption, then they provide direct observational evidence for $\gtrsim 10^{12}$-gauss magnetic fields in the burst sources.

The principal processes suggested for producing the gamma-ray continuum in burst sources are bremsstrahlung (Gilman et al 1980), Compton scattering (Liang 1981, Bussard & Lamb 1982, Fenimore et al 1982), and synchrotron radiation (Ramaty, Lingenfelter & Bussard 1981). To account for the observed gamma-ray burst spectra, bremsstrahlung requires a hot plasma with $T \gtrsim 10^9$K, Comptonization requires electrons of similar temperature and a copious supply of cooler photons that gain energy from the electrons in Compton collisions, and gyrosynchrotron radiation is produced in strong magnetic fields (10^{11} to 10^{12} gauss) by MeV electrons. In certain cases these mechanisms can operate simultaneously, as in the model of Liang (1981) where MeV electrons produce gyrosynchrotron photons of energies $\lesssim 100$ keV and subsequently Compton scatter them up to energies of an MeV. Figure 1, shows the observed spectrum of the March 5 event (Mazets et al 1979) and theoretical calculations that involve annihilation, synchrotron radiation, and Comptonization of the synchrotron photons.

An important property of gamma-ray burst spectra is that they appear to be optically thin (Gilman et al 1980), especially at the higher energies ($\gtrsim 100$ keV). An optically thin emission region is also required to produce the ~ 0.43-MeV emission line, except in the case where grasar action is important (Ramaty, McKinley & Jones 1982). An optically thin source requires a sufficiently small ratio of source depth to source area,

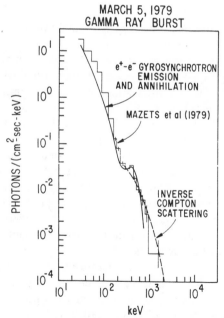

Figure 1 The observed and calculated energy spectrum of the March 5, 1979, gamma-ray burst. The observations are from Mazets et al (1979); the solid curve is the synchrotron and annihilation spectrum of an e^+e^- plasma (Ramaty, Lingenfelter & Bussard 1981) and the dashed curve is the spectrum resulting from the Compton scattering of the synchrotron photons by the energetic e^+e^- pairs (Liang 1981).

so that the small opacity can be consistent with the high observed luminosity. The gamma-ray emission should therefore be produced in a thin layer containing a high density of radiating matter. The most extreme conditions are found in the March 5 event, where in the model of Ramaty, Lingenfelter & Bussard (1981) the observed radiation comes from a magnetically confined thin layer (~ 0.1 mm) of dense ($\sim 10^{26}$ cm^{-3}) e^+e^- pairs covering the surface of a neutron star. The instantaneous energy content of this layer is orders of magnitude smaller than the total energy of the burst, so that energy must be supplied continuously to the layer. This is achieved by the neutron star vibrations discussed above. An attractive consequence of the continuous particle acceleration by vibrations is that the duration of the burst is determined by the damping time (0.15 s) of the vibrations. Indeed, the neutron star mass-to-radius ratio, deduced from the observed gravitational redshift, implies a vibrational damping time almost exactly the same as the duration of the main emission spike of the burst (Ramaty et al 1980).

3.2 Gamma-Ray Line Transients

There are apparently two other types of gamma-ray transients in which all of the radiation observed so far is in emission lines. One such gamma-ray line transient was discovered (Jacobson et al 1978, Ling et al 1982) with a high-resolution Ge detector on June 10, 1974, from an unknown source. This event, lasting about twenty minutes, was characterized by strong emission in four relatively narrow energy bands at 0.40–0.42 MeV, 1.74–1.86 MeV, 2.18–2.26 MeV, and 5.94–5.96 MeV with no detectable continuum. Subsequent searches (Heslin et al 1981), however, failed to observe similar line transients and therefore imply that their frequency is less than 30 per year.

Lingenfelter, Higdon & Ramaty (1978) suggested that the gamma-ray line transient observed by Jacobson et al (1978) could result from episodic accretion onto a neutron star from a binary companion leading to redshifted lines from the neutron star surface and unshifted lines from the atmosphere of the companion star. The observations could then be understood in terms of neutron capture and positron annihilation. Specifically, positron annihilation and neutron capture on hydrogen and iron at and near the surface of the neutron star with a surface redshift of ~ 0.28 would produce the observed redshifted line emission at about 0.41, 1.79, and 5.95 MeV, respectively. The same processes in the atmosphere of the companion star would produce unshifted lines, of which only the 2.223-MeV line from neutron capture on hydrogen was observed. The unshifted 0.511-MeV positron annihilation line could not have been seen because of the large atmospheric and detector background at this energy, while the line emission from neutron capture on iron should be significant only from the iron-rich surface of the neutron star but not from the companion star.

The other type of transient line emission is observed in the pulsed spectrum of the Crab pulsar. This very narrow (FWHM < 4.9 keV) emission line, which may vary slightly in energy from 73 to 77 keV, was first observed by Ling et al (1979) from the Crab nebula. The line was subsequently shown (Strickman, Kurfess & Johnson 1982) to be pulsed with the Crab pulsar period of 0.033 s and to persist only for about 20 minutes and then turn off. The most likely source of this line is cyclotron emission at the gyrofrequency of an intense ($\sim 8 \times 10^{12}$ gauss) magnetic field at the polar cap of a neutron star. In addition, a very narrow 0.4-MeV line was observed (Leventhal, MacCallum & Watts 1977) from a broad field of view that included both the Crab nebula and the source direction of the June 10, 1974, transient.

4. GALACTIC GAMMA RAYS

In addition to the transients, a rich variety of other more steady sources of galactic gamma-ray emission has been observed. These include an intense source of electron-positron annihilation radiation at the Galactic Center, the Crab and Vela pulsars, the binary source Cygnus X-3, a number of unidentified discrete sources, and diffuse emission resulting from cosmic-ray interactions in the interstellar medium.

We discuss first the gamma-ray line emission from the Galactic Center and other potential sources, then turn to the continuum emission from both diffuse regions and localized sources.

4.1 Galactic Center

Intense positron annihilation radiation at 0.511 MeV has been observed from the direction of the Galactic Center for over a decade. This emission was first seen in a series of balloon observations with low-resolution NaI detectors, starting in 1970 (Johnson, Harnden & Haymes 1972, Johnson & Haymes 1973, Haymes et al 1975). But it was not until 1977 that the annihilation line energy of 0.511 MeV was clearly identified with high-resolution Ge detectors flown by Leventhal, MacCallum & Stang (1978). The latter observation also revealed that the line is very narrow (FWHM $\lesssim 3.2$ keV) and that it shows evidence for three-photon positronium continuum emission below 0.511 MeV, which implies that $\sim 90\%$ of the positrons annihilate via positronium. Thus, the observed intensity of $\sim 2 \times 10^{-3}$ photons per $cm^2 \cdot s$ implies an annihilation rate of $\sim 4 \times 10^{43}$ positrons s^{-1} or an annihilation radiation luminosity of $\sim 6 \times 10^{37}$ ergs s^{-1} at the 10-kpc distance of the Galactic Center. The gamma-ray line at 511 keV and the continuum at lower energies are shown in Figure 2.

Recent Ge detector observations (Riegler et al 1981) on HEAO-3 confirm the narrowness (FWHM <2.5 keV) of the line and provide more precise information on the location of the source and strong constraints on the size of the emission region. These measurements showed that the direction of the source is coincident with that of the Galactic Center (within the ±4° observational uncertainty) and that the line intensity varies with time, decreasing by a factor of three in six months from the Autumn of 1979 to the Spring of 1980. This six-month variability implies that the sizes of both the annihilation region and the positron source are less than the light-travel distance of 10^{18} cm.

The nature of the positron annihilation region is further constrained by the observed line width and intensity variations. The line width (FWHM <2.5 keV) requires a gas temperature in the annihilation region less than 5×10^4 K and an ionization fraction greater than 10%

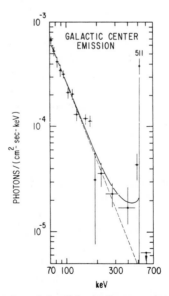

Figure 2 The energy spectrum of the Galactic Center region observed by Leventhal, MacCallum & Stang (1978). The solid curve is the continuum produced by triplet positronium annihilation in addition to a power-law x-ray continuum as indicated by the dashed curve.

(Bussard, Ramaty & Drachman 1979). If the gas were neutral, the line width would be larger than observed, because it would be Doppler broadened, not by the thermal motion of the gas, but by the velocity of energetic positrons forming positronium in flight by charge exchange with neutral hydrogen. In a partially ionized gas, however, the positrons lose energy to the plasma fast enough that they thermalize before they annihilate or form positronium. The line width thus reflects the temperature of the medium, requiring it to be $\lesssim 5 \times 10^4$ K. The intensity variation not only constrains the size of the annihilation region to be less than 10^{18} cm, but it requires that the density of gas in it be high enough that the positrons can slow down and annihilate in less than half a year. If the positrons are produced with kinetic energies on the order of their rest mass, then the time it takes for them to slow down by Coulomb collisions is longer than the time it takes for them to form positronium in such a gas once they have slowed down. Both times are inversely proportional to the gas density. A slowing-down time of $\lesssim 1.5 \times 10^7$ s requires a density of $\gtrsim 10^5$ H cm^{-3}. Such regions appear to exist in both the peculiar warm clouds (Lacy et al 1980) and the compact nonthermal source (Kellermann et al 1977) within the central parsec of the Galaxy.

The nature of the positron source is also strongly constrained by the observed variation of the 0.511-MeV intensity and by observations at other wavelengths. The decrease of a factor of three in the line intensity in six months clearly excludes any of the multiple, extended sources previously proposed, such as cosmic rays, pulsars (Sturrock & Baker 1979), supernovae (Ramaty & Lingenfelter 1979), or primordial black holes (Okeke & Rees 1980). Instead, it essentially requires a single, compact ($<10^{18}$ cm) source that is apparently located either at or close to the Galactic Center and that inherently varies on time scales of six months or less. With a luminosity of at least 6×10^{37} ergs s^{-1}, this source is the most luminous gamma-ray source in the Galaxy.

The various possible positron production processes and the observational constraints on them were recently reviewed by Lingenfelter & Ramaty (1982). They find that the observational (Matteson et al 1979, Matteson 1982) upper limits on accompanying continuum emission at energies $> m_e c^2$ appear to set the strongest constraints on the positron production process, requiring high efficiency such that more than 30% of the total radiated energy $> m_e c^2$ goes into electron-positron pairs. Under the conditions of positron production on time scales comparable to that of the observed variation and in an optically thin, isotropically emitting region, only photon-photon pair production among ~MeV photons can provide the required high efficiency. Moreover, the absolute luminosity of the annihilation line requires that the photon-photon collisions take place in a very compact source ($d < 5 \times 10^8$ cm). Pair production in an intense radiation field around an accreting black hole of $<10^3 M_\odot$ appears to be a possible source. Other mechanisms (Lingenfelter & Ramaty 1982), such as pair production in an electromagnetic cascade in a strong electric field of an accreting and rotating black hole, would be possible if the above constraints were relaxed.

4.2 Other Sources of Line Emission

Thermonuclear burning in supernovae and novae (e.g. Woosley, Axelrod & Weaver 1981, Clayton 1982) and nuclear interactions of low-energy cosmic rays with interstellar gas (Ramaty, Kozlovsky & Lingenfelter 1979) are all expected (Ramaty & Lingenfelter 1981) to produce throughout the Galaxy a variety of nuclear deexcitation lines, as well as additional positron-annihilation line emission. Observations (Albernhe et al 1981, Gardner et al 1982) of the galactic 0.511-MeV emission with wide ($\gtrsim 50°$) field-of-view detectors reveal considerably higher line intensities than would be expected from the Galactic Center source alone, which suggests that there may be a spatially diffuse source of 0.511-MeV line emission in the Galaxy. Only upper limits have been set (Mahoney et al 1982) on the intensities of other lines from processes of nucleosynthesis, but these already significantly constrain some of the theoretical models (Clayton 1982).

The most abundant radionuclide expected from explosive nucleosynthesis in supernovae is ^{56}Ni (Clayton, Colgate & Fishman 1969), which decays with a 6.1-day half-life to ^{56}Co, which, in turn, decays with a half-life of 78.8 days to ^{56}Fe; 20% of the ^{56}Co decays are via positron emission. Nucleosynthesis of ^{56}Ni in supernovae is thought to be the primary source of galactic ^{56}Fe (e.g. Woosley, Axelrod & Weaver 1981). The bulk of the gamma rays (Colgate & McKee 1969) and positrons (Arnett 1979) from the ^{56}Ni decay chain, however, are absorbed in the expanding nebula and their energy emerges only as lower energy radiation. The characteristic light curves of Type I supernovae, in fact, appear to follow the ^{56}Ni and ^{56}Co decay (Colgate & McKee 1969) and optical lines from both ^{56}Co and the resulting ^{56}Fe have recently been detected in the spectrum of an extragalactic supernova, SN 1972e (Axelrod 1980). Any such direct gamma-ray line emission escaping from the nebula would be detectable for only a few years after the supernova explosion. But a fraction of the positrons from ^{56}Co decay are expected to escape into the interstellar medium (Arnett 1979). Since in the tenuous interstellar gas the positron lifetime against annihilation is quite long (10^5 yr in a density of 1H cm^{-3}), positrons should accumulate from several thousand supernovae, assuming that galactic supernovae occur about once every 30 years. Their annihilation should thus produce diffuse galactic gamma-ray line emission at 511 keV (Ramaty & Lingenfelter 1979). Conclusive measurements of such diffuse line emission can put constraints on the fraction of positrons that escape from supernovae and on the average rate of galactic nucleosynthesis during the last 10^5 years.

Similarly, the long-lived radionuclei ^{60}Fe (half-life $\sim 3 \times 10^5$ yr) and ^{26}Al (half-life $\sim 7.2 \times 10^5$ yr), which are also expected from explosive nucleosynthesis, should accumulate from $\sim 10^4$ or more supernovae and be well distributed through the interstellar medium before they decay. Diffuse galactic line emission is thus expected at 1.809 MeV from ^{26}Al decay to ^{26}Mg (Ramaty & Lingenfelter 1977, Arnett 1977) and at 1.332 MeV, 1.173 MeV, and 0.059 MeV from ^{60}Fe decay to ^{60}Co and its subsequent decay to ^{60}Ni (Clayton 1971).

Another important radionuclide from explosive nucleosynthesis in supernovae is ^{44}Ti (Clayton, Colgate & Fishman 1969). This isotope decays with a half-life of 47 years into ^{44}Sc, producing lines at 0.078 and 0.068 MeV. ^{44}Sc subsequently decays into ^{44}Ca with line emission at 1.156 MeV. The ^{44}Ti half-life is comparable to the average time between galactic supernova explosions and therefore gamma-ray lines from this decay chain could be observed from the few youngest galactic supernova remnants.

Explosive nucleosynthesis in novae is expected to produce ^{22}Na (Clayton & Holye 1974) and ^{26}Al (Woosley & Weaver 1980). Since about 40 novae occur in the Galaxy every year, the 1.275-MeV line emission from ^{22}Na with a 2.6-yr half-life should be observable from $>10^2$ novae at any particular time. Thus, both ^{22}Na and ^{26}Al from novae can also provide diffuse galactic line emission, and observational limits on their intensity can constrain nucleosynthetic models of novae.

The most intense deexcitation lines resulting from low-energy (< 100 MeV per nucleon) cosmic-ray interactions are expected at 6.129 MeV from ^{16}O*, at 4.438 MeV from ^{12}C*, and at 0.847 MeV from ^{56}Fe*. Of special interest are the very narrow lines (FWHM $\sim$ 5 keV), such as that at 6.129 MeV from ^{16}O*, resulting from deexcitation of nuclei in interstellar grains (Lingenfelter & Ramaty 1977). The line broadening, which in gases is caused by the recoil velocities of the excited nuclei, is greatly reduced in solids where these nuclei or their radioactive progenitors can come to rest before deexcitation if the lifetime of the level is $>10^{-11}$ s. The detection of gamma-ray lines from low-energy cosmic-ray interactions in the interstellar medium would measure the unknown interstellar density of these cosmic rays, and provide information on the distribution, motion, composition, and size of interstellar dust grains.

4.3 *Diffuse Continuum Emission*

Diffuse gamma-ray continuum emission from the Galaxy was first observed by Clark, Garmire & Kraushaar (1968) with a high-energy (>50 MeV) detector on OSO-3. These observations showed (Kraushaar et al 1972) a clearly defined band (±15° latitude) of enhanced intensity lying along the galactic equator, resulting from emission from the galactic disk. The observations also showed a strong longitudinal dependence of the intensity with a broad peak extending from −30° to +30° longitude around the Galactic Center, resulting from enhanced emission from the region within about 5 kpc of the Galactic Center. The intensity in this direction was $(1.3 \pm 0.3) \times 10^{-4}$ photons per cm^2. s. rad of longitude, approximately three times that from other directions in the disk.

Subsequent surveys in this energy range with spark chambers on SAS-2 (Fichtel et al 1975, Hartman et al 1979) and COS-B (Mayer-Hasselwander et al 1982) have mapped the sky with an angular resolution of a few degrees. They resolved a number of particularly intense discrete sources, including the Crab and Vela pulsars (Kniffen et al 1974, Thompson et al 1975). Calculations by Higdon & Lingenfelter (1976) suggested that as much as half of the observed galactic gamma-ray continuum emission may come from young ($<10^4$ yr), distant (>1 kpc), and as yet undiscovered pulsars. More recent studies give both lower (Harding 1981a) and higher (Salvati & Massaro 1982) values of the pulsar contributions. Surveys with much higher sensitivity and angular resolution are needed to determine what fraction of the galactic continuum emission comes from such sources and what fraction is truly diffuse emission from the interstellar medium.

Several possible sources of gamma-ray continuum emission, resulting from cosmic-ray interactions in the interstellar medium, have been suggested: Compton scattering of starlight (Feenberg & Primakoff 1948) and cosmic blackbody (Gould 1965) photons by cosmic-ray electrons; bremsstrahlung from cosmic-ray electron interactions with interstellar gas (Hutchinson 1952); and decay of π^0 mesons produced by cosmic-ray nucleon interactions with interstellar gas (Hayakawa 1952). Whatever the emission process may be, diffuse gamma-ray continuum observations should give new information on the galactic distribution of cosmic rays.

The unique energy spectrum of gamma rays expected from the decay of π^0 mesons, resulting from cosmic-ray interactions, can be calculated (Stecker 1970, 1971, Cavallo & Gould 1971, Higdon 1974, Badhwar & Stephens 1977) with reasonable accuracy since the cosmic-ray nucleon spectrum above the effective pion production threshold is fairly well known. But even though the other emission processes are well understood (e.g. Blumenthal & Gould 1970), their relative contributions as a function of gamma-ray energy are harder to determine because the spectrum and intensity of both the low-energy (<500 MeV) cosmic-ray electrons, responsible for the bremsstrahlung, and the ambient interstellar photons, responsible for the Compton scattering, are only poorly known.

Comparative calculations (Fichtel et al 1976, Stecker 1977, Cesarsky, Paul & Shukla 1978, Higdon 1979), however, all suggest that π^0 decay is the principal source above 100 MeV, while at lower energies bremsstrahlung is the dominant mechanism, except at high galactic latitudes (Kniffen & Fichtel 1981) where Compton scattering is more important. These conclusions are also consistent with the spectrum of the galactic continuum measured from 50 MeV to 3 GeV on SAS-2 (Hartman et al 1979) and COS-B (Paul et al 1978), as well as with the lower-energy emission (0.06–5 MeV) observed from the galactic disk by a detector on Apollo 16 (Gilman et al 1979). The differential gamma-ray production rates at the solar position in the Galaxy, including the contribution of pulsars, are shown in Figure 3, together with the observed spectrum above 50 MeV.

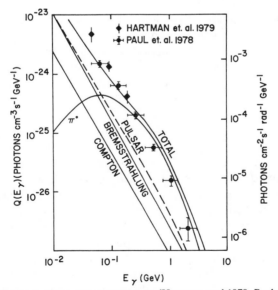

Figure 3 Diffuse galactic gamma-ray spectrum (Hartman et al 1979, Paul et al 1978) and gamma-ray production rates by galactic cosmic rays in the vicinity of the solar system together with the contribution of pulsars (from Harding & Stecker 1981).

Finally, since the absolute intensity of diffuse gamma-rays above 100 MeV depends primarily on the product of the cosmic-ray and ambient gas densities along the line of sight, the variation of that product as a function of galactocentric distance can be deduced from the variation of the diffuse intensity as a function of galactic longitude. Assuming interstellar gas density distributions, based on either spiral arm structure or large-scale density gradients in molecular hydrogen, and further assuming that the observed intensity is entirely of diffuse origin, Stecker et al (1975), Fichtel et al (1976), Hartman et al (1979), and Higdon (1979) calculated the implied cosmic-ray distribution. These calculations imply a cosmic-ray gradient in the Galaxy with higher densities in the inner part of the Galaxy and lower densities in the outer part, which is consistent with current ideas of a galactic origin for the cosmic rays.

4.4 *Discrete Galactic Sources*

A number of discrete galactic sources of gamma-ray continuum have been discovered by balloon- and satellite-borne detectors at energies of 1 MeV to several GeV and by ground-based detectors at very high energies (>100 GeV). These sources appear to encompass a variety of objects: dense interstellar clouds, pulsars, accreting neutron stars, and several as yet unidentified objects. The proposed gamma-ray emission processes are equally diverse, including not only π^0 meson decay, electron bremsstrahlung, Compton scattering and synchrotron emission, but also curvature radiation of electrons in intense ($\sim 10^{12}$ gauss) magnetic fields. Curvature radiation is produced by high-energy electrons moving along curved magnetic field lines (e.g. Sturrock 1971). The observations and models of these sources were recently reviewed by Salvati (1980), Pinkau (1980), Sreekantan (1981), and Bignami & Hermsen (1983).

Gamma-ray emission from the relatively close interstellar cloud near the star ρOph was first reported from balloon observations at >100 MeV by Frye et al (1972) and Dahlbacka, Freier & Waddington (1973). Subsequent observations on COS-B confirmed (Mayer-Hasselwander et al 1980) these measurements and also showed (Caraveo et al 1980) high-energy gamma-ray emission from the Orion cloud complex. Black & Fazio (1973) first suggested that the gamma rays from the ρOph cloud were produced by cosmic-ray interactions with the dense gas in the cloud. Subsequent studies indicate that cosmic-ray electron bremsstrahlung and decay of π^0 mesons produced by cosmic-ray nucleons can indeed account for the observed emission for both the ρOph and Orion clouds (see Bignami & Hermsen 1983 for review).

Pulsed gamma-ray emission has been observed only from the Crab and Vela pulsars and their spectra and light curves differ significantly from each other.

Observations of the Crab pulsar emission from 15 keV to 10 MeV were made from HEAO-1 (Knight 1981) and from 30 MeV to several GeV from SAS-2 (Kniffen et al 1974) and COS-B (Lichti et al 1980). Conflicting reports of very high-energy (>500 GeV) emission are reviewed by Sreekantan (1981). The light curves of the Crab pulsed emission are remarkably similar at all wavelengths, radio, optical, x-ray and gamma-ray, showing two-peaks separated in phase by about 140°. The relative intensity of the peaks, however, varies with wavelength and also with time (Wills et al 1982) at gamma-ray energies >50 MeV. The overall spectrum of the Crab pulsar indicates that its peak luminosity occurs at several MeV (Knight 1981).

Gamma-ray emission from the Vela pulsar has been confirmed only in the energy range from 50 MeV to several GeV by detectors on SAS-2 (Thompson et al 1975) and COS-B (Lichti et al 1980). No pulsed emission was detected at 15 KeV to 10 MeV by HEAO-1 (Knight 1981) and there are conflicting reports of emission at >500 GeV (see Sreekantan 1981). Unlike the Crab, the Vela light curves in the radio, optical, and gamma-ray bands differ greatly from one another and the resultant spectrum is more strongly peaked in the gamma-ray range with a peak luminosity at $\gtrsim 1$ GeV.

The emission process responsible for the pulsed gamma rays is still uncertain, but the most likely, and certainly the best studied, process is curvature radiation. The emission is produced by charged particles moving along intense ($\sim 10^{12}$ gauss) curved magnetic field lines near the polar caps of the neutron star. The particles are accelerated by electric fields induced by the star's rotation. This source of pulsar gamma-ray emission was first suggested by Sturrock (1971) and more detailed models have since been developed (e.g. Ayasli 1981, Harding 1981b) that give good fits to the observed spectra from about 50 MeV to several GeV.

Alternative emission processes for the pulsed gamma rays have been suggested. These include synchrotron emission (Hardee 1979) and Compton scattering (Schlickeiser 1980, Kundt & Krotscheck 1980) of synchrotron photons by ultrarelativistic electrons. These particles could be accelerated (Hardee 1979) in electric fields associated with the breakdown of co-rotation in the magnetosphere at the light circle where the co-rotation velocity approaches the speed of light. Details of the geometry responsible for the light curves and their energy and time variation remain a problem in all of the models.

The peculiar source Cygnus X-3 has been studied in the radio, infrared, x-ray, and gamma-ray bands (see Stepanian 1981 for review). The emission shows a 4.8-hour modulation, attributed to eclipsing of the source or to rotation or precession of an emission beam. Gamma-ray emission at >40 MeV with this modulation period was detected (Galper et al 1977, Lamb et al 1977) from the source for about six months following a giant radio outburst in September, 1972, but not subsequently (Bennett et al 1977). At much higher energies ($>10^3$ GeV), however, gamma-ray emission modulated with the 4.8-hour period has been steadily observed since 1972 (e.g. Neshpor et al 1981, Lamb et al 1982).

There is no generally accepted model of Cygnus X-3 (see Stepanian 1981 and Bignami & Hermsen 1983 for review), but most involve a rapidly rotating neutron star with a close binary companion from which gas may be accreted. The proposed gamma-ray emission processes include those suggested for pulsars as well as decay of π^0 mesons produced by accelerated accreting matter.

In addition to these known objects, several other discrete sources of gamma-ray emission have been reported (Swanenburg et al 1981) but not yet identified with any known astronomical object. The most intense of these unidentified sources, CG195+4, is in fact the second brightest source in the sky at energies >10 MeV. Some if not all of these may also be as yet unidentified pulsars or dense clouds.

5.   EXTRAGALACTIC GAMMA RAYS

Gamma rays have been observed from a few extragalactic objects (see Dean & Ramsden 1981 for review). In addition, there also is a diffuse background that on a coarse scale appears to be isotropic (Trombka et al 1977, Fichtel, Simpson & Thompson 1978). The discrete extragalactic sources are nearby active galaxies of various types, and the diffuse background could be, at least in part, unresolved emission from similar galaxies at cosmological distances. We first discuss the observations of discrete sources and their implications.

5.1 Discrete Extragalactic Sources

There is a variety of very luminous extragalactic objects generally referred to as active galaxies (e.g. Hazard & Mitton 1979). This class of objects contains radio galaxies, Seyfert galaxies, BL Lacertae objects, and quasars. Gamma rays have been observed from some of the brightest of these objects: the radio galaxy Centaurus A (Grindlay et al

1975, Hall et al 1976, Baity et al 1981), the Seyfert galaxy NGC 4151 (Perotti et al 1979, 1981), and the quasar 3C273 (Swanenburg et al 1978, Bignami et al 1981).

Active galaxies have been extensively observed in the radio, infrared, optical, ultraviolet, and x-ray bands. In Figure 4 we combined such observations with the gamma-ray observations for the three active galaxies from which gamma rays were seen. For Centaurus A, the radio measurements are from Kellermann (1974), Price & Stull (1975), and Beall et al (1978); the infrared data are from Grasdalen & Joyce (1976); the optical measurements are from Kunkel & Bradt (1971); the x-ray and gamma-ray results up to a few MeV are from the HEAO-1 observations of Baity et al (1981); the gamma-ray upper limits around 100 MeV are from the SAS-2 observations of Bignami et al (1979); and the gamma-ray measurement above a few hundred GeV is from the atmospheric Čerenkov light observations of Grindlay et al (1975). For NGC 4151 the radio observations are from Haynes et al (1975), the IR and visible data are from Rieke & Lebofsky (1979); the x-ray and gamma-ray data and limits up to several MeV are from the summary of White et al (1980);

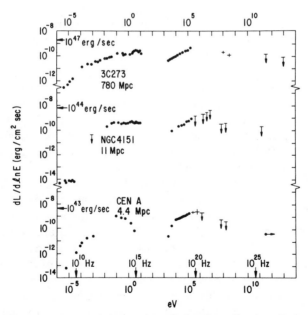

Figure 4 The energy spectra of three active galaxies, the radiogalaxy Centaurus A, the Seyfert galaxy NGC4151, and the quasar 3C273, given in units of luminosity per logarithmic energy interval. The sources of the data are given in the text.

and the gamma-ray limits at $\sim$100 MeV and $\sim$100 GeV are, respectively, from Bignami et al (1979) and Porter & Weekes (1979). For 3C273, the radio data are from Kellermann & Pauliny-Toth (1969); the infrared and visible measurements are from Rieke & Lebofsky (1979); the ultraviolet observations are from Boggess et al (1979); the x-ray data are from Worrall et al (1979) and Bradt et al (1979); the $\sim$100 MeV gamma-ray observations are from Bignami et al (1981); and the high-energy gamma-ray upper limits are from Porter & Weekes (1979).

The spectra are plotted as $dL/d\ln E$ in Figure 4 in order that the relative luminosity at various photon energies may be easily compared. As can be seen, the peak luminosities in the radio galaxy Centaurus A, the Seyfert galaxy NGC4151, and the quasar 3C273 all peak at gamma-ray energies somewhat above 0.1 MeV, which suggests that observations in these energy regions can directly probe the central source of power of these objects. Moreover, this result is apparently not limited to just these three objects. A sample (Boldt 1981) of nearly 20 active galaxies observed in the 3–50-keV range all show differential luminosities, $dL/d\ln E$, that increase with increasing photon energy, demonstrating that the luminosities of these active galaxies also peak at energies at least as high as 50 keV.

A likely source of energy in active galaxies is accretion onto a massive black hole (Lynden-Bell 1969). The luminosity that can be extracted from such accreting matter is not expected to exceed significantly the Eddington limit, $L_E = 4\pi GMm_p c/\sigma_T \simeq 1.2 \times 10^{38} \, M/M_\odot$ erg s^{-1}, where G is the gravitational constant, m_p is the proton mass, σ_T is the Thompson cross section, and $M_\odot$ is the mass of the Sun. If the isotropic luminosity of an accreting object were to exceed L_E substantially, the radiation pressure due to the emergent radiation would be larger than the gravitational attraction on the infalling gas and accretion would stop. The luminosities of 10^{47}, 10^{44}, and 10^{43} erg s^{-1} for 3C273, NGC4151, and Centaurus A, respectively, would thus imply accreting black holes with masses in excess of 10^9, 10^6, and $10^5 \, M_\odot$.

Upper limits on the sizes of the x-ray emitting regions in active galaxies can be obtained from observed time variations (Marshall, Warwick & Pounds 1981, Tennant et al 1981). The source sizes should be less than $c\Delta t$, where Δt is the minimum time scale of the variability. Using such arguments, Bassani & Dean (1981) set upper limits on the sizes of the x-ray sources in a number of active galaxies, including 3C273 and NGC4151. For both these objects they find $c\Delta t \lesssim 10^{15}$ cm. These limits are consistent with the minimal size of the emitting region, which clearly is the Schwarzschild radius of the hole, $r_s = 2GM/c^2 \simeq$

$3 \times 10^5 (M/M_\odot)$ cm. Using the lower limits set on M by the Eddington limit, we see that at least for 3C273 the size of the x-ray source is not much larger than the minimal Schwarzschild radius, i.e. $c\Delta t < 3r_s$. This is also consistent with the large observed luminosities of active galaxies, which require the very efficient release of energy from accreted matter. Such energy release is possible in the deep potential well of the black hole close to its Schwarzschild radius.

Similar constraints could be set on the size of the gamma-ray emitting region, but the time variability of gamma-ray emission from active galaxies is only very poorly known. The fact that finite gamma-ray fluxes up to about one MeV were reported from NGC4151 on two occasions (Perotti et al 1979, 1981), but only lower upper limits were reported at essentially the same energies on another occasion (White et al 1980), can be interpreted as a time variation. On the other hand, 3C273 was observed twice with the instrument on COS B (Swanenburg et al 1978, Bignami et al 1981), and both observations yielded essentially the same gamma-ray flux at ~100 MeV.

Independent information on the nature of the gamma-ray sources, however, can be obtained from considerations of opacity due to photon-photon pair production. The optical depth to gamma rays of the x-ray source region can be estimated from the observed x-ray luminosity and the upper limits on the source size obtained from the observed variability. Thus, Bassani & Dean (1981) find that for isotropic x-ray emission, 3C273, as well as several other quasars, should be opaque to essentially all gamma-rays of energies greater than the pair-production threshold (0.511 MeV). But since ~100-MeV gamma rays were observed from 3C273 (Swanenburg et al 1978, Bignami et al 1981), either the x-ray emission is beamed or the gamma-ray source is much larger than the x-ray source ($\geq 10^{15}$ cm). A large gamma-ray source region in 3C273 would be consistent with the apparent lack of variability of the ~100-MeV gamma-ray luminosity inferred from the two COS B observations of 3C273 (Swanenburg et al 1978, Bignami et al 1981). On the hand, Bassani & Dean (1981) find that the x-ray sources in Seyfert galaxies are transparent to all gamma rays up to about one GeV, so that for these objects the x-ray and gamma-ray sources could be of the same size. Clearly more data are required on the time variability of the gamma-ray emission from active galaxies.

Several radiation mechanisms could be responsible for gamma-ray production in active galaxies. At least some of the gamma rays could be produced by the same mechanisms that produce the x-rays (e.g. Fabian 1979): bremsstrahlung from a hot ($\sim 10^9$ K) gas, Comptonization of cool photons, and the synchrotron self-Compton model.

In addition, there are mechanisms that operate only in the gamma-ray region. As we have already seen, e^+e^- pair production in photon-photon collisions could be important in active galactic nuclei. If the resultant pairs annihilate in an optically thin region, the annihilation radiation should be observable. In the nucleus of our Galaxy, e^+e^- pairs annihilate in a relatively cool region thereby producing a sharp line at 0.511 MeV (Section 4.1). In an active galaxy, however, the annihilation region could be much hotter in which case the line would be both broadened and blueshifted (Ramaty & Mészáros 1981) and thus could produce a photon excess around one MeV. This effect would also explain the absence of the 0.511-MeV line from the spectrum of Centaurus A (Hall et al 1976, Baity et al 1981).

Gamma rays in the MeV region could also be due to Penrose Compton scattering (Piran & Shaham 1977, Leiter & Kafatos 1978, Leiter 1980). Here the scattering takes place in the ergosphere (Penrose 1969) of a rapidly rotating black hole where blueshifted x-ray photons from an accretion disk could interact with transient matter. If a Compton-scattered electron is knocked into the hole's event horizon, the photon picks up rotational energy from the hole and can emerge into free space as a gamma ray. Both e^+e^- pair annihilation and Penrose Compton scattering predict a break in the photon energy spectrum above a few MeV, consistent with observations (Bignami et al 1979).

The ~100-MeV gamma rays in 3C273 could result from π^0 meson decay. However, the observed photon energy spectrum of 3C273 is much steeper than that predicted from π^0 meson decay produced by galactic cosmic-ray interactions (Bignami et al 1981). This difference could indicate a different energetic particle spectrum in active galactic nuclei from that in our Galaxy, or could be caused by photon-photon absorption in the more compact source region of an active galaxy.

5.2 Diffuse Extragalactic Emission

The diffuse gamma-ray background was first observed in the energy range from about 0.1 to 2 MeV by detectors on the lunar probes, Rangers 3 and 5 (Arnold et al 1962, Metzger et al 1964) and at ~100 MeV by a detector on the OSO-3 spacecraft (Clark, Garmire & Kraushaar 1968, Kraushaar et al 1972). Subsequent observations have been made by a number of people (see Horstman, Cavallo & Moretti-Horstman 1975 for review). In particular, the background has been studied in the 0.3–10-MeV range with detectors on several Apollo missions (Trombka et al 1977) and from about 30 to 150 MeV by the instrument on the SAS-2 spacecraft (Fichtel, Simpson & Thompson 1978).

The spectrum of the diffuse background from a few keV to about 100 MeV is shown in Figure 5 (Trombka et al 1977, Fichtel, Simpson & Thompson 1978, Marshall et al 1980, Rothschild 1982). As can be seen, there is considerable structure to this spectrum. The x-ray background below ~50 keV can be well fitted by a thermal bremsstrahlung spectrum with $kT \simeq 40$ keV (Boldt 1981), but at higher energies much more radiation exists than would be predicted by this spectrum.

Estimates (Marshall et al 1980) of the contribution of Seyfert galaxies to the background up to ~50 keV indicate that these galaxies cannot account for the x-ray background. But if the spectra of the Seyfert galaxies extend into the MeV region, as indicated by the observations of Perotti et al (1979, 1981), then the combined contribution of such galaxies could account (Strong, Wolfendale & Worrall 1976, Bignami et al 1979) for the bulk of the gamma-ray background at least up to a few MeV. Furthermore, if at earlier stages in their evolution Seyfert galaxies were more luminous than at the present epoch and if their spectra were thermal ($kT \sim 200$ keV), then these objects could also account for the x-ray background (Leiter & Boldt 1982). Active galaxies could account for the background at the high energies as well, since the steepening of

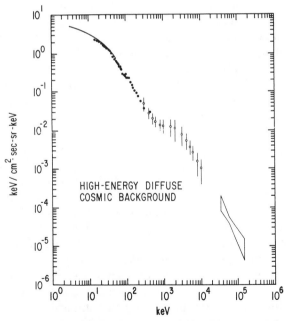

Figure 5 The diffuse x-ray and gamma-ray cosmic background spectrum given in units of photon energy per unit energy interval. The sources of the data are given in the text.

the background spectrum above a few MeV (see Figure 5) is qualitatively similar to the steepening of the energy spectra of individual active galaxies (Bignami et al 1979), and the background spectrum at ~100 MeV has essentially the same spectral index (2.7 ± 0.4, Fichtel, Simpson & Thompson 1978) as that of 3C273 (2.5 ± 0.6, Bignami et al. 1978).

Several other explanations have been put forth for the origin of the diffuse gamma-ray background (e.g. Silk 1970, Horstman, Cavallo & Moretti-Horstman 1975). We mention here, in particular, the possibility of producing the background from matter-antimatter annihilation at the boundaries of superclusters in a baryon-symmetric cosmology (Stecker, Morgan & Bredekamp 1971, Stecker 1978). Here the characteristic peak of the π^0 decay spectrum at 67 MeV is redshifted by the expansion of the universe to about one MeV.

The principal difficulty of a baryon-symmetric cosmology concerns the present photon-to-baryon ratio in the universe (Steigman 1976). In a well-mixed symmetric universe, the bulk of the matter and antimatter is expected to annihilate at a very early stage, which would lead to a present photon-to-baryon ratio of ~10^{18}. This is in contrast to the observed ratio of ~10^9 (e.g. Steigman 1976). Brown & Stecker (1979), however, point out that symmetry breaking in grand unified field theories could produce domains of predominantly matter or antimatter, which, while conserving the overall symmetry of the universe, would greatly reduce the early cosmological annihilation rate and thereby lead to a present photon-to-baryon ratio consistent with observations. The problems of the production and growth of these domains in the early universe were discussed recently by Sato (1981) and by Kuzmin, Shaposhnikov & Tkachev (1981).

6. SUMMARY

We have reviewed observations of cosmic gamma rays, the physical processes responsible for their production, and the astrophysical sites from which they were seen or are expected to be observed in the future. The bulk of the observed gamma-ray emmission is in the photon energy range from about 0.1 MeV to 1 GeV, where observations are made above the atmosphere by instruments carried on balloons and spacecraft. There are also, however, gamma-ray observations at higher energies ($\gtrsim 100$ GeV) obtained by detecting the Čerenkov light produced by the high-energy photons in the atmosphere.

Gamma-ray emission has been observed from sources as close as the Sun and the Moon and as distant as the quasar 3C273, as well as from various other galactic and extragalactic sites. The radiation processes also range from the well understood, e.g. energetic particle interactions with matter, to the still incompletely researched, e.g. radiation transfer in optically thick electron-positron plasmas in intense neutron star magnetic fields. Studies of the gamma-ray sky have already revealed much new information on both the properties of astronomical objects and the high-energy processes that take place in them, and we expect future observations to reveal even more.

ACKNOWLEDGMENTS

We wish to acknowledge financial support from NASA Grant NSG 7541 and the Solar Terrestrial Theory Program. We also acknowledge discussions with E. A. Boldt, E. L. Chupp, A. K. Harding, C. E. Fichtel, J. C. Higdon, D. A. Kniffen, D. Leiter, R. E. Rothschild, and F. W. Stecker.

Albernhe, F., et al. 1981. *Astron. Astrophys.* 94:214

Arnett, W. D. 1977. *Ann. NY Acad. Sci.* 302:90

Arnett, W. D. 1979. *Astrophys. J. Lett.* 230:L32

Arnold, J. R., Metzger, A. E., Anderson, E. C., Van Dilla, M. A. 1962. *J. Geophys. Res.* 67:4876

Axelrod, T. S. 1980. PhD thesis Univ. Calif., Santa Cruz

Ayasli, S. 1981. *Astrophys J.* 249:698

Badhwar, G. D., Stephens, S. A. 1977. *Proc. 15th Int. Cosmic Ray Conf.* 1:198. Plovdiv: Bulg. Acad Sci.

Baity, W. A., et al. 1981. *Astrophys. J.* 244:1981

Bassani, L., Dean, A. J. 1981. *Nature* 294:332

Baym, G., Pethick, C. 1975. *Ann. Rev. Nucl. Sci.* 25:27

Beall, J. H., et al. 1978. *Astrophys. J.* 219:836

Bennett, K., et al. 1977. *Astron. Astrophys.* 59:273

Berko, S., Pendleton, H. N. 1980. *Ann. Rev. Nucl. Part. Sci.* 30:543

Bielefeld, M. J., Reedy, R. C., Metzger, A. E., Trombka, J. T., Arnold, J. R. 1976. *Proc. Lunar Sci. Conf. 7th,* p. 2661. New York: Pergamon

Bignami, G. F., Hermsen, W. 1983, *Ann Rev. Astron. Astrophys.* In press

Bignami, G. F., Fichtel, C. E., Hartman, R. C., Thompson, D. J. 1979. *Astrophys. J.* 232:649

Bignami, G. F., et al. 1981. *Astron. Astrophys.* 93:71

Black, J. H., Fazio, G. G. 1973. *Astrophys. J.* 185:L7

Blumenthal, G. R., Gould, R. J. 1970. *Rev. Mod. Phys.* 42:237

Boggess. A., et al. 1979. *Astrophys. J. Lett.* 230:L131

Boldt, E. A. 1981. *Comments Astrophys.* 9:97

Borner, G., Cohen, J. M. 1973. *Astrophys. J.* 185:959

Bradt, H. V., et al. 1979. *Astrophys. J. Lett.* 230:L5

Brown, R. W., Stecker, F. W. 1979. *Phys. Rev. Lett.* 43:315

Bussard, R. W., Lamb, F. K. 1982. See Lingenfelter, Hudson & Worrall 1982, p. 189

Bussard, R. W., Ramaty, R., Drachman, R. J. 1979. *Astrophys. J.* 228:928

Caraveo, P. A., et al. 1980. *Astron. Astrophys.* 91:L3

Cavallo, G., Gould, R. J. 1971. *Nuovo Cimento B* 2:77

Cesarsky, C. J., Paul, J. A., Shukla, P. G. 1978, *Astrophys. Space Sci.* 59:73

Chupp, E. L. 1982. See Lingenfelter, Hudson & Worral 1982, p. 363

Chupp, E. L., Forrest, D. J. 1981. *Bull. Astron. Soc.* 13:909.

Chupp, E L., et al. 1973. *Nature* 241:333

Chupp, E. L., et al. 1981. *Astrophys J. Lett.* 244:L171

Clark, G. W., Garmire, G. P., Kraushaar, W. L. 1968. *Astrophys. J. Lett.* 153:L203

Clayton, D. D. 1971. *Nature* 234:291
Clayton, D. D. 1982. *Essays in Nuclear Astrophysics*, ed. C. A. Barnes, D. D. Clayton, D. N. Schramm. Cambridge Univ. Press. In press
Clayton, D. D., Hoyle, F. 1974. *Astrophys. J.* 187:L101
Clayton, D. D., Colgate, S. A., Fishman, G. J. 1969. *Astrophys. J.* 155:75
Cline, T. L. 1980. *Comments Astrophys.* 9:13
Cline, T. L. 1981. *Ann. N.Y. Acad. Sci.* 375:314
Cline, T. L. 1982. See Lingenfelter, Hudson & Worrall 1982, p. 17
Cline, T. L., et al. 1981. *Astrophys. J. Lett.* 246:L133
Colgate, S. A., McKee, C. 1969. *Astrophys. J.* 157:623
Colgate, S. A., Petchek, A. G. 1981. *Astrophys. J.* 248:771
Crannell, C. J., Crannell, H., Ramaty, R. 1979. *Astrophys. J.* 229:762
Crannell, C. J., Joyce, G., Ramaty, R., Werntz, C. 1976. *Astrophys. J.* 210:582
Dahlbacka, G. H., Freier, P. S., Waddington, C. J. 1973. *Astrophys. J.* 180:371
Daugherty, J. K., Bussard, R. W. 1980. *Astrophys. J.* 238:296
Dean, A.J., Ramsden, D. 1981. *Philos. Trans. R. Soc. London A* 301:577
Dennis, B. R. et al. 1982. see Lingenfelter, Hudson & Worrall 1982, p. 153
Dyer, P., Bodansky, D., Seamster, A. G., Norman, E. B., Maxson, D. R. 1981. *Phys. Rev.C* 23:1268
Evans, W. D., et al. 1980. *Astrophys, J. Lett.* 237:L7
Fabian, A. C. 1979. *Proc. R. Soc. London A* 366:449
Feenberg, E., Primakoff, H. 1948. *Phys. Rev.* 73;449
Fenimore, E. E., Klebesadel, R. W., Laros, J. G., Stockdale, R. E., Kane, S. R. 1982. *Nature.* 297:665
Fichtel, C. E., Kniffen, D. A., Hartman, R. C. 1973. *Astrophys. J.* 186:L99
Fichtel, C. E., et al. 1975. *Astrophys. J.* 198:163
Fichtel, C. E., et al. 1976. *Astrophys. J.* 208:211
Fichtel, C. E., Simpson, G. A., Thompson, D. J. 1978. *Astrophys. J.* 222:833
Fishman, G. J., Meegan, C. A., Watts, J. W., Derickson, J. H. 1978. *Astrophys. J. Lett.* 223:L13
Frye, G. M., et al. 1972. *Bull. Am. Phys. Soc.* 17:524
Galper, A. M., et al. 1977. *15th Int. Cosmic Ray Conf. Pap.* 1:131 Plovdiv: Bulg. Acad. Sci.
Gardner, B. M., Forrest, D. J., Dunphy, P. P., Chupp, E. L. 1982. *Galactic Cen-*

ter, ed. G. Riegler, R. Blandford. New York: AIP, p. 144
Geiss, J., Reeves, H. 1972. *Astron. Astrophys.* 18:126
Gilman, D., Metzger, A. E., Parker, R. H., Evans, L. G., Trombka, J. I. 1980. *Astrophys. J.* 236:951
Gilman, D., Metzger, A. E., Parker, R. H., Trombka, J. I. 1979. *Astrophys. J.* 229:753
Gould, R. J. 1965. *Phys. Rev. Lett.* 12:511
Grasdalen, G. L., Joyce, R. R. 1976. *Astrophys. J. Lett.* 201:L133
Grindlay, J. E., Helmken, H. F., Hanbury-Brown, R., Davis, J., Allen, L. R. 1975. *Astrophys. J. Lett.* 197:L9
Hall, D. N. B. 1975. *Astrophys. J.* 197:509
Hall, R. D., Meegan, C. A., Walraven, G. D., Djuth, F. T., Haymes, R. C. 1976. *Astrophys. J.* 210:631
Hardee, P. E. 1979. *Astrophys. J.* 227:958
Harding, A. K. 1981a. *Astrophys. J.* 247:639
Harding, A. K. 1981b. *Astrophys. J.* 245:267
Harding, A. K., Stecker, F. W. 1981. *Nature* 290:316
Hartman, R. C., et al. 1979. *Astrophys. J.* 230:597
Harwit, M., Salpeter, E. E. 1973. *Astrophys. J. Lett.* 187:L97
Hayakawa, S. 1952. *Prog. Theor. Phys.* 8:517
Haymes, R. C., et al. 1975. *Astrophys. J.* 201:593
Haynes, R. S., et al. 1975. *A Compendium of Radio Measurement of Bright Galaxies.* Melbourne: CSIRO, Div. Radiophys
Hazard, C., Mitton, S. 1979. *Active Galactic Nuclei.* Cambridge Univ. Press
Heslin, J. P., et al. 1981. *Bull. Am. Astron. Soc.* 13:901
Higdon, J. C. 1974. PhD thesis. Univ. Calif., Los Angeles
Higdon, J. C. 1979. *Astrophys. J.* 232:113
Higdon, J. C. 1982. See Lingenfelter, Hudson & Worrall 1982, p. 115
Higdon, J. C., Lingenfelter, R. E. 1976. *Astrophys. J. Lett.* 208:L107
Hjellming, R. M., Ewald, S. P. 1981. *Astrophys. J. Lett.* 246:L137
Horstman, H. M., Cavallo, G., Moretti-Horstman, E. 1975. *Riv. Nuovo Cimento* 5:255
Hudson, H. S., et al. 1980. *Astrophys. J. Lett.* 236:L91
Hurley, K. 1980. *Non-Solar Gamma Rays*, ed. R Cowsik, R. D. Wills, P. 123. Oxford: Pergamon
Hutchinson, G. W. 1952. *Philos. Mag.* 43:847
Ibragimov, I. A., Kocharov, G. E. 1977. *Sov. Astron. Lett.* 3(5):221

Jacobson, A. S., Ling, J. C., Mahoney, W. A., Willett, J. B. 1978. *Gamma-Ray Spectroscopy in Astrophysics,* ed. T. L. Cline, R. Ramaty, P. 228, Greenbelt, Md: Goddard NASA
Jennings, M. C. 1982. See Lingenfelter, Hudson & Worrall 1982, p. 107
Johnson, W. N., Haymes, R. C. 1973. *Astrophys. J.* 184:103
Johnson, W. N., Harnden, F. R., Haymes, R. C. 1972. *Astrophys. J.* 172:L1
Kane, S. R., et al. 1980. See Sturrock 1980, P. 187
Katz, J. 1982. *Astrophys. J.* In press
Kellermann, K. I. 1974. *Astrophys. J. Lett.* 194:L135
Kellermann, K. I., Pauliny-Toth, I. I. K. 1969. *Astrophys. J. Lett.* 155:L71
Kellermann, K. I., Shaffer, D. B., Clark, B. G., Geldzahler, B. J. 1977. *Astrophys. J.* 214:L61
Klebesadel, R. W., Fenimore, E. E., Laros, J. G., Terrell, J. 1982. See Lingenfelter, Hudson & Worrall 1982, P. 1
Klebesadel, R. W., Strong, I. B., Olson, R. A. 1973. *Astrophy. J. Lett.* 182:L85
Kniffen, D. A., Fichtel, C. E. 1981. *Astrophys. J.* 250:389
Kniffen, D. A., et al. 1974, *Nature* 251:397
Knight, F. K. 1981. PhD thesis. Univ. Calif., San Diego
Kozlovsky, B., Ramaty, R. 1974. *Astrophys. J. Lett.* 191:L43
Kozlovsky, B., Ramaty, R. 1977. *Astrophys. Lett.* 19:19
Kraushaar, W. L., et al. 1972. *Astrophys. J.* 177:341
Kundt, W., Krotscheck, E. 1980. *Astron. Astrophys.* 83:1
Kunkel, W. E., Bradt, H. V. 1971. Astrophys. J. Lett. 170:L7
Kuzmin, V. A., Shaposhnikov, M. E., Tkachev, I. I. 1981. *Phys. Lett.* 105B:165
Lacy, J. H., Townes, C. H., Geballe, T. R., Hollenbach, D. J. 1980. *Astrophys. J.* 241:132
Lamb, D.Q. 1982 See Lingenfelter, Hudson & Worrall 1982, p. 249
Lamb, R. C., Godfrey, C. P., Wheaton, W. A., Turner, P. 1982. *Nature* 296:543
Lamb, R. C., et al. 1977. *Atrophys. J.* 212:L63
Laros, J. G., et al. 1981. *Astrophys. J. Lett.* 245:L63
Leiter, D. 1980. *Astron. Astrophys.* 89:370
Leiter, D., Boldt, E. A. 1982. *Astrophys. J.* In press
Leiter, D., Kafatos, M. 1978. *Astrophys. J.* 226:32
Leventhal, M., MacCallum, C. J., Stang, P. D. 1978. *Astrophys. J.* 225:L11

Leventhal, M., MacCallum, C. J., Watts, A. C. 1977. *Astrophys. J.* 216:491
Lewin, W. H. G., Joss, P. C. 1981. *Space Sci. Rev.* 28:3
Liang, E. P. T. 1981. *Nature* 292:319
Lichti, G. G., et al. 1980. See Hurley 1980, p. 49
Ling, J. C., Mahoney, W.A., Willett, J. B., Jacobson, A. S. 1979. *Astrophys. J.* 231:896
Ling, J. C., Mahoney, W. A., Willett, J. B., Jacobson, A. S. 1982 See, Lingenfelter, Hudson & Worrall 1982, p. 143
Lingenfelter, R. E., Ramaty, R. 1967. *High-Energy Nuclear Reactions in Astrophysics,* ed. B. S. P. Shen, p. 99. New York: Benjamin
Lingenfelter, R. E., Ramaty, R. 1977. *Astrophys. J.* 211:L19
Lingenfelter, R. E., Ramaty, R. 1982. See Gardner et al 1982, p. 148
Lingenfelter, R. E., Flamm, E. J., Canfield, E. H., Kellman, S. 1965. *J. Geophys. Res.* 70:4077, 4087
Lingenfelter, R. E., Higdon, J. C., Ramaty, R. 1978. See Jacobson et al 1978, p. 252
Lingenfelter, R. E., Hudson, H. S., Worrall, D. M. 1982. *Gamma-Ray Transients and Related Astrophysical Phenomena.* New York: AIP
Lynden-Bell, D. 1969. *Nature* 233:690
Mahoney, W. A., et al. 1982. *Astrophys. J.* 262: In press
Marshall, F. E., et al. 1980. *Astrophys. J.* 235:4
Marshall, N., Warwick, R. S., Pounds, K. A. 1981. *Mon. Not. R. Astron. Soc.* 194:987
Matteson, J. L. 1982. See Gardner et al 1982 p. 109
Matteson, J. L., Nolan, P. L., Peterson, L. E. 1979. *X-Ray Astronomy*, ed. W.A. Baity, L.E. Peterson, p. 543 Oxford: Pergamon
Mayer-Hasselwander, H. A., et al. 1980. *Ann. NY Acad. Sci.* 336:211
Mayer-Hasselwander, H. A., et al. 1982. *Astron. Astrophys.* 105:164
Mazets, E. P., Golenetskii, S. V. 1981. *Astrophys. Space Sci.* 75:47
Mazets, E. P., Golenetskii, S. V., Ilyinskii, V. N., Aptekar, R. L., Guryan, Yu. A. 1979. *Nature* 282:587
Mazets, E. P., et al. 1981a. *Astrophys. Space Sci.* 80:1
Mazets, E. P., Golenetskii, S. V., Aptekar, R. L., Guryan, Yu. A., Ilyinskii, V. N. 1981b. *Nature* 290:378
Metzger, A. E., Anderson, E. C., Van Dilla, M. A., Arnold, J. R. 1964. *Nature* 204:766

Metzger, A. E., Trombka, J. I., Reedy, R. C., Arnold, J. R. 1974. *Proc. Lunar Sci. Conf. 5th,* p. 1067. New York: Pergamon

Neshpor, Yu. I., et al. 1981. *Philos. Trans. R. Soc. London Ser. A* 301:633

Okeke, P. N., Rees, M. J. 1980. *Astron. Astrophys.* 81:263

Paul, J. A., et al. 1978. *Astron. Astrophys.* 63:L31

Penrose, R. 1969. *Riv. Nuovo Cimento,* Vol. 1, Num. Spec. 252

Perotti, F., et al. 1979. *Nature* 282:484

Perotti, F., et al. 1981. *Astrophys. J. Lett.* 247:L63

Peterson, L. E., Winckler, J. R. 1959. *J. Geophys. Res.* 64:697

Pinkau, K. 1980. *Ann. NY Acad. Sci.* 336:234

Piran, T., Shaham, J. 1977. *Phys. Rev. D* 17:1615

Porter, N. A., Weekes, T. C. 1979. *Smithson. Astrophys. Observ. Spec. Rep. No.* 381

Price, K. M., Stull, M. A. 1975. *Nature* 255:467

Prince, T. A., Ling, J. C., Mahoney, W. A., Riegler, G. R., Jacobson, A. S. 1982. *Astrophys. J. Lett.* 255: L81

Ramaty, R., Crannell, C. J. 1976. *Astrophys. J.* 203:766

Ramaty, R., Lingenfelter, R. E. 1977. *Astrophys. J. Lett.* 213:L5

Ramaty, R., Lingenfelter, R. E. 1973. *13th Int. Cosmic-Ray Conf. Pap.* 2:1590. Univ. Denver, Colo.

Ramaty, R., Lingenfelter, R. E. 1979. *Nature* 278:127

Ramaty, R., Lingenfelter, R. E. 1981. *Philos. Trans. R. Soc. London Ser. A* 301:671

Ramaty, R., Mészáros, P. 1981. *Astrophys. J.* 250:384

Ramaty, R., Kozlovsky, B., Lingenfelter, R. E. 1975. *Space Sci. Rev.* 18:341

Ramaty, R., Kozlovsky, B., Lingenfelter, R. E. 1979. *Astrophys. J Suppl.* 40:487

Ramaty, R., Kozlovsky, B., Suri, A. N. 1977. *Astrophys. J.* 214:617

Ramaty, R., et al. 1980. *Nature* 122

Ramaty, R., Lingenfelter, R. E., Bussard, R. W. 1981. *Astrophys. Space Sci.* 75:193

Ramaty, R., Lingenfelter, R. E., Kozlovsky, B. 1982. See Lingenfelter, Hudson & Worrall 1982, p. 211

Ramaty, R., McKinley, J. M., Jones, F. C. 1982. *Astrophys. J.* 256: 238

Reedy, R. C. 1978. *Proc. Lunar Planet. Sci. Conf. 9th,* P. 2961. New York: Pergamon

Reedy, R. C., Arnold, J. R., Trombka, J. T. 1973. *J. Geophys. Res.* 78:5847

Riegler, G. R., et al. 1981. *Astrophys. J. Lett.* 248:L13

Rieke, G. H., Lebofsky, M. J. 1979. *Ann. Rev. Astroy. Astrophys.* 17:477

Rothschild, R. E. 1982. *Proc. Workshop X-Ray Astron. Spectrosc.* Goddard Space Flight Center, Md: NASA, p. 599

Ruderman, M. 1975. *Ann. NY Acad. Sci.* 262:164

Salvati, M. 1980. See Hurley 1980, p. 29

Salvati, M., Massaro, E. 1982. *Mon. Not. R. Astron. Soc.* 198:11

Sato, K. 1981. *Phys. Lett.* 99B:66

Schaefer, B. E. 1981. *Nature* 294:722

Schlickeiser, R. 1980. *Astrophys. J.* 236:945

Silk, J. 1970. *Space Sci. Rev.* 11:671

Sreekantan, B. V. 1981. *Philos. Trans. R. Soc. London Ser. A* 301:629

Stecker, F. W. 1969. *Astrophys. Space Sci.* 3:579

Stecker, F. W. 1970. *Astrophys. Space Sci.* 6:377

Stecker, F. W. 1971. *Cosmic Gamma Rays, NASA SP-249* Washington, DC: NASA

Stecker, F. W. 1977. *Astrophys. J.* 212:60

Stecker, F. W. 1978. *Nature* 273:493

Stecker, F. W., Morgan, D. L., Bredekamp, J. 1971. *Phys. Rev. Lett.* 27:1469

Stecker, F. W., Solomon, P. M., Scoville, N. Z., Ryter, C. E. 1975. *Astrophys. J.* 201:90

Steigman, G. 1976. *Ann. Rev. Astron. Astrophys.* 14:339

Stepanian, A. A. 1981. *17th Int. Cosmic Ray Conf. Pap.* 1:50. Paris: French Atomic Energy Commission

Strickman, M. S., Kurfess, J. D., Johnson, W. N. 1982. *Astrophys. J. Lett.* 253: L23

Strong, A W., Wolfendale, A. W., Worrall, D. M. 1976. *J. Phys. A* 9:1553

Sturrock, P. A. 1971. *Astrophys. J.* 164:529

Sturrock, P. A. 1980. *Solar Flares.* Boulder: Colo. Assoc. Univ. Press

Sturrock, P. A., Baker, K. B. 1979. *Astrophys. J.* 234:612

Swanenburg, B. N., et al. 1978. *Nature* 275:298

Swanenburg, B. N., et al. 1981. *Astrophys. J.* 243:L69

Teegarden, B. J. 1982. See Lingenfelter, Hudson & Worrall 1982, p. 123

Teegarden, B. J., Cline, T. L. 1980. *Astrophys. J. Lett.* 236:L67

Teegarden, B. J., Cline, T. L. 1981. *Astrophys. Space Sci.* 75:181

Tennant, A. F., Mushotzky, R. F., Boldt, E. A., Swank, J. H. 1981. *Astrophys. J.* 251:15

Thompson, D J., et al. 1975. *Astrophys. J. Lett.* 200:L79

Trombka, J. I., et al. 1977. *Astrophys. J.* 212:925

Tsygan, A. I. 1975. *Astron. Astrophys.* 44:21; 49:159

Vedrenne, G. 1981. *Philos. Trans. R. Soc. London Ser. A* 301:645

Von Rosenvinge, T. T., Ramaty, R., Reames, D. V. 1981. *17th Int. Cosmic-Ray Conf. Pap.* 3:28. Paris: French Atomic Energy Commission

Wang, H. T., Ramaty, R. 1974. *Solar Phys.* 36:129

White, R. S., et al. 1980. *Nature* 284, 608

Wills, R. E., et. al. 1982. *Nature* 296:723

Woosley, S. E. 1982. See Lingenfelter, Hudson & Worrall 1982, p. 273

Woosley, S. E., Taam, R. E. 1976. *Nature* 263:101

Woosley, S. E., Weaver, T. A. 1980. *Astrophys. J.* 238:1017

Woosley, S. E., Axelrod, T. S., Weaver, T. A. 1981. *Comments Nucl. Part. Phys.* 9:185

Worrall, D. M., et al. 1979. *Astrophys. J.* 232:683

Ann. Rev. Nucl. Part. Sci. 1982. 32:271–308
Copyright © 1982 by Annual Reviews Inc. All rights reserved

RESONANCES IN HEAVY-ION NUCLEAR REACTIONS[1]

T. M. Cormier[2]

Department of Physics and Astronomy and Nuclear Structure Research Laboratory, University of Rochester, Rochester, New York 14627

CONTENTS

1. INTRODUCTION

With the advent of tandem electrostatic accelerators in the early 1960s, nuclear physicists began the exploration of nuclear properties under extreme conditions of shape, angular momentum, temperature, density, and N/Z with heavy-ion reactions. This exploration continues today with ever-increasing vigor. In this review we examine some of the new and unexpected aspects of the nuclear structure of light nuclei that have been revealed in this exploration. We are primarily concerned with a

[1] Supported in part by The National Science Foundation under Grant No. PHY-79-23307.
[2] Partial support from the Alfred P. Sloan Foundation.

271

new class of nuclear eigenstates observed experimentally in heavy-ion resonance reactions. These states are now known to exist in nuclei ranging from the s–d shell through the $f_{7/2}$ shell and are characterized by unexpectedly large partial decay widths for decay into fragments of comparable mass. These fission-like decays are usually taken as evidence for extreme deformations, and the term "nuclear molecular state" has become almost universal in the literature. It should be pointed out, however, that the terminology, although universally accepted, implies more than is actually known about these states. In the following I give a brief account of the development of resonance studies with heavy-ion reactions, including current theoretical efforts.

The examples I use come mostly from the $^{12}C + ^{12}C \rightarrow ^{24}Mg$ system simply because I am most familiar with work in this area. Other less restrictive reviews can be found in the literature (see for example Cindro 1978).

2. RESONANCES NEAR THE BARRIER

Figure 1 shows the first reported detailed study of the energy dependence of a heavy-ion nuclear reaction. These data of Almqvist, Bromley & Kuehner (1960) show the energy variation of the inclusive yields of neutrons, protons, α particles, and γ rays. The energy range shown here spans the $^{12}C - ^{12}C$ Coulomb barrier ($E_{cm} \sim 6.5\,MeV$), which accounts for the steep rise of all of the yield curves. At least three very narrow features were observed in these measurements at ~ 5.7, 6.0, and $6.3\,MeV$ with total widths of the order of $100\,keV$. The observation of a simultaneous enhancement in all reaction channels at these energies identifies these structures as proper states of the ^{24}Mg compound system at excitation energies of $\sim 20\,MeV$. The total level density in ^{24}Mg at these energies is in the range of $100–1000$ levels per MeV, and thus the observation here of strong isolated resonances points to the special structure and/or symmetry of these states. In the intervening two decades since their discovery, exhaustive experimental and theoretical efforts have focused on these and other more recently discovered molecular states in ^{24}Mg in this excitation energy region.

Figure 2 shows an up-to-date compilation of the ^{24}Mg molecular spectrum from $E_x(^{24}Mg) \cong 17–24\,MeV$. Plotted here is the so-called nuclear structure factor, which is essentially the total $^{12}C + ^{12}C$ reaction cross section divided by an energy-dependent penetrability factor that removes the very strong energy variation due to the Coulomb barrier. It is worth mentioning that the cross sections displayed here become extremely small below $E_{cm} \sim 5 \, MeV$, and it is a tribute to the ingenuity of experimenters in this field that the spectrum is so well determined.

In most instances the spins of these states are firmly established, and their rather low values in this energy range place them well inside the ^{24}Mg yrast line where the background level density of states with the same spin is quite high.

Spin assignments for states high in the continuum such as these are most easily established from the observation of a resonant α decay to the ^{20}Ne ground state. In this process the channel spin is zero, and the angular distributions show a characteristic $P_J^2 (\cos \theta)$ shape for a state of spin J (Erb et al 1976).

Elastic scattering measurements across a resonance potentially contain significantly more information. In addition to spin, one hopes to

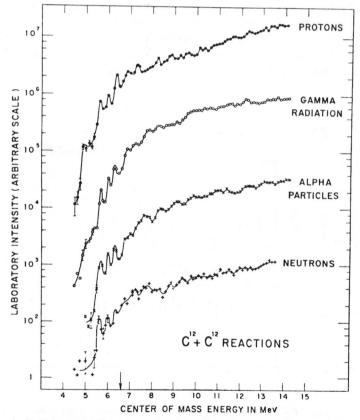

Figure 1 Inclusive yield of n, p, α, and γ's from $^{12}C + ^{12}C$ reactions in the vicinity of the Coulomb barrier. Narrow resonances are visible at a few energies.

learn the partial width Γ_c for decay into $^{12}C + {}^{12}C$. It is the magnitude of this number that distinguishes molecular states from the average compound nucleus state. Elastic scattering measurements close to the Coulomb barrier are difficult, however, where even a very pronounced resonance will produce only minor deviations of the angular distribution from Mott scattering. Recently, however, precision elastic scattering measurements were completed at Yale in the range of $E_{cm} = 6.44$–$6.99\,MeV$. A sample of the data is shown in Figure 3. The solid lines are phase shift fits for an s matrix $s_L = \eta\,\exp(2i\delta_L)$. The fits were constrained by the experimental total reaction cross section and give an essentially perfect description of the data. The left panel of Figure 4 shows the phase shifts δ_L and reflectivities obtained from these fits along with a comparison of the predicted and experimental nuclear structure factor. The $E_{cm} = 6.64$ and $6.83\,MeV$ resonances are seen to have $J^{\pi} = 2^+$ and 4^+, respectively, in agreement with other independent determinations. The right-hand panel shows a Breit-Wigner parameterization of the s matrix from which it is deduced that $\Gamma_c/\Gamma = 0.29$ and 0.09 for the 2^+ and 4^+ states, respectively. These partial widths imply an enhancement factor of ~ 100 compared to statistical compound nucleus

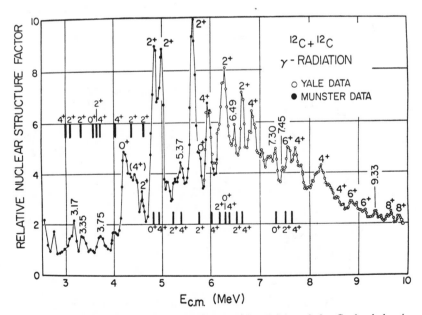

Figure 2 The $^{12}C + {}^{12}C$ molecular spectrum in the vicinity of the Coulomb barrier showing several firm spin assignments. The vertical bars are the result of a calculation discussed in the text. The data come from the Yale and Munster groups (Bromley 1978).

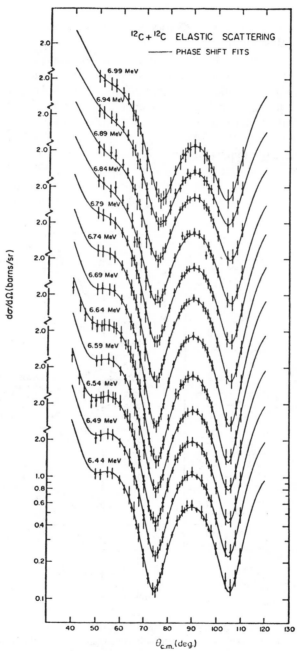

Figure 3 Detailed $^{12}C + ^{12}C$ elastic scattering angular distributions at finely spaced energies from 6.44 to 6.99 MeV in the center of mass. The solid lines are phase shift fits to the data.

decay. Such enhancement factors are typical of those measured or inferred in this region.

Having seen that $^{12}C + ^{12}C$ reactions at low energies can proceed through exotic compound nucleus states, it is reasonable to question whether these molecular states are sufficient in number to alter significantly the average energy dependence of the total $^{12}C + ^{12}C$ cross section below the barrier. This question is of fundamental importance to our understanding of carbon burning in stellar interiors. Some of the earliest measurements (Mazarkis & Stephens 1973) of the $^{12}C + ^{12}C$ total cross section actually suggested that the $^{12}C + ^{12}C$ fusion mechanism was anomalous, and that "absorption under the barrier" was operating, which would drastically alter the extrapolation to stellar

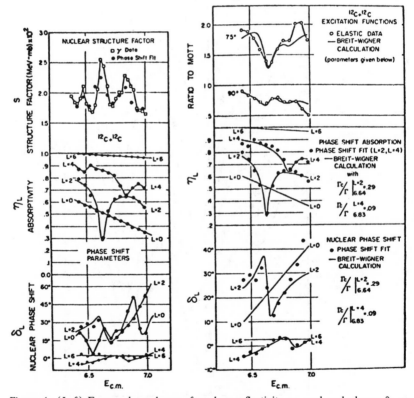

Figure 4 (*Left*) Energy dependence of nuclear reflectivity η_L and real phases δ_L as determined in fits to elastic angular distributions and nuclear structure factor calculated from fitted s-matrix elements compared with experiment. (*Right*) Breit-Wigner fits to the fitted s-matrix elements (η_L and δ_L) and the resulting agreement with elastic scattering excitation functions.

energies. More recent measurements have shown, however, that the average behavior of the ^{12}C + ^{12}C total reaction cross section is quite regular at low energies, obviating the need for absorption under the barrier. Figure 5, taken from Erb et al (1980), underscores this point by comparing the ^{12}C + ^{12}C total reaction cross section to a simple incoming wave boundary condition model.

3. EXPERIMENTS ABOVE THE BARRIER

Early experiments (for example, Almqvist et al 1964) at energies well above the barrier largely reported that cross sections to individual channels fluctuated randomly with energy in a manner consistent with the model of Ericson (1963). More exhaustive studies in recent years, based frequently on very large data sets, revealed in addition a rich spectrum of resonances whose widths increase relatively slowly with increasing excitation energy. The problem of isolating molecular states

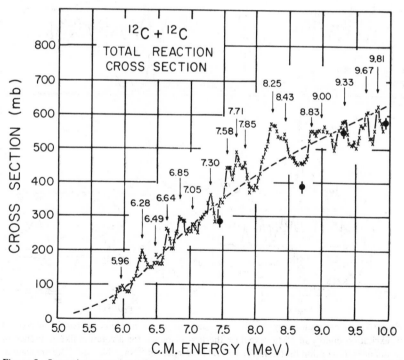

Figure 5 Incoming wave boundary condition model (*dashed line*) fitted to the ^{12}C + ^{12}C total reaction cross section in the vicinity of the Coulomb barrier.

in the higher energy region is essentially one of signal-to-noise ratio, where the noise is random Ericson fluctuations frequently compounded by large nonresonant background processes.

An early systematic study of the ^{12}C + ^{12}C system in the higher energy region by Cosman et al (1975) is shown in Figure 6. This work reported the first evidence for a systematic spacing of resonances of increasing spin suggestive of a highly deformed rotational band. Shown here are excitation functions for light-particle decay to mostly high-spin states in ^{22}Na, ^{23}Na, and ^{20}Ne. The neutron channel not shown in Figure 6, but

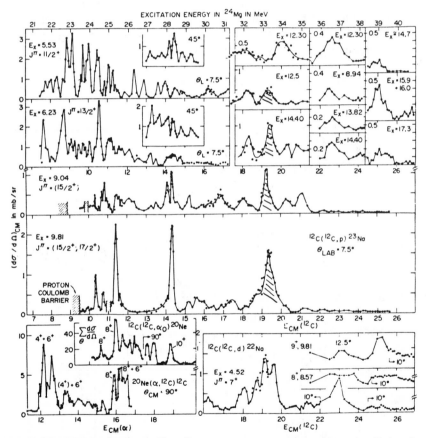

Figure 6   An early summary of ^{12}C + ^{12}C reactions above the Coulomb barrier all plotted vs excitation energy in ^{24}Mg revealing the first evidence for a rotation-like sequence of resonances. Excitation functions for several final states in ^{22}Na, ^{23}Na, and ^{20}Ne are shown. The cross-hatched peak is the first prominent resonance observed well above the barrier at E_{cm} = 19.3 MeV [$E_x(^{24}$Mg$)$ = 33.3 MeV].

studied by Evers et al (1977), has an energy dependence that mirrors that of the proton channel. The cross-hatched peak was the first $^{12}C + ^{12}C$ resonance to be discovered well above the barrier (Van Bibber et al 1974) ($E_{cm} = 19.3$ MeV), and in the following we concentrate on the properties of this state.

To demonstrate the molecular character of this resonance we must establish a Γ_c value in excess of compound nucleus expectation. Figures 7 and 8 show elastic and inelastic scattering excitation functions measured across the same range (Cormier et al 1978). The $E_{cm} = 19.3$-MeV resonance is prominent in both data sets.

Early studies (Shapira et al 1974) of $^{12}C + ^{12}C$ elastic scattering showed that the strong energy dependence seen here is consistent with statistical fluctuations. The newer, more extensive data, however, show substantial cross-channel correlations that are inconsistent with statistical fluctuations. These data are adequate to identify safely the strongest resonances, but the weaker states will require very careful study. The extent of cross correlations present in the inelastic scattering data is exhibited in Figure 9.

For a pair of excitation functions i and j one defines

$$C_{ij} = \frac{2}{N(N-1)} \sum_{i>j} \frac{\left\langle \left(\frac{\sigma_i}{\langle \sigma_i \rangle_\varepsilon} - 1 \right) \left(\frac{\sigma_j}{\langle \sigma_j \rangle_\varepsilon} - 1 \right) \right\rangle_E}{\sqrt{R_i(0)R_j(0)}}.$$

In this expression N is the number of excitation functions, $\langle \sigma_i \rangle_\varepsilon$ is a running average (Pappalardo 1964) over an interval of width ε, $\langle \ \rangle_E$ indicates averaging over the entire energy interval, and the $R(0)$ factors normalize C_{ij}. The frequency distribution of C_{ij} for the inelastic scattering data of Figure 8 is shown in Figure 9. For a purely random set of excitation functions, the mean value of C_{ij} is zero. The set of 16 inelastic scattering excitation functions analyzed here (120 distinct C_{ij}'s) have a mean C_{ij} value of 0.29. The probability that a random data set will produce an average correlation this large is less than 1%.

To demonstrate that the resonances seen here are molecular (i.e. preferentially decay into fission-like channels), we consider the measurement of Γ_c for the 19.3 MeV(12^+) resonance. There are three fission-like channels to consider, all of which couple strongly to this (12^+) resonance: the elastic channel 0^+0^+, the single inelastic channel 2^+0^+, and the mutual inelastic channel 2^+2^+ (denoted 00, 20, and 22 below). In addition, the resonance is damped into the compound nucleus (CN), a fact that accounts for its observation in the n, p, d, etc channels. Figure 10 shows isolated Breit-Wigner plus optical model fits

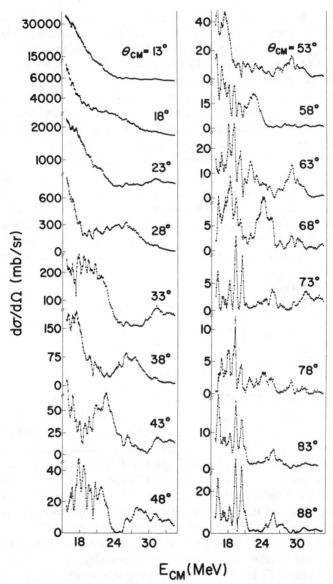

Figure 7 Detailed $^{12}C + ^{12}C$ elastic scattering excitation functions revealing a dense spectrum of narrow peaks.

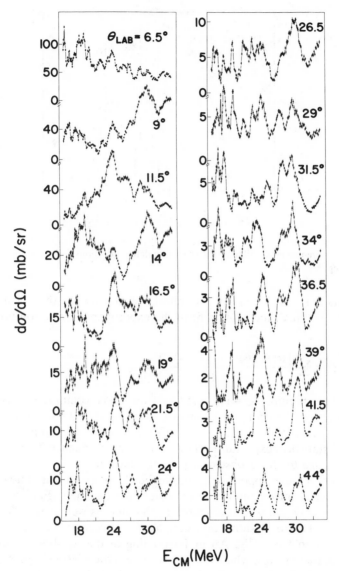

Figure 8 Detailed ^{12}C + ^{12}C inelastic scattering excitation functions to the first excited state $J^{\pi} = 2^+$, $E_x = 4.43$ MeV. Most of the strong peaks visible in the elastic data are also visible here.

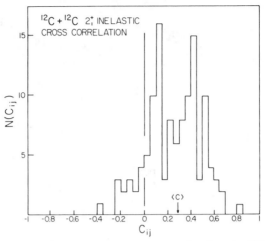

Figure 9 Frequency distribution of cross correlation functions for $^{12}C + ^{12}C$ inelastic scattering.

(Cormier et al 1978) to the elastic scattering in the vicinity of $E_{cm} = 19.3$ MeV, including a fit to the experimental deviation function $D(E)$:

$$D(E) = \sum_{\theta} \left| \frac{\sigma(E,\ \theta) - \langle \sigma(E,\ \theta) \rangle_{\varepsilon}}{\langle \sigma(E,\ \theta) \rangle_{\varepsilon}} \right|.$$

The data require $\Gamma_c = \Gamma_{00} = 75$ keV, and this combined with total cross-section data for this resonance gives $\Gamma_{00}/\Gamma = 0.18$, $\Gamma_{20}/\Gamma = 0.32$, $\Gamma_{22}/\Gamma \leq 0.05$, and $\Gamma_{CN}/\Gamma = 0.45$. The Γ_{00} and Γ_{20} numbers greatly exceed statistical expectation showing that this resonance is very similar in structure to those observed near the barrier.

A new feature appearing in the higher-lying molecular states is the direct decay into a multitude of fission-like channels. For $^{12}C + ^{12}C$, these are all open inelastic scattering channels as well as the $^{10}B + ^{14}N$ channels. Figure 11 shows an example of a molecular state that decays strongly into $^{10}B + ^{14}N$. An understanding of the two-body branching ratios of these states, which ultimately reflects their intrinsic microscopic structure, presents an exciting theoretical challenge. For example, the resonance shown in Figure 11 does not decay within detectable limits into $^{11}B + ^{13}N$, a channel that is kinematically virtually equivalent to $^{10}B + ^{14}N$.

A grand summary of the $^{12}C + ^{12}C$ resonances in the high-energy region (Cosman et al 1981) is shown in Figure 12. Although the spectrum looks hopelessly complicated, there *is* still hope. Figure 13*a*

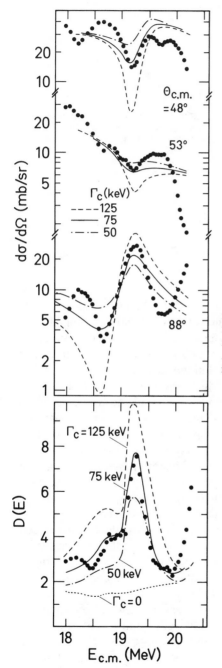

Figure 10 Breit-Wigner plus optical model fits to $^{12}C + ^{12}C$ elastic scattering functions near the $E_{cm} = 19.3$-Me V resonance; the fit to the experimental function D (E) is included. The value $\Gamma_c \cong 75$ ke V is indicated.

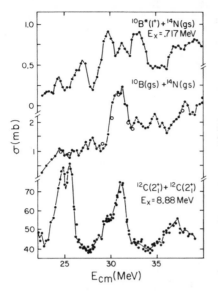

Figure 11 The energy dependence of total cross sections in the $^{10}B + ^{14}N$ and $^{10}B^* + ^{14}N$ channels compared with $^{12}C^* + ^{12}C^*$.

shows the same data plotted as E_x vs $J(J+1)$ including all of the near-barrier resonances as well. The solid points have reasonably firm spin assignments while the open points are less certain. In spite of the few uncertain spin assignments, the pattern is quite clear. States of a given spin cluster together in a limited excitation energy region. Remarkably, the total width over which any cluster is spread does not increase dramatically from spin 0 to spin 16, and the trajectory along which all these states lie is straight. This figure suggests that *total* cross sections for various processes might display gross oscillations resulting from the cumulative effect of the clustered resonances.

Figure 13b compares the total $^{12}C + ^{12}C$ fusion cross section measured at Argonne National Laboratory (Sperr et al 1976) with the locations of the resonances from Figure 13a. The comparison is quite suggestive. It seems possible that the broad oscillations in the fusion cross section may merely reflect the cumulative effect of many individual narrow resonances. Of course, it might equally well be true that the resonances are simply more visible within the confines of an enhancement in the fusion cross section, and that the enhancements have an independent origin. This is discussed in greater detail below.

Other total cross sections have been measured for specific channels. Figure 14 shows data taken by my colleagues and me at Rochester, Brookhaven National Laboratory, and Stony Brook (Cormier et al 1977, 1978) for single and mutual inelastic scattering to the 2_1^+ state and single inelastic scattering to the 3_1^-, 4_1^+, and 0_2^+ of ^{12}C (Fulton et al

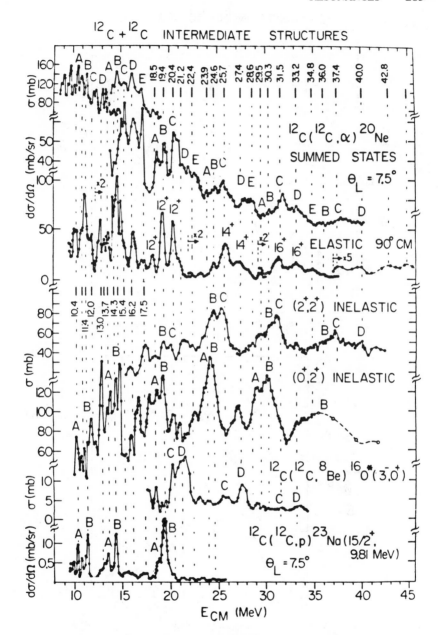

Figure 12   A grand compilation of ^{12}C + ^{12}C excitation functions at energies above the barrier. A complete set of energies at which nonstatistical features are seen is given.

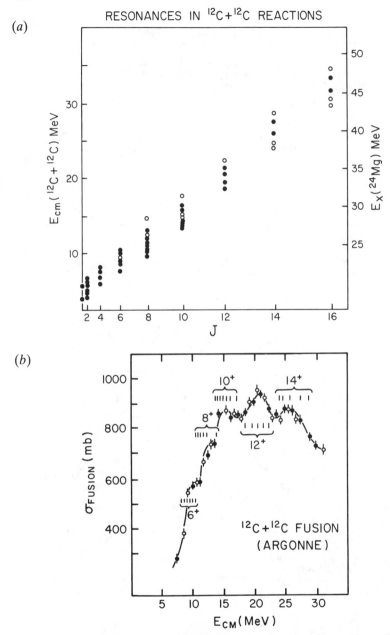

Figure 13   (*Top*) E_x vs $J(J+1)$ plot for all of the resonances from Figure 12 as well as all resonances known from sub-Coulomb studies. Open circles indicate non-firm spin assignments. (*Bottom*) The energy dependence of the $^{12}C + ^{12}C$ total fusion cross section. The observed structure is suggestively correlated with the positions of known intermediate structure resonances (*vertical lines*).

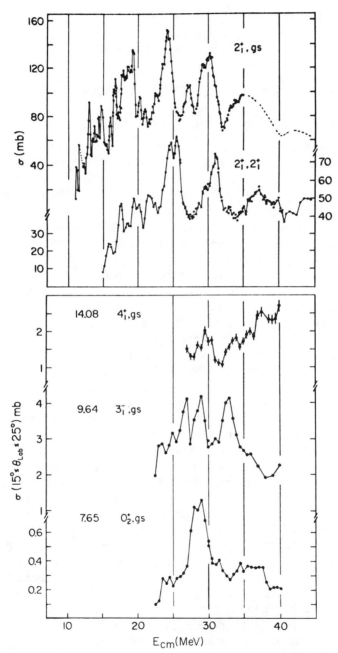

Figure 14 The energy dependence of total cross sections in various $^{12}C + ^{12}C$ inelastic scattering channels.

1980). All of these show broad oscillations that could conceivably be simply the result of the resonance clusters, but as we discuss in the next section alternative explanations are also possible.

4. MODELS OF MOLECULAR STATES

4.1 *Fragmentation of Giant Resonances*

The clustering of the narrow resonances into a limited energy region is reminiscent of a giant resonance phenomenon. This point was emphasized by Feshbach (1976) in a recent review. The situation envisaged assumes a single simple eigenstate $|\alpha\rangle$ at each J fragmented by some weak residual interaction and spread through an energy width $\Gamma^{\downarrow}$ over the more complex states $|\beta_i\rangle$ of the system. The simple state is called an isolated doorway state (Feshbach et al 1967). Through the residual interaction the characteristic "strength" of the doorway is spread over several more complex states. Given this situation, the most relevant experimental quantity becomes the strength function. This function displays the distribution in energy of some characteristic of the doorway state. An example from some work done at Yale has been discussed by Bromley (1978) for the distribution of Γ_c for $J = 2$ and 4. In principle, when all of the fragments are known, the excitation energy of the doorway is determined from the center of gravity of the strength for given J.

Various theoretical discussions of the $^{12}C + ^{12}C$ resonances differ mainly in their prescriptions for the states $|\alpha\rangle$ and $|\beta_i\rangle$. The simplest imaginable situation occurs when $|\alpha\rangle$ is a shape resonance, i.e. a single particle state in the ion-ion mean field, and $|\beta_i\rangle$ are the statistical compound nucleus states. This possibility is quickly ruled out, however, by level density arguments since there are far too few fragments observed at each J compared to the compound nucleus level density. Ultimately, however, some damping to the statistical compound states is required to explain the observation of resonant strength in channels fed by compound nucleus evaporation. This process is usually pictured as shown in Figure 15. This is a schematic view of the time evolution of the scattering state wave function. The doorway state is formed in the first step and is responsible for gross structure in the energy dependence of cross sections. In the second step, the doorway state couples to some slightly more complicated degree of freedom of the system, splitting the gross structure into a number of intermediate structure fragments. Higher-order processes are possible in which more complicated degrees of freedom are involved. The series approaches, in the limit, the statistical compound nucleus, a thermally equilibrated state composed

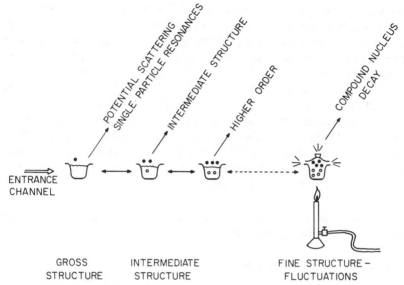

Figure 15 A schematic illustration of the time evolution of a scattering state showing a doorway state, intermediate structure states, etc through the statistical compound nucleus.

of a mixture of very complex configurations. Figure 15 shows how the series can be interpreted in the case of nucleon-induced reactions as the sequential excitation of particle-hole states. In practical calculations it is usually not possible to treat more than the first two steps explicitly. The remaining steps are, of course, normally accounted for by the imaginary part of the optical potential.

4.2 *Intermediate Structure Schemes Based on Intrinsic Excitations*

Several more or less successful models for ^{24}Mg molecular states are based on an idea of Nogami (1968, unpublished) and Imanishi (1968). Their idea is shown schematically in Figure 16. At an energy E a single particle resonance of the mean field with spin J is coupled to an intrinsic excitation of one or both of the incoming ions. In this way the relative kinetic energy is reduced by an amount ΔE_x while the orbital angular momentum is reduced by the vector sum of the spins of the intrinsic states $I_1 + I_2$. If the energy is low enough, the excited molecular state may be quasi-bound and narrow intermediate structure resonances will be observed. At higher incident energies, where the excited molecules are at an energy over their outgoing channel barrier, strong resonant yields are expected to appear in the relevant inelastic channel. In the notation of the preceding discussion, $|\alpha\rangle$ represents the elastic shape resonance and the set $|\beta_i\rangle$ represents various couplings of the intrinsic states of the interacting nuclei. Clearly the set $|\beta_i\rangle$ is rather limited in practice to those states that couple easily to the ground state. That is to say, in the sense of Figure 15, they should be only one or two steps removed from the ground state. For $^{12}C + {}^{12}C$, these are the 2_1^+ at 4.43 MeV, the 3_1^- at 9.6 MeV, and the 4_1^+ at 14.1 MeV.

Practical calculations based essentially on this picture have been performed by Imanishi (1968, 1969), Fink, Scheid & Greiner (1972), Abe, Kondo & Matsuse (1980), and Tanimura (1978 and references therein). A particularly lucid scheme for visualizing what happens in these complicated coupled-channel calculations was devised by Abe et al (1980).

The single particle states of a reasonable potential at not too low an energy are accurately described by

$$E_x = \frac{\hbar^2}{2\mu R^2} L_i(L_i + 1) + E_0$$

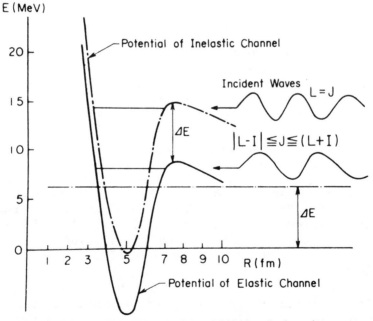

Figure 16 The physical basis of the Nogami-Imanishi idea. An incoming wave can be delayed by exciting an inelastic state (excitation energy $= \Delta E$), which increases the outgoing barrier.

where E_0, the band head, is approximately the compound nucleus Q value plus the inter-ion Coulomb potential, μ is the reduced mass, R is the inter-ion distance, and L_i is the incoming orbital angular momentum. If we take ions with $I^\pi = 0^+$ ground states, then $J = L_i$ where J is the compound nucleus spin.

Shape resonances also occur in outgoing inelastic channels, and, neglecting any change in the moment of inertia, they follow

$$E_x = \frac{\hbar^2}{2\mu R^2} \, L_f \, (L_f + 1) + E_0 + \Delta E_x.$$

Now $\mathbf{L}_i = \mathbf{J} = \mathbf{L}_f + \mathbf{I}$

whence $L_f = J + I, \ldots, |J - I|,$

where I is the outgoing channel spin. Thus there will be a number of inelastic bands, one for each coupling combination allowed by parity conservation.

An example is shown in Figure 17a for the 3_1^- excitation in ^{16}O. One of the bands, the so-called aligned band, will cross the elastic band at

$$J_{\text{cross}} = \frac{1}{2} \frac{\Delta E_x}{(\hbar^2/2\mu R^2)I} + \frac{1}{2} \, (I - 1).$$

Clearly, for J close to J_{cross}, the scattering becomes double resonant, i.e. the outgoing channel resonates simultaneously with the incoming channel. In terms of our simple picture of $|\alpha\rangle$ and $|\beta_i\rangle$ at J values close to J_{cross}, the mixing becomes a degenerate perturbation calculation with two states. Thus one expects to see doublets, each of which shares approximately half of the elastic and inelastic widths. The influence of the other inelastic bands will be much smaller in the vicinity of J_{cross}.

The possibility that aligned inelastic states can become degenerate with elastic states for certain J means that realistic coupled-channel calculations must include all relevant inelastic excitations. Figure 17b shows the pertinent aligned bands for $^{12}C + ^{12}C$. It is clear that a rotational sequence of resonances will be predicted by the full calculation, following, more or less, the trajectory of the elastic band, depending somewhat on the strength of the coupling matrix. It is also clear, and this is an important difference from normal low-lying rotational bands in nuclei, that the intrinsic structure of the molecular state, which results from the $|\alpha\rangle \otimes |\beta_i\rangle$ mixing, changes dramatically from state to state up the band. Thus as a function of increasing J, Figure 17b suggests a steady evolution of the intrinsic structure from mostly $0^+ \otimes 2^+$ to $0^+ \otimes [2^+ \otimes 2^+2^+]$ to $0^+ \otimes 3^-$ to $0^+ \otimes 4^+$ etc. This complexity of the model implies that general properties of the states such as the decay pattern say to $\alpha + ^{20}Ne$, as a function of J, may not

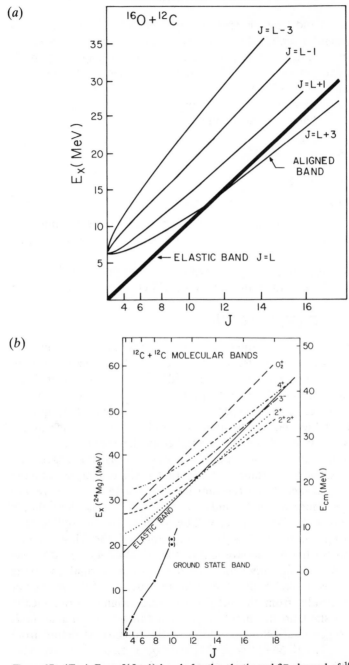

Figure 17 (*Top*) E_x vs $J(J+1)$ bands for the elastic and 3^- channel of $^{16}O + ^{12}C$. The aligned 3^- band crosses the elastic band. (*Bottom*) Band-crossing diagram for $^{12}C + ^{12}C$ showing the elastic and all relevant aligned inelastic bands. It is noted that the excited 0^+ band is parallel to the elastic band.

reveal any simple systematic trends. Probably more significantly the model suggests that electromagnetic transitions between molecular states may be inhibited because of intrinsic structural differences; this makes it difficult to interpret recent experimental searches for these transitions.

The spectrum of the near-barrier molecular states has been calculated as described above by Kondo et al (1978), including the single and mutual 2_1^+ inelastic channels. These channels are effectively closed at these energies by the outgoing Coulomb barrier but the effect of their inclusion in the calculation is enormous. A detailed comparison of the experimental and theoretical spectrum is shown with solid bars in Figure 2. Remarkably, the calculation does a reasonable job in reproducing the approximate density of molecular states, and one is almost, but not quite, tempted to try a one-to-one matching between peaks. Furthermore, theoretical values for Γ_c are in qualitative agreement with experimental values. For example, the precision elastic scattering data of Figure 3 yielded $\Gamma_c = 29_{-9}^{+6}$ keV for the $J^\pi = 2^+$ state at $E_{cm} = 6.64$ MeV, while the model 2^+ states in this energy range have $\Gamma_c = 16$–50 keV.

Finally, the average total cross section near the barrier is well described by the model, which indicates that the major parameters of the optical potentials are reasonable. This comparison is shown in Figure 18, where the energy-smoothed nuclear structure factor from experiment is compared to the model prediction. A major criticism of the original Imanishi calculations of the sub-barrier resonances was that the total reaction cross section was not reproduced. These more recent calculations by Kondo et al solve that problem.

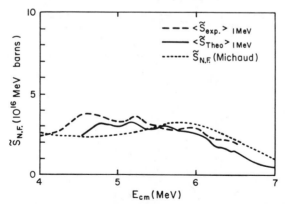

Figure 18 Comparison of experimental (*dashed line*) and theoretical nuclear structure factors averaged to smooth out resonances (Kondo et al 1978, Michaud & Vogt 1972).

Given the remarkable success of this simple theory in producing a qualitative and even semi-quantitative description of the low-energy molecular spectrum in ^{24}Mg, the next natural step is to examine its applicability in the higher energy region. Kondo, Abe & Matsuse (1979) reported an extension of their earlier calculations into the high-energy range. The basis of the model was expanded to include all of the relevant inelastic channels and an angular momentum dependent absorption was devised so as to guarantee surface transparency of the optical potential. A comparison of their calculations with the energy dependence of the total inelastic cross section in the 2^+, 2^+2^+, and 3^- channels is shown in Figure 19. The calculations (*solid line*) are, in general, in good accord with both the magnitude of the cross sections and the spacing of the gross structures. The major failing of the calculation, however, is its inability to reproduce the fragmentation of the gross structures, particularly in the vicinity of the 12^+ resonances. This suggests that important degrees of freedom may still be missing from the calculation (Abe et al 1980). The fact that the inelastic strength is reasonably predicted suggests that these partial widths are being qualitatively reproduced. However, Kondo et al (1979) do not include a calculation of the elastic or total fusion cross sections ($\sigma_{fusion} \cong \sigma_{reaction} - \sigma_{inelastic}$) so it is difficult to appraise the overall success of the theory.

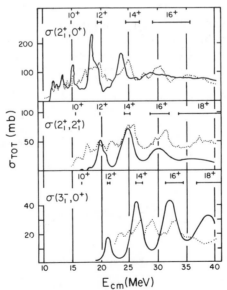

Figure 19   Comparison of total ^{12}C + ^{12}C inelastic cross sections in the 2^+, 2^+2^+, and 3^- channels (*dotted lines*) with the band-crossing model calculations (*solid lines*).

Tanimura (1978 and references therein) performed similar calculations using a double-folding prescription for both the diagonal and
off-diagonal elements of the potential. In this case the coupling interactions are characteristically stronger than the phenomenological potentials used by Kondo et al (1979). The gross features of the elastic and
inelastic angular distributions and excitation functions, as well as the
energy dependence of the total fusion cross section, are described
extremely well by these calculations. Again, however, the intermediate
structure is not reproduced, and it appears, therefore, that while the
Nogami-Imanishi picture of $^{12}C + ^{12}C$ molecular states may contain
much of the essential physics, the full dynamics of the fragmentation of
the gross structures is not in hand.

4.3 Dynamical Symmetries in Nuclear Molecular Spectra

Recently Iachello (1981) and Erb & Bromley (1981) studied the
molecular spectrum of ^{24}Mg near the barrier with the intention of
searching for the long-known pattern of excited states characteristic of
diatomic molecules. Iachello noted that although the potential appropriate to nuclear molecular binding is at present unknown, it is nonetheless
true that any Hamiltonian possessing $U(4) \supset O(4) \supset O(3)$ symmetry
(e.g. a diatomic molecule) will yield an eigenspectrum:

$$E(v, L) = -D + a(v + \tfrac{1}{2}) + b(v + \tfrac{1}{2})^2 + cL(L + 1),$$

where D, a, b, and c are constants that depend on the details of the
Hamiltonian, and v and L are the vibrational and rotational quantum
numbers, respectively. Since the spectrum-generating algebra is of
complete generality, it is interesting to study the nuclear molecular
spectrum of ^{24}Mg for evidence of the above dynamical symmetry.

The fit of the above equation to the molecular spectrum near the
barrier is shown in Figure 20. The parameters D, a, b, and c have been
adjusted to produce this fit, which describes 28 known states with a
rootmean square error of 44 keV. States with uncertain J^π (open circles)
circles) have not been included in the fitting procedure.

Several parameter sets give equivalent fits to the spectrum, and there
is a discrete ambiguity in the assignment of the vibrational quantum
numbers. States with unknown J^π have been inserted in the diagram
where they fit best, and thus the model is essentially making on the
order of ten spin predictions. Obviously, the experimental determination of these spins will constitute an extremely stringent test of the
proposed symmetry.

Independent of ambiguities in the parameterization of the spectrum, the vibrational dissociation limit of the molecule for $L = 0$ is

well determined, $E_d \cong 7\,\mathrm{MeV}$. This is tantalizingly close to the $^{12}\mathrm{C} \rightarrow {}^{8}\mathrm{Be} + \alpha$ threshold and suggests to Erb & Bromley the possible importance of α-particle degrees of freedom in the molecular interaction.

The value of the rotational parameter $c = 76\,\mathrm{keV}$ corresponds to a $^{12}\mathrm{C} - {}^{12}\mathrm{C}$ dumbbell consisting of two touching $^{12}\mathrm{C}$ nuclei or an inter-ion separation of 6.75 fm. This is a rather large separation if one thinks in terms of conventional optical potentials. Erb & Bromley (1981) suggested that this fact, taken with the inferred dissociation energy, implies that the 0^+ $E_x = 7.6$-MeV state in $^{12}\mathrm{C}$, with its very extended 3α structure, may constitute the internal degree of freedom coupling to the entrance wave.

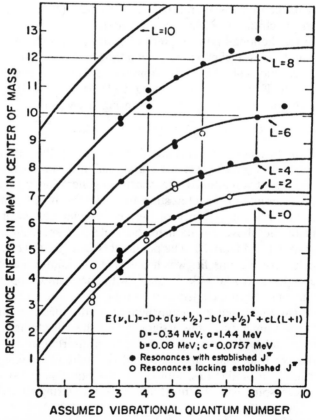

Figure 20 Comparison of experimental molecular spectrum with fitted spectrum based on the U(4) symmetry, (Erb & Bromley 1981).

This possibility (though not in these terms) was actually considered by Michaud & Vogt (1972) very early in the development of molecular theories. The early success of the Imanishi calculations has clearly directed theoretical effort into coupled-channel studies based on the collective states in ^{12}C. A detailed theory including coupling to α degrees of freedom is yet to be formulated.

4.4   Diffraction or Resonances?

A recent paper by Phillips et al (1979) raised some interest by questioning whether the gross oscillations seen in inelastic scattering were, in fact, resonances at all. Their point was that the appearance of strong energy variation in a surface-peaked reaction will naturally result when the nuclear interior is black, and the nuclear surface is transparent. One normally believes that these conditions obtain for ^{12}C + ^{12}C scattering. The question is then a quantitative one: does diffractive inelastic scattering have the correct energy dependence and is the strength correct?

To investigate this question, Phillips et al (1979) formulated the inelastic scattering problem in the Blair-Austern formalism (Austern & Blair 1965). Beginning with a strong absorption elastic s matrix

$$s_1 = \eta_1 \exp(2i\delta_1)$$

where
$$\eta_1 = \left[1 + \exp\left(\frac{E - E_G}{\Delta}\right) \right]^{-1},$$

$$\delta_1 = \delta_{\max}(1 - \eta_1),$$

and
$$E_G = E_O + \frac{\hbar^2}{2\mathscr{I}} l(l + 1).$$

The inelastic amplitude becomes

$$f^{IM}(\theta) = \frac{1}{2i} \left(\frac{K'}{K}\right)^{1/2} \beta R \sum_{ll'} \sqrt{2l' + 1} \, \exp[i(\sigma_1 + \sigma_{1'})] \left(\frac{\partial s_l}{\partial l} \frac{\partial s_{l'}}{\partial l'}\right)^{1/2}$$
$$\times (l'\mathrm{I}00|l0) \ (l'\mathrm{I} - m \cdot m | l0) \ Y_{l'}^{-m}(\theta).$$

Strong structures will appear when the $\partial s_l / \partial l$ and $\partial s_{l'} / \partial l'$ have good overlap (i.e. a well-matched channel), and when the s_l and $s_{l'}$ vary quickly with energy. Figure 21a shows the reflection coefficents and derivatives that appear in the calculations of Phillips et al. The most important parameter, which guarantees the pronounced energy dependence of $\partial s_l / \partial l$, is Δ, which is 1 MeV here. Figure 21b shows that pronounced structure in the inelastic scattering excitation function is

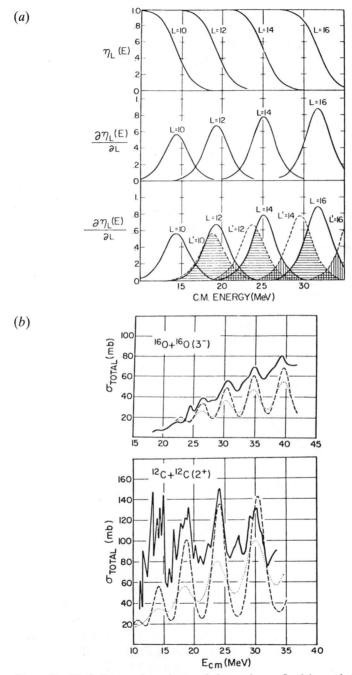

Figure 21 (*Top*) Energy dependence of the nuclear reflectivity and derivatives for
$^{12}C + {}^{12}C(2^+)$ inelastic scatterings as parameterized for diffraction model calculations
(Phillips et al 1979). (*Bottom*) Comparison of experimental (*solid line*) and predicted total
inelastic scattering cross sections using the Austern-Blair diffraction model for $\Delta = 1$ MeV
(*dashed line*) and $\Delta = 1.2$ MeV (*dotted line*).

indeed produced for $\Delta = 1$ MeV (*dashed line*), but it is already suppressed for $\Delta = 1.2$ MeV (*dotted line*).

There has been substantial discussion and interpretation of these calculations in the literature. Friedman, McVoy & Nemes (1979) note that there is nothing intrinsically nonresonant in the calculation. They point out that the physical significance of the narrow l windows is an open question, and that the Blair-Austern formalism in the case of narrow l windows and well-matched channels is identical to a two-Regge-pole problem where one pole is in the entrance channel and one is in the exit channel.

Lee, Chu & Kuo (1981) remarked that the Phillips calculation emphasizes that the reaction is dominated by the perfect kinematic matching of the elastic and 2+ channels, but that the parameter Δ required in a purely diffractive mechanism is too small to be consistent with the proximity potential (Blocki et al 1977, Krappe 1979). They point out that the great bulk of heavy-ion scattering data indicates that the proximity potential is quite reasonable for separations that are not too small. They suggest, therefore, that the choice of $\Delta = 1$ MeV is inconsistent with the conventional potentials normally associated with diffractive scattering.

Recently, Tanimura & Mosel (1981) considered the problem quite generally. Two main points are revealed in their study. First, with regard to well-matched channels, weak coupling approximations (i.e. DWBA, Blair-Austern) readily reproduce the gross structures in inelastic total cross sections when the potentials used support zero-node shape resonances. Well-matched outgoing channels will then have shape resonances simultaneously. The overlap of these incoming and outgoing resonances produces the gross structures. This is precisely the double-resonance mechanism discussed earlier in connection with the band-crossing model. It is now clear, however, that the full coupled-channel calculation is not necessary to produce the gross oscillations.

Tanimura & Mosel's calculations for well-matched channels in ^{12}C + ^{12}C, ^{14}C + ^{14}C, and ^{16}O + ^{16}O are shown in Figure 22. It was already clear that coupling to intrinsic inelastic excitations could not reproduce the multiplicity of narrow resonances seen in the data. It is now clear that the mixing of channels included in the band-crossing model is apparently not essential for the description of the gross features when incoming and outgoing channels are well matched.

The second point made by Tanimura & Mosel (1982) regards poorly matched channels. Recent experimental data revealed gross structure oscillations in total cross section to badly matched channels such as ^{12}C(^{12}C, ^{12}C)^{12}C$^*(0_2^+)$ and ^{12}C(^{12}C, ^{10}B)^{14}N (see Figures 11 and 14) and ^{16}O(^{16}O, ^{16}O)^{16}O$^*(0_2^+)$ (Freeman et al 1980) that are in phase with

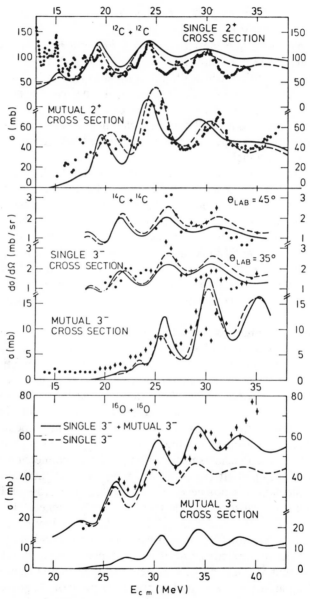

Figure 22 Comparison of various total inelastic scattering cross sections with the one- and two-step weak coupling model.

oscillations in well-matched channels. This discovery is significant because it is now not obvious in terms of, say, the band-crossing model how these gross structures arise. In a badly matched channel there is no simultaneous resonance for the outgoing wave. This is emphasized on the simple band-crossing diagram in the case of 0^+ inelastic scattering where the excited and elastic bands must be parallel (Figure 17b).

The analysis of Tanimura & Mosel shows that a Blair-Austern calculation with small diffuseness mocks up the band-crossing model. We have thus performed such a calculation for the 0_2^+ in ^{12}C. As expected, the calculations reasonably reproduce the phase of the oscillations in the well-matched 3^- excitation, but they are completely out of phase with the 0_2^+ data. Freeman et al (1980) drew a similar conclusion regarding $^{16}O + ^{16}O$ excitation of the $^{16}O(0_2^+)$.

The failure of the Blair-Austern formalism (or DWBA) to reproduce the excitation functions in mismatched channels indicates that these channels are sensitive to the reaction mechanism. Tanimura & Mosel succeeded in reproducing the correlation between poorly matched and well-matched channels using the strong coupling that results from the double-folding procedure. Their result for $^{16}O + ^{16}O$ is compared with the data of Freeman et al in Figure 23.

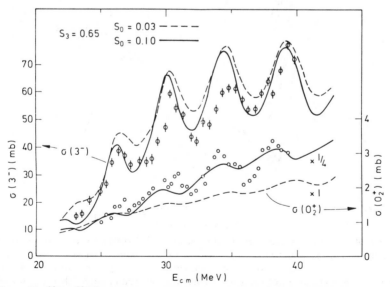

Figure 23 $^{16}O + ^{16}O$ inelastic scattering excitation functions to the well-matched 3^- and poorly matched 0_2^+ channels (Tanimura & Mosel 1981). Calculations for two values of the coupling strength (S_0) are shown. The 0_2^+ cross section for the $S_0 = 0.1$ case has been reduced by a factor of 4.

The main difference between these calculations and the band-crossing model is the strength of the off-diagonal elements of the potential. To understand, therefore, how the strong coupling is influencing the calculation of the 0_2^+ excitation function, Tanimura & Mosel calculated the equivalent local potential (TELP) obtained by simply inserting the coupled-channel wave functions into a Schrödinger equation:

$$V_i^{\text{TELP}}(r) = E_{\text{cm}} + \frac{\hbar^2}{2\mu} \frac{1}{\psi_i(r)} \frac{d^2}{dr^2} \psi_i(r).$$

The potentials obtained in this way are compared with the corresponding diagonal terms of the potential in Figure 24. The results are quite

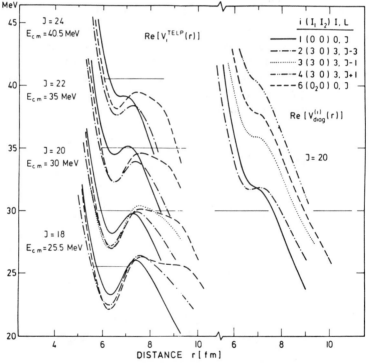

Figure 24 The trivially equivalent local potentials V_i^{TELP} (*left*) for several partial waves and the total diagonal potential for $J = 20\hbar$ (*right*). The various channels ($i = 1$ to 6) are specified in the upper right corner. The elastic channel is $i = 1$, the first excited 0_2^+ is $i = 6$, and $i = 2, 3, 4$ are the three coupling combinations allowed in the 3^- channel. Note in particular the aligned $3^- i = 2$. The diagonal potentials exhibit matching of the elastic and aligned 3^- channel and the mismatch of the 0_2^+ channel. The equivalent local potentials obtained from the coupled channel wave functions, however, exhibit matching of all three channels.

striking. The diagonal potentials (*right side*) clearly uphold the expectation of the band-crossing model (or any weak coupling approach, DWBA, etc) in that the potentials in the elastic channel (*solid curve*) and the aligned component of the 3^- channel (*dashed-dotted curve*) are very similar and will clearly support resonances at the same energy. The diagonal 0_2^+ potential (*dashed curve*), however, is identical in shape but shifted up by the excitation energy of the state and, therefore, is not simultaneously resonant. The situation with the equivalent local potentials (*left side*) is quite different. Here the elastic, the aligned 3^-, *and* the 0_2^+ (*dashed curve*) potentials are very similar, which indicates a good overall matching among all of these channels. The simultaneous resonances, which these potentials support, are the direct result of the mixing of the wave functions through the strong polarization of the potentials seen here (Feshbach 1958, 1962). These large polarization potentials imply significant rearrangement of the nuclear structure of the interacting ions.

4.5   Microscopic Methods

The above result should provide a strong impetus for a microscopic study of the ^{12}C + ^{12}C reaction. There are in the literature a number of attempts to study the nuclear molecular problem from a microscopic point of view. Here I mention only the work of Chandra & Mosel (1978), who studied the high-spin properties of ^{24}Mg using the Strutinsky renormalization and a self-consistent cranking method applied to a nonaxially symmetric shell model Hamiltonian. Figure 25 shows their level diagram as a function of inter-ion distance illustrating the usual continuous evolution from ion + ion to compound nucleus shell structure. Chandra & Mosel calculated the ^{12}C + ^{12}C interaction as a function of R, allowing the nuclear shape complete freedom to adjust itself at each R so as to maintain the minimum energy. The results reveal a sharp discontinuity in the potential at $R \cong 4.5$ fm due to a significant rearrangement of the underlying nuclear structure. This point is indicated in Figure 25 by the dashed lines, where a quasi level crossing occurs. At this point, with R increasing, a rapid transition from a very deformed ^{24}Mg into a dumbbell-shaped ^{24}Mg built of the two nascent ^{12}C nuclei occurs.

At the point of the potential discontinuity there is essentially zero barrier between the molecular and compound nucleus configurations. Nonetheless, the molecular configurations will assume a dynamic stability because of particle transitions at the avoided level crossing. The probability for such a transition is estimated by Chandra & Mosel to be >75% using the Zener-Landau method. It is particularly amusing to

MOLECULAR CONFIGURATIONS

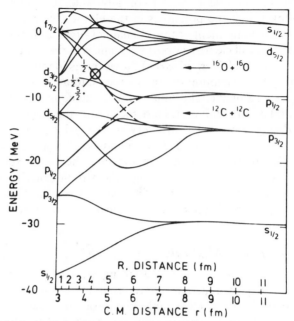

Figure 25 The energy level diagram for a symmetric ion-ion collision. The dashed lines indicate the quasi level crossing relevant to $^{12}C + ^{12}C$.

note (Figure 25) that the molecular configurations owe their stability to a $P_{1/2} \rightarrow P_{3/2}$ promotion. As is well-known, this particular 1p1h state is the primary configuration in the first excited 2^+ state of ^{12}C, and it is just this state that has been used in the phenomenological studies discussed above.

5. SUMMARY

Tremendous progress has been made in recent years toward a theoretical understanding of the large-scale clustering phenomenon in light nuclei known as nuclear molecular structure. Especially for the lower spin states between $E_x(^{24}Mg) = 17$–22 MeV, where the phenomenological models based on the Nogami-Imanishi mechanism appear particularly successful. As we have seen, there may actually be a good microscopic foundation for this model.

Preliminary group-theoretical studies are now indicating that the symmetry properties of the molecular Hamiltonian may eventually be discovered directly from the molecular spectrum once this spectrum is better determined.

The higher energy spectra are, unfortunately, still somewhat of an enigma. While the average properties of the spectrum (i.e. total cross sections, etc) are well described by the models that work well at low energy, the details of the intermediate structure are not. Tanimura & Mosel found convincing evidence for strong polarization of the nuclear wave function during ion-ion contact, thus questioning the validity of the band-crossing model. By inference, then, this casts a shadow on the good agreement obtained for the near-barrier states, since there is no known experimental distinction yet between the structure of the lower and higher energy molecular states.

6. OTHER LIGHT SYSTEMS

Lest I leave the impression that molecular structure is peculiar to ^{24}Mg, I think it best to show Figure 26. This partial summary of the available total inelastic scattering data taken from Haas & Abe (1981, and references therein) should quickly dispel any such notions. All of these cases show gross structure peaks qualitatively similar to those seen in

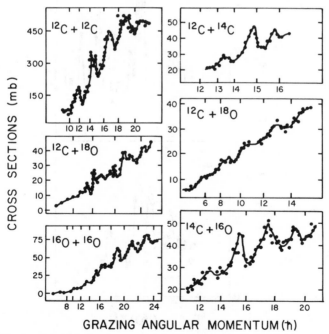

Figure 26 A composite showing some of the existing inelastic scattering total cross sections. Note the magnitude in the ^{12}C + ^{12}C case.

^{12}C + ^{12}C. The main distinction is the magnitude of the cross section, which is almost a factor of 10(!) greater for ^{12}C + ^{12}C. For most of these other systems (except ^{16}O + ^{12}C) it was thought for some time that no intermediate structure states existed to accompany the gross structure phenomena. It is, of course, easy to accommodate such an observation within the framework of the models discussed above by simply recognizing that the widths of the individual intermediate structure states determine their visibility. The high background level density for systems other than ^{16}O + ^{12}C and ^{12}C + ^{12}C could easily broaden individual fragments into an effective continuum. Recently, however, very careful studies of many of these other systems are revealing a rich spectrum of intermediate structure resonances with widths similar to those of the ^{12}C + ^{12}C system. As an example, I show a recent study of the ^{16}O + ^{16}O system by K. Van Bibber (1981, private communication) in Figure 27, which clearly reveals the presence of several narrow resonances.

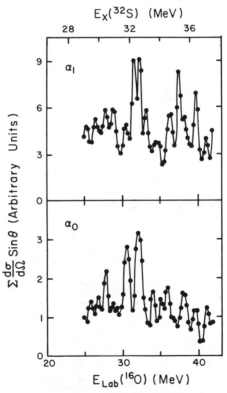

Figure 27   Angle-summed α_0 and α_1 excitation functions showing narrow resonances in the ^{16}O + ^{16}O system previously thought to exhibit only gross structure resonances.

A complete discussion of all of the work in progress on these and other systems would go well beyond the scope of the present pap 'r. I therefore, refer the interested reader to the excellent reviews n this field and especially to the review by Thornton (1980).

7. NEW DEVELOPMENTS AND DIRECTIONS

Space constraints have forced the omission of several of the exciting new developments in this field, such as the possibility of direct electromagnetic transitions between molecular states (McGrath et al 1981), the possibility of direct excitation of molecular states in many-nucleon transfer reactions (Nagatani et al 1979, Szanto de Toledo et al 1981), and the intriguing new prospects for heavy-ion radiative capture through molecular states (Sandorfi & Nathan 1978). These new developments and many of the topics discussed in this review point to a myriad of very basic unanswered questions in this field. The new experimental attacks and recent microscopic theoretical studies are just beginning to shed light on these questions.

Literature Cited

Abe, Y., Kondo, Y., Matsuse, T. 1980. *Prog. Theor. Phys. Suppl.* 68:303–58
Almqvist, E., Bromley, D. A., Kuehner, J. A. 1960. *Phys, Rev. Lett.* 4:515
Almqvist, E., et al. 1964. *Phys. Rev.* 136:1384
Austern, N., Blair, J. S. 1965. *Ann. Phys.* 33:15
Blocki, J., Randrup, J., Swiatecki, W. J., Tsang, C. F. 1977. *Ann. Phys.* 105:427
Bromley, D. A. 1978. See Cindro 1978, pp. 3–60
Chandra, H., Mosel, U. 1978. *Nucl. Phys. A* 298:151
Cindro, N., ed. 1978. *Nuclear Molecular Phenomen.* Amsterdam:North Holland
Cormier, T. M., Applegate, J., Berkowitz, G. M., Braun-Munzinger, P., Cormier, P. M., Harris, J. W., Jachcinski, C. M., Lee, L. L. Jr., Barrette, J., Wegner, H. E. 1977. *Phys. Rev. Lett.* 38:940
Cormier, T. M., Jachcinski, C. M., Berkowitz, G. M., Braun-Munzinger, P., Cormier, P. M., Gai, M., Harris, J. W., Barrette, J., Wegner, H. E. 1978. *Phys, Rev. Lett.* 40:924
Cosman, E. R., Cormier, T. M., Van Bibber, K., Sperduto, A., Young, G., Erskine, J., Greenwood, L. R., Hansen, O. 1975. *Phys. Rev. Lett.* 35:265
Cosman, E., et al. 1981. MIT Preprint (unpublished)

Erb, K. A., Betts, R. R., Hansen, D. L., Sachs, M. W., White, R. L., Tung, P. P., Bromley, D. A. 1976. *Phys. Rev. Lett.* 37:670
Erb, K. A., Betts. R. R., Korotky, S. K., Hindi, M. M., Tung, P. P., Sachs, M. W., Willett, S. J., Bromley, D. A. 1980. *Phys. Rev. C* 22:507
Erb, K. A., Bromley, D. A. 1981. *Phys. Rev. C* 23:2781
Ericson, T. 1963. *Ann. Phys.* 23:390
Evers, D., Denhofer, G., Assmann, W., Harasim, A., Konrad, P., Ley, C., Rudolph, K., Sperr, P. 1977. *Z. Phys. A* 280:287
Feshbach, H. 1976. *J. Phys. C* 5:Suppl. 11, p. 37
Feshbach, H. 1958. *Ann. Phys.* 5:357
Feshbach, H. 1962. *Ann. Phys.* 19:287
Feshbach. H., Kerman, A. K., Lemmer, R. H. 1967. *Ann. Phys.* 41:230
Fink, H. J., Scheid, W., Greiner, W. 1972. *Nucl. Phys. A* 188:259
Freeman, W. S., Wilschut, H. W., Chapuran, T., Piel, W. F. Jr., Paul, P. 1980. *Phys. Rev. Lett.* 45:1479
Friedman, W. A., McVoy, K. W., Nemes, M. C. 1979. *Phys. Lett.* 87B:179
Fulton, B. R., Cormier, T. M., Herman, B. J. 1980. *Phys. Rev. C* 21:198
Haas, F., Abe, Y. 1981. *Phys. Rev. Lett.* 46:1667

Iachello, F. 1981. *Phys. Rev. C* 23:3778

Imanishi, B. 1968. *Phys. Lett.* 27B:267

Imanishi, B. 1969. *Nucl. Phys. A* 125:33

Kondo, Y., Abe, Y., Matsuse, T. 1979. *Phys. Rev. C* 19:1356

Kondo, Y., Matsuse, T., Abe, Y. 1978. *Prog. Theor. Phys.* 59:465

Krappe, H. J. 1979. *Phys. Rev. Lett.* 42:215

Lee, S. Y. Chu, S., Kuo, T. 1981. SUNY, Stony Brook, Preprint (unpublished)

Mazarkis, M., Stephens, W. E. 1973. *Phys. Rev. C* 7:1230

McGrath, R. L., et al. 1981. *Phys. Rev. C* 24:2374

Michaud, G. J., Vogt, E. W. 1972. *Phys. Rev. C* 5:350

Nagatani, K., Shimoda, T., Tanner, D., Tribble, R., Yamaya, T. 1979. *Phys. Rev. Lett.* 43:1480

Pappalardo, G. 1964. *Phys. Lett.* 13:320

Phillips, R. L., Erb, K. A., Bromley, D. A., Weneser, J. 1979. *Phys. Rev. Lett.* 42:566

Sandorfi, A., Nathan, A. M. 1978. *Phys. Rev. Lett.* 40:1252

Shapira, D., Stokstad, R. G., Bromley, D. A. 1974. *Phys. Rev. C* 10:1063

Sperr, P., et al. 1976. *Phys. Rev. Lett.* 37:321

Szanto de Toledo, A., Coimbra, M. M., Carlin Filho, N., Cormier, T. M., Stwertka, P. M. 1981. *Phys. Rev. Lett* 47:632

Tanimura, O. 1978. *Nucl. Phys. A* 309:233

Tanimura, O., Mosel, U. 1981. *Phys. Rev. C* 24:321

Tanimura, O., Mosel, U. 1982. Unpublished

Thornton, S. T. 1980. *Proc. 18th Int. Winter Meet. Nucl. Phys., Bormio, Italy.* Univ. Milan Press

Van Bibber, K., Cosman, E. R., Spertudo, A., Cormier, T. M., Chin, T. N., Hansen, O. 1974. *Phys. Rev. Lett.* 32:687

Ann. Rev. Nucl. Part. Sci. 1982. 32:309–334

THE MACROSCOPIC APPROACH TO NUCLEAR MASSES AND DEFORMATIONS

W. D. Myers and W. J. Swiatecki[1]

Nuclear Science Division, Lawrence Berkeley Laboratory, University of California, Berkeley, California 94720

CONTENTS

1. INTRODUCTION

The mass of the neutron is $M_N = 939.5731\,\mathrm{MeV}/c^2$ and of the proton $M_Z = 938.2796\,\mathrm{MeV}/c^2$. The mass of an atomic nucleus with A nucleons (N neutrons and Z protons) consists of the sum ($NM_N + ZM_Z$), reduced by the mass associated with the binding energy between the nucleons, as required by the relativistic equivalence between mass and energy. These binding energies vary from 2.2 MeV for the deuteron to 1900 MeV for ^{256}Fm. After sixty years of atomic

[1] This work was supported by the Director, Office of Energy Research, Division of Nuclear Physics of the Office of High Energy and Nuclear Physics of the US Department of Energy under Contract W-7405-ENG-48. The US Government has the right to retain a nonexclusive royalty-free license in and to any copyright covering this paper.

309

mass measurements, the binding energies are known today experimentally for ~1900 nuclei in their ground-state equilibrium configurations, as well as for dozens of nuclei in deformed, fission-barrier saddle-point shapes, and for hundreds of interaction-barrier shapes corresponding to pairs of nuclei in contact. These binding-energy (or mass) measurements are often made with a precision corresponding to a small fraction of an MeV and, together, they represent an immense amount of information of practical relevance for many branches of nuclear physics, nuclear engineering, and astrophysics. They also represent an exacting challenge to theoretical efforts at understanding the basic properties of the unique many-body problem presented to us by an atomic nucleus.

In the macroscopic approach to nuclear structure, one attempts to simplify the description of certain aspects of the nuclear many-body problem, with its $3A$ individual-particle degrees of freedom (not to mention possible quark degrees of freedom), by focusing attention on a number of suitably chosen macroscopic features. First and foremost among these are the degrees of freedom describing the *shape* of the nuclear surface (which, although not perfectly sharp, is known experimentally to be fairly well defined, except for very small nuclei). The subject of the present review, the macroscopic approach to nuclear masses and deformations, is then the description of theories, formulae, and techniques for the calculation of nuclear masses (or binding energies) in their dependence on macroscopic (shape) degrees of freedom.

The major part of the binding energy of nuclei may be accounted for by a simple "liquid-drop" formula, consisting of a volume energy [assumed to depend quadratically on the relative neutron excess I, defined as $(N - Z)/A$], a surface energy proportional to the surface area, and the electrostatic energy of a uniform distribution of electric charge inside the nucleus. The effectiveness of this simple liquid-drop-model treatment of nuclear ground-state energies is illustrated in Figure 1. Originally conceived more than 45 years ago (1, 2) for the purpose of calculating only such ground-state masses, the model began to assume a wider range of applicability when it was recognized that the gross properties of nuclear fission could be understood in terms of the shape dependence of the surface and electrostatic energies of the nuclear drop (3). Unfortunately, however, there was an historically understandable tendency to associate the liquid-drop model (even in its static aspects) with a system of strongly interacting particles characterized by short mean-free paths (and treated according to classical mechanics). Because of this misconception and the failure of idealized versions of the liquid-drop model to account for nuclear *dynamics* (e.g. excited nuclear states), the soundness of the liquid-drop model, even in its description of the gross, static aspects of nuclear binding energies, began to be questioned when the nuclear shell model was found to be a good approximation to nuclear structure (4). According to this model, one could think of nuclei as consisting of *weakly* interacting constituents in a common potential well, with quantization of the particle orbits playing an essential role. How then could the liquid-drop model of nuclear binding energies be taken seriously? We discuss this question in a later section. Notwithstanding these reservations, refinements to the liquid-drop model formula for nuclear masses continued to be made. They did

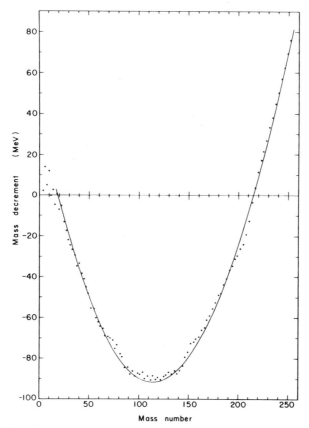

Figure 1 The mass decrements (closely related to nuclear binding energies) are plotted for 97 beta-stable nuclei. The curve is a liquid-drop fit based on the "local" part of the nuclear potential-energy expression. The deviations are due mostly to shell effects.

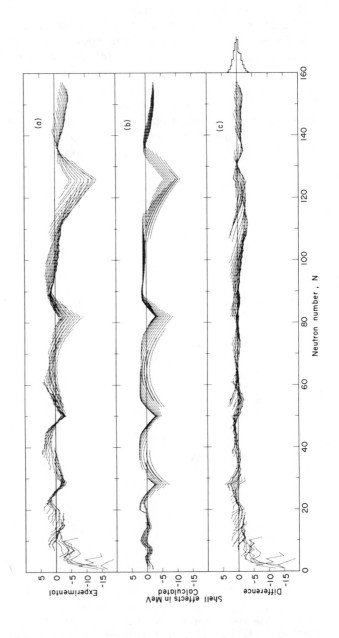

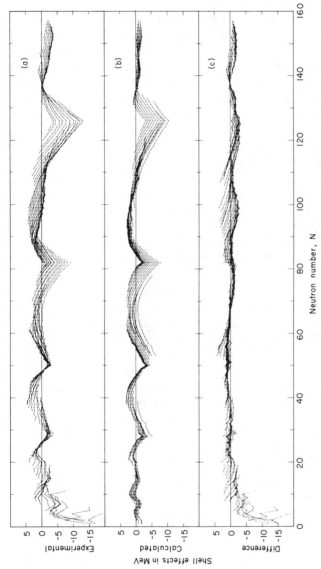

Figure 2 In the two separate parts of this figure, the shell correction to nuclear binding energies (*i.e.*, the experimental mass minus a droplet model fit) is displayed as a function of neutron number in lines (*a*). Line (*b*) is a theoretically calculated shell correction, using a schematic model of bunched levels in the upper figure and the Strutinsky shell-correction method in the lower figure. Line (*c*) is the remaining deviation in each case (30).

not, however, reduce substantially the remaining discrepancies between theory and experiment (up to about 10 MeV), which were soon recognized qualitatively as "shell effects," associated, indeed, with the quantization of the nucleon orbits.

Major advances in treating the shell effects took place about fifteen years ago. [Before this time the shell effect deviations had usually been treated in an ad hoc way by use of tabulated empirical correction functions (5–8).] In References (9–11) it was recognized that the main features of the shell effect deviations (see Figure 2) could be understood in terms of the bunching of the quantized nucleon levels into bands, the bunching being governed by the symmetries of the nuclear shape in question and thus disappearing when the symmetries were destroyed by a deformation. A semi-empirical algebraic treatment of shell effects, made possible by these insights (11–14), was soon followed by quantitative calculations (requiring, however, numerical solutions of the Schrödinger equation in an appropriate potential well) (15–18). The result of these calculations was not only a dramatic reduction of the discrepancies between theoretical and experimental masses (from around 10 MeV to around 1 MeV) but also the explanation of the long-standing puzzle of the mass asymmetry of nuclear fission in terms of shell effects at the fission barrier, as well as the explanation of the existence of relatively stable, strongly deformed nuclei with axes in the ratio of about 2 : 1 (the fission isomers). Finally, and potentially most significant, these shell effect calculations predicted the possible existence of an island of relatively stable nuclei beyond the known limits of the periodic table of elements.

Concurrently with this conquest of the nuclear shell effects, there followed a substantial further improvement in the treatment of the liquid-drop model [the so-called droplet model (19–31)] and the development of successful "proximity" and "folding" techniques for calculating the nuclear interaction between approaching nuclei (32–39), essential for a description of the energies of interaction-barrier configurations. Taking together these three contributions, liquid-drop, proximity, and shell effects, one can now theoretically estimate the binding and deformation energies of known or hypothetical nuclei with an accuracy often approaching or even exceeding 1 MeV.

An independent advance has been the development of techniques that solve the nuclear many-body problem within a self-consistent mean-field approximation [Hartree-Fock calculations with simplified effective interactions (40)]. These potentially most powerful techniques are still somewhat limited by the computational effort they require.

Finally, a major advance in interpolation/extrapolation methods took place around 1966 (41–46), resulting in a very elegant and generally accurate way of predicting nuclear ground-state masses from known neighboring masses (see Section 6) (47–53).

2. FRAMEWORK

The importance of the macroscopic description of nuclear deformation energies can be appreciated by viewing the problem in the wider context of the macroscopic description of nuclear dynamics. In order to discuss a macroscopic dynamical problem, one often needs three components in the equations of motion, corresponding to inertial, dissipative, and conservative forces, each given as a function of the macroscopic (e.g. shape) degrees of freedom (and their time derivatives) (54, 55). The conservative forces follow from the potential energy (expressed as a function of shape). In the case of nuclei, the local or absolute minima in this potential-energy landscape give the ground-state masses. Saddle-point masses are related to (fission) barriers and, generally, the land-scape provides the stage on which dynamical evolutions (to be treated classically or quantally) will be taking place.

In the context of nuclear deformation energies, the problem is then to write down the potential energy V of a nucleus of arbitrary shape — a diffuse nuclear blob consisting of A nucleons — as a function (more precisely, a functional, $V[\Sigma]$) of its shape Σ. [The blob may be in the form of one or more deformed diffuse pieces, but the contour Σ is, by definition, a sharply defined figure. The diffuseness of the surface region of the blob may be specified by a "width" b, of the order of the range of nuclear forces. The size of the blob (not spherical, in general) may be specified by a radius R (or volume $4\pi R^3/3$).]

In a direct attack on this question one may simply attempt a solution of the many-body problem as a whole, using a suitable approximation. In the nuclear context, the mean-field Hartree-Fock approach (using simplified effective interactions) has been particularly successful in recent years and promises to provide eventually the most reliable estimates of certain features of the potential-energy landscape.

The indirect approach, which up to now has provided the main tools for accounting quantitatively for nuclear binding and deformation energies, relies on splitting the total potential energy into three parts and treating them separately. The physical reasons for the split have to do with the fact that the nuclear blob Σ is made up of elements (nucleons) that can feel each other over finite distances by virtue of nuclear interactions (of range $\sim b$) and that inside the blob there are individual-particle wave functions that can feel out the shape of the blob as a whole. (The eigensolution of the Schrödinger equation in a cavity is sensitive to the shape and size of the cavity as a whole.) If it were not for certain specific effects of the finite range of nuclear forces and the global character of the eigenvalue problem, the total energy could be written as a sum of "local" contributions, but the finite range adds a specific "proximity part" and the global character of the wave functions adds a "global part." Thus

$$V[\Sigma] = \text{local part} + \text{proximity part} + \text{global part}.$$

The shape dependence of the local part is made up of contributions from different points on Σ, each contribution being a function only of the local conditions at that point. The proximity part is made up of contributions that depend also on conditions a finite distance of order b away from the point in question. The global part cannot be written as a sum of local contributions — it knows about the shape as a whole and, in particular, about the symmetries of the shape.

In less formal language the local part is, essentially, the liquid-drop or droplet contribution to nuclear masses (it is typically of the order of hundreds of MeV). The proximity part or proximity potential shows up most strikingly in the attraction (of range $\sim b$) between the surfaces of two approaching nuclei (it is typically of the order of tens of MeV). The global part, in particular since it is sensitive to symmetries, contains the shell effects (typically of the order of a few MeV). (The Coulomb energy, which may range from tens to hundreds of MeV, is also part of the global contribution, but it is not specifically sensitive to the symmetries of the shape.)

We describe in the following sections the techniques used to treat these three parts of the nuclear potential energy.

3.   LOCAL PART

For any "saturating" system, such as a nucleus or a drop of water, the main deviations from bulk behavior are confined to a surface layer (of width $\sim b$, say) that is small compared to the size of the system ($b/R \ll 1$). A leptodermous potential-energy theorem may then be derived (32, 54), according to which the local part of the potential energy can be written as the following expansion in powers of b/R:

Relative Order

$$V = c_1 (4/3)\pi R^3 \qquad \text{volume energy} \qquad\qquad 1 \qquad A$$

$$+ c_2 \oint d\sigma \qquad \text{surface energy} \qquad\qquad b/R \qquad A^{2/3}$$

$$+ c_3 \oint \kappa\, d\sigma \qquad \text{curvature energy} \qquad\qquad (b/R)^2 \quad A^{1/3}$$

$$+ c_4 \oint \Gamma\, d\sigma$$
$$\qquad\qquad\qquad\qquad \left\{\begin{matrix}\text{higher-order curvature}\\ \text{corrections}\end{matrix}\right\} \quad (b/R)^3 \quad A^0$$
$$+ c_4' \oint \kappa^2\, d\sigma$$

$$+ \text{corrections that go to zero as } A \to \infty \qquad\qquad A^{-n}.$$

1.

In the above, the integrals are over the surface Σ defining the shape of the system, κ is the total curvature, and Γ the Gaussian curvature at a point on the surface ($\kappa = R_1^{-1} + R_2^{-1}$, $\Gamma = R_1^{-1}R_2^{-1}$, where R_1, R_2 are the principal radii of curvature at the point in question).

The coefficients $c_1 \ldots c_4'$ are independent of the shape and size of the system, but they may depend, in general, on the bulk density and composition (neutron excess). With respect to the leading volume-energy term, the constants are of the relative order 1, b, b^2, b^3, b^3, which implies energy contributions of order A, $A^{2/3}$, $A^{1/3}$, A^0, A^0.

The derivation of Equation 1 may be found in an article by Blocki et al (32); the important point to stress is that it does not rest on assumptions that the particles constituting the system are classical objects with short mean-free paths. The crucial assumption is that the deviations from bulk behavior should be confined to a (relatively thin) surface layer, an assumption that is found to be satisfied quite accurately also for systems of quantized, weakly interacting (or even noninteracting) particles (22, 54, 56).

In particular, it is now well established that, when the nuclear problem of quantized individual-particle orbits in a common potential is treated by the statistical (nuclear) Thomas-Fermi method, the resulting energy reduces, in the appropriate limit of large systems with relatively thin surfaces, to a volume energy, a surface energy, and curvature corrections, as predicted by Equation 1 (19, 57–59).

It follows that the structure of the liquid-drop formula, Equation 1, is not an ad hoc parametrization, but a well-defined approximation exploiting the smallness of the expansion parameter (b/R), and accurate to within the nonlocal effects to be described in Sections 4 and 5.

Apart from the electrostatic energy, the standard liquid-drop mass formula follows from Equation 1 by writing $R = r_0 A^{1/3}$ (r_0 = the nuclear radius constant, about 1.18 fm), and assuming the volume- and surface-energy coefficients c_1, c_2 to depend quadratically on the relative neutron excess I. Thus

$$V = -a_1(1 - \kappa I^2)A + a_2(1 - \kappa_s I^2)A^{2/3}B_s + \text{higher-order terms,} \qquad 2.$$

where a_1 and a_2 are new constants (with the dimensions of energy), κ is the "symmetry energy coefficient," and κ_s is the "surface symmetry energy coefficient." The quantity B_s, a dimensionless functional of the shape Σ, is the surface area of Σ divided by $4\pi R^2$. (Thus $B_s = 1$ for the spherical shape.) In older treatments, the poorly determined coefficient κ_s was usually set equal to zero. A somewhat more reasonable choice is to put $\kappa_s = \kappa$. If this is done (13), then Equation 2 predicts that, if the measured nuclear binding energies per particle (V/A) are corrected for the neutron excess, shell effects, and the electrostatic energy, and are then plotted against $A^{-1/3}$, a straight line should result with $-a_1$ as the intercept and a_2 as the slope. How well this expectation is borne out is shown in Figure 3 taken from Reference (13).

A refinement of the liquid-drop model may be achieved by retaining in Equation 1 higher-order terms in the small expansion parameters $A^{-1/3}$ and I^2. This is indicated in Figure 4. A theory of nuclear binding energies retaining only terms of order A corresponds to the study of standard nuclear matter. Including the terms of order $A^{2/3}$ (surface energy) and $I^2 A$ (volume symmetry energy) corresponds to the liquid-drop model. Retention of the terms in $A^{1/3}$, $I^2 A^{2/3}$, and $I^4 A$ defines the nuclear "droplet model" (19, 24, 30). It turns out that in order to work consistently to this order, it is necessary to include in the theory degrees of freedom corresponding to compressibility and polarizability (i.e. deviations of the neutron and proton densities from uniform values in the bulk) as well as a "neutron-skin" degree of freedom (i.e. the introduction of separate neutron and proton effective surfaces Σ_n, Σ_p). The droplet model thus becomes very much richer than the liquid-drop model and establishes contact with many nuclear phenomena such as details of nuclear rms radii, charge distributions, isotope shifts, and giant dipole resonances. Regarding nuclear ground-state binding energies, the droplet model formula is still a closed algebraic expression.

Another extension of the liquid-drop model can be found in the work of Weiss & Cameron (60, 61), who consider a number of higher-order terms in the symmetry energy. Truran, Cameron & Hilf (62) used these higher-order terms in an actual fit to masses. Similar factors enter in the work of Baym, Bethe & Pethick (63) and Mackie & Baym (64), who are

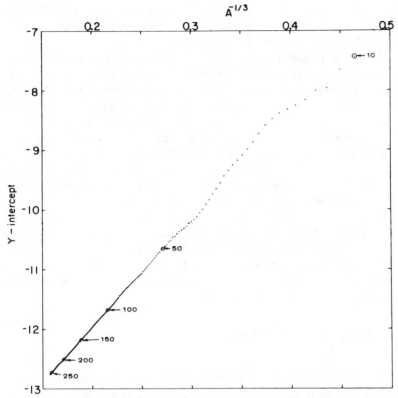

Figure 3 The experimental nuclear binding energy per particle, corrected for the neutron excess, shell effects, and the electrostatic energy, is plotted versus $A^{-1/3}$. The conformation of the experimental points to a linear trend down to mass numbers as low as 10 (see the labels along the data points) confirms the validity of the leptodermous expansion and suggests a relatively small value for correction terms beyond the surface energy.

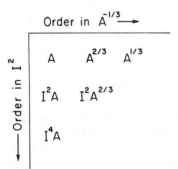

Figure 4 The orders of various terms in the expansion of the energy of a nucleus in powers of $A^{-1/3}$ and I^2. The liquid-drop model includes terms of order A, $A^{2/3}$, and $I^2 A$. The droplet model is defined by the requirement that it should include, in addition, *all* terms of order $A^{1/3}$, $I^2 A^{2/3}$, and $I^4 A$.

concerned mainly with formulating a binding-energy expression that goes over correctly into an equation of state for pure neutron matter when the neutron excess is increased.

When the liquid-drop or droplet model formulae are used for the specifically nuclear part of the binding energy, a term representing the electrostatic interaction of the protons must be added. (Although the electrostatic energy is, strictly speaking, an example of a global contribution, it is more logical to discuss it along with the liquid-drop or droplet model formula.) In the simplest approximation the electrostatic energy is taken to be that of a uniform distribution of charge Ze inside the sharp surface Σ. Closed expressions for this energy are available for slightly distorted spheres, spheroids of any eccentricity, slightly distorted spheroids, and some other special cases (36, 65–67). In general, however, the Coulomb energy must be calculated by numerical quadratures (68, 69). Corrections to the electrostatic energy for the diffuseness of the charge distribution and for the anticorrelation of the protons (due to the exclusion principle) are easily estimated. Their inclusion in a mass formula is trivial since, to lowest order in b/R, they turn out to be constants independent of shape (14, 30).

4. PROXIMITY PART

It may come as a surprise that the leptodermous expansion, Equation 1, even if carried to an infinite order in the small parameter $\varepsilon = A^{-1/3}$, is bound to miss an important piece of even the smooth part of the nuclear energy (quite apart from the oscillating global shell effects). This has to do with the circumstance that, in a system made up of particles interacting through finite-range forces, the interaction energy contains, in general, a part that "knows about" the conditions at two finitely separated points (for example, two surface elements of approaching nuclei, or the front and back sides of a single nucleus). This part cannot be reduced to a sum of local contributions, each a function of local conditions on the surface Σ, and this invalidates the assumption underlying the local leptodermous potential-energy theorem. The mathematical feature of this elusive contribution that evades even an infinite power expansion is its non-analyticity, which means that the contribution in question cannot be expanded in a Taylor series. [A typical example of such a term is $\exp(-A^{1/3})$, i.e. $\exp(-1/\varepsilon)$. See p. 454 in Reference (32).] This type of contribution may be only a fraction of an MeV for a single undeformed nucleus, but it reaches 20–30 MeV for two nuclei near contact and is of decisive importance for the discussion of such configurations. It is also of considerable importance for the fission saddle-point configurations of the lighter nuclei, in the shape of two pieces connected by a small neck.

Krappe, Nix & Sierk (34–36) developed a method of calculating potential energies that, in addition to the local part, generates also a proximity part. It consists of folding an effective short-range interaction [of Yukawa type, $\exp(-x)/x$, or, more recently, a special mixture of a Yukawa and an exponential] into a sharp (or diffuse) density distribution representing the nuclear shape. A particularly elegant version of this method (36) uses the special two-parameter folding function $C(1 - 2x^{-1})\exp(-x)$, where $x = r_{12}/a$ and r_{12} is the separation between two points. This effective interaction has the property of leaving the volume energy unaffected (its average in uniform matter is zero) and it also has the desirable property (required by nuclear saturation) that the interaction energy between two semi-infinite slabs should be stationary when the slabs are in contact. By adjusting the parameters C and a it is then possible to reproduce the empirical surface energy, as well as to give a useful approximation to the interaction energy between two nuclei. As in the case of the Coulomb energy, closed formulae for the folding energy may be derived in several cases but, in general, numerical quadratures are required.

A less comprehensive but algebraic method of treating the proximity contribution in certain cases was developed by Blocki et al (32, 33), where one can also find references to the earlier literature. It rests on the seemingly trivial observation that the interaction energy between two curved (nuclear) surfaces with least separation s may be approximately written as

$$V_{\mathrm{p}}(s) \approx \iint e(D)\,\mathrm{d}x\,\mathrm{d}y, \qquad\qquad 3.$$

where $e(D)$ is the interaction energy per unit area between two *flat*, parallel surfaces at separation D, and the integral is over the transverse dimensions of the gap between the curved surfaces, the gap being specified by the function $D(x, y)$. A change of variables (32, p. 430) leads to

$$V_{\mathrm{p}} = 2\pi\bar{R}\int_{D=s}^{\infty} e(D)\,\mathrm{d}D, \qquad\qquad 4.$$

where $\bar{R} = (R_x R_y)^{1/2}$, R_x and R_y being the radii of curvature (evaluated at the point of least separation) of the surface obtained by plotting D versus x and y. Differentiation with respect to s gives the "proximity force theorem":

$$F(s) \equiv -(\partial V_p / \partial s) = 2\pi \bar{R} \, e(s). \tag{5}$$

That is, "The force between two gently *curved* surfaces as a function of the separation degree of freedom, s, is proportional to the interaction potential per unit area, $e(s)$, between two *flat* surfaces, the constant of proportionality being 2π times the reciprocal of the square root of the Gaussian curvature of the gap width function at the point of least separation between the surfaces" (32).

The theorem reduces the calculation of the force (or potential energy) for approaching nuclei to the calculation of the geometrical quantity $\bar{R}$ [which, for two spherical nuclei with radii R_1 and R_2, turns out to be equal to the reduced radius, $R_1 R_2 / (R_1 + R_2)$] and a universal function $e(s)$ that has been calculated and tabulated, together with its integral, for nuclear surfaces described by the nuclear Thomas-Fermi approximation (32).

It is a trivial matter to estimate the nuclear interaction energy V_p between two approaching nuclei using the simple cubic-exponential approximation given in Reference (32) for the dimensionless quantity $\Phi(\zeta)$, where

$$\Phi(\zeta) \equiv \int_\zeta^\infty \frac{e(D)}{2\gamma} \frac{\mathrm{d}D}{b}. \tag{6}$$

In this expression $\zeta = s/b$, γ is the surface-energy coefficient, and b is the surface width (~ 1 fm). Equation 6 can be used to rewrite Equation 4 in the standard form,

$$V_p(\zeta) = 4\pi\gamma b \bar{R} \Phi(\zeta). \tag{7}$$

5. GLOBAL PART (SHELL EFFECTS)

We mentioned in Section 3 that the local, liquid-drop, or droplet model behavior of the potential energy is a very general property of thin-skinned systems but that, in the nuclear context, it may also be regarded as the result of applying the Thomas-Fermi method to the nuclear many-body problem of (weakly) interacting quantized particles in a common potential. The Thomas-Fermi method is based on the statistical assumption that, for a large system, the density of states in phase space (coordinate space plus momentum space) is, on the average, one per h^3, where h is Planck's constant. It follows at once that, on the average, the energy ε_n of the nth eigenvalue of the solution of the Schrödinger equation in a large, deep potential cavity is proportional to

$n^{2/3}$, the total energy E $(= \sum_1^{n_{max}} \varepsilon_n)$ is proportional to $n_{max}^{5/3}$, where n_{max} is the total number of sequentially filled eigenvalues, and the level density $g(\varepsilon_n)$ (number of eigenvalues, dn, per interval of energy, $d\varepsilon_n$) is proportional to $\varepsilon_n^{1/2}$. Thus

$$\varepsilon_n \propto n^{2/3} \tag{8.}$$

$$E = \sum_1^{n_{max}} \varepsilon_n \propto n_{max}^{5/3} \tag{9.}$$

$$g(\varepsilon_n) = \frac{dn}{d\varepsilon_n} \propto \varepsilon_n^{1/2}. \tag{10.}$$

For a cavity that is not very deep or very large, surface-layer corrections to the above formulae may be readily derived [following, for example, the method of Toke & Swiatecki (70)]. In any case, insofar as the statistical assumption is valid, the level density $g(\varepsilon_n)$ and related quantities are smooth functions of ε_n or n. In the case of an irregular cavity devoid of any symmetries, the eigenvalues would in general be nondegenerate and the inaccuracy of the statistical assumption would be relatively small, reflecting only the discrete spacing of the eigenvalues and random fluctuations around the average. The presence of degeneracies will cause deviations from the statistical (liquid-drop or droplet model) behavior, the deviations being proportional to the strength of the degeneracies in question. Thus the approximate 2×2 spin-isospin degeneracy of the nuclear problem has long been known to contribute to the special stability of light "alpha-particle" nuclei and was early suggested (71, p. 7) as a factor in the even-odd staggering of nuclear masses (corresponding to the special stability of even-even nuclei). Much more drastic are the deviations associated with the $(2l + 1)$ degeneracy of eigenvalues in a spherically symmetric potential (l is the angular momentum quantum number), or the even stronger degeneracies associated with the isotropic harmonic oscillator potential or the inverse-distance potential. In all these cases, instead of an almost smooth spectrum of levels one has a "bunched" spectrum, with several levels per bunch and with gaps in between that would otherwise be populated by the debunched levels. The reason for the relative stability of systems with a particle number corresponding to a closed shell (a particle number corresponding to a filled bunch of levels) is clear: when one begins to fill a new bunch of levels, the eigenvalues are at first anomalously high compared to an average (one has to overcome an anomalously large gap) so the first particles are relatively poorly bound. On the other hand, when completing the filling of a bunch, one is putting particles into eigenvalues that are by now lower than debunched eigenvalues would be (they would be halfway up the next gap). Thus, the eigenvalues ε_n considered as a function of the particle number n go up and down like a zig-zag (a vertical zig followed by a sloping zag) crossing and recrossing the average. The total energy is a running integral over this zig-zag, and its deviation from the average will be a series of arches, with deepest cusped points corresponding to the especially well-bound closed shells. The tops of the arches — half-filled shells — will have anomalously poor bindings (high masses). Mathematically, the binding energy anomaly — the shell effect — can thus be written (13, Equation 2) as

$$V_{\text{shells}} = \sum_{n=1}^{n_{\max}} \varepsilon_n(\text{bunched}) - \sum_{n=1}^{n_{\max}} \varepsilon_n(\text{unbunched}) \qquad 11.$$

$$\approx \int_0^{n_{\max}} \varepsilon(\text{bunched}) \; dn - \int_0^{n_{\max}} \varepsilon(\text{unbunched}) \; dn. \qquad 12.$$

This formula underlies both the treatments of shell effects in References (9–13) and (15–18, 72–76).

In the semi-empirical treatment of Reference (13), the unbunched spectrum of the separate proton and neutron eigenvalues was taken to be that of the Thomas-Fermi method (proportional to $n^{2/3}$). This spectrum was then imagined cut up into bands corresponding to the known neutron or proton magic numbers for spherical nuclei: N, $Z = M_i$, with $M_i = 2, 8, 14$ (or 20), 28, 50, 126, 184. The bands were squeezed by a (common) adjustable factor and moved slightly down together by a second adjustable factor to form the bunched spectrum. Insertion in Equation 12 then gave the proton and neutron shell corrections for the spherical shape. Since the bunching responsible for the known magic numbers is associated with the spherical shape, this bunching and the resulting shell effects should be damped out as the shape is distorted from the sphere. This was achieved in the semi-empirical method by multiplying the shell effect for the spherical shape by a shell damping function in the form of a Gaussian, $\exp(-\theta^2)$, where θ is the root-mean-square deviation of the nuclear surface from the sphere, in units of an adjustable range parameter a. When the resulting three-parameter algebraic shell correction is added to a liquid-drop or droplet model formula, a fair account can be given of the nuclear ground-state masses, equilibrium deformations, and fission-barrier heights. (The comparison with the ground-state masses is shown in the upper part of Figure 2.)

The Strutinsky method (15–18, 75) may be considered as resulting from Equation 12 by changing the variable of integration from n to ε:

$$V_{\text{shells}} = \int_{-\infty}^{\varepsilon_F} g(\varepsilon)\varepsilon \; d\varepsilon - \int_{-\infty}^{\varepsilon_F} \tilde{g}(\varepsilon)\varepsilon \; d\varepsilon, \qquad 13.$$

where $g(\varepsilon)$ is the actual level density, $dn/d\varepsilon$, of the bunched eigenvalues, and $\tilde{g}(\varepsilon)$ is the level density of the unbunched eigenvalues. The Fermi energy ε_F is determined by particle-number normalization,

$$N \text{ or } Z = \int_{-\infty}^{\varepsilon_F} g(\varepsilon) \; d\varepsilon. \qquad 14.$$

In practice, the bunched level density $g(\varepsilon)$ (a series of delta functions) is obtained by calculating numerically all the eigenvalues in a suitable shell model potential with spin-orbit coupling [a modified oscillator (73), a Woods-Saxon well (18), or a potential obtained by folding a Yukawa interaction into a sharp-surfaced generating density (75)]. The level density $\tilde{g}(\varepsilon)$ is most often obtained by smoothing $g(\varepsilon)$ by means of an essentially Gaussian smearing function of suitable range c, i.e.

$$\tilde{g}(\varepsilon) = \int_{-\infty}^{+\infty} g(\varepsilon) \frac{\exp[-(\varepsilon - \varepsilon')^2/c^2]}{c\sqrt{\pi}} [1 + \text{modification}] \; d\varepsilon'. \qquad 15.$$

The "modification" is a polynomial in the argument $(\varepsilon - \varepsilon')/c$, chosen to maximize the smoothing of the rapid oscillations in g, while doing the least damage to the long-range smooth dependence of g on energy.

An additional feature of the Strutinsky method is the inclusion of a correction to the binding energy arising from the pairing of nucleons moving in time-reversed orbits. This pairing correction is important for a realistic description of nuclear energies and is relatively easily treated by means of the Bardeen-Cooper-Schrieffer method, once the eigenvalues ε_n have been calculated with the aid of a computer (18, 73, 75). The final correction to the smooth, local binding energy consists then of separate shell and pairing corrections for the neutrons and protons. (A comparison with ground-state masses is shown in the lower part of Figure 2.)

The Strutinsky method has essentially no adjustable parameters and, given the requisite computational effort, provides the shell correction for any nuclear shape. When combined with a suitable liquid-drop or droplet formula, it has been spectacularly successful in accounting for known features of nuclear deformation energies and in making predictions of new phenomena.

For examples, the reader is referred to Section 6 and to the vast literature reviewed by Nix (76). The foundation of the method and, in particular, its relation to self-consistent Hartree-Fock treatments of the nuclear problem, are discussed by Brack and others (18, 77) and by Bohr & Mottelson (78, Vol. II, pp. 367–71).

6. SPECIFIC TREATMENTS

In the foregoing sections we described the principal physical ingredients that go into typical macroscopic theories of nuclear binding and deformation energies. In the present section we review briefly the relevant literature and provide some details in a few cases. The various developments can be traced in the proceedings of a long series of conferences on atomic masses (79–85), a series on nuclei far from stability (86–90), and a series of conferences on the physics and chemistry of fission (91–94).

Myers & Swiatecki (12) were among the first to make extensive tabulations of predicted masses that combined the liquid-drop model with semi-empirical algebraic shell corrections (11, 13). They also stressed the dependence of nuclear binding energies, including shell effects, on the *shape* of the nucleus (rather than concentrating on the ground-state masses only). We give some details of this mass formula as an example of this type of treatment.

The (atomic) mass was written as a sum of a liquid-drop part and a shell correction, as follows:

$$M(N, Z, \text{shape}) = M_{\text{LD}}(N, Z, \text{shape}) + M_{\text{shells}}(N, Z, \text{shape}). \qquad 16.$$

The liquid-drop part is given by

$$M_{\text{LD}} = M_{\text{N}}N + M_{\text{H}}Z - a_1(1 - \kappa I^2)A + a_2(1 - \kappa I^2)A^{2/3}B_{\text{s}} + c_3 Z^2 A^{-1/3}B_{\text{c}} - c_4 Z^2 A^{-1} + \delta. \qquad 17.$$

The first two terms are the masses of the neutron and of the hydrogen *atom* (this then allows for the masses of the Z atomic electrons), the next two are the volume and surface energies, with a common quadratic dependence on the relative neutron excess I. The shape dependence of the surface energy is contained in B_{s}, the ratio of the area of the shape in question to that of a sphere of equal volume. The next term is the electrostatic energy, whose shape dependence is contained in B_{c} (the ratio of the Coulomb energy of the shape in question to that of the sphere). The next term is a (shape-independent) correction to the electrostatic energy due to the diffuseness of the charge distribution.

The last term is the even-odd correction, taken empirically as $\pm 11A^{1/2}$ MeV for odd-odd or even-even nuclei and zero for odd-mass nuclei. The quantities a_1, a_2, κ, and c_3 are four adjustable parameters (c_4 is related to c_3).

The shell correction, which results from cutting up the nuclear energy spectrum into bands defined by the magic numbers M_i, as described in Section 5, is given by

$$M_{\text{shells}} = Cs(N, Z) \exp(-\theta^2),\qquad\qquad 18.$$

where $\theta = $ (rms deviation of shape from sphere)$/a$,

$$s(N, Z) = \frac{F(N) + F(Z)}{(A/2)^{2/3}} - cA^{1/3}\qquad\qquad 19.$$

$$F(N) = q_i(N - M_{i-1}) - \tfrac{3}{5}(N^{5/3} - M_{i-1}^{5/3}), \text{ for } M_{i-1} \leqslant N \leqslant M_i,$$

with

$$q_i = \tfrac{3}{5}(M_i^{5/3} - M_{i-1}^{5/3})/(M_i - M_{i-1}),$$

$$M_i = 2, 8, 14 \text{ (or 20)}, 28, 50, 82, 126, 184.\qquad\qquad 20.$$

The quantities C, a, and c are three adjustable parameters.

For a given N and Z the ground-state mass and deformation were calculated from Equation 16 by minimizing the total energy with respect to ellipsoidal distortions. The mass table generated in this way (12) listed the liquid-drop mass, the shell correction, the deformation, various decay energies, and the predicted fission barrier for about 8000 different combinations of N and Z. (The inclusion of fission barriers is necessary for a firm determination of the adjustable surface- and Coulomb-energy coefficients.)

In later versions of this type of treatment, the shape dependence of the shell correction was modified, on physical grounds, to $(1 - 2\theta^2)$ $\exp(-\theta^2)$ (14). Later this shell correction was combined with the droplet model to create another table of nuclear masses and other properties (30). A feature of this table is that it includes predicted (effective sharp) radii of the neutron and proton distributions. These followed from minimizing the energy with respect to the droplet model degrees of freedom describing the neutron and proton density distributions. The droplet model formula has a structure similar to Equation 16 but is more complicated and has four additional shape-dependent functionals in addition to B_s and B_c. It also includes the exchange correction (linear in Z) for the anticorrelation of the protons and a semi-empirical "Wigner term," proportional to $|I|$, believed to reflect the tighter binding of particles in identical (or closely similar) quantized orbitals (30).

Part of the droplet model mass table just mentioned (30) also appears in the extremely useful collection of calculated masses that was assembled by Maripuu (95). This same collection also contains a droplet model calculation by von Groote, Hilf & Takahashi (28) that improves the agreement with the measured masses by introducing additional flexibility into the schematic shell correction function. In another contribution to this collection, Seeger & Howard (96) combine a liquid-drop model (modified along droplet model lines) with shell corrections calculated using the Strutinsky method.

At this point it is probably good to remind the reader that the adoption of Strutinsky shell corrections (15–18, 72, 77, 97–99) constitutes a major advance but also a break from the previous semi-empirical shell corrections, which were algebraic in nature. The improved predictive power of the Strutinsky method and its essentially unlimited range of applicability are obtained at the cost of a major increase in calculational complexity. No aspect of this method is amenable to hand calculation, and large electronic computer calculations are required for every result.

An enormous literature based on applications of the Strutinsky method has grown up around the prediction of the various nuclear properties, including the properties of superheavy elements that are thought to form an island of stability beyond the end of the known periodic table. Besides the early work by Strutinsky (15–17), some of the landmark papers in this area are those of Nilsson et al (73), Brack et al (18), Bolsterli et al (75), and the review in this series by Nix (76). Figure 5 gives an illustration of results obtained with the Strutinsky method.

A significant recent compilation of nuclear properties that makes use of the Strutinsky method is that of Möller & Nix (37, 38). The liquid-drop part of their energy expression employs a surface energy determined by folding a particular combination of a Yukawa and an exponential interaction into the specified nuclear density distribution, as explained in Section 4. This treatment of the proximity part, in addition to approximating the interaction energy of approaching nuclei, appears to have a very significant effect on lowering the fission barriers of medium and light nuclei. Several other refinements are included in the treatment (corrections for the proton form factor, an exact treatment of the diffuseness correction to the electrostatic energy, the effect of a slight charge asymmetry in the nuclear force, corrections for zero-point motions in the ground state). But the correction that produced a really significant improvement in the fit to nuclear ground-state masses resulted from the simple inclusion of a shape-independent A^0 term (i.e. a constant) in the liquid-drop part of the formula. This cut the rms deviations in the ground-state masses from 1.93 to 0.97 MeV and seemed to remove virtually all systematic smooth deviations between theory and experiment. This may be the first time that significant contact between experiment and the leptodermous expansion has taken place at the A^0 level.

The treatment of the proximity part of the macroscopic energy with the aid of a folding technique (36) requires numerical integrations similar to those involved in the evaluation of the electrostatic energy. The more restricted but algebraic treatment that follows from the proximity force theorem in Section 4 has been found useful for discussing the interaction energy between nuclei. The expression for V_{p} given in Section 4, when combined with the electrostatic interaction energy of two rigid spheres, overestimates the experimental interaction barriers between two nuclei (as measured for projectiles and targets throughout

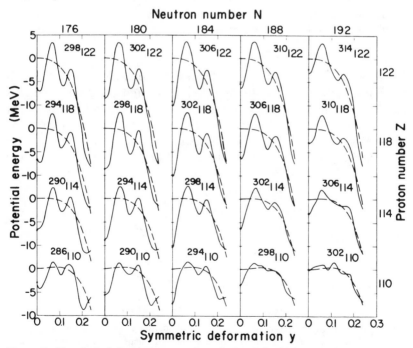

Figure 5 The dashed lines represent the calculated liquid-drop model deformation energies of a number of superheavy nuclei. The solid lines show how the deformation energy is changed when Strutinsky shell effects are added (75).

the periodic tables) by about $4 \pm 2\%$ (100). (A discrepancy in this sense is not unexpected, since actual nuclei are not rigid and, by deforming under the influence of the nuclear forces, may lower the interaction barrier.)

In addition to the theories of nuclear binding and deformation energies, which are the subject of this review, we should mention two treatments that do not go beyond the discussion of equilibrium nuclear masses. The first is that of Zeldes and co-workers (101–105). They employ shell model considerations in constructing algebraic expressions containing a large number of parameters adjusted to account for various aspects of the nuclear level scheme. A table of mass predictions based on this approach was prepared by Liran & Zeldes (106). It agrees extremely well with the known masses on which it is based and is most reliable for short-range extrapolations.

Another approach is that of nuclear mass relations, which was reviewed in this series by Garvey (43). For example, there is the isobaric-multiplet mass equation (107–109), which is obtained from the (more general) Wigner supermultiplet theory (110, 111). This equation asserts that the $(2T + 1)$ masses of an isobaric multiplet with isospin T should be related by the expression,

$$M(T_z) = a + bT_z + cT_z^2. \qquad\qquad 21.$$

In this same category of mass relations, but of wider applicability, is the Garvey & Kelson (41–46) approach illustrated in Figure 6, which is based on adding and subtracting masses so that, within the framework of a shell model treatment, the various interactions between the particles will cancel. The masses of six nuclei should sum to zero when combined according to these patterns. If five masses are known, these

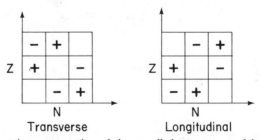

Figure 6 Schematic representation of the so-called transverse and longitudinal mass relations. The boxes represent nuclei from the nuclidic chart with N horizontal and Z vertical. The presence of a plus or a minus sign in a box indicates that the mass value of the respective nucleus is to be added or subtracted. If the Garvey-Kelson mass relations were exact, the sum of the six masses would be zero.

relations can be used to predict the mass of the missing member. In this way tables of predicted masses can be built up by successive application of the basic relations. However, in their simplest forms, the predictions can go badly astray if naively applied to long-range extrapolations (48, 112). The advantages and limitations of this type of mass relations were investigated by Jänecke & others (47–53), who find a number of ways of enhancing their long-range predictive power. The method has also been applied by Monahan & Serduke (113–115).

Finally, as we mentioned in Section 2, a bold attempt is being made to go beyond the indirect approach and actually address the full many-body problem of nuclear structure with the aid of the Hartree-Fock approximation. In spite of the difficult numerical calculations required, this area has seen a great deal of growth in the last decade; current developments were recently reviewed in this series by Quentin & Flocard (40). A survey of nuclear radii throughout the periodic table using Hartree-Fock methods has been undertaken by Beiner, Lombard & Mas (116), who have also prepared a table of nuclear masses (117). A number of related studies is currently underway by Tondeur (118–126) and by Pearson and co-workers (123–126).

While still somewhat phenomenological in nature, because of the effective force (chosen for calculational convenience) and the Hartree-Fock approximation, this approach gives the local, proximity, and global contributions to the energy within a single unified self-consistent approach. This may be essential for the proper treatment of the problem when large excursions away from known nuclei (in shape, particle number, or N/Z) are being considered.

7. CONCLUDING REMARKS

After approximately half a century of nuclear physics there is available today a large amount of experimental information on the masses and deformation energies of nuclei relatively close to the line of beta stability. The interpretation of these measurements, which began in the 1930s with semi-empirical fits to ground-state masses, has improved over the years and has been integrated into detailed theories of nuclear structure and deformabilities. The resulting understanding of both the gross features, as approximated by a liquid-drop or droplet model formula, and of the fine shell effect details, calculated using the Strutinsky method, is generally adequate to account for the binding energies with an accuracy that may be considerably better than 1 MeV (better than 0.1%) for short-range extrapolations, but becomes uncertain for more distant extrapolations. The relatively recent breakthrough in describing quantitatively the shell effects was associated with the indirect approach, in which shell corrections are added to a smooth background, provided by the liquid-drop model. In the future, especially when very distant extrapolations come into consideration (as in astrophysics), the direct approach of attacking the full many-body problem (in a suitable approximation) may become relatively more important. However, even then, the indirect macroscopic approach, suitably enriched to take into account the new situations (e.g. very neutron-rich neutron-star matter, or the bubble or foam topologies of nuclear matter in collapsing supernovae) should continue to be a valuable tool for understanding these complex processes.

Literature cited

1. von Weizsäcker, C. F. 1935. *Z. Phys.* 96:431
2. Bethe, H. A., Bacher, R. F. 1936. *Rev. Mod. Phys.* 8:82–229
3. Bohr, N., Wheeler, J. A. 1939. *Phys. Rev.* 56:426–50
4. Mayer, M. G., Jensen, J. H. D. 1955. *Elementary Theory of Nuclear Shell Structure*, New York: Wiley. 269 pp.
5. Green, A. E. S., Edwards, D. F. 1953. *Phys. Rev.* 91:46–53
6. Cameron, A. G. W. 1957. *Can. J. Phys.* 35:1021
7. Cameron, A. G. W., Elkin, R. M. 1965. *Can. J. Phys.* 43:1288–1311
8. Seeger, P. A. 1967. See Ref. 86, pp. 495–508
9. Mozer, F. S. 1959. *Phys. Rev.* 116:970–75
10. Kümmel, H., Mattauch, J. H. E., Thiele, W., Wapstra, A. H. 1964. See Ref. 81, pp. 42–57
11. Swiatecki, W. J. 1964. See Ref. 81, pp. 58–66
12. Myers, W. D., Swiatecki, W. J. 1965. *Nuclear Masses and Deformations*. Univ. Calif. Lawrence Berkeley Lab. Rep., UCRL–11980. 395 pp.
13. Myers, W. D., Swiatecki, W. J. 1966. *Nucl. Phys.* 81:1–60
14. Myers, W. D., Swiatecki, W. J. 1966. *Ark. Fys.* 36:343–52
15. Strutinsky, V. M. 1967. *Nucl. Phys. A* 95:420–42
16. Strutinsky, V. M. 1967. See Ref. 86, pp. 629–32
17. Strutinsky, V. M. 1968. *Nucl. Phys. A* 122:1–33
18. Brack, M., Damgaard, J., Jensen, A. S., Pauli, H. C., Strutinsky, V. M., Wong, C. Y. 1972. *Rev. Mod. Phys.* 44:320–406
19. Myers, W. D., Swiatecki, W. J. 1969. *Ann. Phys. NY* 55:395–505
20. Myers, W. D. 1969. *Phys. Lett.* 30B:451–4
21. Myers, W. D. 1970. *Nucl. Phys. A* 145:387–400
22. Myers, W. D. 1972. *Dynamic Structure of Nuclear States*, ed. D. J. Rowe, p. 233. Univ. Toronto Press. 585 pp.
23. Ludwig, S., von Groote, H., Hilf, E., Cameron, A. G. W., Truran, J. 1973. *Nucl. Phys. A* 203:627–40
24. Myers, W. D., Swiatecki, W. J. 1974. *Ann. Phys. NY* 84:186–210
25. Möller, P., Nix, J. R. 1974. *Nucl. Phys. A* 229:269–91
26. Hilf, E. R., von Groote, H., Takahashi, K. 1975. *Atomic Masses and Fundamental Constants 5*, p. 293. New York: Plenum
27. Takahashi, K., von Groote, H., Hilf, E. R. 1975. See Ref. 84, pp. 250–56
28. von Groote, H., Hilf, E. R., Takahashi, K. 1976. *At. Data Nucl. Data Tables* 17:418–27
29. Myers, W. D. 1976. *At. Data Nucl. Data Tables* 17:411–17
30. Myers, W. D. 1977. *Droplet Model of Atomic Nuclei*, New York: IFI/Plenum Data Co. 150 pp.
31. Myers, W. D., Swiatecki, W. J. 1980. *Nucl. Phys. A* 336:267–78
32. Blocki, J., Randrup, J., Swiatecki, W. J., Tsang, C. F. 1977. *Ann Phys. N.Y.* 105:427–62
33. Blocki, J., Swiatecki, W. J. 1981. *Ann. Phys. NY* 132:53–65
34. Krappe, H. J., Nix, J. R. 1974. See Ref. 93, paper IAEA-SM-174-12
35. Krappe, H. J., Nix, J. R., Sierk, A. J. 1979. *Phys. Rev. Lett.* 42:215–18

36. Krappe, H. J., Nix, J. R., Sierk, A. J. 1979. *Phys. Rev. C* 20:992–1013
37. Möller, P., Nix, J. R. 1981. *Nucl. Phys. A* 361:117–46
38. Möller, P., Nix, J. R. 1981. *At. Data Nucl. Data Tables* 26:165–96
39. Krappe, H. J. 1981. *Proc. 14th Masurian Summer Sch., Mikolajki, Poland.* To be published in *Nukleonika*
40. Quentin, P., Flocard, H. 1978. *Ann. Rev. Nucl. Part. Sci.* 28:523–96
41. Garvey, G. T., Kelson, I. 1966. *Phys. Rev. Lett.* 16:197–98
42. Garvey, G. T., Gerace, W. J., Jaffe, R. L., Talmi, I., Kelson, I. 1969. *Rev. Mod. Phys.* 41:S1–S80 (Suppl.)
43. Garvey, G. T. 1969. *Ann. Rev. Nucl. Sci.* 19:433–70
44. Garvey, G. T. 1972. *Comments Nucl. Part. Phys.* 5:85–88
45. Comay, E., Kelson, I. 1975. See Ref. 84, pp. 272–78
46. Comay, E., Kelson, I, 1976. *At. Data Nucl. Data Tables* 17:463–66
47. Jänecke, J. 1972. *Phys. Rev. C* 6:467–68
48. Jänecke, J., Behrens, H. 1972. *Z. Phys.* 256:236–42
49. Jänecke, J., Behrens, H. 1974. *Phys. Rev. C* 9:1276–91
50. Jänecke, J., Eynon, B. P. 1975. *Nucl. Phys. A* 243:326–48
51. Jänecke, J., Eynon, B. P. 1975. See Ref. 84, pp. 279–85
52. Jänecke, J., 1976. *At. Data Nucl. Data Tables* 17:455–62
53. Jänecke, J., Eynon, B. P. 1976. *At. Data Nucl. Data Tables* 17:467–73
54. Swiatecki, W. J. 1980. *Prog. Part. Nucl. Phys.* 4:383–450
55. Swiatecki, W. J. 1981. *Phys. Scr.* 24:113–22
56. Tsang, C.-F. 1969. PhD thesis. *Univ. Calif. Lawrence Radiat. Lab. Rep. UCRL–18899.* 209 pp.
57. Strutinsky, V. M., Tyapin, A. S. 1963. *Zh. Eksp. Teor. Fiz.* 45:960 (Transl: *Sov. Phys.-JETP* 18:664–67)
58. Brueckner, K. A., Chirico, J. H., Meldner, H. W. 1971. *Phys. Rev. C* 4:732–40
59. Chu, Y. H., Jennings, B. K., Brack, M. 1977. *Phys. Lett. B* 68:407–11
60. Weiss, R. A., Cameron, A. G. W. 1969. *Can. J. Phys.* 47:2172–209
61. Weiss, R. A., Cameron, A. G. W. 1969. *Can. J. Phys.* 47:2211–54
62. Truran, J. W., Cameron, A. G. W.
63. Baym, G., Bethe, H. A., Pethick, C. J. 1971. *Nucl. Phys. A* 175:225
64. Mackie, F. D., Baym, G. 1977. *Nucl. Phys. A* 285:332–48
65. Swiatecki, W. J. 1956. *Phys. Rev.* 104:993–1005
66. Swiatecki, W. J. 1958. In *Proc. 2nd Int. Conf. Peaceful Uses At. Energy, Geneva,* pp. 248–72. London: Pergamon
67. Hasse, R. W. 1971. *Ann. Phys. NY* 68:377–461
68. Davies, K. T. R., Sierk, A. J. 1975. *J. Comp. Phys.* 18:311–25
69. Davies, K. T. R., Nix, J. R. 1976. *Phys. Rev. C* 141977–94
70. Toke, J., Swiatecki, W. J. 1981. *Nucl. Phys. A* 372:141–50
71. Fermi, E. 1951. *Nuclear Physics.* Univ. Chicago Press. 248 pp.
72. Bjoernholm, S., Strutinsky, V. M. 1969. *Nucl. Phys. A* 136:1–24
73. Nilsson, S. G., Tsang, C. F., Sobiczewski, A., Szymanski, Z., Wycech, S., Gustafson, C., Lamm, I.-L., Möller, P., Nilsson, B. 1969. *Nucl. Phys. A* 131:1–66
74. Seeger, P. A. 1970. See Ref. 87, pp. 217–62
75. Bolsterli, M., Fiset, E. O., Nix, J. R., Norton, J. L. 1972. *Phys Rev. C* 5:1050–77
76. Nix, J. R. 1972. *Ann. Rev. Nucl. Part. Sci.* 22:65–120
77. Brack, M., Quentin, P. 1974. See Ref. 93, pp. 231–48
78. Bohr, A., Mottelson, B. R. 1969. *Nuclear Structure,* New York/Amsterdam: Benjamin. 471 pp.
79. Hintenberger, H., ed. 1957. *Nuclear Masses and Their Determination,* London: Pergamon. 265 pp.
80. Duckworth, H. E., ed. 1960. *Proc. Int. Conf. Nuclidic Masses.* Univ. Toronto Press, 539 pp.
81. Johnson, W. H. Jr., ed. 1964. *Proc. 2nd Int. Conf. Nuclidic Masses.* Vienna: Springer-Verlag. 473 pp.
82. Barber, R. C., ed. 1967. *Proc. 3rd Int. Conf. At. Masses,* Winnepeg: Univ. Manitoba Press. 901 pp.
83. Sanders, J. H., Wapstra, A. H., eds. 1972. *Atomic Masses and Fundamental Constants 4.* London: Plenum. 571 pp.
84. Sanders, J. H., Wapstra, A. H., eds. 1975. *Atomic Masses and Fundamental Constants 5.* New York: Plenum. 681 pp.
85. Nolen, J. A. Jr., Benenson, W.,

Hilf, E. 1970. See Ref. 87, pp. 275–306

eds. 1979. *Atomic Masses and Fundamental Constants 6.* New York: Plenum. 572 pp.

86. Forsling, W., Herrlander, C. J., Ryde, H., eds. 1967. *Proc. Int. Conf. Properties of Nuclei far from Region of Beta-Stability, Lysekil, Sweden, Ark. Fys.* 36:1–686

87. Rudstam, G., ed. 1970. *Proc. Int. Conf. Properties of Nuclei far from Region of Beta-Stability, Leysin, Switzerland, CERN 70–30.* 1151 pp.

88. Klapisch, R., ed. 1976. *3rd Int. Conf. Nuclei far from Stability, CERN 76–13.* 608 pp.

89. Hamilton, J. H., Spejewski, E. H., Bingham, C. R., Zganjar, E. F., eds. 1980. *Future Directions in Studies of Nuclei Far from Stability.* Amsterdam: North-Holland. 424 pp.

90. Hansen, P. G., Nielsen, O. B., eds. 1981. *4th Int. Conf. Nuclei far from Stability, CERN 81–09,* 809 pp.

91. IAEA. 1965. *Physics and Chemistry of Fission 1.* Vienna, Vol. 1, 633 pp., Vol. 2, 469 pp.

92. IAEA. 1969. *Physics and Chemistry of Fission 2.* Vienna, 983 pp.

93. IAEA. 1974. *Physics and Chemistry of Fission 3,* Vienna, Vol. 1, 579 pp., Vol. 2, 525 pp.

94. IAEA. 1980. *Physics and Chemistry of Fission 4.* Vienna Vol. 1, 629 pp., Vol. 2, 501 pp.

95. Maripuu, S., ed. 1976. 1975 Mass Predictions. *At. Data Nucl. Data Tables* 17:411–608

96. Seeger, P. A., Howard, W.M. 1976. *At. Data Nucl. Data Tables* 17:428–30

97. Bunatian, G. G., Kolomiets, V. M., Strutinsky, V. M. 1972. *Nucl. Phys. A* 188:225–58

98. Strutinsky, V. M. 1975. *Nucl. Phys. A* 254:197–210

99. Strutinsky, V. M., Ivanjuk, F. A. 1975. *Nucl. Phys. A* 255:405–18

100. Vaz, L. C., Alexander, J. M., Satchler, G. R. 1981. *Phys. Rep.* 69:373–99

101. Zeldes, N., Grònau, M., Lev, A. 1965. *Nucl. Phys.* 63:1–75

102. Zeldes, N., Grill, A., Simievic, A.

1967. Mat. Fys. Skr. Dan. Vid. Selsk. Vol. 3, No. 5, 163 pp.

103. Zeldes, N. 1967. *Ark. Fys.* 36:361–83

104. Comay, E., Liran, S., Wagman. J., Zeldes, N. 1970. See Ref. 87, pp. 165–216

105. Zeldes, N. 1972. See Ref. 83, pp. 245–54

106. Liran, S., Zeldes, N. 1976. *At. Data Nucl. Data Tables* 17:431–41

107. Robson, D. 1966. *Ann. Rev. Nucl. Sci.* 16:119–49

108. Cerny, J. 1968. *Ann. Rev. Nucl. Sci.* 18:27–52

109. Benenson, W., Kashy, E. 1979. *Rev. Mod. Phys.* 51:527–40

110. Wigner, E. P., Feenberg, E. 1941. *Rep. Prog. Phys.* 8:274

111. Franzini, P., Radicati, L. A. 1963. *Phys. Lett.* 6:322–4

112. Sorensen, R. A. 1971. *Phys. Lett.* 34B:21–3

113. Monahan, J. E., Serduke, F. J. D. 1977. *Phys. Rev. C* 15:1080–84

114. Monahan, J. E., Serduke, F. J. D. 1978. *Phys. Rev. C* 17:1196–1204

115. Monahan, J. E., Serduke, F. J. D. 1979. See Ref. 85, pp. 151–59

116. Beiner, M., Lombard, R. J., Mas, D. 1976. See Ref. 88, pp. 212–21

117. Beiner, M., Lombard, R. J., Mas, D. 1976. *At. Data Nucl. Data Tables* 17:450–54

118. Tondeur, F. 1978. *Nucl. Phys. A* 303:185–98

119. Tondeur, F. 1978. *Nucl. Phys. A* 311:51–60

120. Tondeur, F. 1978. *Z. Phys. A* 288:97–101

121. Tondeur, F. 1979. *Nucl. Phys. A* 315:353–69

122. Tondeur, F. 1979. *Astron. Astrophys.* 72:88–91

123. Cote, J., Pearson, J. M. 1978. *Nucl. Phys. A* 304:104–26

124. Farine, M., Pearson, J. M., Rouben, B. 1978. *Nucl. Phys. A* 304:317–26

125. Pearson, J. M., Rouben, B., Saunier, G., Brut, F, 1979. *Nucl. Phys. A* 317:447–59

126. Farine, M., Cote, J., Pearson, J. M. 1980. *Nucl. Phys. A* 338:86–96

Ann. Rev. Nucl. Part. Sci. 1982. 32:335–89
Copyright © 1982 by Annual Reviews Inc. All rights reserved

CALORIMETRY IN HIGH-ENERGY PHYSICS

C. W. Fabjan

CERN, Geneva, Switzerland

T. Ludlam

Brookhaven National Laboratory, Upton, Long Island, New York 11973, USA

CONTENTS

1. INTRODUCTION

The detection of particles is the experimental tool of high-energy physics research, and the evolution of the field has been closely coupled

0163-8998/82/1201-0335$02.00

335

to the development of improved methods for detecting and measuring an ever-widening spectrum of particle properties.

Much of our present knowledge about the physics of elementary particles has been the result of a continuing refinement of techniques for measuring the trajectories of individual charged particles — from the early hodoscopes, cloud chambers, and emulsion stacks, to today's extraordinarily precise bubble and streamer chambers and sophisticated magnetic spectrometers, equipped with multiwire proportional or drift chambers (Charpak et al 1978, Fabjan & Fischer 1980).

In this article we examine an intrinsically different class of detector that has come to the fore over the past decade and promises to greatly influence the scope of future experiments — total absorption detectors, or calorimeters.

Conceptually, a calorimeter is a block of matter, which intercepts the primary particle and is of sufficient thickness to cause it to interact and deposit all its energy inside the detector volume in a subsequent cascade or "shower" of increasingly lower-energy particles. Eventually most of the incident energy is dissipated and appears in the form of heat. Some (usually a very small) fraction of the deposited energy goes into the production of a more practical signal (e.g. scintillation light, Čerenkov light, or ionization charge), which is proportional to the initial energy. In principle, the uncertainty in the energy measurement is governed by statistical fluctuations in the shower development, and the fractional resolution σ/E improves with increasing energy E as $E^{-1/2}$.

The first large-scale detectors were used in cosmic-ray studies (Murzin 1967). Interest in calorimeters grew in the late 1960s and early 1970s in view of the coming machines (CERN-ISR, Fermi National Accelerator Laboratory FNAL, CERN-SPS) with their greatly changed experimental environments, making, for example, magnetic momentum analysis in large-solid-angle experiments increasingly difficult (Atač 1975).

This period corresponded also to the advent of intense, high-energy neutrino beams with the need for very massive detectors to study their interactions. Not surprisingly, this detector development was paralleled by the rapid growth of analogue-signal processing techniques: during the last decade the typical number of analogue-signal channels of nuclear spectroscopy quality has increased from about 10 to 10^4 in high-energy physics experiments!

At the outset it was noted that calorimetric detectors offer many other attractive capabilities, aside from the energy response, all of which have since been exploited in varying degrees:

1. they are sensitive to neutral as well as charged particles;
2. the size of the detector scales logarithmically with particle energy E,

whereas for magnetic spectrometers the size scales with $p^{1/2}$, for a given relative momentum resolution $\Delta p/p$;

3. with segmented detectors, information on the shower development allows precise measurements of position and angle of the incident particle;

4. the differences in response to electrons, muons, and hadrons can be exploited for particle identification;

5. their fast time response allows operation at high particle rates, and the patterns of energy deposition can be used for real-time event selection.

We are now entering an era in which high-energy colliding-beam machines may reveal a "new" physics of supermassive particles (Ellis 1981) in extremely rare and complex events, and whose signatures include the production of very energetic leptons and the fragmentation of massive quarks. In these experiments the traditional momentum analysis of a few charged particles will be replaced by measurements of momentum and energy flow among multiple jets of particles. Calorimetric detectors are uniquely suited to such tasks, provided the intrinsic potential of these instruments as primary detectors can be fully realized.

In Section 2 we survey the calorimeter installations that have been introduced into a variety of high-energy physics experiments. We describe the physics requirements leading to finely instrumented total absorption detectors for neutrino experiments and investigations of nucleon stability, followed by a summary of the uses of calorimeters for the detection of multiparticle final states in experiments with hadron and electron beams.

In Section 3 we review the physics of the cascade development for electromagnetic and hadronic showers in dense materials, emphasizing the measurements and calculational techniques of properties that govern the performance of calorimeters. This is followed, in Section 4, by a discussion of the detector parameters relating to the intrinsic performance and implementation for specific experimental applications: the resolving power for energy; position and angle measurements; timing properties; the capability for particle identification; and factors such as calibration and monitoring of large detector systems. In Section 5 we examine the methods of readout employed for calorimetric measurements. Finally, we summarize the state of the art, point to specific trends of development, and assess the potential of the calorimetric methods for future physics programs.

We have aimed at presenting a broad view of the role of calorimeters in high-energy physics and have preferred to illustrate our important points with representative examples. This review is therefore not rigorously complete, and it is suggested that recent excellent discussions be consulted (Iwata 1979, Albrow 1981, Amaldi 1981, Astbury 1981, Gordon et al 1981).

2. CALORIMETER SYSTEMS FOR PHYSICS APPLICATIONS

2.1 Neutrino Physics and Nucleon Stability

Detectors for these studies share a number of common features: the event rate is low and proportional to the total instrumented mass; the physics requires a fine-grained readout system, which permits detailed three-dimensional pattern reconstruction. Bubble chambers have therefore been used extensively but lack sensitivity owing to their limited mass. The present generation of neutrino detectors makes extensive use of wire-chamber techniques to approach the intrinsic spatial and angular resolution of calorimeters. New "visual" electronic techniques are being developed for proton decay experiments, for which the most massive detectors with the highest density of signal channels are required.

2.1.1 NEUTRINO DETECTORS   The size of these detectors is governed by the cross section for neutrino interactions, $\sim 10^{10}$ times smaller than hadronic cross sections. Even with intense neutrino beams, the detector must have a very large mass (typically many hundreds of tons), and its volume must be uniformly sensitive to the signature that an interaction has occurred and to the characteristics of the reaction products. These requirements explain the modular construction typical of modern electronic neutrino detectors, of which two examples are shown in Figure 1. Each calorimeter module consists of several planes of dense absorber interleaved with planes of detectors to record the pattern of charged particles emerging from the absorber layers. The detector planes are subdivided into a sufficiently fine-grained array of independent readout cells to allow adequate pattern recognition. The general form of the interaction is $\nu_{(\mu, e)} + \text{nucleon} = \ell_{(\mu, e)} + X$, where ℓ is a charged lepton (charged-current interaction) or a neutrino (neutral-current interaction), and X represents the system of recoiling hadrons projected forward into a collimated jet by relativistic kinematics. In the simplest instance the signature for a neutrino interaction is the sudden appearance of a large amount of energy in a few layers, deep in the detector. If the scattered lepton is a muon, it leaves in each detector layer the characteristic signal of a single minimum-ionizing particle. In some detectors the absorber layers are magnetized iron, which makes possible a determination of the muon momentum from the curvature of its trajectory. (For an example, see Figure 22.) For the detailed study of neutral-current interactions, a very fine-grained subdivision of the calorimeter system is required for measuring the energy and direction of the hadronic system X and for reconstructing the "missing" momentum of the final-state neutrino.

Clearly the scope and sensitivity of neutrino experiments would be much improved if even the most massive detectors could resolve final-state particles with the reliability and precision typical of a bubble chamber. With this goal in mind, some schemes are currently being investigated that use drift chamber methods with large volumes of compressed gas, which provides a visual quality characteristic of homogeneously sensitive detectors. In one case (Vishnevskii et al 1979) a cylindrical detector 3.5 m in diameter and 35 m long is envisioned, containing about 100 tons of argon gas at 150 atm pressure, with ionization electrons collected on planes of anode wires. The idea has also been advanced (Bouclier et al 1980) of using compressed mixtures of more common gases (air or freon) or room-temperature liquid

Figure 1   The two counter neutrino facilities at CERN. The ν beam enters first the CDHS experiment (in the background) and subsequently the apparatus of the CHARM Collaboration. Both experiments are based on very massive, highly modular scintillation calorimeters, and measure particle position with wire chamber techniques (photo CERN).

hydrocarbons (W. J. Willis, 1980, private communication) and detecting the ions that migrate away from charged-particle tracks (positive and/or negative ions produced by electron attachment). Many technical problems need further study before these ideas can be realized in practical detectors.

Closely connected with neutrino physics is a class of investigations referred to as "beam dump" experiments. Here, intense beams of protons are directed at an inert, massive absorber (the beam dump) capable of fully containing the hadronic reaction products. A detector downstream from this absorber is used to compare the fluxes of emerging muons and neutrinos with that of the incident hadrons. The observed leptons are presumed to be products of the weak decay of particles produced by the interacting protons. The results can be corrected for the lepton flux due to pions and kaons, which decay before interacting in the absorber. An excess of leptons is observed, which gives evidence for additional families of particles with much shorter lifetimes. Such experiments gave an early indication of the hadronic production of charmed particles (Dydak 1980, Derrick 1981).

2.1.2 NUCLEON STABILITY   The possibility of nucleon decay has been studied experimentally for several decades. Recently, these efforts received a strong impetus from "grand unified theories" of weak, electromagnetic, and strong interactions, which provide definite predictions of the nucleon lifetime (Georgi et al 1974, Pati & Salam 1974, Ellis 1981). Present estimates place the proton lifetime around $\tau_p \simeq 10^{31\pm2}$ y, about three orders of magnitude above the current experimental limit of $\tau_p \simeq 10^{28}$ y (Krishnaswamy et al 1981) and barely within reach of present experimental finesse. To achieve a limit of about 10^{32} y, approximately 1000 tons, containing 10^{33} nucleons, have to be instrumented to search sensitively for some of the expected decay modes such as: $p \rightarrow e^+ + h^0$, where h^0 is a neutral meson (π^0, η, ρ^0, ω^0, etc). The signature of such a decay is rather striking but requires a detector with sufficient subdivision to recognize the back-to-back decay into a lepton and a hadron with the relatively low energy deposit of about 1 GeV. The sensitivity is limited by the flux of muons and neutrinos originating from atmospheric showers. Only μ's can be shielded by placing the experiments deep underground in mine shafts or in road tunnels beneath Alpine mountains. The ν-induced rate simulating nucleon decay is estimated at $\sim 10^{-2}$ events per ton per year if energy deposition alone is measured. If complete event reconstruction is possible, a further suppression of about 100 may be achieved and one may reach an experimental limit of $\tau \gtrsim 10^{33}$ y (Goldhaber & Sulak 1981, Treille 1981). Section 2.3 gives further information.

2.2   *Storage Ring Experiments*

The apparent lack of driving physics motivations paired with major technical difficulties was not propitious for a vigorous development of calorimeters at storage rings: such detectors are large, complex, and

Figure 2   View of the hadron calorimeter of the UA2 Collaboration at the CERN p̄p collider. This spherical detector is made from 24 "orange"-like segments, containing 10 cells with threefold longitudinal subdivision (Photo CERN).

costly, because most of the solid angle has to be covered in a complicated (approximately spherical) geometry and adequately instrumented (Figures 2 and 3). Only now have electromagnetic calorimeters become relatively "standard" detectors, whereas hadronic calorimeter facilities are still quite rare. Beautiful Gedanken experi-

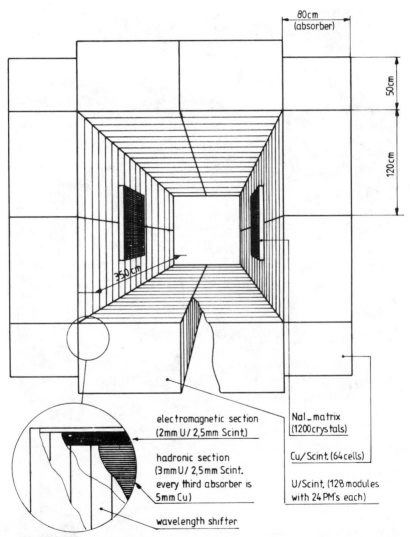

Figure 3 Isometric view of the AFS uranium calorimeter at the CERN ISR. The 8-sr detector is subdivided into 800 cells with four readout channels each. Also shown are two high-granularity NaI "crystal walls," with a total of 1200 readout channels.

ments have therefore preceded reality by more than a decade (Willis 1972).

Most of our comments, which we illustrate with storage ring applications, apply also to the use of calorimeters for fixed-target experiments. Several pioneering studies at FNAL (Bromberg 1979, Dris et al 1979) and at the CERN-SPS (Pretzl 1981) have contributed to improving the understanding of calorimetry in hadronic physics and have found unexpected and striking results on the production cross section and topology of events with a large transverse energy.

2.2.1 PHYSICS APPLICATIONS   At *hadron machines* the studies focus on reactions that are characterized by a large transverse energy flow, as a signal for a highly inelastic interaction between the nucleon constituents. The signature appears in many different characteristic event structures and may therefore be efficiently selected with hadron calorimeters: examples are single high-p_T particles, "jets" of particles, or events exhibiting large transverse energy E_T, irrespective of its detailed structure. Topical applications include invariant mass studies of multijet systems. This is considered as one possible strategy that could be used to search for a $t\bar{t}$ signal, because two-jet or three-jet decays may dominate over the $\mu^+\mu^-$ channel by a factor $\gtrsim 100$. Further motivation for this "jet spectroscopy" is provided by the gauge theories of elementary particles (Bég 1981), stipulating new particles in the 10–100 GeV mass range, which are expected to decay with a characteristic pattern of several jets. Apart from these studies, which are being pursued with present detectors at the CERN ISR and the $p\bar{p}$ collider, and which influence the design of future detectors for ISABELLE and the $p\bar{p}$ collider at FNAL, hadron calorimeters are considered to be most essential in studies of ultrarelativistic heavy-ion collisions (Willis 1981). As an example, in collisions between nuclei with $A \gtrsim 100$ and ~20 GeV per nucleon, several thousands of particles may be produced, carrying a total of several thousand GeV. Except for small-solid-angle, special-purpose detectors, probably only calorimeters will provide meaningful measures of such events, which are too complicated at the individual-particle level although suitable averages may reveal striking signatures.

At *electron-positron colliders*, electromagnetic detectors are frequently used to measure the dominant fraction of neutral particles, π^0's. They are also the ideal tool for detecting electrons, which may signal decays of particles with one or more "heavy" (c, b, . . .) quarks. Unique investigations of $c\bar{c}$ and $b\bar{b}$ quark spectroscopy were accomplished with very high resolution NaI-type shower detectors (Scharre 1981, Schamberger 1981).

Many features of the physics program at the next generation of e^+e^- colliders [LEP (Picasso 1981), SLAC Linear Collider (Panofsky 1981), CESR II (McDaniel 1981)] will require the extensive use of hadronic calorimeters (Cundy et al 1976, Willis & Winter 1976, Fabjan 1980). In particular, the characteristic feature of e^+e^- annihilation — production of particles at relatively large angles with a total energy equal to the center-of-mass (c.m.) collision energy — makes the experimental technique of total energy measurement particularly appropriate. For future e^+e^- physics, this method will be important because

1. the level of neutron and K_L^0 production, measurable only with hadronic calorimeters, is expected to increase with energy;
2. a large and most interesting fraction of events will contain neutrinos in the final state; missing energy and momentum analysis provides the sole handle on such reactions;
3. a considerable fraction of events will show good momentum balance but large missing energy; these may be two-photon events or, above the Z^0 pole, events on the radiative tail. Total energy will provide the cleanest signature;
4. hadronic calorimeters will be the most powerful tool for measuring the reaction $e^+e^- \rightarrow W^+W^-$, either in channels where each W decays hadronically (total of four jets) or through the rarer leptonic decay channels.

2.2.2 CONSEQUENCES OF DETECTOR LIMITATIONS   Energy and space resolution are the principal calorimeter properties affecting physics studies. Consider as an example a typical multijet event of 200 GeV that can be measured with an accuracy of $\sigma \approx 6$–8 GeV. Likewise, the total momentum balance can be checked at a level of $\sigma \approx 4$ GeV/c (Cundy et al 1976). These are intrinsic performance figures, disregarding possible instrumental effects (see Section 5) and are well matched to the next generation of e^+e^- programs. While such a performance is also adequate for work at hadronic machines, a further serious difficulty arises from the convolution of the *energy response function* with the steeply falling p_T distribution of hadronically produced secondaries (Selove 1972, Almehed et al 1977, Dris 1979; M. Block, 1977, unpublished note UA1-26, UA1 Collaboration, CERN). As a consequence (see Figure 4), the measured energy deposit E_m in the detector will originate predominantly from incident particles with energy $E < E_m$; therefore count rates and trigger rates are higher than the true physics rates. The result can be devastating for detectors with poor energy resolution or a non-Gaussian response function ("tails" in energy resolution), making the deconvolution almost impossible. The problem is compounded by the tendency to respond to π^0's with a signal larger than that from charged pions (see Sections 3 and 4). Without adequate precautions, these detectors would preferentially select π^0's, which would render them useless for general trigger applications.

The *granularity* of the detector characterizes the spatial separation between two particles in an event. The required degree depends on the

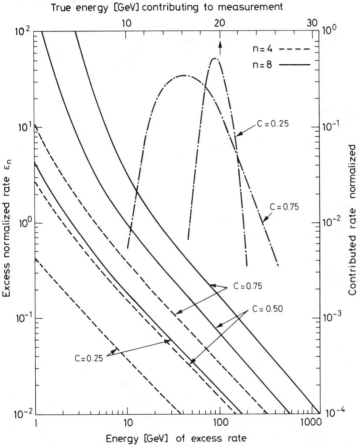

Figure 4 The results (left ordinate versus bottom abscissa) of the convolution $\bar{S}(E_0) = \int_{E_1}^{\infty} S(E) \cdot R(E, E_0) \, dE$ of a Gaussian energy response function $R(E, E_0) = [1/\sqrt{2\pi}\,\sigma(E)] \exp[-(E - E_0)^2/2\sigma^2(E)], \sigma(E) = cE^{1/2}$, with a spectrum $S(E) = A/E^n$. The excess normalized rate is given by $\varepsilon_n = [\bar{S}_n(E_0) - S_n(E_0)]/S_n(E_0)$. Also shown (dashed-dotted curve corresponding to right ordinate versus top abscissa for $n = 8$) is an example of the corresponding energy distribution, which contributes to an energy measurement of $E_0 = 20$ GeV. Due to the steep fall $\sim E^{-n}$, the most probable energy E contributing may be substantially lower than E_0 (adapted from M. Block, 1977, unpublished note UA1-26, UA1 Collaboration, CERN).

Table 1 Triggering with hadron calorimeters

Experiment	Trigger

Single-particle
  inclusive distribution
  correlations
	Localized energy deposit in spatial coincidence
	  with matching track; several thresholds used
	  concurrently

Jet studies	Extended ($\sim 1\,\mathrm{sr}$) energy deposit; several thresholds and multiplicities

Inclusive leptons,
  multileptons
	Electromagnetic deposit in spatial coincidence with
	  matching track; several thresholds and multiplicities

Heavy flavor jets	Jet trigger with additional lepton required
Correlations	Various combinations of above triggers

minimum angle θ to be detected between particles and is therefore imposed by the physics. An example is the individual detection of some or most of the particles inside a jet, e.g. the detection of an electron inside such an event. Showers have to be separated by approximately one shower diameter d in order to be recognized individually, which therefore defines the minimum distance D of the detector from the interaction point as $D \gtrsim d/\theta$. As the total instrumented volume of the calorimeter scales approximately with $\sim D^3 \sim d^3$, one might pay a premium for a high-density detector and still achieve the overall most economical design, provided the distance scaling is correctly implemented.

The granularity requirement dominates the *size* consideration for calorimeters. In contrast, the c.m. energy of the collider imposes a weaker scaling of the size: for full shower containment the thickness of the detector scales logarithmically with energy and increases by a factor of two from $E = 10\,\mathrm{GeV}$ to $100\,\mathrm{GeV}$.

The *trigger capability* is a unique, and perhaps the most important, requirement of hadron calorimeters employed at hadron machines. For satisfactory operation, one needs uniform response irrespective of event topology and particle composition; good energy resolution at the trigger level to minimize effects of the response function; and adequate granularity at the trigger level for selection of specific event topologies. For high selectivity, rather complex analogue computations are required, as may be seen from the examples in Table 1 (Rosselet 1981).

2.3 Calorimeter Systems

In Table 2 we have summarized information on some calorimeter facilities. This incomplete selection emphasizes the great variety of applications, techniques, and performances. (CDHS: Abramowicz et al

1981; CHARM: Diddens et al 1980; FNAL Exp 594: Bogert et al 1982; BNL Exp. 734: Amako 1981; Fréjus: Barloutaud 1981; IMB: Van der Velde 1981; NA5: Eckardt et al 1978; MARK J: Barber et al 1980; CELLO: Behrend et al 1981; AFS: Botner et al 1981a; UA1: Corden et al 1982; UA2: Clark et al 1982; MAC: Anderson et al 1978).

3.   ENERGY LOSS IN DENSE ABSORBERS

3.1   *Electromagnetic and Hadronic Showers*

In this section we describe the properties of electromagnetic and hadronic showers. We emphasize the mechanisms contributing to fluctuations in the cascading process, which impose the fundamental limitations on calorimetric determination of the energy, position, and direction of particles. Characteristic properties describing the average behavior of electromagnetic and hadronic showers are summarized in Table 3. One notes the remarkably different dependence on the absorber material, which reflects the very different nature of the two cascading processes.

For electrons and photons, the shower develops predominantly through bremsstrahlung and pair production. The scale for the longitudinal distribution is set by the "radiation length" (X_0) related to the mean path length of an electron in a material (A, Z). Through the multiplication process, the largest number of secondary particles is reached at a depth t_{max}, after which the number and energy of secondaries decrease with a characteristic attenuation length λ_{att}, which is determined by the mean free path of photons with energy corresponding to minimum absorption in the material (Figure 5). Eventually, the particles reach a "critical" energy, $\varepsilon \sim 550$ MeV$/Z$, below which no further multiplication occurs (ε is defined as the energy at which the rate of energy loss per radiation length equals the total energy of the electron). The depth of containment of the shower depends logarithmically on energy. The lateral distribution of the secondaries shows a marked dependence on the depth along the shower axis (Figure 5). Close to the shower maximum, the swarm of still rather energetic secondaries is quite collimated and contained in a cylinder with radius $R_{max} \simeq 1X_0$; deeper in the cascade, the distribution is dominated by multiple scattering of low-energy electrons, which no longer radiate and which travel far from the shower axis. For this part of the shower, the lateral distribution scales in units of the "Molière" radius ρ_M.

Some of the characteristic features of electromagnetic showers are conveniently described in analytic form using "Rossi's Approximation

Table 2 Examples of calorimeter facilities

Logo	Major physics goals	Size (tons) (approx.)	Technique
CDHS	ν physics: emphasis on structure functions	1400	2.5-cm or 5-cm Fe plates sampled with scintillators and drift chambers
CHARM	ν physics: emphasis on neutral-current events	150	8-cm marble plates sampled with scintillator and proportional tubes
FNAL Exp. 594	ν physics	340	Flash tubes between layers of sand and steel pellets
BNL Exp. 734	ν physics	120	Homogeneous liquid scintillator, proportional tubes
Fréjus	Nucleon stability	1500	Iron, flash tubes
IMB	Nucleon stability	7000	Homogeneous water Čerenkov detector
NA5	Hadronic high-p_T particle production	70	0.55-cm Pb/scintillator for e.m. part 5-cm Fe/scintillator for H part
MARK J	Study of e^+e^- annihilation	180	0.5-cm Pb/scintillator for e.m. part Iron/scintillator for H part
CELLO	Study of e^+e^- annihilation	50	0.12-cm Pb/0.36-cm liquid argon
AFS	Hadronic production of very high E_T states	300	0.2-cm U/0.25-cm scint. for e.m. part 0.3-cm U/0.25-cm scint. for H part
UA1	Hadronic production at $s^{1/2} = 540$ GeV	1000	0.2–0.4-cm Pb/scint. for e.m. part 5-cm Fe/scintillator for H part
UA2	Hadronic production at $s^{1/2} = 540$ GeV	100	3.5-mm Pb/scint. for e.m. part 1.5-cm Fe/scint. for H part
MAC	Study of e^+e^- annihilation	400	0.28-cm Pb/proportional tubes for e.m. part 2.7-cm Fe/proportional tubes for H part

No. of read-out channels	σ_E (for 1 GeV)		Comment
	Hadrons	Photons	
3600 PMs 4000 drift wires	58(70)% 2.5(5)cm	23%	
~1600 scintillators ~10^4 proportional tubes	53%	20%	Optimized for angular resolution for missing momentum determination
~4×10^5 flash tubes	80%	10%	Very fine-grained readout for optimum pattern recognition
3800 PMs 13,000 proportional tubes	Not available		Active absorber for optimum sensitivity to elastic scattering events
~10^6	$\sigma \approx 15\%$ expected for $p \rightarrow e^+ \pi^0$		
1350 8 in. diameter PMs	$\sigma \approx 20\%$ expected for $p \rightarrow e^+ \pi^0$		Patterns of Čerenkov light give adequate spatial and energy resolution with largest feasible detector mass
500 PMs on wavelength shifters	86%	16%	Complete azimuthal coverage for bias-free triggering
Coarse readout with PMs Drift chamber	20% for 30-GeV events	10%	Coarse sampling, but almost complete 4π coverage
8000 charge amplifiers	—	9%	4π coverage with high-performance e.m. calorimeter
3200 PMs on wavelength shifters	35%	15%	Complete azimuthal coverage for bias-free triggering
2200 PMs	75%	15%	4π coverage Hadron calorimeter integrated with spectrometer magnet
1700 PMs on wavelength shifters	60%	15%	Spherical geometry!
10^5 wires grouped in 6000 signal channels	75%	17%	Calorimeter iron is magnetized for μ spectroscopy; nearly spherical geometry

Table 3 Average properties of electromagnetic and hadronic showers

Quantity	Electromagnetic showers	Hadronic showers
Mean free path	$9X_0/7$ for γ's $X_0 \simeq 180A/Z^2 [\text{g cm}^{-2}]$	$\lambda = A/(N_{\text{Avogadro}} \cdot \sigma_{\text{Abs}})$ $\propto A^{1/4}$
Secondary particles	e^+, e^-, γ; below critical energy $\varepsilon \simeq 550\,\text{MeV}/Z$ ionization loss only; inelasticity $\kappa = 1$ (all energy given to particle production).	Fast nucleons, pions; medium-energy ($\sim 100\,\text{MeV}$) p's and n's; low-energy ($\sim 10\,\text{MeV}$) p, n, γ; nuclear fragments; inelasticity $\kappa \sim 0.5$.

Average shower dimensions

Shower maximum[a]	$t_{\text{max}}[X_0] \simeq \ln(E/\varepsilon) - \alpha$ ($\alpha \simeq 1$ for e's, $\simeq 0.5$ for γ's)	$t_{\text{max}}[\lambda] \simeq 0.6 \ln E[\text{GeV}] - 0.2$
Depth for $\simeq 95\%$ longitudinal containment[a]	$L_{0.95}[X_0] \simeq t_{\text{max}} + 0.08Z + 9.6$	$L_{0.95}[\lambda] \simeq t_{\text{max}} + 4E^{0.15}[\text{GeV}]$
Radius for $\simeq 95\%$ radial containment	$R \simeq 2\rho_M \simeq 14A/Z[\text{g cm}^{-2}]$	$R \simeq \lambda$

[a] Measured from face of calorimeter.

B'' (Rossi 1964). In this one-dimensional treatment (multiple scattering is ignored), the energy loss of electrons is $\Delta E = \varepsilon/X_0 \simeq 3Z/A$ [MeV g^{-1} cm^{-2}], and the interactions of the secondaries of all energies are approximated by their asymptotic expressions. Under these assumptions the shower distributions in different materials scale in units of X_0 and ε. The total track length T, which is the sum of all track segments of charged tracks, is given in units of radiation lengths, as $T/X_0 \approx E/\varepsilon$. The measurable value T_m of the total track length T depends on the minimum detectable energy η in the calorimeter, and one finds that T_m can be expressed with a universal function $F(\eta)$ in the form $T_m = F(\eta) \cdot E/\varepsilon$ (see, for example, Amaldi 1981). Calorimetry works because T_m is also proportional to the incident energy. Furthermore, statistical fluctuations $\sigma(T_m)$ of the measurable track length are found to give the intrinsic limitation to the energy resolution $\sigma(E)$ (see discussion in Section 4). A striking display of the energy profile of a high-energy electromagnetic shower is presented in Figure 6 (P. Sergiampietri, 1981, private communication).

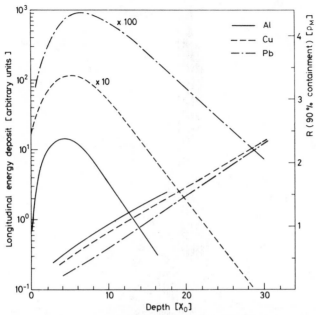

Figure 5   Longitudinal shower development (left ordinate) of 6 GeV/c electrons in three very different materials, showing the scaling in unit of radiation lengths X_0. On the right ordinate the shower radius for 95% containment of the shower is given as a function of the shower depth. In the later development of the cascade, the radial shower dimensions scale with the Molière radius $\rho_M \sim 7A/Z$ (data adapted from Bathow et al 1970).

The description of showers induced by strongly interacting particles is considerably more complex: a wide spectrum of secondary particles is produced, and nuclear physics effects associated with the excitation of the absorber nuclei are so large that they may dominate effects due to technical features of the detector. The *average* properties of hadronic showers can be parametrized as shown in Table 3, with the characteristic dimensions determined by the nuclear absorption length λ. Examples of longitudinal and transverse distributions are given in Figure 7.

Typical of hadronic interactions is the multiple particle production with limited transverse momentum, $\langle p_T \rangle \approx 0.35$ GeV/c, for which about half of the incident energy is consumed (the inelasticity $\kappa \approx 0.5$). The remainder of the energy is carried by fast forward-going (leading) particles. The secondaries are mostly pions and nucleons, with a multiplicity composition only weakly energy dependent above the resonance region ($E \gtrsim 1$ GeV). Two features specific to the propagation of hadronic showers constitute the principal limitations to hadronic

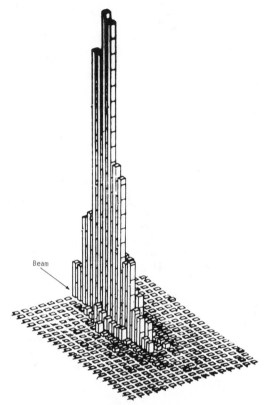

Figure 6 Two-dimensional measurement of an electromagnetic shower induced by a 27-GeV/c electron. The beam direction is indicated by the arrow. The calorimeter is an essentially homogeneous detector of liquid argon. The very fine transverse ($\sim 1 X_0$) and longitudinal ($\sim 1 X_0$) granularity was achieved with a suitable geometry of charge-collecting electrodes (courtesy P. Sergiampietri, Pisa).

calorimeters (Table 4):

1. A major component of the secondaries are π^0's, which will propagate electromagnetically without any further nuclear interactions; the average fraction converted into π^0's is $f_{\pi^0} \simeq 0.1 \ln E[\text{GeV}]$. The fluctuations in π^0 production from one cascade to the next are mostly determined by the nature of the first inelastic interaction. They are therefore large and approximately at the 50% level.

2. In the hadronic interactions of the cascade, a sizeable amount of the available energy is converted into excitation or breakup of the nuclei, of which only a fraction will eventually appear as detectable energy (see Table 4) with concomitant large event-to-event fluctuations.

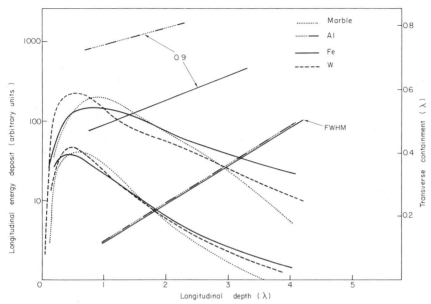

Figure 7 Longitudinal shower development (left ordinate) induced by hadrons in three different materials, showing approximate scaling in absorption length λ. The shower distributions are measured from the vertex of the shower and are therefore more peaked than those measured with respect to the face of the calorimeter. For the transverse distributions as a function of shower depth, scaling in λ is found for the narrow core (FWHM) of the showers. The radius of the cylinder for 90% lateral containment is much larger and does not scale in λ. Note that marble and aluminum have almost identical absorption and radiation length (Marble: Jonker et al 1982; Al: Friend et al 1976; Fe: Holder et al 1978; W: Cheshire et al 1977).

Clearly, these two processes are strongly correlated, and together they represent the intrinsic limit to the performance of hadronic calorimeters (Fabjan et al 1975, Fabjan & Willis 1975).

3.2 *Calculational Techniques*

Detailed understanding of the shower development in various materials, and its dependence on the nature and energy of the incident particle, is essential to the design of calorimeters and the development of improved detector techniques. For these purposes, analytical parametrizations of the average shower behavior are generally not sufficient, since the most critical measures of calorimeter performance — the resolving power for energy and position measurements — are dictated by event-to-event fluctuations in the shower cascade. For this reason the simulation of

Table 4 Characteristic stages of hadronic showers

Reaction product	Characteristic time (s)	Characteristic properties	Effects on energy resolution
Secondary hadrons	$\sim 10^{-22}$	Multiplicity $\sim A^{0.1} \ln E[\text{GeV}]$	Fluctuations in $\pi^{\pm}$ versus π^0 production
Nuclear excitation	$10^{-18}\text{–}10^{-13}$	Emission of p's and n's ($\sim 100\,\text{MeV}$); "evaporation" of n's, γ's ($\sim 10\,\text{MeV}$)	$\sim 15\%$ of hadronic cascade energy converted into nuclear binding energy losses; large fluctuations and vastly different detection efficiencies
Pion and muon decay	$10^{-8}\text{–}10^{-6}$	Fractional "invisible" energy $\sim 0.04/\ln E[\text{GeV}]$	Negligible contribution due to small loss of μ's and ν's

showers by Monte Carlo techniques has been developed. In these calculations, which generally have their origins in programs developed to study radiation-shielding questions, stochastic models for the elementary electromagnetic and hadronic scattering processes are employed to generate individual cascades and to follow their progress in considerable detail.

Needless to say, the reliability of these results depends on the fidelity of the models for primordial interactions, and on the quality of available data for atomic and nuclear structure of absorber materials. Widely used programs become increasingly predictive as they are refined and checked against measurements. This is particularly true for electromagnetic showers, for which a very general and well-documented system of computer codes called EGS (electron gamma shower) (Ford & Nelson 1978) has been almost universally adopted.

With EGS, one may specify an arbitrary detector geometry; each shower particle is transported, in small steps, through the absorbing media, using cross sections and branching ratios for the various energy loss processes (Compton scattering, pair production, and photoelectric effect for photons; multiple Coulomb scattering, annihilation, Bhabha scattering, Møller scattering, and bremsstrahlung for electrons and positrons). The complete history of each generated cascade is available for further analysis, including the energy deposited by individual shower particles in each volume element of material. One example, showing the detailed agreement with measured data, is given in Figure 8.

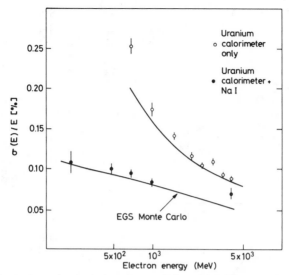

Figure 8 Comparison of measured energy resolution with EGS-Monte Carlo evaluation (*solid lines*). The high-resolution device consists of $6X_0$ deep NaI crystals, backed up by the AFS hadron calorimeter. The open circles give the energy resolution measured in the uranium calorimeter alone (P. Jeffreys & T. Jensen, 1982, unpublished note, AFS collaboration, CERN).

The simulation of hadronic cascades is much more complicated. Many approximations are necessary, and these must be tailored to the specific application. The principal difficulty stems from a lack of experimental data on the energy-dependent interaction cross sections for all the different secondary particles. The details of the nuclear break up, evaporation, and de-excitation must be accounted for, as well as the different response of the active detector medium to the different particles.

A number of hadron shower calculations appear in the literature (Jones 1969, Ranft 1972, Goebel et al 1973, Gabriel & Amburgey 1974, Baroncelli 1974, Grant 1975, Gabriel & Schmidt 1976, Gabriel 1978). Usually these agree well with the average behavior of experimental data, but often differ in the relative contributions of different energy loss mechanisms, particularly for high-energy (>10 GeV) showers (Figure 9). Figure 10 shows a computer simulation of a 100 GeV π^- meson incident on a block of iron and illustrates the large fluctuations in the spatial development of hadronic showers.

A practical limitation to the use of such Monte Carlo techniques is the necessary computer time. Even the existing limited programs are very costly and become more so in proportion with the shower energy. This is

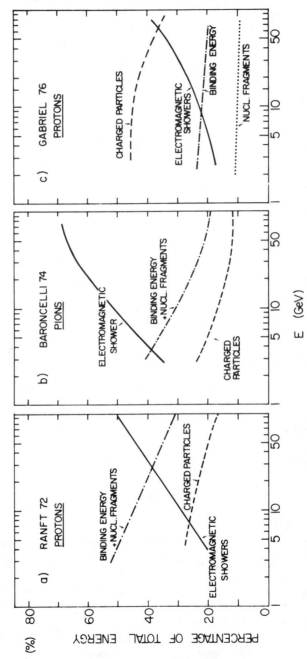

Figure 9 Relative contributions of the most important processes to the energy dissipated by hadronic showers, as evaluated by three representative Monte Carlo calculations (from Amaldi 1981).

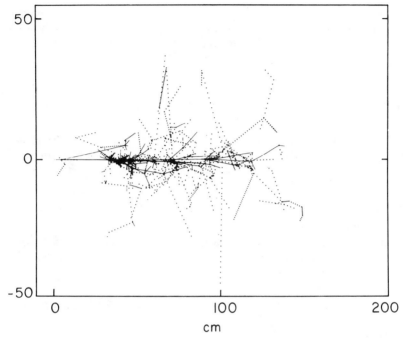

Figure 10   Computer simulation of a hadronic cascade induced by a 100-GeV/c π^- in iron. The dotted tracks represent neutrons that are seen to propagate far away from the shower axis (Grant 1975).

particularly troublesome, because frequently large numbers of showers need to be generated to examine effects in the tails of statistical distributions.

As an alternative to such studies, the considerable body of existing shower data is used to develop parametric approximations to the differential energy deposit over the volume of the cascade. One example (Böck et al 1981) for such a parametrization of the energy deposit as a function of shower path lengths, s, is

$$dE = k[wt^{a-1}e^{-bt} + (1-w)l^{c-1}e^{-dl}]\,ds,$$

where t is the depth, starting from the shower origin, in radiation lengths, and l is the same depth in units of absorption lengths. The parameters a, b, c, d, w are energy-dependent fits to the data. Crude shower fluctuations are simulated by randomly varying the depth of the shower origin; smearing the incident particle energy (to simulate the energy resolution of the calorimeter); and randomly varying the length of the shower by scaling the values of t and l. This approach, while obviously not satisfactory for evaluating basic questions of energy resolution, can be very effective for investigating spatial resolution and for studying problems of pattern recognition in multiparticle events, jet reconstruction, etc (Della Negra 1981, Gordon et al 1981).

4. ELEMENTS OF DETECTOR DESIGN

While the primary goal in designing a calorimeter is usually to achieve the best possible energy resolution for the particles of interest, this is not the sole consideration. Specific experiments generally impose additional requirements. For instance,

1. the necessity to cover a large solid angle may render the best techniques prohibitively expensive;
2. in large detector systems, particularly for colliding-beam experiments, the calorimeter must be tightly integrated with other types of detectors; it may have to function in the presence of strong magnetic fields;
3. the necessity to sustain high particle rates or the use of the calorimeter information in fast trigger decisions requires special consideration;
4. the degree to which the calorimeter is to be exploited for position measurement and particle identification profoundly influences the nature and complexity of its structure.

In this section we discuss the detector parameters bearing on these issues.

4.1 *Energy Resolution*

The ultimate limit for the energy resolution of a homogeneously sensitive calorimeter is determined by fluctuations intrinsic to the mechanisms for the development of showers, as described in Section 3. For electromagnetic showers, this limitation results from variations in the net track length of charged particles in the cascade. For hadron showers, fluctuations in the fractional energy loss accounted for by each of several interaction mechanisms set a much higher limit. In both cases, however, the underlying phenomena are statistical processes, whose effects grow in magnitude as $E^{1/2}$. Hence the limiting accuracy, expressed as a fraction of the total energy, *improves* with increasing energy as $E^{-1/2}$.

In Table 5 we list the major contributions to the final resolution of practical detectors. For electromagnetic showers, the intrinsic fluctuations set a limit so small as to be seldom approached in practice. (Indeed, the value given in the table is derived from calculations, not measurements; see Longo & Sestili 1975.)

The most precise electromagnetic measurements are achieved with homogeneous shower counters, in which the entire volume responds with a measurable signal to the passage of charged particles. One such detector with excellent performance is sodium iodide (NaI), a high-density scintillator, with which resolutions close to the shower fluctuation limit have been obtained for energies of $\simeq 1$ GeV. However, over most of the energy range explored in high-energy physics above 1 GeV, the measurement of electromagnetic showers in this and other detectors is dominated by various instrumental limitations.

This is not the case for hadron showers, as Table 5 indicates. Here the inherent shower fluctuations are the major determinant of the final performance. For this reason there is little practical motivation to

Table 5 Limitations to energy resolution

Contributing mechanisms (add in quadrature)	Electromagnetic showers	Hadronic showers
Intrinsic shower fluctuations	Track length fluctuations: $\sigma/E \simeq 0.005\, E^{-1/2}$ [GeV]	Fluctuations in the mechanism of energy loss: $\sigma/E \simeq 0.50\, E^{-1/2}$ [GeV]; With compensation for nuclear effects: $\sigma/E \simeq 0.20\, E^{-1/2}$ [GeV]
Sampling fluctuations[a]	$\sigma/E \simeq 0.04\,(\Delta E/E)^{1/2}$	$\sigma/E \simeq 0.09\,(\Delta E/E)^{1/2}$
Instrumental effects	Noise and pedestal width: $\sigma/E \sim 1/E$ determine minimum detectable signal; limit low-energy performance Calibration errors; nonuniformities: $\sigma/E \sim$ constant limit high-energy performance	
Incomplete containment of shower (energy leakage)	$\sigma/E \sim \log E$ For leakage fraction $\gtrsim$ few %: nonlinear response and non-Gaussian "tail"	

[a] ΔE = energy loss by a single charged particle in one sampling layer, measured in MeV; E = total energy, measured in GeV.

employ homogeneous detectors for such measurements. Given the high cost of dense scintillating material and the present highly developed techniques of sampling calorimetry, it is mostly the latter approach that is used for the measurement of hadron showers.

With sampling calorimeters, the shower is developed mainly in an inert absorber material interspersed with an active medium that samples the energy loss at fixed intervals as the cascade develops through the depth of the calorimeter. The typical configuration is a stack of many thick plates of a dense metallic absorber interleaved with layers of active material. The sampled energy measures the ionization loss of shower particles entering the active layers, representing a small but (on the average) fixed fraction of the total cascade. Such constructions are less costly than homogeneous detectors. In addition, one gains the flexibility to optimize the energy and position measurement capability for specific applications by decoupling the absorber and readout functions. Typical choices for the sampling layers are plastic scintillators, liquid argon (with the metal absorber plates acting as electrodes of an ion chamber), and multiwire proportional planes (the active medium being a gas, usually at atmospheric pressure). A detailed discussion of readout techniques is given in Section 5.

For electromagnetic showers in a well-designed sampling calorimeter, the largest contribution to the energy resolution is given by statistical fluctuations in the sampling process, as determined by the energy loss of shower electrons in the sampling layers and, ultimately, by the number of photons or ionization events detected by the readout system.

To first order, the sampling fluctuations are determined by the number of sampled electrons n_e in the shower: $\sigma/E \approx n_e^{-1/2}$. In Rossi's Approximation B, $n_e = E/\Delta E$, where ΔE is the energy loss of a single charged particle in one layer of the calorimeter. For a sampling layer of thickness Δx, one has (cf Section 2) $\Delta E = (\varepsilon \, \Delta x/X_0)$ and therefore

$$\sigma/E \simeq (\varepsilon \cdot t/E)^{1/2},$$

where t is the thickness of each sampling layer, in units of radiation lengths.

This dependence on the sampling interval and the energy is a well-established characteristic of electromagnetic shower measurements with sampling calorimeters. A more refined parametrization (Amaldi 1981), which is quite accurate for dense (i.e. nongaseous) sampling materials, is

$$\sigma/E = 3.2\% \ (\varepsilon[\mathrm{MeV}]/F(Z)\langle\cos\,\theta\rangle)^{1/2}(t/E[\mathrm{GeV}])^{1/2}$$
$$= 3.2\% \ (\Delta E[\mathrm{MeV}]/E[\mathrm{GeV}])^{1/2}[F(Z)\langle\cos\,\theta\rangle]^{-1/2}.$$

Values for $F(Z)^{-1/2}$ and $\langle \cos \theta \rangle^{-1/2}$ are typically in the range of 1.0 to 1.25. The factor $F(Z)$ accounts for the shortening of the path length due to electrons slowing down and stopping in the sampling medium; $\langle \cos \theta \rangle$ is the mean angular spread of cascade particles about the shower axis.

Some measured values of the resolution for several calorimeter structures are shown in Figure 11. The results with dense sampling media are well described by the above parametrization, while the gas-sampling devices give performance figures that are poorer by about a factor of two. The additional factors that broaden the resolution in the gas-sampling case are related to the signal collection properties in very thin sampling media — a matter we take up in Section 5.

As indicated in Table 5, the energy resolution for hadrons is not determined by sampling fluctuations. These have been measured for hadronic showers (Fabjan et al 1977) and, for a given sampling

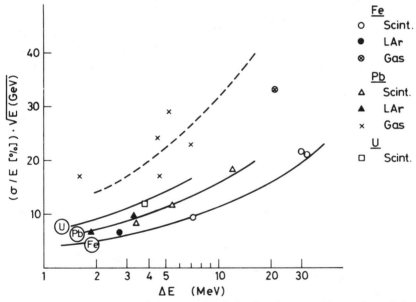

Figure 11   Measured values of the energy resolution for electrons with sampling calorimeters of different constructions: Fe/scint. (Stone et al 1978, Abramowicz et al 1981); Fe/LAr (Hitlin et al 1976); Pb/scint. (Stone et al 1978); Pb/LAr (Hitlin et al 1976, Asano et al 1980); U/scint. (Botner 1981); Pb/gas (Mueller et al 1981, Price & Ambats 1981, Atač et al 1981a, Anderson et al 1978); Fe/gas (Ludlam et al 1981). ΔE is the energy loss in a single sample for a minimum-ionizing particle. The solid curves are estimates as given in Table 5 for the indicated materials and geometries. The dashed curve gives twice the predicted value, representative of the performance of gas-sampling devices.

geometry ΔE, are roughly twice as large as those observed in electromagnetic showers. The dominant contribution to the resolution width comes from fluctuations in the nuclear processes and the correlated loss in detectable energy due to nuclear binding effects, and from the production of particles whose energy goes undetected in the sampling medium. The effect of this invisible energy is clearly seen in the response of a calorimeter to electron and hadron beams (Figure 12). On the average, the ratio of detected signals for electrons/hadrons is $S_{e/h} \simeq 1.4$. Unless event-to-event fluctuations in the electromagnetic component of hadron cascades are somehow corrected for or compensated, the resolution for hadrons (Table 5) is

$$\sigma/E \simeq \{(0.5E^{-1/2})^2 + [0.09(\Delta E/E)^{1/2}]^2\}^{1/2}.$$

If the calorimeter is instrumented so that the signal from each sampling layer is separately recorded to provide detailed longitudinal shower information (as, for example, in the neutrino detectors described in Section 2), then some compensation for the electromagnetic/hadronic fluctuations on a shower-by-shower basis is possible. Relative improvements of about 20% in the resolution for hadronic energies have been obtained by application of weighting procedures based on observed correlations between the total shower signal and the

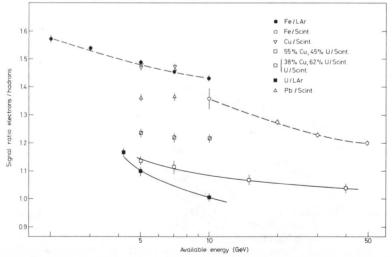

Figure 12 Ratio of average energy deposit for electrons and hadrons as a function of available energy in various materials. The lines through some of the data are drawn to guide the eye (Fe/LAr, U/LAr: Fabjan et al 1977; Fe/scint.: Abramowicz et al 1981; U/Cu/scint.:Botner et al 1981a; other data: Botner 1981).

maximum single-layer signal (Rishan 1979, Selove et al 1980, Abramowicz et al 1981).

A more direct method for improving the energy resolution of
hadronic calorimeters by correcting the nuclear fluctuations is based on
the use of ^{238}U as absorber material (Fabjan & Willis 1975, Fabjan et al
1977). In the uranium absorber some of the normally invisible energy
expended in nuclear breakup leads to neutron-induced fission, which in
turn produces detectable energy in the calorimeter, mostly in the form
of photons of typically 10 MeV. This mechanism of fission compensation for the unseen energy in nuclear cascades is, coincidentally, almost
exact: in tests of a uranium/liquid-argon calorimeter with 1.7 mm thick
uranium plates (Fabjan et al 1977), the electron/hadron ratio $S_{e/h}$ was
measured to be 1.05 (Figure 12) and the measured energy resolution for
hadronic showers was 30% $E^{-1/2}$. The energy resolution obtained with
several representative hadron calorimeters is summarized in Figure 13.

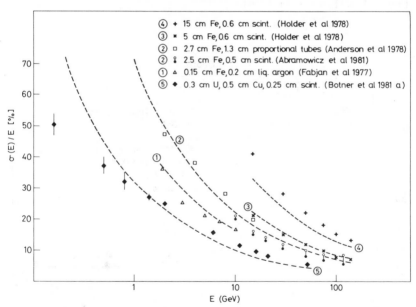

Figure 13 Energy resolution for hadrons measured with iron and uranium-sampling
calorimeters. Curves 1–4 are calculated with the values of intrinsic and sampling
fluctuations as given in Table 5. For the data of Abramowicz et al (1981), the open circles
are the raw data; the solid circles are the results of the off-line analysis, using the
longitudinal shower information to correct for fluctuations in the electromagnetic/
hadronic energy ratio. For curve 5 the intrinsic fluctuation is assumed to be $0.2E^{-1/2}$, and
does not take account of the 35% (in units of λ) admixture of Cu. Below 1 GeV the
resolution improves over the expected value and indicates the influences of mechanisms
such as ranging and reduced nuclear effects.

The resolution figures determined by intrinsic shower and sampling fluctuations will not be realized if showers are not adequately contained within the calorimeter volume. In any practical detector some average fraction f of the shower energy escapes through the sides (lateral leakage) or back (longitudinal leakage). (For early-developing hadron showers, some energy may be lost through the front face, a phenomenon referred to as albedo.) The fraction f increases logarithmically with energy, and it is found that lateral containment is less critical for the resolution than longitudinal containment (Amaldi 1981). In either case, values of f greater than a few percent yield a measurable degradation of performance (Figure 14).

For low-energy showers the resolution may become worse as the calorimeter depth is increased beyond the optimum for containment, owing to larger-than-necessary levels of noise and systematic errors in the readout. This point is illustrated by the measurements shown in Figure 14 (Hitlin et al 1976).

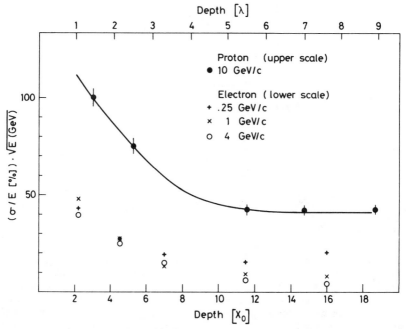

Figure 14 Effect of longitudinal leakage on energy resolution for electromagnetic (Hitlin et al 1976) and hardronic showers (Fabjan et al 1977).

4.2 Position and Angle Measurement:
Resolution and Granularity

If the readout is segmented along one or more directions transverse to the shower axis, the location of the shower can be accurately determined by measuring the centroid of the deposited energy. This requires that the transverse cell size be comparable to the lateral shower dimension. For homogeneous detectors this may be achieved by assembling a close-packed array of independent counters. The Crystal Ball detector (Chan et al 1978) is an elegant example of this approach (Figure 15). In sampling calorimeters, the most straightforward

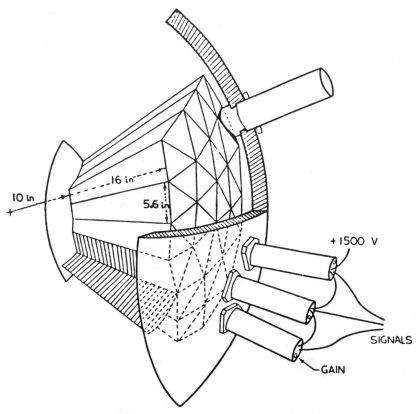

Figure 15 A section of the Crystal Ball detector, containing 54 sodium-iodide elements. Each NaI crystal is viewed by a photomultiplier tube, some of which are shown. Almost complete spherical coverage around the e^+e^- collision region is obtained with 672 NaI elements (Chan et al 1978).

approach is to subdivide the active layers into individually read out strips, with successive layers measuring the position along different coordinates.

For electromagnetic showers, the RMS lateral dimension is of the order of a radiation length (typically a few centimeters) and is slowly varying with energy (see Section 3). Spatial resolutions of a few millimeters are achieved for high-energy electromagnetic showers, using lead-glass blocks of lateral dimension ~5 cm (Akopdjanov et al 1977, Amendolia et al 1980). Similar resolution values are achieved in sampling calorimeters by subdividing the readout of active layers into 1–2 cm wide segments.

The transverse distribution of hadron showers is approximately one absorption length, substantially broader than for electromagnetic showers, and fluctuations in the distribution of deposited energy through the volume of the cascade are larger. For a calorimeter segmentation s, a spatial resolution $\sigma_x \approx 1.0 \exp(1.2s/\lambda)$ (all dimensions in cm) is obtained (Amendolia et al 1980). Position accuracies $\lesssim 1$ cm can be achieved (Binon et al 1981), but figures of merit of a few centimeters are more typical. The use of shower measurements to determine the direction of the initiating particle has been carefully studied in connection with the large neutrino detectors. For the detector of FNAL experiment 594 (see Table 2), the energy dependence of angular resolution for electrons and hadrons was measured to be $\sigma(\theta_e)$ [mrad] $= 3.5 + 53/E$[GeV] and $\sigma(\theta_h)$ [mrad] $= 6 + 640/E$[GeV] (Bogert et al 1982). With the CHARM detector (Figure 1) an angular resolution $\sigma(\theta_h)$ [mrad] $= 160/E^{1/2}$[GeV] $+ 560/E$[GeV] was obtained for hadrons and about an eight times better resolution for electrons (Diddens et al 1980). In both cases the angular fluctuations were minimized by choosing an absorber material in which electromagnetic and hadronic showers have approximately equal length ($12X_0$[cm] $\approx 3\lambda$ [cm]).

In most applications it is important that the calorimeter be able to measure simultaneously the position and energy of two or more particles, and several important types of calorimetric measurements are limited by the ability to resolve close pairs of showers. One example is the signature of high-energy π^0 particles. With well-segmented calorimeters, the limit for distinguishing nearby showers is determined by the lateral size and shape of the cascade. Electromagnetic showers can be distinguished down to separations of $\gtrsim 1X_0$ (Kourkoumelis et al 1980). For hadron showers the minimum separation of resolved pairs is $\gtrsim 1\lambda$.

For events with high multiplicities of particles striking the calorimeter, the pattern recognition problem becomes severe if the lateral segmentation is in strips and projections have to be used for the reconstruction of space points. In present colliding-beam experiments, the interesting events tend to be just those in which a very large number of particles enter the detectors. While for some measurements it is sufficient to record the angular distribution of particle energies rather coarsely, classifying events in terms of patterns of energy flow, it is more frequently required to measure the energies of individual particles in events of high particle density. In this case it is necessary to segment the readout into cells that are small in both lateral dimensions, as in the Crystal Ball configuration (Figure 15) or in the TASSO shower counters (PETRA Bulletin No. 14, 1979).

An example of the motivation for fine-grained spatial segmentation is shown in Figure 16. This is a simulation (Gordon et al 1981) of an event in which two jets of particles result from the hadronization of a pair of heavy quarks produced in 400-GeV colliding proton beams. Such events are extremely rare and difficult to isolate, yet their detection and measurement is one of the primary reasons for constructing high-energy colliding-beam machines. By recording the energy deposited in each of the eight calorimeter modules, one could recognize the presence of the two jets in this event and measure the energy carried by each. But this is not enough. Figure 16 (bottom) shows the energy deposited in module 3, as seen by a fine-grained readout organized into towers of 20-cm sides. With this information it is possible to reconstruct detailed correlations between energy and angle within the jet of particles (e.g. to measure the invariant mass of the jet).

4.3 Particle Identification

Such measurements, which are not usually associated with calorimeter applications, have traditionally implied identification of protons, charged pions, and kaons. With the discovery of the "new" quarks, emphasis has shifted to the identification of signatures from c, b, and possibly heavier quarks. Various techniques are applied to these "modern" identification requirements and we summarize here the role of calorimetry. Table 6 gives an overview.

The art of identifying electrons is usually based on a coarse measure of the longitudinal shower profile. Dominant background comes from charged π's producing a leading π^0 system through charge exchange in the calorimeter, with a probability at the percent level for few GeV/c π's. Typically, therefore, a π rejection of about 100 is achieved if information on the momenta of the incident particles is not available, but may approach about 1000 if momentum measurements are included (Apel et al 1975, Lederman et al 1975, Cobb et al 1979, Basile

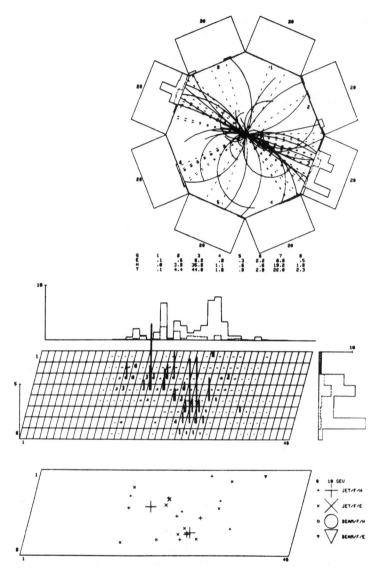

Figure 16 (*Top*) Monte Carlo simulation of an event producing high-energy jets of particles in the collision of two 400-GeV/c proton beams. Solid trajectories are charged particles in a 1-T magnetic field; neutrals are dashed. The energy deposition in the eight calorimeters is indicated. (*Bottom*) Response of octant 3 for the same event. It is subdivided into 8×46 towers of 20 cm sides. The signal recorded in each tower, of which the first 10 radiation lengths are read out separately, is indicated by the length of the two bars (energy in GeV with scale on lower left). For comparison, the point of impact of the particles is also shown (size of indicator is proportional to energy) (Gordon et al 1981).

Table 6 Particle identification with calorimeters

Particle produced	Calorimeter technique	Comment
Electron, e	Charged particle initiating the electromagnetic shower	Background from charge exchange $\pi^{\pm} N \rightarrow \pi^0 + X$ in calorimeter; π discrimination of ~ 10–1000 possible
Photon, γ	Neutral particle initiating the electromagnetic shower	Background from photons from meson decays
$\pi^0, \eta, \ldots \rightarrow \gamma\gamma\, \rho, \phi,$ $J/\psi, Y, \ldots \rightarrow e^+ e^-$	Invariant mass obtained from measurement of energy and angle	Classical application for electromagnetic calorimeters
Protons, deuterons, tritons, . . . and their antiparticles	Comparison of visible energy E_{vis} in calorimeter with momentum of particle	$E_{\text{vis}}^{b(\bar{b})} = (\mathbf{p}_b^2 + m_b^2)^{1/2} - (+)m_b$ Protons (antiprotons) identified up to 4 (5) GeV/c; deuterons (antideuterons) correspondingly higher
(Anti)neutrino	Visible energy E_{vis} in calorimeter compared with missing momentum	Important tool for $e^+ e^- \rightarrow \nu(\bar{\nu}) + X$ and at CERN collider (ISABELLE, . . .) $pp(p\bar{p}) \rightarrow \nu(\bar{\nu}) + X$
Muon	E_{vis} compared to $\mathbf{p}$; range	Background from noninteracting pions
Neutron or $K_L^0 (\bar{n}, \bar{K}_L^0)$	Neutral particle initiating hadronic shower	Some discrimination perhaps possible based on detailed (longitudinal) shower information

et al 1979). As the charge-exchange cross section drops with increasing energy, a π rejection factor approaching 10^4 is expected for momenta $p \simeq 100$ GeV/c. The effect of this hadronic background is minimized by building detectors from absorbers with a large ratio of $\lambda/X_0 \sim Z^2/A^{3/4}$ Lead is therefore a popular choice for the electromagnetic part of calorimeters. A further small (about 2- to 3-fold) improvement in rejection is achieved if transverse shower dimensions are measured (Lederman et al 1975, Cobb et al 1979). Longitudinal information on early shower development is provided by the "converter" method. A combination of a $\sim 1X_0$ thick Pb plate followed by a scintillator or a multiwire proportional chamber (MWPC) ("passive" converter) or by instrumented layers of lead-glass Čerenkovs ("active" converter) can signal an early interaction, predominantly of electromagnetic origin (Lederman et al 1975, Gabathuler et al 1978). The ultimate "converter" for relativistic electrons is not based on bremsstrahlung emitted from one X_0 of a high-Z material, but rather exploits radiation emitted from ~$0.01X_0$ of very-low-Z substances: transition radiation detectors (Cobb et al 1977). These "converters" will be particularly valuable for the identification of electrons amidst densely collimated jets of hadrons (Fabjan et al 1981).

In a similar way, photons are identified that are either produced directly (Diakonou el at 1979) or from meson decay. It has been found that a transverse granularity in $2X_0$ allows discrimination between one and two photons (e.g. from π^0 decay) if they are separated by more than one radiation length.

Identification of particles through the determination of the invariant mass of their decay products is a rather standard calorimetric method, on which a number of interesting recent physics results were based: charmonium spectroscopy; discovery of the η_c (Coyne 1981) and the hadronic production of χ states (Kourkoumelis et al 1979); the possible detection of a state $\theta(1640) \rightarrow \eta\eta$ (Scharre 1981), considered a glueball candidate; the discovery of directly produced high-p_T photons and their separation from high-p π^0's — all illustrate the power of this method and its role in the "new" physics.

An interesting method for identifying protons, deuterons, etc, and their antiparticles over a limited momentum range is to compare the particle's momentum with its characteristic energy deposition, its "available energy." For electrons and protons this is the kinetic energy of the particle; for mesons, it is the total energy; while for antiprotons it is the total energy plus one proton mass (Fabjan et al 1977, Corden et al 1982). Proton and antiproton identification up to about 4 GeV/c is possible. Hadronic high-p_T production of antibaryons may also be studied, as the increased available energy at a given p_T dominates the $\pi^\pm$ trigger rate over a limited kinematical region (W. J. Willis, 1979, and O. Bortner & C. W. Fabjan, 1981, both unpublished notes, AFS Collaboration, CERN). Recently, the calorimeter response to nuclei with A in the range 2 to 20 has been measured and found to be proportional to the kinetic energy (Stevenson et al 1981).

Muon identification with calorimeters is an important technique for neutrino physics and for the detection of uncommon particle decays (charm, W, etc). In calorimeters with frequent longitudinal subdivision (e.g. neutrino detectors), the multiple ionization measurement consistent with the passage of fast μ's discriminates against pions, which penetrate with a probability exp $(-d/\tilde{\lambda})$. The observed path length d is measured in units of "detectable" absorption length $\tilde{\lambda}$, which is found to agree closely with tabulated values (Baum et al 1975, Grant 1975, Holder et al 1978). With purely passive absorbers, μ's may be separated from π's to a lesser degree, based on their different "punch-through" probability. Spatial consistency between the tracks entering and exiting the absorber within multiple scattering limits enhances the discrimination (Grant 1975). Events containing high-p_T neutrinos will probably be studied with a calorimetric technique for the first time at the hadron collider. The method is based on a precise measurement of the missing momentum in an event. It might be comparatively easy for cases such as $W \rightarrow \ell\nu$ decay, where the ν is reflected in the characteristic spectrum of the charged lepton. As already mentioned in Section 2, missing-momentum techniques are expected to find their widest ranging applications at future e^+e^- colliders.

4.4 Rate Capability and Trigger Selection

The event rates at hadron machines, either fixed-target or colliding beams, can be enormously high. The collision rate at the ISR, operating in the luminosity range 10^{31}–10^{32} cm^{-2}s^{-1}, is $\gtrsim 1$ MHz, and luminosities higher by an order of magnitude are contemplated for ISABELLE. The power of these machines is not the high collision rate in itself, but the ability to produce useful numbers of very rare events over periods of months. A very high degree of trigger selectivity is required, rejecting most events seen by the detectors. Indeed, a large detector system with thousands of readout channels generally cannot record more than a few events per second.

The rate and trigger capability depends on the intrinsic resolving time of the calorimeter, which is determined by the time required to develop a measurable signal in the readout medium and by the pulse-shaping time required for noise filtering. If this time is comparable to the mean time between events, then the energy recorded for a particular event will be influenced by "pile-up" of spurious energy deposited by nearby events in time. The typical resolving times for calorimeters range from a few tens of nanoseconds (for scintillator readout) to several hundred nanoseconds (for charge collection readout)(see Section 5).

For most high-energy physics applications, the effect of energy and time resolution on trigger selectivity is intimately coupled with the rapid decrease in the particle production spectrum as a function of the transverse momentum p_T. As an example, we show in Figure 17 the p_T spectrum of incident particles and the result of the measurement, which is distorted by such pile-up effects according to the following assumption: 20% probability for two events to occur within the resolving time

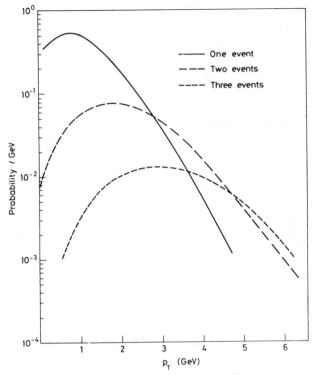

Figure 17 Influence of detector resolving time on the measured transverse momentum spectrum. This example assumes a 20% (4%) probability for two (three) collisions to occur within the resolving time.

of the calorimeter (and, by Poisson statistics, a 4% probability for three events, etc). In this case, most of the events above 3 GeV result from multiple event pile-up, depicted by the dashed curves in Figure 17 (Carithers 1982).

Other factors related to high rates include space-charge saturation effects, which may be important in the case of gas sampling with avalanche gain (see Section 5), and limited detector lifetime. Scintillators, lead glass, wire chambers, and the components of sensitive preamplifiers all suffer a measurable degree of radiation damage after integrated particle fluxes $\gtrsim 10^{12}$ cm^{-2} (Schönbacher & Witzeling 1979, Hilke 1981).

4.5   *Calibration and Monitoring*

A large calorimeter consists of many individual detector elements, each of which must be continually and precisely adjusted if the performance of the device is not to be dominated by systematic errors. This usually implies gain control of each analogue readout channel to about 1%.

With the calibration and monitoring scheme, one needs to (a) establish absolute energy scale; (b) compensate for time variations of the response of individual channels (e.g. thermal drifts); and (c) detect malfunctioning channels. For these purposes a mechanism is required for injecting a known calibration signal into each channel. For lead-glass and scintillation counters this is achieved by distributing precisely controlled pulses of light to each detector element. In large systems it is current practice to illuminate a bundle of optical fibers with a single light source, each fiber being connected to an individual detector element. Lasers or gas discharge lamps are used as light sources (Powell et al 1981, A. Clark, 1981, unpublished note, UA2 Collaboration, CERN).

Alternatively, radioactive sources are used to excite the scintillator. This provides a very stable calibration signal, completely monitoring the active system. Ideally, such sources are uniformly distributed in the calorimeter (Botner et al 1981a) or are permanently introduced into certain areas (Botner et al 1981b), or they are mounted externally for dedicated calibration runs (Corden et al 1982). For charge-collection devices (e.g. liquid-argon and proportional wire calorimeters) a voltage pulse across a fixed capacitor injects a measured calibration charge at the input of each amplifier in the system (Cobb et al 1979).

The absolute energy scale is determined by exposing each detector element to particles of known energy (e.g. monoenergetic electron beams), or known energy deposit (e.g. muons). In fixed-target experiments this step can be accomplished by scanning the calorimeter array across the beam with a system of rails and stepping motors. For colliding-beam experiments the initial energy calibration must be done in test beams. Once the detector is in place, an absolute energy scale can be obtained from cosmic-ray muons or, in the case of e^+e^- machines, from elastic (Bhabha) scattering events.

In most experiments a final adjustment of the overall energy scale is made during the later stages of data analysis, using the known masses of reconstructed particles and narrow resonance states (π^0, η, J/ψ, Υ, . . .). For today's very large detector systems, the task of injecting calibration pulses into each channel, adjusting voltage settings to compensate for gain changes, and preparing a parameter list for off-line corrections is done exclusively under computer control (Breidenbach 1981).

We stress that the quality of the calibration system will have the most pronounced influence on the performance of a calorimeter facility. In the conceptual design stage of such a detector, various alternatives must therefore be evaluated in conjunction with a careful assessment of the respective calibration methods.

5.   READOUT TECHNIQUES

In the previous sections we discussed the physics and detector require- ments of large experimental facilities, as well as the properties of electromagnetic and hadronic showers and their consequences for the performance of calorimeters. Much of the recent calorimeter develop- ments concentrated on instrumental techniques for converting the signature of the showers into a measurable signal, which would be well matched to the experimental requirements. These "readout" techniques are described in this section.

Two different methods have found wide applications and are schema- tically shown in Figure 18.

1. "Photon-collecting" techniques are based on scintillation or Čeren- kov light in the active medium. A system of light guides is usually required, and light is converted into charge with photomultipliers. This technique can be mechanically engineered into large systems with relative ease. However, light collection is intrinsically nonisotropic, light guides introduce nonuniform areas, and calibration is complex.

2. "Charge-collecting" techniques measure the ionization produced in the active absorber; the charge may be amplified with external circuitry ("ionization chamber" mode) or to various degrees internally at the charge-collecting electrodes (proportional mode, Geiger mode, etc). Charge-collection uniformity is good, segmentation and calibration easy.

We first discuss homogeneously sensitive detectors, which offer attractive possibilities for the energy measurement of electromagnetic showers. Next, sampling devices are discussed, which are widely used for electromagnetic calorimetry and offer the only practical method for hadron detectors.

5.1 Homogeneous Calorimeters

The intrinsically superior energy resolution for electromagnetic showers is the principal attraction of homogeneous absorbers. Practical detectors may be realized given the relatively small size of electromagnetic showers. For hadronic calorimeters, homogeneous systems offer no interest and would be impractically large.

A summary of homogeneous detector properties is given in Table 7.

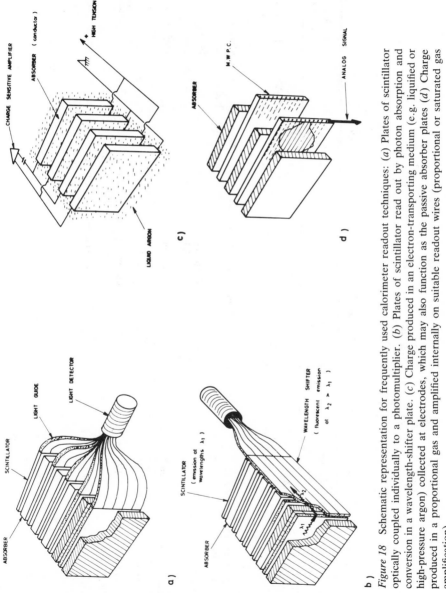

Figure 18 Schematic representation for frequently used calorimeter readout techniques: (*a*) Plates of scintillator optically coupled individually to a photomultiplier. (*b*) Plates of scintillator read out by photon absorption and conversion in a wavelength-shifter plate. (*c*) Charge produced in an electron-transporting medium (e.g. liquified or high-pressure argon) collected at electrodes, which may also function as the passive absorber plates (*d*) Charge produced in a proportional gas and amplified internally on suitable readout wires (proportional or saturated gas amplification).

Table 7 Properties and performances of electromagnetic shower detectors

Type	Radiation length (cm)	Density (g cm^{-3})	Detection mechanism	Energy resolution (E_{GeV})	Principle limitation to $\sigma(E)$
Sodium iodide NaI (Tl)	2.59	3.7	Scintillation	$\sim 0.015E^{-1/2} < 1\,\text{GeV}$ $\sim 0.015E^{-1/4} > 1\,\text{GeV}$	Shower fluctuations; optical nonuniformity
Bismuth germanate Bi$_4$Ge$_3$O$_{12}$ (BGO)	1.12	7.13	Primary scintillation	Below 1 GeV comparable to NaI	Similar to NaI
Scintillating glass	~ 4	~ 3.5	Scintillation some Čerenkov light	$\sim 0.02E^{-1/2}$	Photon statistics
Lead-glass (55% PbO, 45% SiO$_2$)	2.36	4.08	Čerenkov light	$\sim 0.04E^{-1/2}$	Photon statistics
Tl (HCO$_2$)-liquid "Helicon"	~ 1.9	~ 4.3	Čerenkov light	Comparable to Pb-glass	Photon statistics
Liquid argon	14	1.4	Ionization charge	$0.02E^{-1/2}$	Effect of shower fluctuation on electron collection

Type	Signal (photoel/GeV)	Characteristic time (ns)	Radiation damage (rad)	Mechanical stability	References
Sodium iodide NaI (Tl)	10^6	250	$\sim 10^4$	Hygroscopic, fragile	Beron et al 1973 Chan et al 1978
Bismuth germanate $Bi_4Ge_3O_{12}$ (BGO)	$\sim 10^5$	350	~ 1000	Good	Blenar et al 1981 Kobayashi et al 1981a
Scintillating glass	few $\times 10^3$	~ 70	$\sim 10^6$	Good	Yoshimura et al 1976 Bartalucci et al 1980 Kobayashi et al 1981b
Lead-glass (55%PbO, 45%SiO$_2$)	10^3	~ 20	$\sim 10^4$	Good	Bartel et al 1979
Tl(HCO$_2$)-liquid "Helicon"	$\lesssim 10^3$ (?)	~ 20	$\gtrsim 10^6$	Liquid toxic (?)	Kusumegi et al 1981
Liquid argon	$\lesssim 2 \times 10^6$ (electrons)	$\gtrsim 100$	Not measured	Cryogenic liquid	Chen et al 1978 Radeka 1977

The following parameters are of primary interest:

1. Energy resolution: NaI and BGO show the best performance.
2. Short radiation length: this permits compact detectors and very high granularity. BGO is most attractive.
3. Light output: it is highest for NaI and permits detection with vacuum photodiodes and external amplifiers. Very compact detectors, insensitive to strong magnetic fields, can be constructed (Anassontzis et al 1981).
4. Radiation resistance: lead-glass and scintillators are known to be affected. Helicon appears to be more resistant.
5. Mechanical ruggedness: lead-glass and BGO excel.
6. Price: for correctly designed detectors the price per X_0^3 is relevant. Expressed in these units, NaI and BGO are rather comparable. Surface treatment of the absorbers is expensive but may be minimized by employing novel production or assembly techniques without significant loss in performance (Grannis et al 1981, Bartalucci et al 1980, Anassontzis et al 1981).

5.2   Readout Systems for Sampling Calorimeters

5.2.1   LIGHT-COLLECTING SAMPLING CALORIMETERS   The renaissance of such calorimeters started with the introduction of cheap "plastic scintillators" and elegant light-readout techniques using "wavelength shifters" (WLSs), replacing the cumbersome scheme of scintillator plates individually coupled to a light guide (Figure 18a). The principle is indicated in Figure 18b (Garwin 1960, Erwin et al 1975, Selove et al 1979, Botner et al 1981a). Some scintillation light crosses an air gap and enters the WLS, where it is absorbed and subsequently re-emitted at longer wavelengths; a fraction of this "wavelength-shifted" light is then internally reflected to the light detector. This scheme avoids complicated and costly optical contacts between the scintillators and light collectors, and minimizes dead spaces. A variety of scintillators have been developed for the large calorimeter facilities. They are based on PMMA (polymethyl methacrylate) (Kienzle 1975) or polystyrene (Corden et al 1982) as the matrix for the primary scintillating agent. The light yield is close to that of more conventional organic scintillators (usually based on a polyvinyl toluene solvent), if certain aromatic compounds, e.g. up to about 20% naphthalene, are added. These new scintillators are more easily mass produced, hence cheaper, and have superior mechanical properties. Some of the limitations of the WLS method may be removed by further developments: better spectral matching between the scintillator emission and the WLS absorption, and also between the WLS emission and the photocathode sensitivity will increase the number of detected photons, which is marginal in present systems. Related developments might result in the use of thinner yet more uniform WLSs; increased granularity might be achieved with WLSs having spatially different spectral sensitivities (Eckardt et al 1978). Potentially the most promising developments concern scintillators: they are still rather inefficient (a small percentage of the energy loss is converted into visible photons), and reduced saturation of the response to densely ionizing nuclear fragments should improve the energy resolution of hadron calorimeters (see Section 4). This is an important scintillator property for calorimeter applications and it should be carefully investigated and specified.

5.2.2 CHARGE-COLLECTION READOUT   The ionization charge produced from the passage of the charged particles of the shower may be collected from solids, liquids, or gases. Solids (Brisson et al 1981) and liquids can only be used in an ionization chamber mode with no internal amplification. The best known and to date only practical example is based on the use of liquid argon (Willis & Radeka 1974). In specific cases, liquid xenon may be used (Alvarez 1968, Derenzo et al 1974, Masuda et al 1980, Doke 1981). The use of room-temperature liquids has also been advocated (private communication, C. R. Gruhn 1974), but with increasing operating temperature the tolerable level of impurities decreases strongly: for liquid-argon operation, impurity concentrations at the ppm level are acceptable; liquid-xenon operation requires control close to the ppb level; while for room-temperature operation, even lower values are required. If gas is used as the active sampling medium, internal amplification to various degrees is usually exploited: proportional chambers or tubes provide a signal proportional to the energy loss. At higher gas gain, with devices operating in a controlled streamer or Geiger mode, the measured signal is related to the number of shower particles that traverse the active medium ("digital readout")(Conversi 1973).

The principal advantage common to all these charge-collection methods is seen in the ease of segmentation of the readout and the capability to operate in magnetic fields. Some features specific to the various types are

1. Operation in the ionization mode, i.e. liquid-argon calorimeters, provides the best energy resolution and control of systematic effects (cf Figure 11)(Cobb et al 1979, Abrams et al 1980, Kadansky et al 1981, Behrend et al 1981, Slattery et al 1981);

2. Gas proportional devices offer a wide variety of relatively inexpensive construction methods (Anderson et al 1978, Bosio et al 1978, Stone 1981);
3. Digital operation, in the Geiger or streamer mode, allows for very simple and cheap signal-processing electronics (Conversi & Federici 1978, Walker et al 1981, Jonker et al 1981).

For ion chambers and proportional wire readout, the measured signal amounts to a few picocoulombs of charge per GeV of shower energy. Since sampling calorimeters are inherently devices of large capacitance, the optimum charge measurement requires careful consideration of the relationships between signal, noise, resolving time, and detector size. A detailed noise analysis (Willis & Radeka 1974) gives

$$ENC_{\mathrm{opt}} = 5.9 \times 10^6 (C_{\mathrm{D}}[\mu\mathrm{F}]/t_{\mathrm{NF}}[\mathrm{ns}])^{1/2}.$$

(ENC "equivalent noise charge," is the input signal level that gives the same output as noise; C_{D} is the detector capacitance, and t_{NF} is the noise filter time). This result gives the fundamental lower limit to the noise, achieved with optimal capacitance matching between the detector and amplifier. The noise figure grows with increasing detector capacitance and can be improved at the expense of increasing resolving time.

Some general features of proportional charge collection devices are illustrated in Figure 19 for two model detectors, each with parameters typical of its type: a liquid-argon calorimeter with ion chamber readout, and a gas-sampling calorimeter with multiwire proportional planes for the readout. For each case the absorber layers are 2–6 mm lead sheets, with sampling geometries giving good energy resolution. For the ion chamber, the measured charge is induced on the collecting electrode by the drifting electrons, and amounts to 50% of the total ionization charge (Rossi & Staub 1949). In the case of proportional wires, the observed signal is due to the motion of positive ions away from the anode wire (Sauli 1977), with about 30% of the total charge collected during the rapidly developing early part of the signal. In both examples a bipolar pulse-shaping wave form is indicated, with noise filter time $t_{\mathrm{NF}} \simeq$ 100 ns. Possible trade-offs between resolving time and noise are exploited more freely in gas-sampling calorimeters, where there is some freedom in the choice of avalanche gain and the inherently poorer resolution reduces the sensitivity to noise. The noise penalty can be severe for very large calorimeters, which must sustain high particle fluxes (Radeka & Williams 1981).

Energy resolution curves for the two model calorimeters of our illustration are shown in Figure 20 for electromagnetic showers. These

Monte Carlo calculations agree well with measured results from real detectors of similar specification. For the liquid-argon detector the resolution is dominated by statistical fluctuations in the number of sampled shower electrons. The remaining component is mainly due to variations in the path length of electrons traversing the sampling gap: soft shower electrons are brought to rest depositing all their kinetic

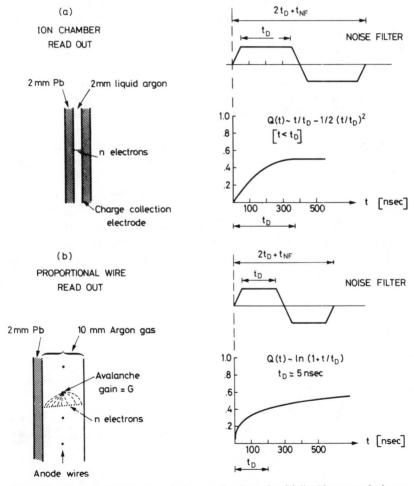

Figure 19 Charge collection in a single sampling layer for (*a*) liquid-argon calorimeter with ion chamber readout; (*b*) gas proportional wire readout. The signal charge $Q(t)$ is shown as a function of time, and t_D is the time required for all ionization electrons to be collected. For each case a bipolar noise filter weighting function is indicated (see text).

energy in the sampling medium, while higher-energy electrons cross the gap, leaving behind a signal that depends on the initial angle and effects of multiple scattering in the sampling gap.

For the gas-sampling calorimeter (Figure 20, *right*), shower electron statistics are dominated by two larger components of nearly equal magnitude: (*a*) path-length variations are much stronger in the gas-sampling medium, because even the very soft shower electrons contribute, which are widely distributed in angle as they enter the sampling gap (Fischer 1978); and (*b*) fluctuations in the rate of energy loss dE/dx (Landau fluctuations) become important in thin layers of gas.

The resolution of gas proportional calorimeters can be somewhat improved by limiting the path-length variations. This is achieved by inserting thin metal walls between anode wires or by using proportional tubes of small cross-sectional area (Fischer 1978, Ludlam et al 1981). Recent investigations of gas mixtures (Atač et al 1981b) demonstrated the feasibility of a "limited streamer" mode in which avalanches remain localized within a few tenths of a millimeter. A calorimeter of this type has given linear response over an extended energy range, with resolution $\sigma/E \simeq 17\% \, E^{-1/2}$ for electrons (M. Atač, 1981, private communication).

While the proportional wire calorimeter gives poorer resolution for electromagnetic showers, the enhanced fluctuations in its response are still small compared to the intrinsic fluctuations in hadron cascades. The choice of gas-sampling calorimeters for hadron measurements can, in principle, be made without compromising energy resolution. This is not expected to be the case for uranium calorimeters, for which, however, no measurements with a gaseous sampling medium have been made so far.

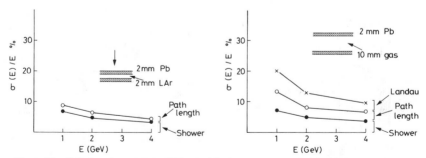

Figure 20 Calculated energy resolution, with electrons at normal incidence, for the two model calorimeters (see text and Figure 19). The lowest curve includes only the effects of fluctuations in shower development, the next curve includes both shower and path-length fluctuations, etc. Landau fluctuations contribute negligibly in the liquid-argon case.

With these types of detectors, spatial segmentation and localization can be easily implemented. This may be achieved in a projective geometry using strips or in a "tower" arrangement, e.g. by measuring the signal charge induced on a pattern of cathode pads. The tower arrangement is valuable for reducing ambiguities and confusion in multiparticle events. For this construction the proportional tubes are made from graphite-impregnated plastic with a sheet resistivity of $\sim 100\,k\Omega$ per pad (Battistoni et al 1980, Ayres et al 1981).

Another technique currently being studied for achieving a very high degree of spatial segmentation is the so-called drift-collection calorimeter (Fischer & Ullaland 1980, Price & Ambats 1981). Shower electrons are collected on a relatively small number of proportional wire planes, with one shower coordinate determined by the drift time of ionization charge onto the collecting wires. A possible configuration is shown in Figure 21.

Gas-sampling calorimetry in a digital mode (using Geiger, streamer, or flash-tube techniques) is a means for greatly simplifying the signal processing circuitry, and offers an expedient method for achieving a very high degree of segmentation. Flash chambers consist of an array of tubular cells filled with a mixture of about 96% neon and 4% helium; a pulsed high voltage is applied across each cell after an external event

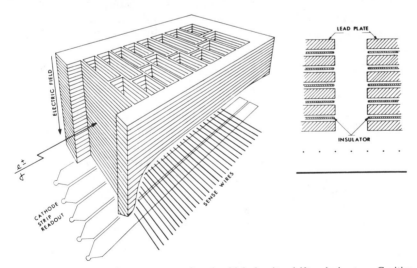

Figure 21 Geometrical arrangement for the high-density drift calorimeter. Cavities between absorber plates allow drifting of ionization electrons over long distances onto MWPC-type detector. Very high spatial granularity can be achieved at the cost of mechanical complexity and rate capability. A very uniform magnetic guidance field parallel to the drift direction is usually required (Fischer & Ullaland 1980).

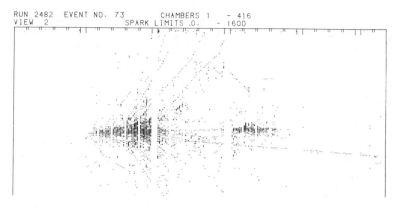

Figure 22   Display of high-energy neutrino event recorded by the flash chamber calorimeter of experiment 594 at FNAL (Bogert et al 1982).

trigger. In the presence of ionization charge, a signal-producing plasma discharge propagates over the full length of the cell. In one such array for a large neutrino experiment at FNAL, 608 flash-chamber planes with a total of some 400,000 cells are sandwiched between absorber layers of sand and steel shot (Bogert et al 1982). The pattern of struck cells in each plane is read out by sensing induced signals on a pair of magnetostrictive delay lines. Figure 22 shows an on-line display of a high-energy neutrino event in the detector, with each dot representing a struck cell. The high degree of segmentation is evident. The energy response, however, is found to be nonlinear, which indicates severe saturation effects for this type of digital readout at high energies.

Frequently, care is taken to limit the geometrical extension of the discharge region, usually accomplished with various mechanical discontinuities (beads, nylon wires, etc). A calorimeter operated in this mode gave an energy resolution $\sigma \simeq 12\% E^{-1/2}$ for electron energies up to 5 GeV (W. Carithers, 1981, private communication). This is much better than is normally achieved for gas-sampling calorimeters and reflects the absence of Landau and path-length fluctuations. At higher energies the calorimeter showed saturation effects due to the increasing probability for multiple hits over the geometrical extension of the discharge region.

6. OUTLOOK

The use of calorimetric methods in high-energy physics began with rather specialized applications, which capitalized on some unique features not attainable with other techniques: electromagnetic shower detectors for electron and photon measurements, neutrino detectors, and muon identifiers.

The evolution toward a more general detection technique — similar in scope to magnetic momentum analysis — had to wait for the development of hadronic calorimetry. Although the application of this technique to the hadron machines (which were inaugurated in the early 1970s) was delayed, it has shaped the detectors for the new CERN p$\bar{\text{p}}$ Collider and will be central to the detector facilities currently under discussion. At the same time, physics studies have evolved in a direction where measurements based on "classic" magnetic momentum analysis are supplemented or replaced by analyses that are based on very precise global measurements of event structure, frequently requiring extraordinary trigger selectivity, and that are much more suitable to calorimetric detection. For the next decade, physics studies will rely increasingly on these more global studies, with properties averaged over groups of particles and with the distinction between individual particles blurred unless they carry some very specific information. This role of calorimetry is sketched in Table 8.

Despite recent conceptual and technical progress, a number of questions deserve further attention:

1. What is the precise energy dependence of hadronic energy resolution?
2. What improvement in energy, position, and angular resolution could be obtained with complete information on the individual shower distributions? With such information, can we tolerate increased longitudinal leakage without affecting energy resolution?
3. What contributes to the measured energy resolution $\sigma(E) = 0.2E^{1/2}$ of a fission-compensated hadron calorimeter?
4. Can we understand hadronic sampling fluctuations to the same degree as we understand electromagnetic ones?
5. Can particle identification and separation be improved if more detailed shower information is available?
6. Are there advantages in mixing different absorber materials, or in changing the sampling step inside a calorimeter?

The very diverse applications of calorimetric techniques will ensure continued study of these and the many technical questions connected with the readout and signal processing of calorimeter information. There can never be a unique solution, but there should always be a search for the most suitable method. We hope that the information provided in this review will be useful toward this end.

Table 8 Future role of absorptive spectroscopy

Source of particles	Physics emphasis	Calorimeter properties	Technical implications
pp (p$\bar{\text{p}}$) collider collider	Rare processes: high p_T, lepton, photon production manifestations of heavy quarks, $W^\pm, Z^0, \ldots$	4π coverage with e.m. and hadronic detection; high trigger selectivity	Approach intrinsic resolution in multicell device; control of inhomogeneities, stability
e^+e^- collider	Complex, high-multiplicity final states (multijets, electrons in jet, neutrinos)	Precision measurements of total, visible energy and momentum	Very high granularity; particle identification
Secondary beams $p \gtrsim 1\,\text{TeV}/c$	Similar to first entry; with increasing energy, stronger emphasis on global features	Calorimeter becomes primary or sole spectrometer element	High granularity, high rate operation
Penetrating cosmic radiation; proton decay	Detailed final-state analysis of events with extremely low rate	Potentially largest detector systems ($\gtrsim$10,000 tons) with very fine grain readout	Ultra-low-cost instrumentation

ACKNOWLEDGMENTS

We are indebted to many groups for providing us with information about their detectors and, in particular, to our collaborators in the R806/AFS groups at CERN, with whom a number of issues raised here were first discussed and tried. It is a pleasure to acknowledge the stimulation and interest arising from our association with W. J. Willis.

The progress we have reviewed here as well as the promise for the future is due in large measure to the support of calorimeter research by the program directors of many laboratories; we especially acknowledge the conviction and foresight with which such support has been rendered by the Experimental Facilities and Experimental Physics Divisions at CERN.

One of the authors (T.L.) is supported by the US Department of Energy under contract number DE-AC02-76CH00016.

Literature Cited

Abramowicz, H., et al. 1981. *Nucl. Instrum. Methods* 180:429
Abrams, G. S., et al. 1980. *IEEE Trans. Nucl. Sci.* NS–27:59
Akopdjanov, G. A., et al. 1977. *Nucl. Instrum. Methods* 146:441
Albrow, M. 1981. *General Meeting on LEP, Villars*, ed. M. Bourquin, p. 142. ECFA 81/54. Geneva: CERN
Almehed, S., et al. 1977. *CERN/ISRC/76:36*
Alvarez, L. W. 1968. *LRL Physics Note 672*
Amako, Y. 1981. *Proc. 1981 ISABELLE Summer Workshop, BNL Rep.* 51443:1257. Brookhaven Natl. Lab, NY
Amaldi, U. 1981. *Phys. Scr.* 23:409
Amendolia S. R., et al. 1980. *Pisa 80–4*
Anassontzis, E., et al. 1981. *CERN/ISRC/81–16*
Anderson, R. L., et al. 1978. *IEEE Trans. Nucl. Sci.* NS-25:340
Apel, J., et al. 1975. *Nucl. Instrum. Methods* 127:495
Asano, Y., et al. 1980. *Nucl. Instrum. Methods* 174:357
Astbury, A. 1981. *Phys. Scr.* 23:397
Atač, M.,ed. 1975. *Proc. Calorimeter Workshop*, Batavia, Ill: Fermi Natl. Accel. Lab. (FNAL)
Atač, M., et al. 1981a. *IEEE Trans. Nucl. Sci.* NS-28:500
Atač, M., et al. 1981b. *Fermilab preprint FN-339*, to be published in *Nucl. Instrum. Methods*
Ayres, D., et al. 1981. *Design report for Fermilab Collider Detector Facility*. Batavia, Ill: FNAL
Barber, D. P., et al. 1980. (The Mark-J collaboration.) *Phys. Rep.* 63:337
Barloutaud, R. 1981. *Proc. Int. Conf. on High Energy Physics*, Lisbon, to be published
Baroncelli, A. 1974. *Nucl. Instrum. Methods* 118:445
Bartalucci, S., et al. 1980. *Nucl. Instrum. Methods* 178:401
Bartel, W., et al. 1979. *Phys. Lett.* 88B:171
Basile, M., et al. 1979. *Nucl. Instrum. Methods* 163:93
Bathow, G., et al. 1970. *Nucl. Phys. B* 20:592
Battistoni, G., et al. 1980. *Nucl. Instrum. Methods* 176:297
Baum, L., et al. 1975. See Atač 1975, p. 295
Bég, M. A. B. 1981. *Rockefeller Univ. preprint 81-B-9*, and *Proc. EPS Int. Conf. on High Energy Physics*, Lisbon. To be published

Behrend, H. J., et al. 1981. *Phys. Scr.* 23:610
Beron, B. L., et al. 1973. *Proc. 5th Int. Conf. on Instrumentation for High-Energy Physics, Frascati*, p. 362. Frascati: CNEN
Binon, F., et al. 1981. *Nucl. Instrum. Methods* 188:507
Blenar, G., et al. 1981. *Proc. EPS Int. Conf. on High-Energy Physics, Lisbon*, to be published
Böck, R., et al. 1981. *Nucl. Instrum. Methods* 186:533
Bogert, D., et al. 1982. *IEEE Trans. Nucl. Sci.* NS-29:363
Bosio, C., et al. 1978. *Nucl. Instrum Methods* 157:35
Botner, O. 1981. *Phys. Scr.* 23:555
Botner, O., et al. 1981a. *IEEE Trans. Nucl. Sci.* NS-28:510
Botner, O., et al. 1981b. *Nucl. Instrum. Methods* 179:45
Bouclier, R., et al. 1980. *CERN-EP Internal Report 80–07*. Geneva: CERN
Breidenbach, M. 1981. *Phys, Scr.* 23:507
Brisson, V., et al. 1981. *Phys. Scr.* 23:688
Bromberg, C. 1979. *Phys. Rev Lett.* 42:1202
Carithers, W. 1982. *BNL Rep. 30321*. Brookhaven Natl. Lab., NY
Chan, Y., et al. 1978. *IEEE Trans. Nucl. Sci.* NS-25–1:333
Charpak, G., et al. 1978. *Phys. Today* 31(10):23
Chen, H. H., et al. 1978. *Nucl. Instrum. Methods* 150:585
Cheshire, D. L., et al. 1977. *Nucl. Instrum. Methods* 141:219
Clark, A., et al 1982. *The UA2 Central Calorimeter*. Presented at the Int. Conf. Instrumentation for Colliding-Beam Phys, Stanford Univ., Calif.
Cobb, J. H., et al. 1977. *Nucl. Instrum. Methods* 140:413
Cobb, J. H., et al. 1979. *Nucl. Instrum. Methods* 158:93
Conversi, M. 1973. *Nature* 241:160
Conversi, M., Federici, L. 1978. *Nucl. Instrum. Methods* 151:193
Corden, M. J., et al 1982. *Phys. Scr.* 25:5
Coyne, D. G. 1981. *SLAC-PUB 2809*. Stanford, Calif: SLAC
Cundy, D., et al. 1976. In *Physics with very high energy e^+e^- colliding beams*, CERN 76-18:145
Della Negra, M. 1981. *Phys. Scr.* 23:468
Derenzo, S. E., et al. 1974. *Nucl. Instrum. Methods* 122:319
Derrick, M. 1981. *Preprint ANL-HEP-CP-81–33*. Argonne Natl. Lab, Ill.

388   FABJAN & LUDLAM

Diakonou, M., et al. 1979. *Phys. Lett.* 87B:292
Diddens, A. N., et al. 1980. *Nucl. Instrum. Methods* 178:27
Doke, T. 1981. *Port. Phys.* 12:9
Dris, M. A. 1979. *Nucl. Instrum. Methods* 161:311
Dris, M. A., et al. 1979. *Phys. Rev. D* 19:1361
Dydak, F. 1980. *Preprint CERN-EP/80-224.* Geneva: CERN
Eckardt, V., et al. 1978. *Nucl. Instrum. Methods* 155:353
Ellis, J. 1981. *Phenomenology of Gauge Theories,* CERN-TH 3174; and *Proc. Les Houches Summer School.* To be published
Erwin, A., et al. 1975. See Atač 1975, p. 271
Fabjan, C. W. 1980. In *ECFA-LEP Working Group,* ed. A Zichichi, p. 191. ECFA/79/39
Fabjan, C. W., Fischer, G. H. 1980. *Rep. Prog. Phys.* 43:1003
Fabjan, C. W., Willis, W. J. 1975. See Atač 1975, p. 1
Fabjan, C. W., et al. 1975. *Phys. Lett.* 60B:105
Fabjan, C. W., et al. 1977. *Nucl. Instrum. Methods* 141:61
Fabjan, C. W., et al. 1981. *Nucl. Instrum. Methods* 185:119
Fischer, H. G. 1978. *Nucl. Instrum. Methods* 156:81
Fisher, H. G., Ullaland, O. 1980. *IEEE Trans. Nucl. Sci.* NS-27:38
Ford, R. L., Nelson, W. R. 1978. *SLAC-210.* Stanford, Calif: SLAC
Friend, B., et al. 1976. *Nucl. Instrum. Methods* 136:505
Gabathuler, E., et al. 1978. *Nucl. Instrum. Methods* 157:47
Gabriel, T. A. 1978. *Nucl. Instrum. Methods* 150:145
Gabriel, T. A., Amburgey, J. D. 1974. *Nucl. Instrum. Methods* 116:33
Gabriel, T. A., Schmidt, W. 1976. *Nucl. Instrum. Methods* 134:271
Garwin, R. C. 1960. *Rev. Sci. Instrum.* 31:1010
Georgi, H., et al. 1974. *Phys. Rev. Lett.* 33:451
Goebel, K., et al. 1973. *Nucl. Instrum. Methods* 113:433
Goldhaber, M., Sulak, L. R. 1981. *Comments Nucl. Part. Phys.* 10:215
Gordon, H. A., et al. 1981. *Proc. 1981 ISABELLE Summer Workshop, BNL Rep.* 51443:884. Brookhaven Natl. Lab., NY
Grannis, P. D., et al. 1981. *Nucl. Instrum. Methods* 188:239

Grant, A. 1975. *Nucl. Instrum. Methods* 131:167
Hilke, H. J. 1981. *Proc. 1981 ISABELLE Summer Workshop, BNL Rep.* 51443:1275. Brookhaven Natl. Lab., NY
Hitlin, D., et al. 1976. *Nucl. Instrum. Methods* 137:225
Holder, M., et al. 1978. *Nucl. Instrum. Methods* 151:69
Iwata, S. 1979. *Report DPNU-3-79.* Nagoya Univ., Japan
Jones, W. W. 1969. *Phys. Rev.* 187:1868
Jonker, M., et al. 1981. *Phys. Scr.* 23:677
Jonker, M., et al. 1982. *Nucl. Instrum. Methods.* In press
Kadansky, V., et al. 1981. *Phys. Scri.* 23:680
Kienzle, W. 1975. *CERN-NP Int. Rep. 75-12.* Geneva: CERN
Kobayashi, M., et al. 1981a. *Nucl. Instrum. Methods* 189:629
Kobayashi, M., et al. 1981b. *Proc. Int. Symp. on Nucl. Radiat. Detectors, Tokyo,* p. 465. Univ. Tokyo
Kourkoumelis, C., et al. 1979. *Phys. Lett.* 81B:405
Kourkoumelis, C., et al. 1980. *Z. Phys.* C5:95
Krishnaswamy, M. R., et al. 1981. *Phys. Lett.* 106B:339
Kusumegi, A., et al. 1981. *Nucl. Instrum. Methods.* In press
Lederman, L., et al. 1975. *Nucl. Instrum. Methods* 129:65
Longo, E., Sestili, I. 1975. *Nucl. Instrum. Methods* 128:283
Ludlam, T., et al. 1981. *IEEE Trans. Nucl. Sci* NS-28:517
Masuda, K., et al. 1980. *Nucl. Instrum. Methods* 174:439
McDaniel, B. D. 1981. *Proc. 1981 Int Symp. on Lepton and Photon Interactions at High Energies,* ed. W. Pfeil., p. 921. Univ. Bonn
Mueller, J. J., et al. 1981. *IEEE Trans. Nucl. Sci.* NS-28:496
Murzin, V. S. 1967. *Progress in Elementary Particle and Cosmic-Ray Physics,* ed. J. G. Wilson, I. A. Wouthuysen, Vol. IX, p. 247. Amsterdam: North-Holland
Panofsky, W. K. H. 1981. See McDaniel 1981, p. 957
Pati, H. C., Salam, A. 1974. *Phys. Rev D* 10:275
Picasso, E. 1981. See Albrow 1981, p. 32
Powell, B., et al. 1981 *Nucl. Instrum. Methods.* In press
Pretzl, K. 1981. *MPI-PAE/Exp. El. 95;* and *Proc. SLAC Summer Inst. Part. Phys.,* Stanford Univ.

Price, L., Ambats, I. 1981. *IEEE Trans. Nucl. Sci.* NS-28:506
Radeka, V. 1977. *IEEE Trans. Nucl. Sci.* NS-24:293
Radeka, V., Williams, H. 1981. *Proc. 1981 ISABELLE Summer Workshop,* BNL 51443:1153. Brookhaven Natl. Lab, NY
Ranft, J. 1972. *Part. Accel.* 3:129
Rishan, J. P. 1979. *SLAC 216.* Standford, Calif: SLAC
Rosselet, L. 1981. *CERN 81-07:316.* Geneva: CERN
Rossi, B. 1964. *High Energy Particles.* New York: Prentice Hall
Rossi, B., Staub, H. 1949. *Ionization-Chambers and Counters.* New York: McGraw-Hill
Sauli, F. 1977. *CERN 77-09.* Geneva: CERN
Schamberger, R. D. 1981. See McDaniel 1981, p. 217
Scharre, D. 1981. See McDaniel 1981, P. 163
Schönbacher, H., Witzeling, W. 1979. *Nucl. Instrum. Methods* 165:517
Selove, W. 1972. *CERN-NP Internal Report 72-25.* Geneva: CERN
Selove, W., et al. 1979. *Nucl. Instrum. Methods* 161:233
Selove, W., et al. 1980. *Univ. Penn. Preprint UPR-75E*
Slattery, P., et al. 1981. *Proc. Int. Conf. on High Energy Physics,* Lisbon. To be published
Stevenson, J., et al. 1981. *Nucl. Instrum. Methods* 188:41
Stone, S. L. 1981. *Phys. Scr.* 23:605
Stone, S. L., et al. 1978. *Nucl. Instrum. Methods* 151:387
Treille, D. 1981. *Preprint CERN-EP/81-134.* Geneva: CERN
Van der Velde, J. 1981. *SLAC* 239:457. Stanford: SLAC
Vishnevskii, A. V., et al. 1979. *Preprint ITEP-53*
Walker, R. L., et al. 1981. *Preprint CALT-68-823.* Pasadena: Calif. Inst. Technol.
Willis, W. J. 1972. *BNL 17522:207.* Upton, NY: Brookhaven Natl. Lab.
Willis, W. J. 1981. *Preprint CERN-EP/81-21;* and *Proc. Workshop Future Relativistic Heavy Ion Experiments,* Darmstadt
Willis, W. J., Radeka, V. 1974. *Nucl. Instrum. Methods* 120:221
Willis, W. J., Winter, K. 1976. See Cundy et al 1976, p. 131
Yoshimura, Y., et al. 1976. *Nucl. Instrum. Methods* 137:57

Ann. Rev. Nucl. Part. Sci. 1982. 32:391–441

INERTIAL CONFINEMENT FUSION[1]

Denis Keefe

Lawrence Berkeley Laboratory, University of California, Berkeley, California 94720

CONTENTS

1. INTRODUCTION

1.1 *Thermonuclear Fusion for Electricity Generation*

The attraction of using thermonuclear fusion of light nuclei as a source of electrical energy lies in the small investment needed in the kinetic

[1] This work was supported by the Assistant Secretary for Defense Programs, Office of Inertial Fusion, Laser Fusion Division, US Department of Energy, under Contract No. DE-AC03-76SF00098. The US Government has the right to retain a nonexclusive royalty-free license in and to any copyright covering this paper.

energy of the ions to achieve a significant number of exothermic fusion reactions, each resulting in a very much larger kinetic energy of the reaction fragments. A variety of candidate reactions that have been considered are shown in Table 1 (Dean 1981).

Immediate plans for fusion development rely exclusively on the first reaction, between deuterium and tritium:

$$D + T \rightarrow n + He^4 + 17.6\,MeV$$

because it requires the least initial energy investment and provides the greatest energy amplification. Despite the theoretically large factor for energy amplification shown in Table 1, much of it will be whittled down by a host of inevitable inefficiency factors in any practical system and by the need to operate at energies substantially above the threshold shown. While "scientific breakeven" experiments are expected to occur within a few years at the Tokamak Fusion Test Reactor (TFTR) at Princeton, it is a long step from that to engineering breakeven (a genuinely energy self-sufficient system) and still a longer one to net output power production. Hence it is natural to concentrate now on the least difficult option — trying to burn deuterium-tritium (D-T) fuel. When that has been successfully achieved, other more difficult fuel choices will, no doubt, be explored for reactors of a later generation; for example, the D-D reaction, which would avoid the need to breed the radioactive isotope tritium, or the p-B reaction, which produces only charged particles in the final state and so offers the possibility of direct electrical energy conversion without the interpolation of an inefficient thermo-electrical conversion stage.

Table 1 Fusion reactions

	Reaction energy (MeV)	Threshold plasma temperature (keV)	Maximum energy gain per fusion
$D + T \rightarrow {}^4He + N$	17.6	4	1800
$D + D \rightarrow {}^3He + N$	3.2	50	70
$D + D \rightarrow T + P$	4.0	50	80
$D + {}^3He \rightarrow {}^4He + P$	18.3	100	180
${}^6Li + P \rightarrow {}^3He + {}^4He$	4.0	900	6
${}^6Li + D \rightarrow {}^7Li + P$	5.0	>900	6
${}^6Li + D \rightarrow T + {}^4He + P$	2.6	>900	3
${}^6Li + D \rightarrow 2({}^4He)$	22.0	>900	22
${}^7Li + P \rightarrow 2({}^4He)$	17.5	>900	18
${}^{11}B + P \rightarrow 3({}^4He)$	8.7	300	30

For fusion to work as a practical source of electricity, physics requires that two conditions be achieved simultaneously for a deuterium-tritium mixture:

1. The temperature should be in the region of $kT = 20\,\text{keV}$.
2. The hot plasma should be adequately confined; that is, the product, $n\tau$, of the number density n and the "confinement time" τ should lie close to $10^{15}\,\text{s}\,\text{cm}^{-3}$. This corresponds to the so-called Lawson criterion and is discussed in Section 1.2 (Lawson 1957).

In short, these conditions will ensure that a sufficiently large number of interactions occur (i.e. enough fuel will be burned), in the time for which the plasma density and temperature are both adequately high, to produce useful amounts of output energy.

Three methods are known for confinement of thermonuclear plasmas. *Gravitational* confinement, as in the sun and other stars, is quite successful and has a long lifetime but, because the confining force of gravity is so weak, it has a system scale length, (the solar radius) that is unacceptably large for earthbound fusion. Systems that use *magnetic fields*, in either a toroidal or mirror configuration, are being steadily advanced and seem virtually assured of scientific breakeven within a matter of years. For example, the Alcator A tokamak at MIT has already achieved an $n\tau$ product of 3×10^{13} s cm^{-3}; this would have been adequate for a breakeven demonstration had a hot enough plasma been contained. Finally, *inertial* confinement refers to the situation where a small volume of thermonuclear fuel — a sphere or pellet of frozen D-T mixture, for instance — is rapidly heated and promptly begins to fly apart on a very short time scale, τ, corresponding to the ratio of the pellet radius divided by the thermal speed of the ions $(2kT/M)^{1/2}$. While strictly this represents an unconfined system, there is a small but nonzero time in which the plasma density remains adequately large. For a pellet size suitable for electricity generation, this time is about 50 ps and one must arrange conditions to achieve enough nuclear interactions in that time to produce useful amounts of energy. The "confinement" time can be increased somewhat by inclusion of tampers of higher-density material to lengthen the time of disassembly.

Like its gravitational counterpart, inertial confinement fusion is known to work — in the form of the hydrogen bomb — again on an unacceptable scale. A major thrust of inertial confinement fusion research, therefore, is to show that very small amounts of D-T can be used successfully. For reference, the complete burning of one milligram of D-T will produce about 350 MJ; thus a reactor scenario based on inertial fusion might call for 10 microexplosions, each burning 1 mg of

D-T, to take place in a reactor vessel every second. If the output energy of the reactions (3.5 GJ per second) is absorbed in a thermal blanket, then — allowing for the inefficiency of converting the heat to electricity — one arrives at an electrical power output of ≈ 1 GWe. Also for reference, 1 MJ is roughly the energy released when half a pound of TNT explodes.

A magnetically confined plasma can be heated successfully by beams of neutral deuterium atoms or by radio-frequency waves or, perhaps, by beams of heavier atoms such as carbon. For inertially confined plasmas, energy supplied in the surface layers of a pellet by laser light or short-range ions causes ablative implosion of the pellet and can result in satisfactory compressional heating. As discussed in Section 1.2, volume compression of the fuel is in any case required for other reasons.

1.2   The Lawson Criterion for Magnetic and Inertial Confinement

While the condition for the product, $n\tau$, is roughly the same for magnetic and inertial fusion, the individual values of n and of τ needed are dramatically different for the two cases. A simple version of the criterion for ideal energy breakeven follows straightforwardly from energetics. If the number densities for deuterium ions and tritium ions are each denoted by n (thus total number density $= 2n$) then the reaction rate is

$$R = n^2 \,\overline{\sigma v}\, \text{cm}^{-3}\,\text{s}^{-1},\tag{1.}$$

where $\overline{\sigma v}$ is the product of the interaction cross section times the ion thermal velocity averaged over a Maxwellian velocity distribution. Figure 1 shows that $\overline{\sigma v}$ for the D-T reaction at 20 keV is about $4 \times 10^{-16}\,\text{cm}^3\,\text{s}^{-1}$. If the fusion reactions were to proceed for a time τ (short compared with the depletion time), then

$$\frac{\text{Energy out}}{\text{Energy in}} = \frac{n^2 \overline{\sigma v}\tau(17.6\,\text{MeV})}{2n\,(20\,\text{keV})}$$

$$\approx 2 \times 10^{-13}\, n\tau,\tag{2.}$$

giving a minimum scientific breakeven condition of $n\tau \gtrsim 5 \times 10^{12}\,\text{cm}^{-3}\,\text{s}$.

A more practical interpretation of $n\tau$ is as a measure of the fraction, ϕ, of the heated fuel that is burned in a time τ. At any time the rate of depletion of one fuel component (either D or T) is $-\mathrm{d}n(t)/\mathrm{d}t = \overline{\sigma v}[n(t)]^2$. Integrating for a time τ and setting $\phi = 1 - n(\tau)/n$, we find, for a temperature of 20 keV,

$$\frac{\phi}{1-\phi} = \overline{\sigma v}n\tau = 4 \times 10^{-16}\, n\tau.\tag{3.}$$

In some early experiments on magnetic fusion with pulsed fields, τ was controlled by the pulse length of the equipment. Today, however, experiments are planned with fields maintained for times of the order of a minute; in that case the important time scale is the "energy confinement time" of the plasma, which is determined by the rate of escape of energetic particles and radiation from the hot plasma. This time can be of the order of a second so that at densities of $n \sim 10^{15}\,\mathrm{cm}^{-3}$ a significant fraction of the fuel can be burned. For $\phi \approx 30\%$ fractional burn, an $n\tau$ product of $10^{15}\,\mathrm{cm}^{-3}$ s is needed.

Equations 1–3 also apply, of course, to inertial fusion with some additional constraints. The confinement time, τ, before the pellet has disassembled by virtue of the thermal speed, v, of the ions is of the order of (r/v), where r is the radius of the fuel pellet. (More exact calculation indicates that the effective τ is close to one quarter of this value.) To define an individual scale for either n or τ requires a further

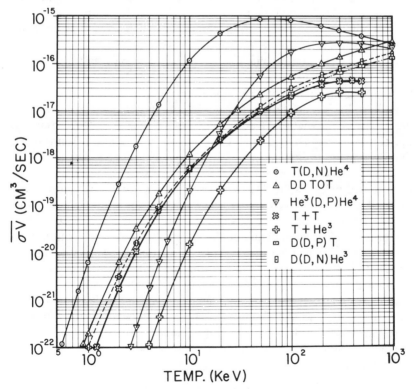

Figure 1 Values of $\overline{\sigma v}$ averaged over a Maxwellian distribution for a variety of fusion reactions.

constraint. If we choose to specify explicitly the pellet mass, M, one can write

$$n\tau = \frac{M}{(m_D + m_T)} \frac{3}{4\pi r^3} \frac{r}{v} \gtrsim 10^{15} \text{ cm}^{-3} \text{ s}$$

or

$$\frac{M}{r^2} \gtrsim 7 \text{ g cm}^{-2}, \qquad\qquad 4.$$

where the ion speed has been taken to be $2 \times 10^8 \text{ cm s}^{-1}$ appropriate to 20 keV. Thus for a pellet of mass $M = 1$ mg, which is in the range of interest for inertial fusion, the fuel radius must be no more than 0.1 mm. This in turn implies, with $n\tau = 10^{15}$, that the number density n should be $2 \times 10^{25} \text{ cm}^{-3}$ or about 1000 times the density in solid D-T. Thus successful burning of an inertial fusion pellet requires that a sphere of solid-density D-T be compressed a factor of 10 in radius; the corresponding tenfold decrease in confinement time, τ, is more than offset by the thousand-fold increase in density, n. (From Equation 4, we note that if we chose a larger value for M less compression would be needed, but the energy yield would become unacceptably large.)

Compression of the fuel one-thousand-fold, needed in any case to meet the Lawson criterion for milligram masses, brings with it an important side benefit. At normal D-T solid density ($\rho = 0.2 \text{ g cm}^{-3}$), the reaction product α particles have a range of about 10 mm — ten times the pellet radius — and escape. Since the range varies as $1/n$, in the compressed fuel it will be 0.01 mm or about one tenth of the compressed pellet radius. Thus most of the α particles will stop in the fuel and help to raise its temperature. Under these "ignition" conditions not all the fuel needs to be heated to 20 keV *ab initio*; it is enough to compress the fuel and adequately heat a small central core, whereupon the α particles will take up the role of heating the surrounding fuel layers, thereby creating a "propagating thermonuclear burn wave."

Among workers in the inertial fusion field the Lawson criterion is usually stated in different units, namely density (ρ in g cm^{-3}) and compressed fuel radius (r in cm). These quantities are directly proportional respectively to n and τ (a temperature of ~20 keV being implicit); a more exact calculation than that used above shows that in these terms

$$\phi = \frac{\rho r}{\rho r + 6}, \qquad\qquad 5.$$

so that a typical rule of thumb for significant fuel burning is

$$\rho r \gtrsim 3 \, \text{g cm}^{-2} \qquad\qquad 6.$$

1.3 Inertial Fusion: Advantages and Issues

Magnetic fusion had already been a subject of substantial experimental study by the time the field emerged from beneath the umbrella of classification in 1956. Now it has reached the point where adequately high temperature and adequate confinement have been separately demonstrated in two different tokamak experiments, and we are on the threshold of a further experiment (TFTR) in which both conditions are expected to be achieved simultaneously and scientific breakeven established. After the physics is proved will come the time to address the formidable engineering problems of making a realistic reactor system. The first wall, the thermal blanket (and breeder blanket, if there is one), must be incorporated close to the plasma within the complicated magnetic confinement coil system and must be designed so that it can be serviced and maintained after it has become radioactive. The complicated topology of toroidal, as opposed to mirror, devices makes this especially difficult.

The history of controlled inertial fusion is younger than that of magnetic fusion by some ten years and it is probably fair to say that our physics understanding of it continues to lag about five years or so behind our understanding of magnetic fusion. Shortly after Maiman demonstrated a practical working laser in 1960, Nuckolls et al (reported by Emmett et al 1974) realized that such a tool might provide the enormously high surface-heating power needed to compress and burn small volumes of thermonuclear fuel (Nuckolls et al 1972). Declassification of certain ideas that supported "laser-fusion" as an approach to electrical power generation took place early in the 1970s although design details and procedures for certain types of pellet continue to be classified. Using a neodymium-glass laser, Basov et al (1968) observed the first thermonuclear neutrons from a laser-heated plasma. Later, when it was found that particle beams (electrons and, later, ions) could offer special advantages as high-power drivers, the name "laser fusion" was broadened to "inertial confinement fusion" (ICF) or "inertial fusion," for short.

While some years must elapse before the physics issues in inertial confinement can be settled, it could offer practical advantages over magnetic fusion for energy applications. In particular, the reactor vessel, which contains the microexplosions and converts their energy to heat, can be designed with very few constraints imposed by the design of the driver, which supplies the energy to the pellet surface. There is more choice, such as in the nature of the reactor materials or the size and shape of the vessel, than for magnetic systems, and the absence of the high magnetic field allows liquid metals to be used to protect the inner surface of the reactor. Design for maintenance is also eased. A second advantage is the appreciation — following the initial suggestion by Maschke (1974, 1975a) — that, not only would an intense beam of high-energy heavy ions provide a particularly effective way of imploding pellets, but we could also capitalize on several decades of ideas and developments in the accelerator field to shorten the tedious engineering steps between scientific breakeven and a working power plant (Martin 1975, Martin & Arnold 1976, Keefe 1976). Realization of these advantages, however, requires certain physics questions to be resolved for the pellet, and the extrapolation of the accelerator physics from present practice to be shown.

The central issue for the pellet physics is whether a small amount of D-T fuel can be burned successfully under laboratory conditions. While a thermonuclear weapon is known to work, it releases a gigantic amount of energy and requires a high-power driver — a fission device — also of colossal magnitude. It is impossible to extrapolate with certainty any of this experience to the scale of one-milligram pellets. A major issue for the small pellet is whether the large compression can be accomplished in a hydrodynamically stable way and the fuel heated in a delicately prescribed way; if, for instance, premature heating ("preheat") occurs, the fuel mass cannot be compressed enough and will fizzle. The Shiva laser at Livermore has succeeded in compressing a D-T pellet to 50–100 times solid density, albeit with a very small mass and hence with a ρr value orders of magnitude below what is needed (Equation 6). A major issue for the driver, even if it be a single-shot device to test the pellet physics, is the ability to couple energy from the beam to the surface layers of the pellet; as reviewed below, laser and particle beams have quite different properties. Finally, for fusion power generation it is essential that the driver have high repetition rate (1–10 Hz), high efficiency (10–30% or more), long lifetime, and high availability (>80%), i.e. long-term reliability.

2.   PELLET COMPRESSION

2.1   *Target Design Considerations*

To this point the term "pellet" has been used in a simplified sense to denote a tiny sphere of D-T, but the configuration of the fuel and its attendant structure is more complicated. Hence the term "target" to denote the object placed at the focus of the laser or particle beams is used from here on.

Figure 2 illustrates two classes of target design, the "single-shell" and "double-shell" concepts. The outside layer of material in each consists of an ablation layer; when energy from laser light or short-range ions is deposited there, material is ejected and, just as in a rocket engine, an inward force is created by virtue of the outward momentum of the ablated material. In the single-shell design the fuel layer consists of a hollow shell of D-T, which can be created by filling the sphere with high-pressure gas and then freezing it. The object is to maintain a smooth, stable, ablatively driven implosion until the fuel eventually ends up at the center with an appropriate pr value close to $3\,\mathrm{g\,cm^{-2}}$. From a simple one-dimensional rocket equation it is easy to estimate that for an energy-efficient implosion $\sim$80–90% of the mass of the ablator will finally be ejected. If the compression occurs under ideal conditions, it turns out that, while the compressional heating of the fuel can be quite modest (on average $\sim$200 eV), a small volume in the center can be raised to $\sim$20 keV by the converging shock wave created in the fuel. This is exactly the desired condition for ignition to occur at the center and a burn wave due to α-particle heating to propagate into the rest of the fuel. The double-shell design differs in having a small mass of D-T fuel contained in a shell of dense material suspended at the center. In this case the outer shell is driven inward for a while, collides with and

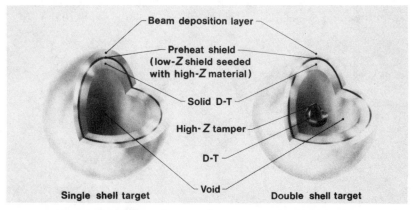

Figure 2 Schematic cross sections of two types of target. The single-shell target (*left*) is simpler to fabricate. The double-shell target (*right*) has two regions containing D-T fuel, ignition occurring first in the inner one. The beam-deposition region may be a single layer of light or dense material or may be composed of several different layers.

transfers energy to the inner shell, and drives the inner fuel to ignition. The burn wave then progagates outward to the main fuel layer and ignites it.

Successful compression is a delicate operation, and it is essential to minimize the energy into the fuel until compression is complete. Direct transfer of energy from the ablator layer can be prevented by a preheat shield (see Figure 2). In addition, however, the implosion speed must be controlled rather carefully so that the compression is isentropic, in which case the pressure-volume variation is described by an adiabatic curve, or adiabat; for minimum heating the compression must take place along the lowest possible adiabat. Ideally, for maximum compression, one wishes to approach a condition where the electrons are in the Fermi-degenerate state, in which case the average fuel temperature can be as little as $\sim200\,$eV for a pr value of $3\,$g cm^{-2}. If premature heating of the fuel occurs it will begin to expand before a suitable value of pr can be reached; alternatively, a very large driving energy would be needed.

Considerable attention is paid by target designers to minimizing the bad effects of fluid instabilities such as the Rayleigh-Taylor instability. This occurs whenever one attempts to accelerate a light fluid against a dense fluid. For example, the buoyancy force of water cannot support a denser liquid above its surface; the interface is unstable and the liquids will begin to interchange. Similar instabilities can occur initially as hot, low-density, ablating plasma pushes cold, high-density material inward and again, later, just before the end of compression when the compressed hot fuel begins to exert an outward force against heavier material. In the double-shell design this instability can occur at both the outer

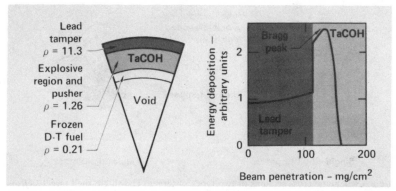

Figure 3 A target designed specifically to be driven by ions. (*Left*) Arrangement of tamper, pusher, and fuel layers. (*Right*) Energy-deposition profile for 10-GeV heavy ions. Most is deposited in polyethylene seeded with tantalum oxide (TaCOH).

and the inner shells. While the occurrence of fluid instabilities cannot be avoided, it is possible by proper choice of materials and geometry to ensure that the characteristic growth time can be adequately long to avoid effects that could be too damaging.

When the input energy is derived from a beam of heavy ions, an alternative to ablative implosion has been proposed by Bangerter & Meeker (1976). Unlike laser light, high-speed heavy ions deposit most of their energy near the end of their range (Bragg peak). Furthermore, specific energy deposition is about twice as great in low-Z materials than in high-Z materials. Figure 3 shows a target design that takes advantage of these properties. Most of the ion energy is deposited in a subsurface layer of polyethylene (CH_2) that is seeded with a very small amount of tantalum oxide to inhibit preheating of the fuel. The thin outside layer of lead acts as a relatively immobile tamper that retards the outward motion of the expanding plasma. The action can be compared to that of a cannon — the gun barrel containing the explosion — as opposed to the rocket-like action of ablatively driven targets.

Finally, the discussion of the target physics in this paper is treated entirely in the context of spherically symmetric illumination. In fact, that is not necessary and target designs are possible for which the energy is supplied by just two beams (or two bundles of beams) impinging 180° apart. (From a practical point of view, two-sided illumination from opposite poles is much easier to arrange for in a reactor scenario than is spherical illumination.) In this case, one possibility is to convert the focused energy of the laser or particle beams to soft x rays, which, if contained in a black-body cavity ("hohlraum"), can provide spherically symmetric compression of a physically separate component containing the thermonuclear fuel. While such indirectly driven target designs lie beyond the scope of this paper, the same sort of physics issues discussed herein are also involved.

2.2   Energy and Power Needed for Successful Compression

Results of elaborate computer programs (such as LASNEX at Lawrence Livermore National Laboratory) show that the desired implosion velocity should be close to $2 \times 10^7 \, cm \, s^{-1}$. To achieve this the material in the ablator must be raised to a temperature of ~200 eV, which in turn will require a specific energy deposition, w, in the ablator of

$$w \approx 20 \, MJ \, g^{-1}. \qquad\qquad 7.$$

This is one of three fundamental requirements that a laser or particle-beam driver must be able to provide.

A second requirement arises from the total amount of energy that must be supplied to the D-T fuel. Ignoring the relatively small amount of energy needed to ignite the center portion and assuming the average fuel temperature can be kept as low as 200 eV, one finds that the heat energy in a milligram is about 10 kJ. About an order of magnitude more energy must be supplied to the ablator, however, since the ejected ions carry away most of the energy and at best only some can be transferred to the remaining part of the ablator, or "payload," in rocket terminology. Thus the driver must be called upon to supply at least 100 kJ in energy. More realistic evaluation of several other effects such as fluid instabilities, imperfections in target fabrication, or illumination asymmetries, all of which affect ignition and burning, leads to about another order of magnitude in the driver energy, Q, requirement. Thus

$$Q \approx 1 \text{ MJ} \qquad\qquad 8.$$

A third requirement arises through the short time scale of the compression process. If, for instance, the targets in Figure 2 were initially 3 mm in radius they would be compressed to 0.3 mm in radius in a time of $0.27/(2 \times 10^7)\,\text{s} \approx 14\,\text{ns}$. From the previous requirement (Equation 8) it is clear that the power, P, needed is given by

$$P = 1\,\text{MJ}/14\,\text{ns} \approx 7 \times 10^{13}\,\text{W} = 70\,\text{TW}.$$

A more relevant quantity is the power per square centimeter, or irradiance, S, which for this example can be seen to be of order $10^{14}\,\text{W cm}^{-2}$. Another argument to establish the order of magnitude of the power density required arises from the fact that the ablator surface layers are extremely hot ($\sim 200\,\text{eV}$) so that if one wishes to avoid significant cooling due to black-body radiation, an irradiance $S \approx \sigma T^4$ is needed, where σ is the Stefan-Boltzmann constant, $5.67 \times 10^{-12}\,\text{W cm}^{-2}\,\text{deg}^{-4}$; this, too, demands that $S \gtrsim 10^{14}\,\text{W cm}^{-2}$. This argument, while helpful in setting a general scale, is in practice not strictly applicable; the plasma layer, in fact, is optically thin and the radiation loss will be well below the black-body level. In summary, careful considerations lead to the following condition — probably the most demanding of all:

$$S \approx 2 \times 10^{14}\,\text{W cm}^{-2} = 200\,\text{TW cm}^{-2}. \qquad\qquad 9.$$

Finally, successful target compression, i.e. maintaining the fuel conditions on a low adiabat, requires that the supplied energy pulse be shaped in time rising from a low power value at first to the peak value for the last 10 ns or so. The performance of single-shell targets is somewhat more sensitive to maintenance of the proper pulse shape than that of double-shell targets.

3. APPLICATION OF INERTIAL FUSION TO ELECTRIC POWER PRODUCTION

3.1 *Target and Driver Requirements*

Figure 4 summarizes the results of computer calculations of target gain versus input driver energy, gain being defined as the ratio of the energy arising from the microexplosion to the energy supplied to the target by the driver (Coleman et al 1981a). It is clear that double-shell targets can achieve higher gains than single-shell targets when a sufficiently large driver energy is available ($\gtrsim 5$ MJ). The rapid decrease in the double-shell gain curve as one proceeds downwards in input energy below 3 MJ seems an inevitable consequence of the two-step nature of that design.

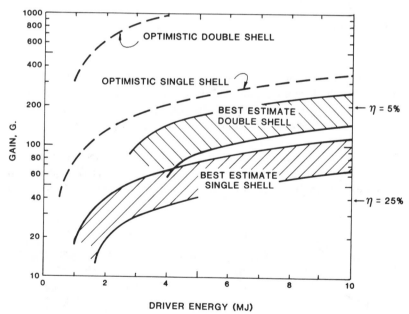

Figure 4 Calculated gain for single- and double-shell targets as a function of driver input energy (Coleman et al 1981a). Correction factors related to realistic experimental uncertainties reduce the gain estimates below those obtainable under ideal conditions (labeled "optimistic"). For double-shell targets, the gain drops rapidly for driver energies below the range plotted. The lines marked $\eta = 5\%$, 25% indicate the gain needed for drivers with these respective efficiencies (Equation 11).

What value of target gain may be needed for electric power production is intimately connected with the properties of the driver, as can be seen by reference to a simplified power-flow diagram (Figure 5). The driver efficiency, η, is defined as the energy per pulse delivered by the laser or particle beam to the target divided by the total electric energy required to produce that pulse. Of the gross electric power produced, an amount fP is recirculated to operate the driver and the remaining amount, $(1-f)P$, represents the net generating capacity of the system. If the thermoelectrical conversion efficiency of the turbine generators is denoted by ε ($\approx 33\%$), the total power can be written

$$P = \eta(fP)G\varepsilon,$$ 10.

where G is the target gain. If one demands for economic and other reasons that the recirculating power fraction be fairly small ($f = 1$ corresponds to no net power) then

$$1 \gg f = \frac{1}{\eta G \varepsilon} \approx \frac{3}{\eta G},$$

or, for example,

$$\eta G \gtrsim 10.$$ 11.

Thus the efficiency of a proposed driver system plays a crucial role in determining the target gain needed. Reference to Figure 4 shows that to achieve an ηG product of 10 a driver with 5% efficiency (characteristic of a good laser system) requires a double-shell target to achieve gain $G = 200$ and an input energy per pulse of 7 MJ. If the driver efficiency is as much as 25% (characteristic of a particle-beam accelerator), the situation is greatly eased; the minimum gain needed is $G = 40$, which permits either single- or double-shell targets to be used, and the beam

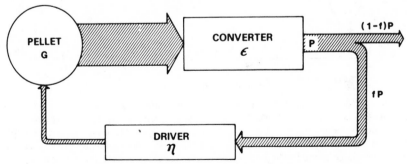

Figure 5 Electrical power flow in a reactor. The recirculating power fraction is f.

energy can be as low as 3 MJ per pulse. Because they are easier to make and cost less, single-shell targets are to be preferred.

To achieve significant amounts of electric power requires several microexplosions per second inside a containment vessel (reactor). Although the repetition rate needed is less for the higher-gain targets, the impulsive wall-loading per event will be an order of magnitude larger for the example just given, namely $(200 \times 7)/(40 \times 3) \approx 12$ times.

While the energy release per pulse in the high-gain case corresponds to that from several hundred pounds of TNT, the nature of the explosion is quite different from a chemical detonation. Most of the energy is carried to the walls by neutrons, x rays, and ions. Since the effect of a blast wave varies directly as the square root of mass, for a given energy, the small target mass results in a shock to the wall some two orders of magnitude less than that due to a chemical explosion. Nonetheless, the more rapid succession of smaller explosions appropriate to a high-efficiency driver makes it easier to ensure that the containment vessel can be kept to a reasonable size (radius ≈ 5–10 m). Table 2 summarizes the requirements that a driver for an electric generating system must fulfill.

Historically, lasers have provided the major tools for studies of the target physics problems in inertial fusion largely because they could supply the enormous irradiance that is needed. As candidates for a power-plant driver, however, they seem unpromising. Solid-state lasers (e.g. neodymium-glass) have low efficiency ($<1\%$), and thermal cool-down requirements restrict them to single-pulse operation. Gas lasers offer the possibility of the needed repetition rate since the active medium can be circulated and cooled. The carbon dioxide laser may have too long a wavelength ($10~\mu$m) to be satisfactory for the conventional targets discussed here, although its efficiency can approach 10%. Certain lasers that use excimers such as krypton fluoride have the advantage of suitable wavelength ($0.25~\mu$m) but would require heroic

Table 2 Driver requirements for power production

Energy — 1 to 10 MJ
Power — 100 to 600 TW
Pulse shape — control needed
(Driver efficiency) $\times$ (target gain) $= \eta G > 10$
Focusing — to a few millimeters at 5 to 10 m from the reactor wall
Reliability — $>80\%$ on-time
Lifetime — 30 years
Repetition rate — 1 to 10 s^{-1}
Cost — (a few) $\times 10^8$ \$ per GWe of electrical output

optical systems and are limited in efficiency to about 5%. The major problem facing the laser as a power-plant driver, however, is one that has emerged only in the past couple of years, namely, that operation with double-shell targets will demand an energy per pulse approaching 10 MJ unless target gains can approach the ideal curves. This is 500 times the energy of the Shiva glass laser at Livermore and 100 times that of the Nova glass laser now under construction (Coleman et al 1981b) and would cost far too much.

Light-ion systems based on pulsed-power technology, and heavy-ion systems based on high-energy and nuclear physics accelerator technology, share the advantages of experience with multimegajoule beams and can have electrical efficiencies in the region of 25%, perhaps more. Production of intense light-ion beams is, however, still a single-pulse operation; also, achieving the proper irradiance may require an unacceptably short distance between the accelerator and target (<1 m). High-energy accelerator systems that can produce heavy-ion beams have a broad theoretical and technological base and have demonstrated high repetition rate and availability. While the high beam power needed in a heavy-ion beam of good optical quality appears theoretically feasible, it lies far beyond present experience and needs some scaled laboratory demonstration to ensure success.

Other methods that have been suggested for imploding targets include macroparticle acceleration, free-electron lasers, and imploding liners, but all these methods seem conceptually relatively less mature.

3.2 Reactor Considerations

Studies of the likely costs of fusion power plants employing either magnetic or inertial confinement have generally shown that the unit cost of the electrical energy produced decreases as the size of the plant is increased. This result, labelled "economy of scale," leads to the choice of an output capacity of 1 GWe — give or take a factor of two — as suitable for electricity production at a cost per kilowatt-hour comparable to that of today. (Who knows what may still be an acceptable cost fifty years in the future?) An additional design criterion derived from cost studies is that the wall of the fusion reactor should be able to operate at a high neutron loading in excess of 1 MW m^{-2}, preferably 2–4 times that value; the problem of wall survival at such levels is treated below.

A plant to produce 1 GWe requires a thermal input power of some 4 GWt (if one assumes a recirculating power fraction $f = 25\%$), which can be generated by some 5 or 10 microexplosions per second, each with a yield of 800 or 400 MJ. Suitable targets would thus contain just a few milligrams of D-T fuel. A suitable reactor design should

1. for single pulses: provide adequate containment, handle the conversion of the energy in the neutrons, x rays, and target debris to a thermal blanket, and also breed enough tritium to compensate for at least what has been consumed;
2. for multiple-pulse operation at 5–10 Hz: be restored to steady-state conditions in about 100 ms and also have a target injection system with a 5–10-Hz capability.

Recently, Monsler et al (1981) comprehensively reviewed the problems envisioned in reactor design and the possible solutions.

3.2.1 SINGLE-PULSE REQUIREMENTS   About 70% of the energy released from the burning target is in the form of neutrons, distributed in energy with a peak at 14.1 MeV, and a degraded spectrum at lower energies because some neutrons interact before they can escape from the target debris. The remaining 30% of the energy is delivered mainly by x rays and some by the hot ionized target debris. During the brief time ($<$0.1 ns) when thermonuclear interactions are at their peak, a significant amount of energy is emitted as hard x rays with energies of about 100 keV. Shortly thereafter, the target debris that has been heated by absorption of α particles, neutrons, and x rays radiates soft x rays in the 1-keV energy region. As this hot plasma debris expands and cools, the radiation spectrum shifts downward in wavelength to optical and thermal emission. Note that the neutrons and hard x rays will distribute their energy volumetrically, i.e. deep in the thermal blanket material, but because they arrive promptly they cause a thermal shock and thermal gradients that the structure must withstand. While the soft x rays and debris arrive on a somewhat longer time scale, their energy is deposited in a shallow surface layer on the inside of the wall and may produce significant erosion by evaporation and sputtering.

The thermal transients in the bulk material are amenable to conventional engineering design, but wall erosion presents a special concern in a chamber of reasonable size (no more than 10 m in radius) under vacuum. Stainless steel, for example, would be eroded at a rate of several centimeters per year. If required, engineering realizations of the "dry-wall" concept are possible by use of expensive materials, such as pyrolitic graphite or niobium, or by means of a sacrificial liner that could be replaced every few years (Hovingh 1976).

In an effort to design a containment vessel that could last for the life of a power plant (30 years), several schemes for providing wall-protection have been devised. One of these, Solase, calls for the use of a buffer gas, e.g. neon or xenon, that is relatively opaque both to x rays and to the debris ions (Conn et al 1977). If the gas pressure is held at several torr, the burning fuel rapidly creates around it a sphere of hot ionized buffer gas. This "fireball" expands until it cools to the point where its characteristic radiation can no longer be transmitted through the surrounding gas. At the point where the fireball expansion slows down, however, an outward-going shock wave is launched, propagates through the buffer gas, and delivers a severe mechanical shock to the chamber wall, which creates additional structural difficulties. The Solase concept is not applicable for particle-beam drivers, which require quite different gas pressure regimes for propagation, but it could be used for a laser driver, the buffer gas being transparent at optical frequencies.

Alternative ways of protecting the chamber wall to achieve a 30-year life rely on liquid metals, and several variants of this concept have been proposed. In one early version, Hylife (Maniscalco 1977), a curtain or "fall" of liquid lithium is constructed inside the chamber wall. If the curtain thickness is chosen to be 60 cm, the neutrons will deliver ~95% of their energy to the lithium while it will absorb essentially all of the x ray and debris energy. Thus the heat energy can be removed by passing the liquid lithium through the primary of a heat exchanger ahead of the turbines and then returning it to the top of the chamber to continue to supply the fall.

Apart from providing wall protection and allowing for rapid heat extraction from the chamber, the use of liquid lithium fulfills the other condition mentioned earlier for single-pulse operation, namely the breeding of tritium. The breeding ratio, defined as the amount of tritium bred per pulse divided by the amount consumed by thermonuclear burning, clearly should be unity or a little greater for a reactor that is self-sufficient in tritium supply. Two isotopes of lithium are efficient for breeding tritium by the reactions ^{7}Li (n, n'α) T $-$ 2.46 MeV and ^{6}Li (n, α) T $+$ 4.8 MeV, the first reaction being due to fast neutrons and the second to slow neutrons. Neutronics calculations show that there is no difficulty in obtaining a tritium breeding ratio of between one and two. By breeding enough to supply the tritium needs of the targets, the tritium inventory at a power plant can be kept relatively low (several kilograms) and radiation hazards due to accidental tritium release minimized. Because the tritium is bred, it is interesting to note that the raw materials actually consumed are deuterium and lithium, both of which, fortunately, are abundant.

A valuable by-product of a fast-flowing liquid sheet of lithium in the chamber is that operation at high vacuum is possible. The vapor pressure of lithium at a temperature of 660 K (a desirable temperature for reactor operation) is about 10^{-4} torr, which is low enough to allow propagation of the beams from a heavy-ion driver from the wall to the target. Lower pressure, if desired, can be arranged by operating at a slightly reduced temperature; for instance, at 610 K the vapor pressure is 10^{-5} torr. If the liquid lithium flows at a fast enough rate, it also serves as a high-speed vacuum pump to restore rapidly the high-vacuum condition between pulses.

There are other candidates for a low-vapor-pressure fall besides lithium but the need for breeding cannot be ignored. A suitable choice is a lead-lithium alloy Pb_4Li, which would allow a lower lithium inventory (Maniscalco 1977). The lead degrades the neutron energy spectrum but a breeding ratio of unity can be maintained by a modest enrichment with ^6Li. A penalty for the use of this denser material results from the greatly increased pumping capacity needed to recirculate the liquid.

While a big step forward in ensuring long reactor lifetime, the continuous liquid-metal fall does not eliminate some problems due to impulsive mechanical shock on the wall. At this low pressure, there is no fireball occurrence but the combination of evaporated material from the fall and the high central temperature leads to a rapid expansion that drives the fall rapidly and coherently outward to slam into the chamber wall. A variation of the Hylife concept that overcomes this defect replaces the continuous curtain by a forest of some 400 high-speed cylindrical jets, each about 5–15 cm in radius (Monsler et al 1979). The hot expanding gases can find a variety of tortuous paths among the cylindrical jets and the force is primarily taken up in liquid-liquid impact of the colliding jets. The jets are visualized to be arranged in a close-packed hexagonal array, to provide maximum screening of the wall, with a few lines of sight through the forest to allow the laser or ion beams to penetrate freely to the target. Finally, an alternative way of protection by an array of liquid-metal columns is discussed in a recent study for a heavy-ion-driven fusion plant by a Karlsruhe–University of Wisconsin collaboration (Badger et al 1981). Here the high-speed lithium jets are replaced by low-speed streams of lead-lithium alloy guided vertically by flexible tubes made of loosely woven silicon carbide fibers. The main liquid flow is within the tubes, but enough seeps through the weave to provide a substantial protective layer on the outside. Because this concept, called Inport, seems to offer solutions not just for single-pulse wall protection, but also for the many additional problems associated with repetitive pulsing at several hertz, it is discussed in the next section.

3.2.2 REPETITIVE-PULSE REQUIREMENTS   Operation at 5–10 Hz demands that conditions inside the containment chamber be restored to normal within 100–200 ms. (This is not a hard and fast criterion since one can appeal to a design with a small number of separate stacked vessels into which the driver beams can be sequentially switched on successive pulses and the allowed restoration time thereby extended.) In the Solase concept, for example, a large amount of heat is generated in the buffer gas and the thermal transfer to the vessel is so slow that it must be pumped out and replaced by cold gas between pulses (Bohachevsky 1981). In the Hylife concept the disrupted jets must be given time to reassemble into their steady state. As a result, repetition rates more than about 2–4 Hz will be difficult and require high pumping speeds for the liquid metal.

The Inport concept circumvents many of the foregoing problems, provides a solution for operation at 5 Hz, and offers some additional advantages (Figure 6). By conducting the bulk flow of the lead-lithium alloy through woven SiC tubes, the time delay associated with reassembly of a free-flowing jet is avoided. Also the fact that the conducted flow

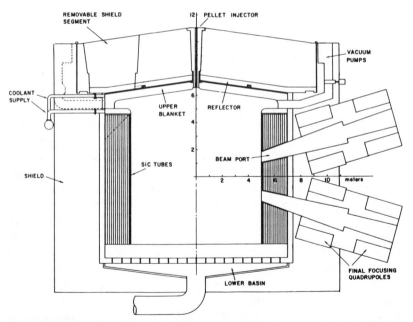

Figure 6 The Inport reactor concept (Badger et al 1981), which incorporates improvements over the Hylife concept (Maniscalco 1977). The coolant is a liquid lead-lithium alloy conducted in open-weave silicon carbide tubes. Ten ports through which the heavy-ion beams enter are arranged in five pairs around the chamber; one pair is shown.

can hence be at a slow rate avoids the need for a comparatively massive pumping system, an early argument against the use of heavy liquid metals. The seepage of material through the coarse weave provides a protective layer of about 1 mm that is quickly evaporated by the x rays and target debris but is reestablished on a time scale of ~100 ms. The lead-lithium alloy has a much lower vapor pressure than lithium and even a working temperature as high as 500°C still allows an operating pressure less than 10^{-5} torr, comfortably low for the propagation of heavy-ion beams to the target.

In a pure lithium liquid-metal system two concerns are (a) the chemical fire hazard due to the large hot lithium inventory and (b) the delicate operation of recovering the tritium from the hot lithium. Both are significantly ameliorated if lead-lithium alloy is used. The amount of lithium is reduced by more than an order of magnitude. Also, tritium has a low solubility in lead-lithium alloy and will predominantly remain behind in gaseous form in the chamber, and thus it can be pumped out and recovered cold by cryogenic pumping systems.

A Westinghouse study (Sucov 1981) on a dry-wall concept using an unprotected tantalum first wall contained the novel suggestion that advantages could be gained by having chemical overlap between the materials used in the target fabrication and those used in the containment vessel. They suggested, for example, that the high-Z tamper be made of tantalum rather than lead. Thus the deposition of material from the target debris could offset the erosion of the tantalum first wall. An analogous approach is advocated in the Hiball study in which the lead tamper is retained but the pusher material, described in Section 2.1 as polyethylene seeded with tantalum oxide, is replaced by lithium seeded with a small fraction of lead with the same average density. Since this provides a quite acceptable target design, one can thereby avoid the need to remove unwanted target debris from the liquid-metal coolant.

Finally, a working reactor requires the delivery of five or more cryogenic targets per second to the center of the chamber and overall system feedback to ensure that the target arrival and the driver pulse occur simultaneously. In a vessel of radius 10 m the time for free fall under gravity exceeds a second, hence the target must be launched at high velocity. Also it is desirable to delay target injection until the chamber conditions have been restored to steady state. Several satisfactory schemes for target injection, for example, by electrostatic, electromagnetic, or pneumatic means, are known and can impart speeds of the order of a few hundred meters per second, which leads to a time for the target to travel the radius of the chamber of 50 ms or less. This time exceeds substantially the time interval between initiation of the driver

pulse and arrival of the beams in the chamber so that the synchronization strategy consists in laser-ranging and tracking the ballistic motion of
the target as it crosses the chamber, predicting its time of arrival at the
beam focus, and firing the driver with a predetermined lead time. To
take the specific example of a heavy-ion induction-linac driver (Section
5.2), the total travel time of the ions from rest at the ion source to arrival
at the center of the chamber is about 100 μs; thus, when the accelerator
is first triggered, the target is only a couple of centimeters from the
desired rendezvous at the center of the chamber. For proper operation
of the accelerator, the switching times of the accelerating modules need
to be controlled to some 10 ns; if this is characteristic of the timing error
in the arrival of the beams it corresponds to a target positional error of
only a few micrometers — negligible compared to the target size. While
the longitudinal synchronization is not a problem, correction measures
may be needed for lateral displacements, i.e. when the pellet trajectory
does not intersect the center of the chamber. The input beam lines
therefore must incorporate deflecting elements — movable mirrors for
laser light beams or pulsed magnets for particle beams — that can be
actuated in response to the target-tracking information. Fortunately,
the needed steering corrections are small, less than 1 mrad in angle, and
pose no serious problems.

4.   LASER DRIVERS FOR INERTIAL FUSION

4.1   *Laser Interaction with Matter*

This is an immense subject (Brueckner & Jorna 1974, Motz 1979,
Bodner 1981, Max 1981) and can only be discussed here in a summary
and qualitative way. Some recent results of relevance to inertial fusion
are given in later sections. It should be noted that many early experiments in laser fusion made use of small glass microballoons filled with a
D-T gas mixture to produce thermonuclear neutrons. In this case the
thin shell is heated all the way through and explodes inward and
outward to produce a rather small compression and a high-temperature
fuel. Experiments with these "exploding-pusher" microballoons have
little relevance to the sort of inertial fusion target conditions discussed
above, where it was seen that an intact ablatively driven shell must
produce large volume compression and, at the same time, maintain the
fuel conditions on a low adiabat.

In considering the laser as a candidate for an inertial-fusion driver one
must examine two basic issues: (*a*) coupling efficiency, and (*b*) hot
electron production. The coupling efficiency, defined as the ratio of the

energy usefully conveyed to the D-T fuel divided by the input energy in the laser light, is made up of two factors, the "implosion efficiency" and the "absorption efficiency." The implosion (or hydrodynamical) efficiency enters in the consideration of any kind of driver and represents the effectiveness of the ablatively driven rocket action in conveying maximum energy, first, to the imploding pusher shell or payload and, second, in conversion of this energy into convergent nonadiabatic shock-heating of the fuel. The question of fluid instabilities in the target is also common to all drivers. The absorption efficiency, however, which measures how well the incoming radiation is converted to mass-ablation of the pusher, depends very much on the nature of the driver. For instance, heavy ions are believed to deposit their energy in well-understood classical fashion and not to create penetrating hot electrons. Laser light, on the other hand, must contend with a multitude of complicated physical processes occurring in the corona region that extends out beyond the ablation surface, and the absorption efficiency is an especially sensitive function of many parameters associated with the laser pulse, e.g. wavelength, irradiance, pulse length, etc. The second issue, hot electron production, can prove fatal if it leads to excessive preheating of the fuel; it also turns out to be critically dependent on the laser light parameters.

Several large research programs in the US and elsewhere are addressed to unravelling the variety of physical processes at play and determining how they scale. While present lasers provide more than adequate irradiance, they fall short by two orders of magnitude in the energy needed for reactor-sized targets (millimeters) and, hence, must conduct experiments on scaled-down sizes and for short pulses. Development of accurate scaling laws that can be extrapolated with confidence is, therefore, of the highest priority. Figure 7 is a schematic of coronal conditions. To first order, absorption of the light occurs predominantly in that part of the corona where the plasma density has the critical value, n_c ($\sim 10^{21}$ cm^{-3} for $\lambda = 1\,\mu$m), such that the plasma frequency, $\omega = (4\pi n_c e^2/m_e)^{1/2}$, matches the angular frequency of the laser light. Light energy is absorbed basically by three mechanisms: (a) inverse bremsstrahlung, (b) resonance absorption, and (c) parametric instabilities.

In the first process, electrons oscillating in the electric field of the light wave collide and exchange energy with the ions. (Its title derives from the change in the energy state of both an electron and an ion following absorption of a photon.) At low laser intensities this collisional process dominates by far and is a desirable one in that only low-temperature electrons are generated. As one proceeds to higher intensities, collective plasma processes begin to predominate, for example the second and third ones above, both collisionless, in which the incoming electromagnetic wave is coupled to a variety of plasma waves through both linear and nonlinear processes. The high electric fields produced by such waves can have the undesirable result of generating hot electrons ($\geqslant 20$ keV) and, consequently, penetrating x rays also. It should be noted that partition of energy among these absorption mechanisms is also a function of wavelength, in particular the collisional loss being enhanced as one proceeds to shorter wavelength.

Not all the laser light energy is absorbed, however, and as one proceeds to high intensity an increasing fraction is lost by reflection through stimulated Brillouin and stimulated Raman backscattering. (Again, these effects are relatively weaker at short wavelengths.) The nomenclature arises by analogy with light scattering from sound waves (Brillouin) or from molecular vibrations (Raman); the corresponding waves in the corona are the ion acoustic oscillations and electron plasma oscillations, respectively, which are pumped, or stimulated, by the laser light. Both Brillouin and Raman scatter represent instability processes in that the scattering contributes energy to the corresponding plasma wave. As a result they can also generate hot electrons. Raman scatter leads to an absolute (i.e. restricted to one location in space) instability at the point where the coronal density is $n_c/4$. Both effects also can lead to convective instabilities that drift away from the region of interaction;

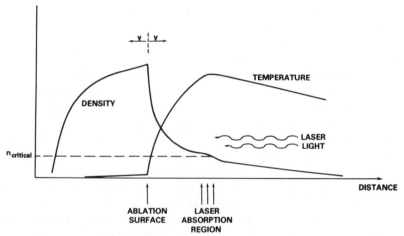

Figure 7 A schematic of the density and temperature distribution near the critical and ablation layers for a laser driver (Bodner 1981). Under ideal conditions the velocities of the payload and the ejecta are equal and opposite.

if such instabilities are growing, their total growth depends on the scale length of the plasma. As mentioned above, limitations in total energy allow laser experiments at appropriately high power only for short times and small sizes, and the magnitudes of the effects remain uncertain for scale lengths and times appropriate for reactor targets. Theoretical estimates for millimeter scale plasmas place the fraction of light reflected by Brillouin backscatter between 30 and 90%, a disturbingly wide range (Bodner 1981).

The mechanism of thermal conduction translating the energy absorbed at the critical layer into the ablation front is not yet properly understood and remains a topic of intensive study. On the one hand, the energy flow may be inhibited and can result in loss of efficiency. On the other hand, thermal diffusion effects can be large and can serve to smooth out the consequences of local hot spots at the critical layer and provide more uniform energy transfer to the ablation face, thus allowing considerable departure from symmetry in the driving beams.

4.2 Laser Facilities and Experiments

4.2.1 AVAILABLE WAVELENGTHS The primary tools in use today for studies of laser fusion are the neodymium-glass laser ("short" wavelength, $\lambda = 1.06$ μm) and the carbon dioxide gas laser ("long" wavelength, $\lambda = 10.6$ μm). In addition, in anticipation of the advantages of operating at very short wavelengths, development of krypton-fluoride gas lasers ($\lambda = 0.25$ μm) is actively underway. One of the most dramatic advances in the last few years has been the demonstration that laser light at higher harmonics (e.g. $\lambda/2$, $\lambda/3$, $\lambda/4$) can be efficiently generated, thus greatly extending the range of experiments on the critical issue of wavelength scaling of the many physical processes occurring in the corona. Most of the Nd-glass facilities have now been modified to include this capability, and experiments have recently been made not just with infrared light ($\lambda = 1.06$ μm) but also with green ($\lambda/2 = 0.53$ μm), blue ($\lambda/3 = 0.35$ μm), and even ultraviolet ($\lambda/4 = 0.26$ μm) light. Frequency conversion can be achieved by passing the infrared light through an optically nonlinear medium, such as a potassium dihydrogen phosphate (KDP) crystal. The polarization density in the crystal responds to the incoming light like an anharmonic oscillator and develops components at the harmonics $\lambda/2$, $\lambda/3$, etc, which can be radiated. High conversion efficiency ($\approx 70\%$) has been shown for $\lambda/2$ and $\lambda/3$ light (Coleman et al 1981c).

4.2.2 SHORT WAVELENGTH LASERS Major inertial fusion programs based on glass lasers are conducted at Lawrence Livermore National

Laboratory (LLNL), KMS Fusion, Naval Research Laboratory (NRL), and the University of Rochester (UR) in the United States and also at Ecole Polytechnique (France), Osaka (Japan), Rutherford-Appleton laboratory (United Kingdom), and the Lebedev Institute (USSR). The largest program and facilities are at LLNL, where recent experiments have been done with the two-beam Argus system (2 kJ, 5 TW) and the 20-beam Shiva system (10 kJ, 20 TW). See Figure 8. Both of these facilities have now ceased operation in preparation for installation of a much larger system, Nova. It will be constructed in two stages, Novette with two beams and Nova with ten beams. Each beam will have the capability of providing 10–15 kJ and 5–15 TW. Novette is scheduled to be completed late in 1982 with the larger system following within a few years. The accomplishments at Shiva include the ablatively driven compression of fuel to 100 times liquid density and the successful implosion of double-shell targets.

The 24-beam system, Omega (4 kJ, 12 TW), at UR has begun operations and is an upgrade of their previous 6-beam system, Zeta. The KMS system (1 kJ, 2 TW) incorporates a unique gas-jet target system to allow experiments on absorption and stimulated Brillouin scatter under conditions of various density gradients. Most of the above systems have some measure of frequency-conversion capability and have contributed greatly to the experimental study of wavelength scaling. A key result is the way the energy absorption efficiency falls off with intensity at different wavelengths. Figure 9 shows clearly the advantage of the shorter wavelength where the absorption mechanism at high irradiance is dominated by inverse bremsstrahlung. Figure 10 shows another important result, the sharp drop-off in hot electron production, inferred from the x-ray spectra, as the wavelength is decreased; it can be seen that a factor of three reduction in wavelength from 1.06 μm reduces the penetrating x-ray flux by more than two orders of magnitude and results in a more favorable, i.e. steeper, slope in the spectrum.

Experiments at NRL have concentrated on detailed studies of ablative acceleration of planar targets by relatively long laser pulses (3 ns). Although the laser energy and power are comparatively low (1 kJ, 0.3 TW), they have been successful in achieving ablation-driven velocities of 10^7 cm s^{-1} with both high absorption and hydrodynamic efficiencies, and with only small heating of the rear surface (relevant for preheating effects). It still remains an issue how these results are to be scaled to the higher power needed for fusion targets.

While Nd-glass lasers continue to provide a valuable research tool for studying the physics of laser light interactions and certain features of fuel compression, it should be noted that they are quite unsuited as a driver for an inertial fusion power plant. After a single high-power pulse the glass must be allowed to cool down and return to thermal equilibrium, which requires about one hour for the largest systems built.

Figure 8 A view along the twenty-beam neodymium-glass laser system Shiva.

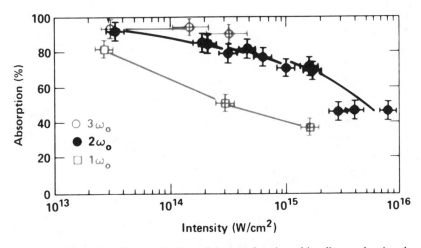

Figure 9 Absorption efficiency for laser light as a function of irradiance, showing the advantage of using wavelengths less than 1 μm (corresponding to an angular frequency, ω_0) at high irradiance.

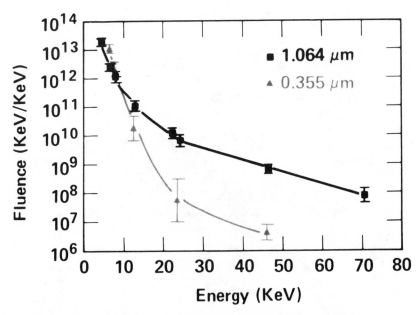

Figure 10 The x-ray spectra produced by hot electrons at $\lambda = 1.06\,\mu m$ and $\lambda = 0.35\,\mu m$. Both the fluxes and mean energies decrease for shorter wavelengths.

Steady operation at many pulses per second is inconceivable. Also, the electrical efficiency is about 0.1%, two orders of magnitude below the region of interest for fusion power.

4.2.3 LONG WAVELENGTH LASERS   Carbon dioxide lasers ($\lambda =$ 10.6 μm) offer a number of practical advantages. The efficiency can be relatively high, as laser systems go, perhaps approaching $\eta \approx 10\%$. Also, the use of gas as an active medium allows the engineering design of a driver with high repetition rate since the gas can be recirculated and cooled on a continuous basis.

Studies with carbon dioxide lasers are centered at the Los Alamos National Laboratory. The two-beam Gemini system operated, until its recent shut down, at 1 kJ and 1 TW, and the larger eight-beam Helios system delivers 8 kJ at 8 TW. Under construction is a still larger system, Antares, which will have 24 beams and deliver 40 kJ at 40 TW; it will be completed in 1983 (see Figure 11). Experiments at Helios have achieved fuel compression to 20 times liquid density in small targets.

The major issue, however, is whether the physics of absorption and hot electron production, at this long wavelength, can in fact allow successful ablative implosion and ignition of reactor targets. Much has been made earlier of the apparent benefits of proceeding to wavelengths shorter than 1.06 μm, and an obvious question is how much less favorable the physics may be for a wavelength ten times longer. The

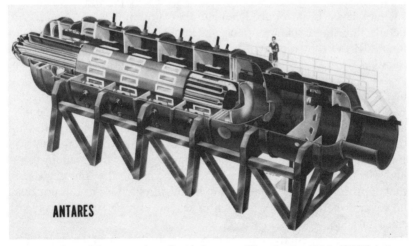

ANTARES

Figure 11   One of the two carbon dioxide laser amplifiers for the 40-kJ, 40-TW Antares facility at Los Alamos. Each simultaneously amplifies twelve laser beams azimuthally arranged near the outer radius (see exit holes in end-plate at right). The medium is pumped by 48 high-current electron beams that emerge radially from the inner cylindrical chamber.

answer seems to point to the need for developing radically different target designs and, indeed, some new concepts have been proposed that could turn to advantage the physics observed for 10.6-μm light.

Experiments on hot electron production at different irradiances at wavelengths from 0.26 up to 10.6 μm are consistent with a scaling law that gives a mean electron temperature proportional to $(S\lambda^2)^k$, where $k \approx 0.33$–0.46. This behavior is inferred from measurements of x-ray and ion fluxes and is therefore dependent on details of the theoretical modelling — hence the uncertainty in the exponent, k. The penetrating electrons generated by CO_2 light are, therefore, five to ten times hotter than those produced by Nd-glass light, for the same irradiance. Priedhorsky et al (1981) report observing a hard x-ray spectrum with a slope of 250 keV at $S = 10^{16}$ W cm^{-2} in an experiment at the Helios facility. They also claim that a substantial fraction, probably between 10 and 100%, of the absorbed laser energy is converted to hot electrons.

Thus the target designer is immediately presented with two challenges: how to avoid preheat without the use of an intolerably massive preheat shield, and how to convert the energy of the electrons into a more useful form. In response, several new concepts have been developed. One interesting possible solution proposed by Lee et al (1979) is called "vacuum insulation." Normally, the inward flux of hot electrons from the coronal region generates an outward return current (to maintain charge neutrality) carried by the cold plasma electrons. If a vacuum layer is introduced behind the outer foil shell (engineering details like maintenance of the vacuum and mechanical support are ignored), the return current is eliminated and the emission of a small number of hot electrons will make the outer layer positively charged and any further electron emission is inhibited by the electric field (Figure 12). Indeed, electrons that start inward can be returned to reflex backward and forward through the outward shell to deposit their energy as useful heat. The envisaged action is similar to that in a reflex ion diode (see Section 5.1). Eventually, however, ions will be emitted from the inner surface of the outer foil and move across the vacuum gap to give electrical closure. The net effect is to change the time scale from that appropriate for fast electrons to a slow-ion time scale. It is not clear, however, whether such a protection scheme can be scaled to reactor-sized targets.

The long wavelength light has been found, also, to lead to substantial fluxes of ions with MeV energies from the coronal region. In seeking to explain many experimental observations of thermal transport inhibition, Forslund & Brackbill (1981) made a major advance in the understanding of not just thermal transport but also the fast-ion production. When laser light is focused to a spot on a thin foil, a density gradient occurs in the direction of the light and a thermal gradient predominantly in the transverse direction toward the spot. Under these circumstances azimuthal magnetic fields in the megagauss range can occur, their strength and rate of rise being greater the hotter the electron temperature. Hot electrons emitted backward from the front surface are trapped by the magnetic field (cf magnetic insulation discussed in Section 5.1) and spread transversely across the foil surface far away from the focal spot by a fast $\mathbf{E} \times \mathbf{B}$ drift motion. This has three important consequences: (a) the energy of the electrons is held trapped in the coronal region, (b) the energy deposited at the focal spot is rapidly convected sideways to provide thermal smoothing over a large area, and (c) the thin sheet of electrons hovering just above the foil surface results in the formation of an ion diode (cf Section 5.1), which extracts ions from the surface plasma and accelerates them back toward the incoming laser light. Because of their higher magnetic rigidity, the ions can cross the shallow region of magnetic field without being trapped.

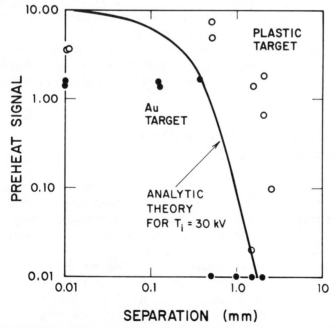

Figure 12 Experimental results of the effectiveness of vacuum insulation of the hot electrons (preheat signal). Measurements indicated even sharper cut-off than the calculated curve based on the model by Lee et al (1979).

Ions of several MeV kinetic energy pose no threat of preheat because of their short range and indeed, as discussed below, provide a much more desirable energy deposition mechanism than do fast electrons. In the example calculation by Forslund & Brackbill (1981) about 20% of the total absorbed energy was converted to ions. This suggests the intriguing possibility of finding suitable conditions such that most of the absorbed energy can be transferred from the electrons to fast ions; preliminary experimental results are very encouraging.

While this may seem, at first sight, an unnecessarily complicated way of heating by light ions compared with a directly accelerated light-ion beam, it circumvents the major problem of propagating a high-current light-ion beam several meters across the reactor (Section 5.1) by using the laser light as an intermediate vehicle to place the ion source within a millimeter or so outside the target.

4.2.4 ADVANCED LASER DEVELOPMENT   Lasers that use gas as an optical medium offer the potential for operation at high repetition rate. Given the recent promising results on absorption efficiency and hot electron spectra as one proceeds to shorter wavelength, it is not surprising that the krypton-fluoride laser ($\lambda = 0.25$ μm) has become the object of attention as a fusion driver. Practically it can reach an efficiency of about 5%. Difficulties still remain with developing optical coatings to handle the power densities in the ultraviolet region in a reasonable way. A discussion of the strategies and laser architecture options can be found in a recent comprehensive review by Holzrichter et al (1982).

5.   PARTICLE-BEAM DRIVERS
FOR INERTIAL FUSION

In many aspects charged-particle-beam drivers represent the antithesis of laser systems. We have seen that laser systems suffer from relatively low efficiency, are costly to scale to the many-megajoule energy range needed, and raise many unanswered questions about absorption efficiency and preheat. Nonetheless, lasers provide a unique way of developing the huge irradiance required because the power can be delivered to a tiny focal spot. Particle-beam drivers, on the other hand, include a variety of accelerator technologies that can have high electrical efficiency (25% and more), have demonstrated multi-megajoule capability, and are believed to deposit energy in a classical and well-understood manner without significant problems of preheat (with the exception of electron beams). In contrast with lasers, the major issues concern reaching adequately high beam power on a small focal spot, 2 or 3 mm in radius. In particular, heavy-ion drivers stand unique in being based on technology developed over many years for accelerators for high-energy and nuclear physics and so can guarantee high repetition rate, long life, and high availability.

Both particle-beam and laser drivers must meet the needs described earlier set by the target performance (see Table 2). For practical reasons, however, the higher electrical efficiency of accelerators permits the use of single-shell targets so that the beam energy per pulse in a fusion power plant can be less than that for lasers by a factor of two or more. The whole class of problems associated with the corona is absent. (In a sense, particle beams represent the ultimate in a short wavelength system having de Broglie wavelengths in the range 10^{-6}–10^{-11} μm!)

As discussed in Section 2.2, successful implosion of a target demands an irradiance $S \sim 2 \times 10^{14}$ W cm^{-2}, a specific energy deposition $w \sim 20$ MJ g^{-1}, and a time scale $t \sim 10^{-8}$ s. Since $w = St/R$, where R is the particle range in g cm^{-2}, one quickly notes that no matter what ion is used its range is constrained to be close to $R = 0.1$ g cm^{-2}. This constraint has dramatically different consequences for the kinetic energy and current of the beam, depending on the mass of the ion. Typical specifications are

1. for electrons: ~1 MeV and 100 megamperes;
2. for protons: ~5 MeV and 20 megamperes;
3. for heavy ions ($A \gtrsim 200$): ~10 GeV and 10 kiloamperes.

5.1   Light-Ion Drivers

Use of light-ion drivers has its roots in a technology for producing intense relativistic electron beams (IREB) that was pioneered more than thirty years ago by C. Martin at Aldermaston, England. In its commonest form an IREB accelerator consists of a slowly pulsed high-voltage supply, such as a Marx generator, which charges a pulse-forming line to a few megavolts; this energy can be switched to a rather simple cold cathode diode in which the electron beam is generated either by field emission or emission from a surface plasma (Nation 1979). Such a pulse-power system is relatively simple and inexpensive to fabricate. Typical IREB currents lie in the range from 10 kA to several megamperes and the largest systems can deliver megajoules of beam energy in a pulse duration of some 100 ns or less. This form of pulse-power technology is useful basically for single-pulse operation and would require extensive modification to achieve high repetition rate. Nonetheless, in the near term it offers the advantage of operating at much higher beam energies (megajoules) than any existing laser system. For testing of target physics the biggest issue is whether beams can be focused to provide adequately high irradiance.

Early experiments were pursued with electron beams principally at Sandia National Laboratory, Albuquerque, and at the Kurchatov Institute, Moscow. Apart from the difficulty of focusing electron beams with such enormous current on a small spot, the electrons have an ill-defined range and also produce x rays; hence preheat can be troublesome to avoid. The realization in the mid-1970s that IREB diodes could be converted to produce huge currents of light ions (protons) led to a dramatic switch in the Sandia program from electrons to ions in 1979. Already under construction at that time was a radial array of 36 separate electron-beam machines, each with its individual diode arranged so that the electron beams would converge at the center of the array. The voltage polarity was quickly reversed, the electron-beam diode designs abandoned and replaced by a design for a single cylindrical ion diode to give an inward-pointing ion beam, and the name of the device changed from EBFA (electron-beam fusion accelerator) to PBFA-I, the first letter now standing for "particle" (Figure 13). This

Figure 13 A cut-away view of PBFA-I showing the arrangement of the Marx generators, pulse-forming lines, and transmission lines that lead to a cylindrical ion diode in a small region (≈ 50 cm) at the center.

system has operated electrically at 1 MJ, 30 TW into a dummy load. Experiments with ion diodes at another Sandia facility, Proto-I, have already achieved an irradiance of 2 TW cm^{-2} on a target.

The brief history of the development of high-current ion diodes has indeed been explosive — proceeding from a few hundred amperes (Humphries et al 1974) to present values of 700,000 A (Cooperstein et al 1979). Also, about 80% of the electrical energy going into the diode can be converted to ion energy. The amazing variety of high-current ion-diode designs and their performance were reviewed by Kuswa et al (1982). The key to successful operation lies in the supression of electron flow. In a conventional IREB diode, electrons emitted from the cold cathode rapidly cross to the anode foil, which thereupon quickly develops a surface plasma that produces ions that stream back toward the cathode. Under these conditions of bipolar Child-Langmuir flow, the backward proton current density is 2.3% of the electron current density, down by the square root of the mass ratio. Thus almost all of the power of the generator goes to the electron beam and little to the ions. A wide variety of techniques has now been devised for inhibiting the electrons from crossing the gap, in which event efficient power transfer from generator to ions can be accomplished.

Three generic methods of electron suppression are shown in Figure 14; many hybrid designs have also been used (Kuswa et al 1982). In the reflex diode (Figure 14a) electrons lose a small amount of energy in passing through the anode foil, fail to reach the second cathode, and reflex backward and forward through the foil many times with decreasing amplitude. Thus the electron current is severely reduced but the electrons, nonetheless, help provide charge neutralization for the ions by virtue of their presence. The reflexing electrons produce an anode plasma which is the source of the ions. Notice that half of the ions in the case depicted will be lost because they are directed unfavorably.

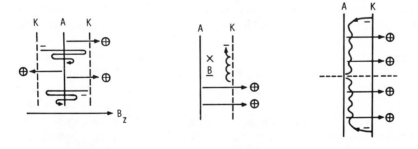

(a) Reflexing-electron   (b) Magnetically-insulated   (c) Pinched electron beam

Figure 14   A schematic of the three general classes of ion diodes discussed (Olson 1982).

Figure 14*b* shows a schematic of the concept of magnetic insulation in which a transverse magnetic field causes the electrons to execute magnetron-type orbits, thus preventing them from crossing the gap. The simplified version shown would work only for a short time, since negative charge accumulation near the top of the cathode would soon lead to electrical breakdown. A wide variety of ingenious designs has been developed to force the magnetic field configuration to have cylindrical symmetry — either about the axis of a planar diode or by developing the diode into a cylindrical diode formed by two co-axial annuli to produce an inward-pointing sheet beam (barrel diode). In either case, the $\mathbf{E} \times \mathbf{B}$ drift carries the electrons around the symmetry axis many times and charge accumulation is avoided. In a magnetically insulated diode the ions come from an anode surface plasma created by an electrical flashover; the cathode can be a thin foil or mesh to allow the ions to escape.

The usual textbook treatment of the Child-Langmuir space-charge limited current in a planar diode makes use of a one-dimensional model. This is adequate for many conventional applications in which the self-magnetic field of the beam is unimportant since the self-field cannot be incorporated in a one-dimensional model in which the beam has no axis and no boundary. The pinched-electron-beam diode (Figure 14*c*) relies heavily on the self-magnetic field of the electrons to create mainly radial flow of the electrons (the more rigid ions will flow almost axially) and thus greatly prolong the electron transit time from cathode to anode. It is well known that an intense electron beam cannot propagate axially through a background of charge-neutralizing positive ions if the current exceeds the Alfvèn-Lawson limit, $I = 17,000 \ \beta\gamma$ amperes, the value at which the self-field at the beam edge can bend an electron into an arc of diameter equal to the beam radius [$\beta = v/c$, $\gamma = (1 - \beta^2)^{-1/2}$]. The pinched-beam diode relies on the electron current being high enough that the self-magnetic bending radius of electrons near the beam edge is somewhat less than the anode-to-cathode gap spacing, d, and electrons flow (in complicated orbits) in the radial direction as sketched in Figure 14*c*. This will occur if the electron current leaving the cathode exceeds $I = 17,000 \ \beta\gamma(r/2d)$ amperes, where r is the cathode radius, and lies beyond the Alfvèn-Lawson limit for the usual geometries where $r/2d > 1$. Although ion flow is mainly axial in the different ion diodes discussed, the ion currents can substantially exceed the Child-Langmuir value computed on the basis of the physical spacing of the anode-cathode gap. This is so simply because the effective diode spacing is really much smaller as a result of the electron cloud (virtual cathode) penetrating into the gap space.

While the Sandia program has included many experiments on ablative implosion of cylindrical foils and successful production of high-temperature (50 eV) imploded plasma, its major efforts are directed at obtaining higher irradiances with the proper time structure at PBFA-I, mainly by testing different diode designs and different methods of focusing and propagating the beams to the target (Yonas 1981). Recent efforts have been devoted to the use of the AMPF ION (auto-magnetic, plasma-filled, ion) diode (see Figure 15). In this, the electric current pulse from the generator into the diode is made to flow through spiral conductors arranged radially to produce an insulating magnetic field with a constant tangential magnetic field at the surface of the anode plasma. The design of the complicated conductor arrangement takes into account the self-field of the ion beam. This design has been found susceptible to damage by the huge particle fluxes, and is costly to fabricate or repair. Also, ions crossing the region of the insulating magnetic field receive small transverse magnetic deflections in various directions and these smear out the energy-deposition region at the target and limit the irradiance. The possibility of proceeding to diode designs that could produce heavier ions, such as carbon, is therefore being examined. A promising prospect for the near future is the "pinch-reflex diode" developed at NRL and it will soon be tried at PBFA (Cooperstein et al 1979). This is a modification of the pinched-beam diode

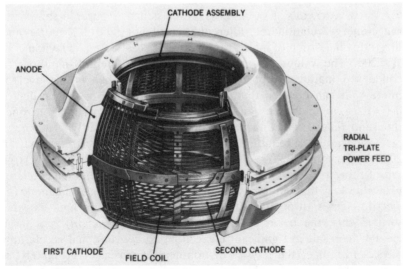

Figure 15   The AMPF ION ion diode for use at PBFA-1 (Kuswa et al 1982). The cylindrical symmetry is about a vertical axis. Current flowing to the diode through complicated spiral conductors provides magnetic insulation that traps electrons.

(Figure 14c) in which the electrons are caused to reflex through a thin anode foil. The geometry is arranged to produce a convergent (ballistically focused) ion beam at 1.3 MeV, which has been propagated through a plasma to produce current densities as high as 200–300 kA cm^{-2} at the focal spot.

Construction of a new facility, PBFA-II, has begun at Sandia, and it should operate in 1986. Like PBFA-I, it will consist of 36 generators arranged radially but each with a higher power capability (3 TW) and a range of voltage options (2–16 MeV). With a total power in excess of 100 TW it is hoped to achieve an irradiance at the target of 50 TW cm^{-2}, which could be enough for breakeven. Just as in the case of PBFA-I, where the advantages of proceeding up the ion mass scale (from electrons to protons) was perceived in the early design stages, plans for PBFA-II now incorporate the possibility of using still heavier ions such as helium, lithium, or carbon.

The research program at the Kurchatov Institute continues to center on the use of electron beams. Construction of a large 50-module radial array of generators, Angara 5, has encountered some technical difficulties and is proceeding slowly (L. I. Rudakov 1981, unpublished). As yet, there has been no indication of any intention or desire to convert the equipment to exploit the more promising ion diode approach.

5.2   Heavy-Ion Drivers

5.2.1   ADVANTAGES, PARAMETERS, AND ISSUES   Figure 16 shows the range-energy relation for different ions in hot (200 eV) matter and illustrates the origin of interest in the use of accelerated heavy ions i.e. $A \gtrsim 200$. The values of range are not very different from those in condensed matter except for very short ranges below the level of interest. It can be seen that, for the desired range in the pusher of $0.1 - 0.2$ g cm^{-2}, the energy of a ^{207}Pb ion (~ 10 GeV) is about one-thousand times greater than that of a proton (~ 10 MeV). Hence, to meet the irradiance requirement, the beam current can be reduced from 20 MA of protons to 20 kA of heavy ions. While this is still a very large ion current, it has opened the possibility of using conventional accelerator technology that employs vacuum transport of beams by magnetic (or electric) lenses through multiple accelerating gaps. At high currents such as this, the beam plasma frequency $\omega = (4\pi ne^2/m)^{1/2}$ is large enough to result in significant corrections to conventional accelerator physics formulae; the physics of non-neutral plasmas, fortunately, is considerably different and simpler to predict than that of conventional neutral plasmas. It can be seen that the major physics issues reduce to the following two questions: (a) Can conventional accelerator technology

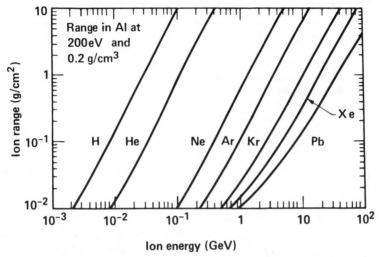

Figure 16   The range-energy relation for several ion species in hot matter (200 eV). The ion range of interest for inertial fusion is about 0.1–0.2 g cm^{-2}.

be reliably scaled to produce 10 kA of ion current? (*b*) If so, can collective or other effects such as those due to nonlinearities, be controlled well enough to ensure that the beam quality (brightness) is good enough to allow it to be focused ten meters away on a spot a few millimeters across? Our present understanding of accelerator physics leads us to believe that both questions can be answered affirmatively for, in fact, more than one type of accelerator system. The design, however, must incorporate some special features: for instance, a beam current in the 20-kA range would require intolerably high magnetic fields for single-channel transport and must be transported near the target in 10 or 20 separately focused channels.

Another feature of the reduction in particle-beam current to 20 kA is the high degree of confidence one can have that the energy deposition in the target is describable in classical terms. Good theoretical arguments exist (Bangerter 1979), but the experimental observation of classical energy deposition for protons at 1 TW cm^{-2} — a particle fluence greater than that needed for heavy ions — provides a further basis of reassurance.

Using the information in Table 2, one can develop the list of parameters for a heavy-ion accelerator driver given in Table 3. For a target gain of $G = 100$, the energy release per pulse is 300 MJ, corresponding to burning 1 mg of D-T fuel per pulse; at a repetition rate of 10 Hz, the electrical output is about 1 GWe.

Table 3   Typical parameters for a ^{207}Pb ion-beam power-plant driver

Beam energy = 3 MJ/pulse
Ion kinetic energy = 10 GeV (50 MeV/amu, β = 0.3)
Ion range = 0.2 g/cm^2
Number of ions = 2×10^{15}/pulse (300 particle μC)
Pulse length needed at target = 20 ns
Power needed at target = 150 TW
Beam current = 15 kA
Focal spot radius = 2.5 mm

The advantages of exploiting conventional accelerator technology are many:

1. There is a substantial physics and engineering base of experience, particularly in the fields of high-energy and nuclear physics where it stretches back for 50 years. The wide repertoire of accelerator tools, concepts, and devices that has been developed and tested can be deployed with a high degree of confidence to meet this new challenge. Techniques developed for rapid beam switching and splitting or combining beams are highly applicable. There is some relevant experience with multi-megajoule beams at Fermilab and CERN; while these are beams of relativistic protons and far from the 100-TW class, there is now considerable experience in such matters as the handling, shaving, aborting, and shielding of beams with this substantial damage potential.

2. Designs that incorporate high repetition rate, high availability, and long lifetime have become commonplace in the conventional accelerator field. In the future development of inertial fusion on the road to a full-scale reactor, the proven feature of high repetition rate is especially crucial. The engineering evaluation of a variety of containment vessel designs demands this ability to bring the vessel up to full temperature under realistic conditions for long test periods. All other proposed driver systems will require many years of difficult engineering to achieve this capability and, indeed, in some cases it may not be achievable.

3. As stressed earlier, high driver efficiency is extremely desirable for economic reasons and also in permitting the use of simpler low-gain targets with relatively low yield per pulse. Hitherto, the design of research accelerators has paid no attention to striving for high efficiency, although some accelerators (SLAC, LAMPF) do operate with an efficiency of several percent. Most designs for a fusion driver accelerator have naturally focused on accelerating as much beam current as possible so that, automatically, the higher beam loading has led to increased electrical efficiency from the mains to the beam. Typical estimates of accelerator driver efficiency lie in the range 20–30%. Higher values

seem possible at the expense of increased capital cost but would have a diminishing effect on the economics of fusion power.

5.2.2 ACCELERATOR PERFORMANCE CRITERIA   The realism of proposed driver schemes has been the subject of extensive review by accelerator scientists at a series of workshops (Bangerter et al 1976, Smith 1977, Arnold 1979, Herrmannsfeldt 1980). At the time of the first workshop, target designers believed that an ion kinetic energy of 40–100 GeV and a beam energy as low as 1 MJ might be suitable; since then, the desired kinetic energy has dropped dramatically and the beam energy increased (see Table 3). Both changes aggravate the accelerator problems significantly in requiring higher beam currents and causing greater difficulty in achieving a small focal spot. The central problem lies in achieving the high beam power, which, as discovered by Maschke (1975b), is limited in a magnetic quadrupole system by practical field strengths; he showed that the transportable beam power scales as $(\gamma - 1)(\beta\gamma)^{5/3}$ (see below) and thus is a steep function of ion kinetic energy. This limit comes into play when the electrostatic self-repulsion of the beam particles (magnetic self-attraction being almost absent at the speeds under consideration) becomes comparable with the magnetic restoring force of the transport lenses.

Another casualty of the changed target parameter specifications has been the synchrotron as a candidate for a driver system. The great attraction of the synchrotron as a tool for high-energy particle physics lies in its ability to amplify proton energies by a factor of about one thousand between injection and extraction. A heavy-ion synchrotron, for several reasons, needs a high injection energy (about 2 GeV at the Berkeley Bevalac, for instance) and, as the gap between the injection energy and the desired extraction energy dwindled to only a small factor, its advantages were soon realized to be outweighed by the disadvantages of introducing the extra beam manipulations involved (e.g. injection, extraction, de- and re-bunching) and it was retired as a driver option (Teng et al 1979).

Nonetheless, two distinct accelerator system designs offer considerable promise of success (Teng 1976). One uses an rf linac patterned on conventional designs (Figure 17a) to accelerate about 100 mA of heavy ions to the full energy of 10 GeV. It differs from today's ion linacs in three regards: first, in the low charge-to-mass ratio of the ions ($q/A \approx 0.01$ compared with 0.1); second, in the high current ($I \gtrsim 100$ mA compared with $I \sim 5$ mA); and, third, in that several parallel injectors are needed to accelerate lower current beams at the front end — because of the velocity dependence of the Maschke limit — to an intermediate energy where they can be combined and safely transported thereafter (Figure 17a). This last operation requires careful manipulations in the transverse and longitudinal phase planes to minimize emittance dilution and so preserve beam brightness. The main rf linac is a constant current device and simply serves to boost the kinetic energy to 10 GeV at a beam current of 150 mA. A three-stage system is then used to amplify the beam current by the required factor of 10^5. The beam is transferred to ten storage rings via an intermediate stacking ring and intermediate condenser rings to allow multi-turn stacking in both vertical and horizontal phase planes. By means of rf bunchers in the rings, the current can be further increased and the pulse length shortened. Next the ten beams are simultaneously extracted from the storage rings, each passed through a further pulsed bunching stage (provided by a ramped-voltage induction linac), and finally carried to the target.

The other scheme is a single-pass induction linac in which amplification of the beam-current takes place continuously during acceleration (Figure 17b). The induction linac consists of a sequence of nonresonant pulsed ferromagnetic cores, each of which supplies an energy increment to the beam by transformer action (Figure 18). This structure is particularly well suited to acceleration of very high beam currents (e.g. 100–100,000 A) in a repetitive pulsed fashion (Faltens & Keefe 1982). The ion injection current is several amperes (one hundred times that in the rf linac) and the entire beam can be accelerated in a single long sausage-like bunch. (Actually, there are cost advantages to subdividing the beam transversely and transporting several long bunches side by side in separate transport channels but all threading the same accelerating cores.) Early on, the voltage pulses to the induction cores are ramped slightly upwards with time, thus differentially accelerating the tail of the bunch with respect to the head. As the velocity is increased and the bunch length decreased, the beam current rises to several kiloamperes at the end of acceleration. A strongly ramped voltage is then applied to initiate a further longitudinal compression (by the needed last factor of five or so) which takes place in the final transport system to the target. An example of how the current and pulse length vary for a particular design is shown in Figure 17b. If a single beam were to be accelerated, it must be split transversely by septum magnets and subdivided into 10–20 channels for transport to the target; obviously, if acceleration of multiple beams (10–20 in number) were to be arranged, this septum-splitting operation could be avoided.

In choosing the detailed design parameters for either system, attention must be paid to several limiting phenomena arising from self-field

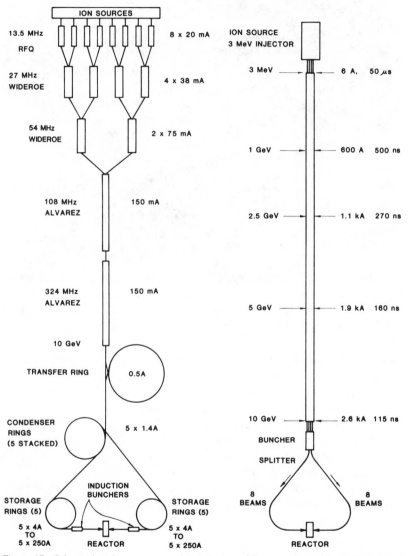

Figure 17   Schematic of two proposed accelerator driver systems: (*a*) The proposed 7-MJ driver for Hiball of the rf/storage-ring type first proposed by Maschke (Teng 1976). Current amplification occurs at full energy by multi-turn stacking in a cascade of rings. A final factor of ten comes from induction bunchers. (*b*) A single-pass four-beam induction linac (3 MJ) in which current amplification occurs continuously to keep pace with the space-charge limit.

Figure 18 A view of the FXR induction linac at the Lawrence Livermore National Laboratory. It accelerates 4 kA of electrons with a pulse width of 60 ns; components toward the high-energy end of an ion induction linac would look very similar.

(non-neutral plasma) effects at high current. Transverse effects limit the current, or lead to undesirable emittance growth; longitudinal effects lead either to beam loss or to an increase in momentum spread, $\Delta p/p$, with consequent chromatic aberration problems in the final focusing stage. When the beam plasma frequency is high, coupling can take place between these degrees of freedom and theoretical description of this coupling is still incomplete. Specific limitations that are known and must be considered in the design are as follows:

1. In the storage ring, the Laslett tune-shift condition (Laslett 1963) applies for transverse stability, i.e. the number N of stored ions is constrained by

$$N < \frac{2\pi \, \Delta \nu}{b r_p} \left(\frac{M}{M_p} \right) \left(\frac{1}{q^2} \right) \varepsilon_N \beta \gamma^2, \qquad 12.$$

where b = bunching factor = (fraction of ring occupied by ions)$^{-1}$, r_p = classical radius of proton = 1.5×10^{-16} cm, q = ionization state of stored ions, $\pi \varepsilon_N = \pi \varepsilon \beta \gamma$ = normalized transverse emittance = constant in ideal accelerator.

For steady-state storage conditions the allowed betatron tune shift, $\Delta\nu$, has the value 0.25. In the final rapid bunching needed just before extraction, this can be exceeded for a transient situation; a bunching experiment by Maschke at the Brookhaven AGS showed that by rapidly passing through betatron resonances $\Delta\nu = 2$ could be attained. For a heavy-ion driver $\beta(\approx 0.3)$, $\gamma(\approx 1)$, and $\varepsilon_N(\approx 2 \times 10^{-5}$ radian meters) are all small; thus the requirements set by Equation 12 demand that the number of storage rings must be large (10–20). Also, it is undesirable to use an ion with a charge state, q, much more than unity.

2. In the storage ring the longitudinal resistive instability can occur and consequently lead to an increase in momentum spread and loss of ions to the walls. This instability will not occur if the injected beam has a sufficiently large momentum spread. Unfortunately, a large momentum spread conflicts with the need to minimize chromatic aberration in the final focusing lenses, and it is found that in most driver scenarios the beam will always be above the instability threshold. Growth times will be of the order of a millisecond and damaging effects need not occur if injection, bunching, and extraction are all completed in a time of the order of a few milliseconds.

3. For linear beam transport systems the maximum current, and hence power, that can be transported in a quadrupole lattice is limited by the maximum attainable focusing (Maschke limit). This limit has been the object of extensive analytical and numerical study (Smith 1982). The limiting beam power, P, in watts is given by

$$P = 1.7 \times 10^{15} \left(\frac{M}{qM_p}\right)^{4/3} (\varepsilon_N B)^{2/3} (\beta\gamma)^{5/3} (\gamma - 1), \qquad 13.$$

where B = quadrupole "pole-tip" field (teslas) averaged along transport line. Delivery of the needed power of 150 TW (Table 3) to the target demands that some 10–20 beams be used if this condition is to be obeyed.

Equation 13 can be rewritten as a limit on beam current by dividing by the beam "voltage," $Mc^2(\gamma - 1)/qe$, and in this form is an important design factor for an induction linac driver. An induction linac will be most compact and efficient if it is accelerating currents in the kiloampere range; the Maschke limit will not, however, permit such currents below a kinetic energy of some 2 GeV (Figure 17b) and the low-velocity section consequently must be less than ideally matched. A way of circumventing this current limit has, however, been discussed by Maschke (1979), who pointed out that transverse subdivision into a number of beamlets each separately focused will allow more total current to be transported. A design for a four-beam induction linac has indeed shown significant cost advantages (Faltens et al 1981). As mentioned earlier, extension of this design to the range 10–20 beams would offer another benefit in avoiding septum splitting after acceleration.

4. The longitudinal resistive instability is not a problem for the rf linac (because the current is low) but must be considered in detail for the induction linac where the current is typically 10,000 times greater. Since one is dealing with a single bunch, the theory is incomplete. Simple theory suggests that two plasma waves are launched forward and backward from the site of a perturbation. The fast forward wave decays; the slow backward wave grows in amplitude as it travels to the back of the bunch but then is reflected into a forward-going decaying wave. Results from more sophisticated analyses give conflicting results, some predicting stability, others instability (Smith 1982). If instability does occur, the growth length λ is given by

$$\frac{1}{\lambda} = \frac{R}{Z_0} \left(\frac{4\pi^2 q^2}{1 + 2 \ln(b/a)} \frac{M_p}{M} \frac{N}{L} r_p \right)^{1/2},$$

where R = real part of impedance/meter, b/a = pipe radius/beam radius ≈ 1, Z_0 = free-space impedance = 377 ohms, and N/L = line density of ions.

Using a conservative value of R (100 ohms/meter), λ turns out to be of the order of one kilometer. In the length of a driver (5–10 km), growth may do little damage. Analysis of a type that indicated instability for a single-beam 3-MJ driver has shown that the situation becomes stable again by resorting to four parallel beams — an added benefit of the multiple-beam approach.

Since its inception in 1977, the heavy-ion fusion program in the US has been seriously hampered by lack of funding and it has not yet been possible to test many of the key issues in the accelerator physics that must be explored before proceeding to build a large device. Nonetheless, experiments at Argonne, Brookhaven, and Berkeley have been able to establish that heavy-ion sources with the needed high current and high brightness can be successfully built, also that the transport of very intense ion beams at low velocity can be handled successfully and conforms to theory. If a funding level comparable to that of the light-ion program (still much below that for the laser programs) is established, it is planned to build an ion accelerator with an energy (2–5 kJ) that will produce plasma temperatures of ~50 eV in a small focal spot. Such a test device would settle many of the accelerator questions, establish the predicted nature of ion energy deposition in hot matter, and also illuminate the problems that could be encountered in propagating intense ion beams in a reactor environment (Bangerter 1981). Based on today's knowledge, the heavy-ion approach seems clearly to offer the most promising driver for a civilian power plant. The lack of proper financial support can be traced partly to its late entry into the field, and partly to the fact that the laser and light-ion approaches are primarily directed at the physics of scientific feasibility and weapons-related physics, for which neither efficiency nor repetition rate is a concern.

5.3 *Focusing of Ion Beams on the Target*

This topic was the subject of a recent extensively referenced review by Olson (1982), who classifies the various methods that have been proposed for final focusing of the beams on the target as follows.

1. Ballistic transport with bare beam: Here the ions move in straight lines converging from the final lens to the target.
2. Ballistic transport with transversely available electrons: In this case electrons are injected radially into the beam to supply neutralization of the self-repulsive space-charge force.
3. Ballistic transport with axially available electrons: This differs from method (2) only in that the electrons are allowed to be dragged along the direction of motion of the ions.
4. Ballistic transport with co-moving electrons: If electrons can be injected at the same speed as the ions and in the same direction, they can, ideally, provide perfect charge and current neutralization.
5. Ballistic transport in gas or plasma: If there is a background gas in the containment vessel, it will rapidly be ionized by the incoming ions to provide some measure of neutralization.
6. Self-pinched transport in gas: If the ion beam can be focused to a spot size of a few millimeters and then passed through a gaseous region at a pressure of a few torr, "self-pinched" propagation can occur. The background plasma produces charge neutralization and partial current neutralization; the net current causes an azimuthal magnetic field to confine the beam to constant size within which the ions oscillate in the transverse direction.
7. Transport in preformed channel: Here a current-carrying conducting plasma channel is created ahead of time by means, for example, of a laser, an electron beam, or an exploding wire; the net current generates an azimuthal magnetic field that may confine the beam.

Two distinct sets of focusing issues must be faced for light ions. First, how does one achieve satisfactory focusing in the present generation of single-pulse experiments where the anode surface subtends a very large solid angle at the target, and the intrinsic spread in angles of the ions forces the stand-off distance from anode to target to be very short (~50 cm)? Second, how can one develop a focusing concept suitable for a reactor vessel where the propagation distance is of order 10 m, and the ports through which the ions enter must subtend a very small solid angle at the target?

For either light-ion application, the simplest focusing scheme in the list, method (1), can be ruled out. For the near-term experiments with PBFA-I and-II, the focusing mechanisms (2), (3), and (5) have received most attention; choice of proper anode curvature ensures that the ions are launched on average to point directly at the target and are thereafter neutralized by electrons or plasma. It should be noted that the proton currents discussed lie near and above the Alfvèn-Lawson limit for protons, $I_A = 31 \beta\gamma$ megamperes ≈ 3 MA, which implies that neutralization conditions must maintained stably to a high degree of precision. Some two dozen plasma instability effects (e.g. two-stream, filamentation, etc) have been examined theoretically and it is not yet clear whether one can stay out of danger from all at once. Also of concern is transient plasma behavior, which may spoil the focusing during parts of the pulse. Some experiments at reduced power and energy have, however, provided encouragement for some approaches (Olson 1982). For a reactor scenario, only mechanisms (6) and (7) are options and the former can probably be ruled out insofar as the azimuthal magnetic fields generated seem too small to contain the beam. Some new instabilities, e.g. hose instabilities, arise in case (7) and are on the verge of being troublesome but probably are not fatal.

For heavy ions, all seven focusing mechanisms are candidates for consideration. The cleanest by far is the first, focusing in vacuum, and is amenable to precise classical calculation including treatment of high-order aberrations (second and third) in the final magnetic quadrupole lenses. Provided the number of final beams is large enough (10–20) to maintain the current in each beam no greater than ~1 kA, a focal spot size of 2.5-mm radius can be achieved for low-charge-state ions. Good vacuum ($<10^{-4}$ torr) is required but reactor concepts have been described that seem to have acceptable properties (Section 3.2). Almost all other options have been considered to some degree; if any were to succeed it could provide further flexibility in the choice of charge state (higher q), or a reduction in the number of final beams entering the reactor, or in exploring a wider variety of reactor concepts. By contrast with light ions, the heavy-ion current per beam entering the reactor is some six orders of magnitude below the ion Alfvèn-Lawson limit, $I_A = 6$

$\beta\gamma$ gigamperes ≈ 2 GA, which eases the performance requirements of the various neutralization schemes. Whether there is a stable window of propagation in gas, or not, at a pressure ~ 1 torr has been the subject of varied theoretical arguments; if it exists, it may or may not be wide enough for reliable use of method (5). Results of numerical simulation of injection of co-moving electrons [cf method (4)] have been extremely encouraging in showing that some four times as much current per beam line as in the vacuum case can be acceptable; thus a smaller number of final beams, perhaps as few as four, might be permissible.

6. SUMMARY AND OUTLOOK

In future years, inertial confinement fusion can offer an inviting alternative to magnetic confinement fusion for electricity generation. In particular, a much wider variety of choice in the design of the reaction containment vessel is allowed since this method of confinement does not require the vessel to be cocooned within intricate field windings. Furthermore, the overlap between reactor design and driver design is minimal.

Whereas the development of magnetic fusion over the years has proceeded by trying to scale up the physics from small experiments to large ones, such as the Princeton TFTR or the Livermore MFTFB, the thrust of inertial fusion research is the opposite, namely trying to scale down the physics known to work for large uncontrolled devices to the tiny scale needed for controlled fusion. A critical issue is whether the physics will allow successful compression and burning of the fuel for a driver energy investment that is not too large—no more than several megajoules. If so, a second critical issue is development of a practicable driver in the megajoule class with high enough efficiency to make electricity economically. It is in this regard that particle-beam drivers offer the most promising solutions; also, they should achieve higher coupling efficiency than lasers because of the absence of coronal phenomena. The additional features of high repetition rate and ability to focus the beams at long stand-off distances make the heavy-ion approach a particularly attractive choice.

ACKNOWLEDGMENTS

I thank R. O. Bangerter, A. Faltens, T. Godlove, W. B. Herrmannsfeldt, R. Johnson, W. Kunkel, and J. M. Peterson for reading early drafts and offering helpful comments and suggestions. My gratitude is due to S. E. Bodner, D. Bruggink, T. Chan, W. F. Krupke, J. W-K.

Mark, C. L. Olson, S. Rockwood, and G. Yonas for providing graphical material. I especially thank Linda Egeberg for her patience and care in the preparation of the text.

This work was supported by the Assistant Secretary for Defense Programs, Office of Inertial Fusion, Laser Fusion Division, US Department of Energy, under Contract No. DE-AC03-76SF00098.

Literature Cited

Arnold, R. C., ed. 1979. *Proc. Heavy-Ion Fusion Workshop, Argonne Natl. Lab., 1978. Argonne Natl. Lab. Rep. ANL-79-31.* 430 pp.

Badger, B., Arendt, F., Becker, K., Beckert, K., Bock, R., Bohne, D., Bozsik, I., Brezina, J., Dalle Donne, M., El-Guebaly, L., Engelstad, R., Eyrich, W., Frohlick, R., Ghoniem, N., Goel, B., Hassanein, A., Henderson, D., Hobel, W., Hofmann, I., Hoyer, E., Keller, R., Kessler, G., Klein, A., Kreutz, R., Kulcinski, G., Larsen, E., Lee, K., Long, K., Lovell, E., Metzler, N., Meyer-ter-Vehn, J., von Mollendorff, U., Moritz, N., Moses, G., Muller, R., O'Brien, K., Peterson, R., Plute, K., Pong, L., Sanders, R., Sapp, J., Sawan, M., Schretzmann, K., Spindler, T., Sviatoslavsky, I., Symon, K., Sze, D., Tahir, N., Vogelsang, W., White, A., Witkowski, S., Wollnik, H. 1981. *HIBALL-A Conceptual Heavy Ion Beam Driven Fusion Reactor Study,* pp. VI.1/1–VI.7/12. *Univ. Wis. Rep. UWFDM-450*

Bangerter, R. O. 1979. See Arnold 1979, pp. 415–20

Bangerter, R. O., compiler. 1981. *Accelerator Fusion—A National Plan for the Development of Heavy-Ion Accelerators for Fusion Power. Los Alamos Natl. Lab. Preprint LA-UR-81-3730.* 32 pp.

Bangerter, R. O., Herrmannsfeldt, W. B., Judd, D. L., Smith, L., eds. 1976. *ERDA Summer Study of Heavy Ions for Inertial Fusion. Lawrence Berkeley Lab. Rep. LBL-5543.* 110 pp.

Bangerter, R. O., Meeker, D. J. 1976. *Lawrence Livermore Lab. Rep. UCRL-78474.* 15 pp.

Basov, N. G., Kriukov, P. G., Zakharov, S. D., Senatski, Yu. V., Tchekalin, S. V. 1968. *IEEE J. Quantum Electron.* 4: 864–67

Bodner, S. E. 1981. *J. Fusion Energy* 1: 221–40

Bohachevsky, I. O. 1981. *Los Alamos Sci. Lab. Rep. LA-8557.* 61 pp.

Brueckner, K. A., Jorna, S. 1974. *Rev. Mod. Phys.* 46: 325–67

Coleman, L. W., Krupke, W. F., Strack,

J. R., eds. 1981a. *1980 Laser Program Ann. Rep.,* Vol. 2, p. 3/18. *Lawrence Livermore Natl. Lab. Rep. UCRL-50021-80*

Coleman, L. W., Krupke, W. F., Strack, J. R., eds. 1981b. See Coleman et al 1981a, Vol. 1, pp. 2/53–2/183

Coleman, L. W., Krupke, W. F., Strack, J. R. 1981c. See Coleman et al 1981a, Vol. 1, pp. 2/254–2/286

Conn, R. W., Abdel-Khalik, S. I., Moses, G. A., Beranek, F., Cheng, E. T., Cooper, G. W., Droll, R. B., Henderson, T., Howard, J., Hunter, T. O., Larson, E. M., Kulcinski, G. L., Lovell, E. G., Magelssen, G. R., Maynard, C. W., Okula, K. R., Ortman, M., Ragheb, M. M. H., Rensel, W. B., Solomon, D., Spencer, R. L., Sviatoslavsky, I. N., Vogelsang, W. F., Watson, R. D., Wolfer, W. G. 1977. *SOLASE, a Conceptual Laser Fusion Reactor Design. Univ. Wis. Rep. UWFDM-220*

Cooperstein, G., Goldstein, S. A., Mosher, D., Oliphant, W. F., Sandel, F. L., Stephanakis, S. J., Young, F. C. 1979. *Proc. 3rd Int. Top. Conf. High Power Electron Ion Beam Res. Technol., Novosibirsk, USSR,* pp. 567–75. Novosibirsk: Inst. Nucl. Phys.

Dean, S. O., ed. 1981. *Prospects for Fusion Power,* p. 3. Elmsford, NY: Pergamon. 90 pp.

Emmett, J. L., Nuckolls, J., Wood, L. 1974. *Sci. Am.* 230 (6): 24–37

Faltens, A., Hoyer, E., Keefe, D. 1981. *Proc. 4th Int. Top. Conf. High-Power Electron Ion Beam Res. Technol. Palaiseau,* ed. H. J. Doucet, J. M. Buzzi, pp. 751–58. Paris: Ecole Polytech. 996 pp.

Faltens, A., Keefe, D. 1982. *Proc. 1981 Linear Accel. Conf., Santa Fe, NM, Oct. 1981,* pp. 205–13. *Los Alamos Natl. Lab. Rep. LA-9234-C.* 381 pp.

Forslund, D. W., Brackbill, J. U. 1981. *Magnetic Field Induced Surface Transport on Laser Irradiated Foils. Los Alamos Natl. Lab. Rep. LA-UR-81-170.* 12 pp.

Herrmannsfeldt, W. B., ed. 1980. *Proc.*

Heavy-Ion Fusion Workshop, Berkeley, 1979. Lawrence Berkeley Lab. Rep. LBL-10301. 515 pp.
Holzrichter, J. F., Eimerl, D., George, E. V., Trenholme, J. B., Simmons, W. W., Hunt, J. T. 1982. *J. Fusion Energy* 2(1):5–45
Hovingh, J. 1976. See Bangerter et al 1976, pp. 41–44
Humphries, S., Lee, J. J., Sudan, R. N. 1974. *Appl. Phys. Lett.* 25:20–22
Keefe, D. 1976. *Lawrence Berkeley Lab. Intern. Rep. BEV-3201.* 7 pp.
Kuswa, G. W., Quintenz, J. P., Freeman, J. R., Chang, J. 1982. Chap. IV, *Applied Charged Particle Optics, Part III, Advances in Electronics and Electron Physics,* Suppl. 13C, ed. A. Septier. New York: Academic. In press
Laslett, L. J. 1963. *Proc. 1963 Summer Study Storage Rings, Accel., Exp. Super-High Energies,* ed. J. W. Bittner, pp. 324–67. *Brookhaven Natl. Lab. Rep. BNL 7534.* 477 pp.
Lawson, J. D. 1957. *Proc. Phys. Soc. (London) Ser. B.* 70:2–10
Lee, K., Forslund, D. W., Kendel, J. M., Lindman, E. L. 1979. *Nucl. Fusion* 19:1447–56
Maniscalco, J. A. 1977. See Smith 1977, pp 73–75
Martin, R. L. 1975. *IEEE Trans. Nucl. Sci.* NS22:1763–64
Martin, R. L., Arnold, R.C. 1976. *Argonne Natl. Lab. Intern. Rep. RLM/RCA-1* and *US Patent 4,069, 457*
Maschke, A. W. 1974. *Brookhaven Natl. Lab. Rep. BNL 19008.* 6 pp.
Maschke, A. W. 1975a. *IEEE Trans. Nucl. Sci.* NS22:1825–27
Maschke, A. W. 1975b. *Brookhaven Natl. Lab. Rep. BNL-20297.* 3 pp.
Maschke, A. W. 1979. *Brookhaven Natl. Lab. Rep. BNL-51029.* 3 pp.
Max, C. E. 1981. *Lawrence Livermore Natl. Lab. Rep. UCRL-53107.* 82 pp.
Monsler, M., Blink, J., Hovingh, J., Meier, W., Walker, P. 1979. See Arnold 1979, pp. 225–36
Monsler, M. J., Hovingh, J., Cook, D. L., Frank, T. G., Moses, G. A. 1981. *Nucl. Technol. Fusion* 1:302–58
Motz, H. 1979. *The Physics of Laser Fusion.* London: Academic. 290 pp.
Nation, J. A. 1979. *Part. Accel.* 10:1–30
Nuckolls, J., Wood, L., Thiessen, A., Zimmerman, G. 1972. *Nature* 239:139–42
Olson, C. L. 1982. *J. Fusion Energy* 1(4):307–39
Priedhorsky, W., Lier, D., Day, R., Gerke, D. 1981. *Phys. Rev. Lett.* 47:1661–64
Smith, L. W., ed. 1977. *Proc. Heavy-Ion Fusion Workshop, Brookhaven Natl. Lab., Brookhaven Natl. Lab. Rep. BNL-50769.* 146 pp.
Smith, L. 1982. See Faltens & Keefe 1982, pp. 111–15
Sucov, E. W., ed. 1981. *Inertial Confinement Fusion Central Station Electric Power Generating Plant,* Vol. 1, pp. 4/75–6, *Westinghouse Fusion Power Systems Rep. WFPS-TME-81-001.* 234 pp.
Teng, L. C. 1976. See Bangerter et al 1976, pp. 13–17
Teng, L. C., Judd, D. L., Mills, F. E., Sutter D. F. 1979. See Arnold 1979, pp. 159–70
Yonas, G. 1981. *Proc. 10th Eur. Conf. Controlled Fusion Plasma Phys., (Moscow, Sept.)* In press

Ann. Rev. Nucl. Part. Sci. 1982. 32:443–97

PHYSICS OF INTERMEDIATE VECTOR BOSONS

John Ellis

CERN, Geneva, Switzerland

Mary K. Gaillard[1]

Lawrence Berkeley Laboratory and Department of Physics,
University of California, Berkeley, California 94720 USA

Georges Girardi and Paul Sorba

L.A.P.P., Annecy-le-Vieux, France

CONTENTS

[1] This work was supported in part by the Director, Office of Energy Research, Office of High Energy and Nuclear Physics, Division of High Energy Physics of the US Department of Energy under Contract W-7405-ENG-48. The US Government has the right to retain a nonexclusive royalty-free license in and to any copyright covering this paper.

1. INTRODUCTION

Since the first accelerator neutrino experiment (Danby et al 1962), searches for the vector bosons hypothesized (Fermi 1933, Yukawa 1935) as mediators of the weak interactions have been uniformly unsuccessful. Nevertheless, most theorists fully expect forthcoming experiments to uncover these particles, and believe that they can accurately predict their masses, decay properties, and production rates. The reason for this confidence is the astonishing success of the now "standard" model (Glashow 1961, Weinberg 1967, Salam 1968) of electromagnetic and weak interactions.

The model was constructed presuming that elementary particle interactions should be describable by a "renormalizable" field theory — that is, one in which all observables are calculable in terms of a finite number of measured parameters. Quantum electrodynamics is the prototype: once the electron's mass and charge are specified, photon and electron scattering amplitudes may be calculated to any desired accuracy. Years of effort to formulate a similarly calculable weak interaction theory finally bore fruit with the work of Gerhard 't Hooft (1971), who demonstrated the renormalizability of a class of theories with massive, electrically charged vector bosons as required to reproduce the observed structure of Fermi weak interaction couplings (see also Lee & Zinn-Justin 1972). A simple example of such a theory had been written down (Glashow 1961, Weinberg 1967, Salam 1968) several years earlier, but it predicted the existence of a type of weak interaction — neutral currents — that had not been observed. Renewed interest in the model spurred dedicated experimental searches that uncovered the predicted phenomena (Hasert et al 1973), and, when the dust settled, all the data appeared consistent with the first simple model, which we describe below.

Within the present state of technology, renormalizable theories must satisfy a tight set of constraints.

1. They include only fields with intrinsic angular momentum (spin) $s \leq 1$.
2. They include only couplings of dimension[2] 4 or less. Bose fields have intrinsic dimension 1 and fermions 3/2, so elementary couplings are at most bilinear in fermions and quadrilinear in bosons.
3. If spin-1 (vector) fields are included, the theory must be invariant under local "gauge" transformations.

---

[2] In mass units. Throughout we set $\hbar = c = 1$.

We illustrate this property of gauge invariance with a simple model, namely quantum electrodynamics (QED), with the Lagrangian

$$\mathscr{L}_{\text{QED}} = -\frac{1}{4} F_{\mu\nu} F^{\mu\nu} + i \sum_a \bar{\psi}_a \gamma^\mu D_\mu \psi_a + \sum_a m_a \bar{\psi}_a \psi_a, \qquad 1.$$

where A_μ is the electromagnetic field (photon) and

$$F_{\mu\nu} = \partial_\mu A_\nu - \partial_\nu A_\mu \qquad 2.$$

is the field strength; ψ_a is a fermion field (quark or lepton) of mass m_a and charge q_a. The "covariant derivative" D_μ is related to the ordinary derivative $\partial_\mu = \partial/\partial x^\mu$ by

$$D_\mu = \partial_\mu - ieA_\mu Q, \qquad 3.$$

where Q is the electric charge operator in units of the positron charge e:

$$Q\psi_a = q_a \psi_a. \qquad 4.$$

It is easy to see that the Lagrangian (Equation 1) is invariant under local (space-time-dependent) phase transformations on fermion fields

$$\psi \rightarrow e^{ie\lambda(x)Q} \psi \qquad 5.$$

if they are accompanied by a corresponding shift (change of "gauge") in the photon field

$$A_\mu \rightarrow A_\mu + \partial_\mu \lambda(x). \qquad 6.$$

This invariance reflects the fact that A_μ is not an observable: the observable field strength $F_{\mu\nu}$ is invariant under Equation 6. Moreover gauge invariance ensures the cancellation of infinites that would otherwise arise in the calculation of multiple photon exchange contributions to scattering amplitudes.

While the photon kinetic energy $(-\frac{1}{4}F_{\mu\nu}F^{\mu\nu})$ is gauge invariant, a photon mass $(\frac{1}{2}m^2 A_\mu A^\mu)$ would not be. This was one of the stumbling blocks to the construction of a theory involving the necessarily massive vector mesons of the weak interactions. A second stumbling block was that they carry electric charge and therefore couple to the photon. The generalization of gauge invariance to the case of self-interacting vector fields was worked out by Yang & Mills (1954), but the proof that such theories are renormalizable was complete only with the work of 't Hooft (1971).

The Lagrangian for a general Yang-Mills theory, including fermions, can be written in the form

$$\mathscr{L}_{\text{YM}} = -\frac{1}{4}G^i_{\mu\nu}G^{\mu\nu}_i + \bar{\psi}\slashed{D}\psi, \qquad \slashed{D} = D_\mu\gamma^\mu, \qquad 7.$$

where ψ is a column matrix with elements ψ_a and the covariant derivative is now a matrix:

$$D_{\mu a}^{b} = \delta_a^b \partial_\mu - i V_\mu^i g_{ia}^b ; \qquad\qquad 8.$$

the fermionic current $\bar{\psi}^a \gamma^\mu \psi_b$ couples to the vector V_μ^i with strength g_{ia}^b. The generalized field strength is

$$G_{\mu\nu}^i = \partial_\mu V_\nu^i - \partial_\nu V_\mu^i + c_{ijk} V_\mu^j V_\nu^k. \qquad\qquad 9.$$

Equation 7 will be invariant under an infinitesimal gauge transformation $[\lambda(x) \ll 1]$:

$$\delta\psi = i \lambda^i(x) g_i \psi$$

$$\delta V_\mu^i = \partial_\mu \lambda^i(x) - c_{ijk} \lambda^j(x) V_\mu^k \qquad\qquad 10.$$

if the c_{ijk} are totally antisymmetric and we impose

$$[g_i, g_j] = i c_{ijk} g_k \qquad\qquad 11.$$

$$c_{ijk} c_{klm} = c_{ikm} c_{kjl} + c_{ilk} c_{kjm}. \qquad\qquad 12.$$

These conditions mean that the g_i represent the infinitesimal generators of a group g with structure constants c_{ijk}. The field strength $G_{\mu\nu}^i$ transforms according to the adjoint representation of g:

$$\delta G_{\mu\nu}^i = -c_{ijk} \lambda^j(x) G_{\mu\nu}^k. \qquad\qquad 13.$$

The rules that emerge for constructing a renormalizable theory with vector fields are thus:

1. The Lagrangian must be invariant under a group of local gauge transformations.
2. The spin-1 fields must transform according to the adjoint representation of g, which determines uniquely their multiplicity.
3. Fermions must transform according to some (reducible) representation of g with their couplings to vectors given by representation matrices for the generators.

We did not include a fermion mass term in Equation 7 because we will consider gauge transformations that depend on helicity (the spin component along the direction of momentum) and that are not respected by mass terms. In such theories fermions, like vectors, acquire their masses from symmetry-breaking effects.

Let us recall the status of weak interactions before the discovery of neutral currents (Hasert et al 1973). Their two distinctive features were that only "left-handed" (negative helicity) fermions participated in weak couplings and that one unit of electric charge was exchanged. The observed weak interactions could be described in lowest order by the Lagrangian

$$\mathscr{L}_{\mathrm{W}} = g \sum_a \bar{\psi}_a \gamma^\mu I^+ \psi_a W_\mu^- + g \sum_a \bar{\psi}_a \gamma^\mu I^- \psi_a W_\mu^+, \qquad 14.$$

where the coupling constant g is related to the Fermi constant G_{F} through the mass m_{W} of the charged gauge bosons $W^\pm$:

$$\frac{G_{\mathrm{F}}}{\sqrt{2}} = \frac{g^2}{8 m_{\mathrm{W}}^2} \qquad 15.$$

and ψ_a are doublets of fermion fields, for example

$$\psi_{\mathrm{e}} = \begin{pmatrix} \nu_e \\ e^- \end{pmatrix}, \quad \psi_\mu = \begin{pmatrix} \nu_\mu \\ \mu^- \end{pmatrix}, \text{ etc.} \qquad 16.$$

The operators $I^\pm$ are represented by the direct product of a Pauli matrix $\tau^\pm$ with the Dirac matrix $L = \frac{1}{2}(1 - \gamma_5)$, which projects out negative helicity:

$$I^\pm = \frac{1}{\sqrt{2}}(I_1 \pm i I_2), \qquad I_i = \frac{1}{2}\tau_i L, \qquad i = 1, 2, 3. \qquad 17.$$

For a gauge-invariant theory we must add a neutral gauge boson W^3 coupled to the neutral current constructed using the commutator I_3 of the operators $I^\pm$:

$$\mathscr{L}_{\mathrm{W}^3} = g \sum_a \bar{\psi}_a \gamma^\mu I_3 \psi_a W_\mu^3. \qquad 18.$$

The matrices I_i generate two-dimensional unitary transformations on left-handed fermions and the gauge group is called $\mathrm{SU(2)_L}$. In a realistic theory we must also include electromagnetic couplings, but they cannot be added directly to Equations 14 and 18 because the charge matrix Q already contains a piece proportional to I_3. We enlarge the gauge group to include the full electromagnetic charge by defining a "hypercharge" operator

$$Y = Q - I_3, \qquad 19.$$

which commutes with the I_i. We then add a coupling

$$\mathscr{L}_{\mathrm{B}} = g' \sum_a \bar{\psi}_a \gamma^\mu Y \psi_a B_\mu \qquad 20.$$

so that the gauge group becomes $\mathrm{SU(2)_L \times U(1)}$, where the generator Y of $\mathrm{U(1)}$, represented by

$$Y = QR + \left(Q - \frac{\tau_3}{2}\right)L, \qquad R = \frac{1}{2}(1 + \gamma_5), \qquad 21.$$

generates helicity-dependent local phase transformations. The electro-magnetic current is therefore split into two components Y and I_3 coupled, respectively, to B and W^3. Defining two orthornormal fields,

$$A_\mu = B_\mu \cos \theta_w + W_\mu^3 \sin \theta_w$$

$$Z_\mu = W_\mu^3 \cos \theta_w - B_\mu \sin \theta_w, \qquad\qquad 22.$$

the sum of the interactions in Equations 18 and 20 can be rewritten as

$$\mathscr{L}_{w^3} + \mathscr{L}_B = A_\mu \, \bar\psi \gamma^\mu [g'Q \cos \theta_w - I_3(g' \cos \theta_w - g \sin \theta_w)]\psi$$
$$+ Z_\mu \, \bar\psi \gamma^\mu [I_3(g \cos \theta_w + g' \sin \theta_w) - Qg' \sin \theta_w]\psi. \qquad 23.$$

Identifying A_μ with the photon, which couples only to charge Q with strength e, gives

$$e = g' \cos \theta_w = g \sin \theta_w \qquad\qquad 24.$$

and the Z coupling reduces to the simple form

$$\mathscr{L}_Z = \frac{g}{\cos \theta_w} Z_\mu \bar\psi \gamma^\mu [I_3 - Q \sin^2 \theta_w]\psi. \qquad\qquad 25.$$

The neutral-current interaction (Equation 25) was obtained by the minimal extension of the known couplings, Equations 1 and 14, to a gauge-invariant theory. At the time when the now "standard" model was first proposed, neutral currents had not been observed, and alternative models were proposed that used the introduction of additional leptons and quarks to eliminate Z couplings to neutrinos (Lee 1972, Prentki & Zumino 1972) or to eliminate the Z entirely (Georgi & Glashow 1972). These models were less appealing on other grounds and were rendered obsolete by the discovery of neutral currents.

Returning to Equation 25, we see that the weak neutral couplings of fermions are completely specified by their charge and (helicity-dependent) weak isospin in terms of a single parameter, the "weak angle" θ_w, which characterizes the singlet-triplet mixing of the neutral vectors. At low energies, transitions involving massive boson exchange appear point-like with strength given by a Fermi constant G. Comparing Equation 25 with 14, we find the relative strength of neutral- and charged-current effects to be

$$G_Z/G_F = m_W^2/\cos^2 \theta_w m_Z^2. \qquad\qquad 26.$$

At this point we must consider how vectors may acquire masses in a gauge-invariant theory. One assumes that particle interactions are governed by a gauge-invariant Lagrangian, but that the lowest energy ("vacuum") state is not gauge invariant. Then the physical spectrum will not appear to satisfy the requirements of the gauge symmetry. According to a general theorem (Goldstone 1961), this situation, known as "spontaneous symmetry breaking," implies the existence of massless scalars called "Goldstone bosons." In the case of a local gauge symmetry, the Goldstone scalars can be removed from the Lagrangian by a gauge transformation and their degrees of freedom reappear as the longitudinal components of those vectors that become massive. This phenomenon is known as the Higgs or Higgs-Kibble mechanism (Higgs 1964, Kibble 1967).

The Higgs mechanism is most easily implemented by the introduction of scalar fields that transform under the gauge group. Let us add to the above model two complex scalar fields (ϕ^+, ϕ^0) that transform as a doublet under $SU(2)_L$ with hypercharge $Y_\phi = \frac{1}{2}$. They couple to gauge vectors through the covariant derivative

$$|D_\mu \Phi|^2 = \left| \left(\partial_\mu - ig \frac{\tau}{2} \, W_\mu - ig' \frac{1}{2} B_\mu \right) \Phi \right|^2, \qquad \Phi = \begin{pmatrix} \phi^+ \\ \phi^0 \end{pmatrix}. \qquad 27.$$

In addition we add a "scalar potential" with $SU(2)_L \times U(1)$ invariant mass and interaction terms:

$$V = -\mathscr{L}_\phi = -\mu^2 \Phi^\dagger \Phi + \lambda (\Phi^\dagger \Phi)^2, \qquad \lambda > 0. \qquad 28.$$

If $\mu^2 < 0$, the bilinear term is an ordinary mass term and V has its minimum at $\Phi = 0$. For $\mu^2 > 0$, the minimum occurs for a value $\Phi \equiv \langle \Phi \rangle$ with

$$|\langle \Phi \rangle|^2 = \frac{\mu^2}{2\lambda} \equiv \frac{v^2}{2}. \qquad 29.$$

In this case we must redefine the scalar field so that scalar excitations correspond to perturbations around the lowest energy state. Since Φ is characterized by four real functions, we may write it as

$$\Phi = \exp\left(\frac{i\theta}{v} \cdot \frac{\tau}{\sqrt{2}} \right) \begin{pmatrix} 0 \\ \dfrac{H(x) + v}{\sqrt{2}} \end{pmatrix}, \qquad 30.$$

where θ_i and H are real and by definition $\langle H(x) \rangle = \langle \theta(x) \rangle = 0$. This choice defines the direction of electromagnetic charge in the internal symmetry space: it is by definition the operator that leaves $\langle \Phi \rangle$ invariant. The potential (Equation 28) is independent of θ; excitation of these degrees of freedom corresponds to a rotation of Φ in isospin space and costs no energy. In the absence of gauge interactions, θ would emerge as a massless Goldstone field with derivative couplings as determined by expanding the kinetic energy term $|\partial_\mu \Phi|^2$. Using instead the covariant derivative, gauge invariance ensures that

$$\mathcal{L}(\Phi, W, B, \psi) = \mathcal{L}(\Phi', W', B', \psi'),\qquad\qquad 31.$$

where the primed fields are obtained from the unprimed ones by a gauge transformation. Choosing as parameters $\lambda_i = -\theta_i$, $\lambda_0 = 0$ removes the fields θ_i from the Lagrangian as expressed in terms of ψ', B' and

$$W_i' = W_i + \frac{\sqrt{2}}{v}\partial_\mu \theta_i + \cdots,\qquad\qquad 32.$$

which now has a longitudinal component. The covariant derivative (Equation 27) (dropping primes on gauge fields) now becomes

$$
\begin{aligned}
|D_\mu \Phi'|^2 &= \frac{1}{2}\left|D_\mu \begin{pmatrix} 0 \\ H+v \end{pmatrix}\right|^2 \\
&= \frac{1}{2}|\partial_\mu H|^2 + \frac{(H+v)^2}{2}\left(\frac{g^2}{2}W^+W^- + \frac{g^2}{4\cos^2\theta}Z^2\right).
\end{aligned}
\qquad 33.
$$

The photon field A does not appear in Equation 33 since it decouples from the neutral scalar; the squared coupling for the Z is taken from Equation 25 with $I_3 = -\frac{1}{2}$, $Q = 0$. The last term in Equation 33 determines the couplings of W and Z to the physical scalar field H and also their masses:

$$m_W^2 = \frac{g^2 v^2}{4} = \cos^2\theta_w m_Z^2.\qquad\qquad 34.$$

Using this result in Equation 26, we find equal effective Fermi constants for charged- and neutral-current couplings. This result follows from the $SU(2)_L$ doublet structure of Φ; a different transformation property would have resulted in a different mass ratio and unequal coupling constants. Fermion masses may be generated by introducing Yukawa couplings of fermions to Φ; redefining the scalar field as in Equation 30 gives rise to a fermion mass matrix proportional to v.

The electroweak gauge model with the specific symmetry mechanism described here is known as the *standard model*. It predicts the strength and structure of all neutral-current phenomena in terms of a single parameter θ_w. Although it has met with remarkable experimental success, many theorists believe that the symmetry-breaking mechanism may be more complicated. These possibilities are mentioned in Section 6, but have little bearing on the physics of vectors, which is our main concern here.

Before proceeding to the discussion of vector phenomenology, we must specify more completely the fermion content of the theory. At the time the electroweak theory was formulated, the data required three $SU(2)_L$ doublets of fermions: the two lepton doublets of Equation 16 and one quark doublet (Gell-Mann & Levy 1960, Cabibbo 1963)

$$\psi_u = \begin{pmatrix} u \\ d_c \end{pmatrix}, \qquad\qquad 35.$$

where u is the "up" quark of charge $\frac{2}{3}e$ and d_c is a superposition of the "down" and "strange" quarks

$$d_c = d \cos \theta_c + s \sin \theta_c \qquad\qquad 36.$$

of charge $-\frac{1}{3}e$. The "Cabibbo angle" θ_c is determined experimentally by comparing rates for semileptonic decays with strangeness change $|\Delta S| = 1$ and 0. The doublet ψ_u (Equation 35) presented a difficulty for the theory since its contribution to the neutral current (Equation 18) should be

$$\bar{\psi}_u \gamma^\mu I_3 \psi_u = \frac{1}{2}(\bar{u}\gamma^\mu Lu - \bar{d}_c \gamma^\mu Ld_c), \qquad\qquad 37.$$

which implied a strangeness-changing neutral-current coupling

$$\mathcal{L}_{\Delta S \neq 0} = \frac{g}{2} \sin \theta_c \cos \theta_c (\bar{d}\gamma^\mu Ls + \bar{s}\gamma^\mu Ld)W_\mu^3. \qquad\qquad 38.$$

This coupling (Equation 38) would induce the decay $K_L \to \mu\mu$ at a rate comparable to that for $K^+ \to \mu^+ \nu_\mu$, while experimentally the $\mu^+\mu^-$ mode is suppressed relative to the $\mu\nu$ mode by a factor 4×10^{-9} in rate. Before the work of 't Hooft (1971), Glashow, Iliopoulos & Maiani (GIM) (1970) had suggested that such strong suppression of strangeness-changing neutral currents, which should occur at higher orders even in a theory with only charged weak currents, could be understood if there were a fourth "charmed" quark c of charge $\frac{2}{3}e$ forming a weak isospin doublet with the superposition s_c of s and d orthogonal to d_c:

$$\psi_c = \begin{pmatrix} c \\ s_c \end{pmatrix}, \, s_c = s \cos \theta_c - d \sin \theta_c. \qquad\qquad 39.$$

Then the strangeness-changing component of the neutral coupling

$$g\bar{\psi}_c \gamma^\mu I_3 \bar{\psi}_c W_\mu^3 \qquad\qquad 40.$$

exactly cancels Equation 38. Although there was no evidence for charmed hadrons, the development of renormalizable theories made a mechanism of the GIM type seem imperative. Studies of higher-order contributions to neutral strangeness-changing currents led to estimates (Vainshtein & Khriplovich 1973, Gaillard & Lee 1974) of the masses of charmed hadrons, and intensive experimental searches eventually established their existence and confirmed their predicted properties (Aubert et al 1974, Augustin et al 1974, Goldhaber et al 1976, Peruzzi et al 1977). This discovery, together with the discovery of neutral currents, cemented the belief in gauge theories for many theorists. Indeed, it is generally believed that the strong nuclear interactions are also described by a gauge theory. In order to understand baryon spectroscopy and other low-energy properties of hadron interactions, it has long seemed necessary to postulate that each of the flavors $(u, d, s, c, ...)$ of quark exist in three different varieties, called colors. This new degree of freedom is believed to be gauged with massless vector bosons called gluons representing an $SU(3)_c$ color group. The resulting gauge theory, called quantum chromodynamics (QCD), has had considerable qualitative success in describing how the strong interactions get weaker at high energies so that the quark constituents of hadrons appear essentially free. One particularly convincing manifestation of this property of asymptotic freedom was in the spectroscopy and decay properties of hadrons containing the relatively massive charm quark.

Along with the discovery of charm, the existence of an unexpected new lepton, the τ, was also revealed (Perl et al 1975). Data suggest that the τ couples weakly to a third neutrino forming a lepton doublet

$$\psi_\tau = \begin{pmatrix} \nu_\tau \\ \tau^- \end{pmatrix}$$ 41.

with the same couplings as ψ_e and ψ_μ. This discovery led to the postulate of yet another quark doublet

$$\psi_t = \begin{pmatrix} t \\ b \end{pmatrix},$$ 42.

again on the grounds of renormalizability. In spite of gauge invariance, helicity-dependent fermion-vector couplings induce infinities in higher-order corrections to the three-vector vertex that can be removed only by cancellations among different fermions. The condition for this cancellation in the standard model is that the sum over fermion charges vanish. Since each quark doublet really represents three (mass- and charge-degenerate) doublets, corresponding to the three color degrees of freedom of the strong interaction gauge group $SU(3)_c$, the charge sum rule is satisfied if there are equal numbers of quark and lepton doublets, which one tends to associate as "generations" of fundamental fermions. Hadrons apparently composed of b and $\bar{b}$ have been identified (Herb et al 1977), while evidence for the t quark is still being sought, and its nonobservation to date means that its mass must be at least 18 GeV. The extension to six quarks had in fact already been proposed by Kobayashi & Maskawa (1973) because the Cabibbo mixing of b with d and s provides a possible source for the CP-violating effects observed in K decays.

To summarize, a large body of data (Hung & Sakurai 1981) supports the belief that weak and electromagnetic interactions can be described together by a renormalizable gauge theory; however, direct confirmation is still lacking. Aside from verifying the existence of the heavy vectors, many theorists believe the crucial test of the gauge theory postulate is the structure of their self-couplings. In the standard model these arise uniquely from the Yang-Mills interaction

$$\mathscr{L}_{\text{YM}} = -\frac{1}{4} G^i_{\mu\nu} G^{\mu\nu}_i, \qquad i = 1, 2, 3 \tag{43.}$$

$$G^i_{\mu\nu} = \partial_\mu W^i_\nu - \partial_\nu W^i_\mu + g \varepsilon_{ijk} W^j_\mu W^k_\nu,$$

which contains trilinear and quartic couplings of the charged

$$W^\pm = \frac{1}{2} (W^1 \pm i W^2) \tag{44.}$$

and neutral

$$W^3 = \cos \theta_w Z - \sin \theta_w A \tag{45.}$$

intermediate bosons. Even accepting a gauge-theory description, uncertainties persist. For example, there is a class of models that mimic the minimal model at low energy but would show up as dramatically different in the spectrum of massive vectors.

This article describes the physics of weakly coupled vector bosons. Section 2 covers their "static" properties: mass, decay widths, and branching ratios. Subsequent sections treat production mechanisms: e^+e^- annihilation, lepton-induced reactions on nucleons, pp and p$\bar{\text{p}}$ collisions. In Section 6, we briefly review the phenomenology of scalar particles associated with the clouded issue of symmetry breaking; the best laboratory for their study may in fact be in associated production with vectors or in their decay. Section 7 summarizes our conclusions.

2. GENERAL PROPERTIES OF THE VECTOR BOSONS

In the previous section, we described the arguments that led to the formulation of the "standard" model of weak and electromagnetic interactions, and we developed much of its structure. In this section we start to focus on the physics of the as-yet unseen vector bosons by listing some of their expected properties. Although experiments to date have been made in restricted kinematic ranges far from those where the vector bosons should appear, the manifold successes of the standard model and its tightly constrained nature tempt us to make the extrapolation and enable us to make relatively detailed predictions for their masses and decay modes.

2.1 Properties of the Charged Vector Bosons

We start with the charged vector bosons $W^{\pm}$, whose masses can be inferred from the known strength of the charged weak interactions:

$$\frac{G_F}{\sqrt{2}} = \frac{g^2}{8m_W^2} \qquad\qquad 15.$$

and the magnitude of the SU(2) coupling constant g:

$$g = \frac{e}{\sin \theta_w} \qquad\qquad 24.$$

in the standard model. Combining these two results, we find

$$m_{W^{\pm}} = \sqrt{\frac{\pi \alpha}{\sqrt{2}\,G_F}} \frac{1}{\sin \theta_w} \approx \frac{37.4\,\text{GeV}}{\sin \theta_w}. \qquad\qquad 46.$$

The experimental value of the neutral weak mixing angle θ_w is such that $\sin^2 \theta_w \approx 0.21$–$0.25$, corresponding to

$$m_W = 75\text{–}82 \text{ GeV} \qquad\qquad 47.$$

if we use Equation 46. This value is calculated directly from the Lagrangian without taking into account radiative corrections. Since the standard model is renormalizable, we are able to compute reliably to any desired order the radiative corrections to Equation 46 and the resulting prediction Equation 47. The one-loop radiative corrections were recently calculated (Bardin & Dokuchaeva 1981, Veltman 1980, Sirlin & Marciano 1981, Llewellyn Smith & Wheater 1981). When used in the analysis of weak neutral-current data, they tend to reduce somewhat the preferred range of $\sin^2 \theta_w$:

$$\sin^2 \theta_w = 0.215 \pm 0.012 \text{ (radiative corrections included)} \qquad 48.$$

and raise the prediction of Equation 47 for the $W^{\pm}$ boson mass by a few percent:

$$m_{W^{\pm}} = 83.0 \pm 0.24 \text{ GeV (radiative corrections included)}. \qquad 49.$$

The principal decay modes of the $W^\pm$ are also very easily deduced from the standard-model Lagrangian of Section 1. One finds, for example, that the leptonic decay mode $W^- \to e^- \bar{\nu}_e$ has the partial decay width

$$\Gamma(W \to e^- \bar{\nu}_e) = \frac{G_F m_{W^\pm}^3}{6\pi \sqrt{2}} . \qquad\qquad 50.$$

The partial rates for other fermion-antifermion decay modes are simply related to Equation 50 if one neglects the effects of finite fermion masses. All such (m_f/m_W) effects are rather small except when they involve the as-yet unseen quark t, whose mass must be at least 18 GeV. One finds that

$$\Gamma(W^- \to e^- \bar{\nu}_e): \Gamma(W^- \to \mu^- \bar{\nu}_\mu): \Gamma(W^- \tau^- \bar{\nu}_\tau): \Gamma(W^- \to \bar{q}q')$$
$$\approx 1: \quad 1: \quad 1: \quad 3|U_{qq'}|^2, \qquad\qquad 51.$$

where the factor 3 in Equation 51 is a counting factor arising because each quark pair $\bar{q}q'$ can be produced in three different colors. The quantity $U_{qq'}$ in Equation 51 parametrizes the mixing (Kobayashi & Maskawa, 1973) of the charged weak current between the quarks q and q′. In the familiar cases of the up, down, and strange quarks, $U_{qq'}$ is essentially given by the well-known Cabibbo angle θ_c:

$$|U_{ud}| = \cos \theta_c, |U_{us}| = \sin \theta_c \qquad\qquad 52.$$

with a natural extension to a world with more quarks and mixing angles. The matrix of couplings $U_{qq'}$ is unitary in a space of dimensionality N_G equal to the total number of fermion generations. Since each doublet of colored quarks is accompanied in the standard model by a doublet of leptons, one expects on the basis of Equation 51 that[3]

$$\frac{\Gamma(W^- \to e^- \bar{\nu}_e)}{\Gamma(W^- \to \text{all})} = \frac{\Gamma(W^- \to \mu^- \bar{\nu}_\mu)}{\Gamma(W^- \to \text{all})} \approx \frac{1}{4N_G} \lesssim \frac{1}{12} \qquad 53.$$

if all the fundamental fermions are much lighter than the $W^\pm$ and if these bosons have no other important decay modes. Indeed, fermion-antifermion pairs are by far the most important decay modes of the $W^\pm$ in the standard model, so that the results expressed in Equation 53 stand as the predictions for the leptonic branching ratios. The total $W^\pm$ decay width can be estimated by substituting the expected mass (Equation 49) into the anticipated decay width for $W^- \to e^- \bar{\nu}_e$:

$$\Gamma(W \to e^- \bar{\nu}_e) \approx 260 \text{ MeV} \qquad\qquad 54.$$

and then multiplying by $4N_G$, as suggested by Equations 51 and 53.

[3] Here, as is customary in this subject, we assume that the partial width for decay into hadrons is approximated by a sum over all energetically allowed $\bar{q}q'$ pairs.

One finds

$$\Gamma(W^- \to \text{all}) \approx N_G \text{ GeV},\qquad 55.$$

which is significantly wider than the mass resolution that should be obtainable in some of the experimental searches discussed in subsequent sections. While comparatively large by hadronic standards, the width (Equation 55) is still only a few percent of the large mass (Equation 49) expected for the $W^\pm$, so that the $W^\pm$ can still be regarded as a "narrow" resonance.

2.2   Properties of the Z^0

Turning now to the Z^0, the expected mass can again be derived simply from the standard model as set out in Section 1:

$$m_{Z^0} = \frac{m_{W^\pm}}{\cos\theta_w} = \sqrt{\frac{\pi\alpha}{\sqrt{2}\,G_F}}\,\frac{1}{\sin\theta_w\cos\theta_w} = \frac{37.4}{\sin\theta_w\cos\theta_w}\text{ GeV.}\qquad 56.$$

As for the $W^\pm$ (Equations 47 and 49), this mass estimate is subject to significant radiative corrections. If one includes the radiative corrections to Equation 56, one finds that

$$m_{Z^0} = 93.8 \pm 2.0 \text{ GeV.}\qquad 57.$$

Also as in the case of the $W^\pm$, one expects the principal decay modes to be fermion-antifermion pairs. We may parametrize the coupling of the Z^0 to a fermion pair $\bar{f}f$ ($f = e^-, \mu^-$, u, d, etc) by

$$\mathcal{L}_{Zf\bar{f}} = -\frac{m_Z}{\sqrt{2}}\left(\frac{G_F}{\sqrt{2}}\right)^{1/2}\bar{f}\gamma_\mu(v_f - a_f\gamma_5)f Z^\mu.\qquad 58.$$

In the standard model, Equation 58 is equivalent to Equation 25 with m_Z determined by Equations 34 and 15, and the vector couplings v_f and axial couplings a_f are

$a_\nu = v_\nu = 1$ for neutrinos

$a_f = \pm 1, v_f = \pm(1 - 4|q_f|\sin^2\theta_w)$ for $q_f \gtrless 0$.   59.

The various partial decay widths can be expressed in terms of the vector and axial couplings of Equation 58:

$$\Gamma(Z^0 \to \bar{f}f) = \frac{G_F m_Z^3}{24\pi\sqrt{2}}(1 \text{ or } 3)(v_f^2 + a_f^2).\qquad 60.$$

In view of the suppression of flavor-changing neutral currents, one expects the decay products to have equal and opposite flavors. The factor of 1 or 3 in Equation 60 depends on whether one is studying a leptonic decay mode or a quark-antiquark mode with its characteristic factor for the number of colors. Using the standard-model neutral-current couplings of Section 1, one finds (Camilleri et al 1976)

$$\Gamma(Z^0 \to \bar{\nu}_e \nu_e): \Gamma(Z^0 \to e^+ e^-): \Gamma(Z^0 \to \bar{u}u): \Gamma(Z^0 \to \bar{d}d)$$

$$= 2: 1 + (1 - 4\sin^2\theta_w)^2: 3\left[1 + \left(1 - \frac{8}{3}\sin^2\theta_w\right)^2\right]:$$

$$3\left[1 + \left(1 - \frac{4}{3}\sin^2\theta_w\right)^2\right] \qquad 61.$$

and similarly for the second- and third-generation flavors if one neglects corrections of order m_f^2/m_W^2. These are not actually negligible (Marciano & Parsa 1981) for the top quark:

$$\Gamma(Z^0 \to \bar{t}t) = \frac{G_F m_{Z^0}^3}{4\sqrt{2}\,\pi}\left(1 - \frac{4m_t^2}{m_Z^2}\right)^{1/2}$$

$$\left[1 - \frac{8}{3}x + \frac{32}{9}x^2 - \frac{m_t^2}{m_W^2}\left(1 + \frac{16}{3}x - \frac{64}{9}x^2\right)\right] \qquad 62.$$

(where $x \equiv \sin^2\theta_w$) implies a reduction of O(20)% by comparison with the decay rate into $\bar{u}u$ if the top quark has a mass of 20 GeV, close to the present experimental limit. If there were equal numbers of light fermions of the different charges (neutrinos, charged leptons, charge $+\frac{2}{3}$ quarks, and charge $-\frac{1}{3}$ quarks), we would expect on the basis of Equation 61 that

$$\frac{\Gamma(Z^0 \to e^+ e^-)}{\Gamma(Z^0 \to \text{all})} = \frac{\Gamma(Z^0 \to \mu^+\mu^-)}{\Gamma(Z^0 \to \text{all})} \approx \frac{1}{11 N_G} \lesssim 3\% \qquad 63.$$

since we know there must be at least three generations. From the expected Z^0 mass (Equation 57) and the formula for the partial decay widths (Equation 60), we deduce

$$\Gamma(Z^0 \to e^+ e^-) \approx 90 \text{ MeV} \qquad 64.$$

and

$$\Gamma(Z^0 \to \text{all}) \approx N_G \text{ GeV}, \qquad 65.$$

which is again considerably wider than the experimental resolution expected in some of the experimental searches. The total width (Equation 65) is slightly reduced by the finite mass correction (Equation 62) for the top quark, but this effect is compensated for by the QCD radiative corrections to the quark-antiquark decay modes, each of which is multiplied by a factor

$$\left[1 + \frac{\alpha_s}{\pi} + \cdots\right] \approx 1.04. \qquad \qquad 66.$$

The end result is to retain the simple formula expressed in Equation 65.

A point of some phenomenological interest and uncertainty is the total decay rate into neutrinos:

$$\Gamma\left(Z^0 \to \sum_\nu \nu\bar{\nu}\right) = 0.18\,\text{GeV} \times N_\nu. \qquad \qquad 67.$$

Neutrinos are, of course, much lighter than the related charged leptons, and we have very meager indications of how many there may be: presumably they could all be produced in Z^0 decays even if their partner leptons were heavier than $\frac{1}{2}m_{Z^0}$. The most direct upper limit on the number of neutrinos from particle physics experiments probably comes from K^+ decays. The limit (Asano et al 1981)

$$\Gamma(K^+ \to \pi^+ \Sigma\, \bar{\nu}\nu)/\Gamma(K^+ \to \text{all}) \lesssim 1.4 \times 10^{-7} \qquad \qquad 68.$$

corresponds (Ellis 1981) to an upper limit of

$$N_\nu \lesssim O(10^5). \qquad \qquad 69.$$

A similar limit can be deduced from the success of simple QCD calculations of the total decay rate and e^+e^- branching ratio of the lightest known meson containing a b quark and its antiquark. By way of contrast, cosmological nucleosynthesis calculations (Steigman 1979) suggest that

$$N_\nu \lesssim 3 \text{ or } 4 \qquad \qquad 70.$$

for neutrinos weighing less than an MeV or so. If the number of light neutrinos is not excessive, a comparison of Equations 65 and 67 suggests that

$$\frac{\Delta\Gamma(Z^0 \to \text{all})}{\Gamma(Z^0 \to \text{all})} \approx 6\% \text{ per neutrino}, \qquad \qquad 71.$$

so that a precision measurement of the Z^0 width should be able to tell us how many neutrinos there are. Provincial particle physicists may trust this more than the nucleosynthesis limit (Equation 70). However, if the Z^0 is significantly wider than the expectation (Equation 65), we must be sure to have excluded the possible existence of other substantial decay modes of the Z^0. None have been found in the standard model: for example, Alles et al (1977) and Albert et al (1980) calculated

$$\frac{\Gamma(Z^0 \to W^\pm + X)}{\Gamma(Z^0 \to \text{all})} \sim 2 \times 10^{-7},$$ 72.

and even decays into the mythical Higgs boson are expected to have relatively small branching ratios:

$$BR(Z^0 \to H^0 + \ell^+ \ell^- \text{ or } \mu^+ \mu^-) \lesssim 10^{-4} \qquad (\text{Bjorken 1976}) \qquad 73.$$

$$BR(Z^0 \to H^0 + \gamma) \lesssim 10^{-6} \qquad (\text{Cahn et al 1979}). \qquad 74.$$

The prospects for counting the number of neutrinos, therefore, seem quite good in the context of the standard model.

What about deviations from the standard model? Alternative theories based on groups larger than the minimal $SU(2) \times U(1)$ of Section 1 generally have more neutral-current interactions and hence more than one neutral vector boson. In quite a large class of theories, the lightest of these is actually lighter than the single Z^0 of the standard model (Georgi & Weinberg 1978). Experiments at $e^+ e^-$ storage rings in particular are gradually increasing the lower bound on the mass of such a light Z^0, and decreasing the phase space available to such models (Branson 1981). An alternative variation on the standard model predicts that the Z^0 should have a significantly higher mass (Abbott & Farhi 1981). In this scenario the $SU(2)$ group of the weak interactions is supposed not to be spontaneously broken, and to be characterized by a strong coupling constant. The Z^0 and $W^\pm$ are then expected to be composite states with strong couplings analogous to the familiar ρ mesons of hadronic physics. Equation 15, relating the masses and couplings of the $W^\pm$ to the observed Fermi constant G_F, still holds as a first approximation. We then see that strongly interacting $W^\pm$ and Z^0 must be heavier than the conventional expectations (Equation 49 and 57). Equations 50 and 60 tell us that the $W^\pm$ and Z^0 must also be considerably wider than the standard-model expectations (Equation 55 and 65). This scenario is certainly very unorthodox, and has several unsolved technical problems (Abbott & Farhi 1981), but it serves to remind us that despite the strong circumstantial evidence for the standard model for the $W^\pm$ and Z^0, experiments may reveal to us something rather different.

3.   VECTOR BOSON PRODUCTION IN $e^+ e^-$ ANNIHILATION

Most of the next generation of electron-positron colliders (see Table 1) are intended to be "Z^0 factories." The determination of the Z^0 mass will

Table 1 Comparison of e^+e^- colliding-beam projects

Name	Location	Maximum center-of-mass energy (GeV)	Expected luminosity $(cm^{-2}s^{-1})$	Status
LEP	CERN, Switz.	140 (260)	10^{32}	approved
SLC	SLAC, USA	100	6×10^{31}	proposed
CESR II	Cornell, USA	100	10^{32}	proposed

provide a precision measurement of the neutral weak mixing angle θ_w and/or a check of the higher-order corrections discussed in Section 2, while the determination of its width counts otherwise undetectable weakly coupled light particles such as neutrinos. The hoped-for production rate of about a million Z^0s per fortnight will allow searches for rare decay modes like those involving the elusive scalar bosons. All of these measurements can provide indirect probes of a still higher energy sector of the theory, just as precision measurements in kaon physics contributed important constraints on the construction of the electroweak theory of Section 1. In addition, the high rate of hadron production on resonance will provide a valuable laboratory for studies of both their weak and strong couplings. Here we emphasize straightforward tests of the electroweak theory at collision energies near the Z^0 pole, which are described in Section 3.1.

Production of charged intermediate bosons will not become significant until we can reach energies above the threshold for $e^+e^- \rightarrow W^+W^-$. The properties of this interaction, which directly probes the structure of the Yang-Mills couplings (Equation 43), are described in Section 3.2.

Most of the results presented in this section are taken from unpublished LEP studies (Camilleri et al 1976, ECFA & CERN 1979, and associated LEP/ECFA reports). Additional unpublished material is contained in the Cornell Z^0 theory workshop (Peskin & Tye 1981).

3.1 Physics Around the Z^0 Pole

In Section 2 we gave in Equation 58 a general parametrization of the Z^0 couplings to fermions. The parameters m_Z, v_f, and a_f are in principle completely determined (up to an overall sign common to all the v's and a's) by measurements of the total cross section as a function of center-of-mass (c.m.) energy, the angular asymmetry of the final-state momentum axis with respect to the beam axis, and polarization-dependent effects in the process

$$e^+e^- \to f\bar{f}. \qquad\qquad 75.$$

Let us first consider the case where the final-state pair is not e^+e^-. Then Reaction 75 will be dominated by single-photon and Z^0 exchange in the direct channel. As is customary, we normalize cross sections with respect to the lowest-order one-photon exchange cross section:

$$\sigma_0 = \sigma(e^+e^- \to \gamma \to \mu^+\mu^-) = \frac{4\pi\alpha^2}{3s} = \frac{87}{s} \text{ GeV}^2 \text{ nb}, \qquad 76.$$

where s is the square of the total c.m. energy, and study the ratio

$$R_f = \sigma(e^+e^- \to f\bar{f})/\sigma_0. \qquad\qquad 77.$$

Then including both photon exchange and Z^0 exchange with the coupling given in Equation 58, we obtain[4]

$$R_f = q_f^2 - 2q_f v_e v_f \rho(s)\chi(s) + (v_e^2 + a_e^2)(v_f^2 + a_f^2)\rho^2(s)\chi(s) \qquad 78.$$

where to simplify notation we have introduced the kinematic factors

$$\rho(s) = \frac{G_F}{8\sqrt{2}\,\pi\alpha} \frac{m_Z^2 s}{s - m_Z^2} \qquad\qquad 79.$$

and

$$\chi(s) = \frac{(s - m_Z^2)^2}{(s - m_Z^2)^2 + \Gamma_Z^2 m_Z^2}. \qquad\qquad 80.$$

The factor $\chi(s)$ gives the Breit-Wigner smearing of the pole and can be set equal to unity in the narrow width approximation. The third term in Equation 78 is the pure Z^0 contribution: the position and shape of the peak reflect the Z^0 mass and width, while the height of the peak determines the product of Ze^+e^- and $Zf\bar{f}$ coupling strengths. The second term arises from Z^0 and photon exchange interference and is sensitive only to the vector couplings of the Z^0; its sign reflects the relative v_e/v_f sign. The relative contribution of vector and axial couplings can be read off from the value of R_f at the minimum, which in the narrow width approximation occurs at

$$\rho(s) = \frac{q_f v_e v_f}{(v_e^2 + a_e^2)(v_f^2 + a_f^2)} \qquad\qquad 81.$$

---

[4] Here and in Equations 82 and 84 one must sum over three colors if f is a quark, so the right-hand side is multiplied by a factor of three for each quark flavor.

and takes the value

$$R_f^{min} = q_f \left[q_f - \frac{v_e v_f}{(v_e^2 + a_e^2)(v_f^2 + a_f^2)} \right].$$ 82.

The ratio $R_\mu(q_f = -1)$ is shown in Figure 1 for $m_Z = 83$ GeV and for several values of v and a where μ-e universality

$$v_e = v_\mu \equiv v, \qquad a_e = a_\mu \equiv a$$ 83.

has been assumed. Note that universality alone predicts that the interference term has the same sign as $s - m_Z$. In the standard model (Equation 59) $v = 0$ for $\sin^2 \theta_w = 0.25$, so that the presently favored value of $\sin^2 \theta_w$ suggests that γ-Z interference effects will be difficult to discern. For hadronic final states, their identification in terms of a primary q$\bar{\text{q}}$ pair will be difficult at best, and realistic measurements of R_q will require an average over quark types of the interference effects, which again are rather small if the parameters of the standard model are used.

More striking effects are expected in angular asymmetries that are more sensitive to axial couplings. Indeed, measurements at PETRA and PEP have already found indications of an axial coupling for the muon that is compatible with the standard model (Branson 1981). Defining θ as the angle between the incident e$^-$ and the outgoing f, the differential cross section for Reaction 75 in the narrow width approximation is given by

$$\frac{d\sigma_f}{d\cos\theta} = \frac{\pi\alpha^2}{2s} \{q_f^2(1 + \cos^2\theta) - 2q_f\rho(s)[vv_f(1 + \cos^2\theta) + 2aa_f \cos\theta]$$

$$+ \rho^2(s)[(v^2 + a^2)(v_f^2 + a_f^2)(1 + \cos^2\theta) + 8vav_fa_f \cos\theta]\}.$$ 84.

The integrated asymmetry, defined by

$$A_f = \frac{\displaystyle\int_0^1 d\cos\theta \frac{d\sigma_f}{d\cos\theta} - \int_{-1}^0 d\cos\theta \frac{d\sigma_f}{d\cos\theta}}{\displaystyle\int_{-1}^1 d\cos\theta \frac{d\sigma_f}{d\cos\theta}},$$ 85.

is given by the expression

$$A_f = \frac{\frac{3}{2}aa_f\rho(s)[-q_f + 2vv_f\rho(s)]}{q_f^2 - 2q_f\rho(s)vv_f + \rho^2(s)(v^2 + a^2)(v_f^2 + a_f^2)},$$ 86.

which vanishes if the axial coupling of either the electron or final-state fermion vanishes, and is bounded by $|A_f| < 0.75$. A_μ is shown

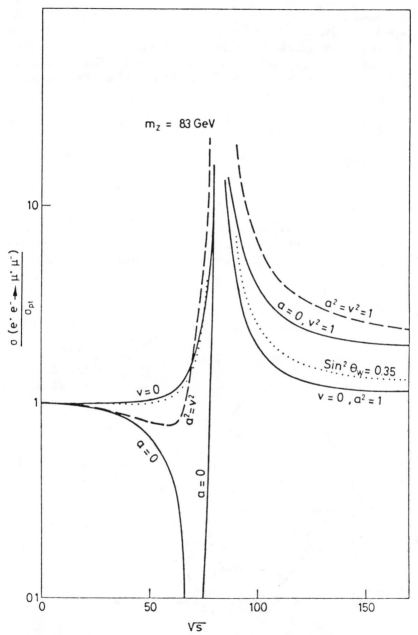

Figure 1 The ratio R_μ of $\sigma(e^+e^- \to \mu^+\mu^-)$ relative to σ_0 (Equation 76) plotted for different values of the vector and axial-vector couplings of e and μ (radiative corrections not included).

in Figure 2, assuming universality (Equation 83), for $m_Z = 83\,\text{GeV}$, $a^2 = 1$, and several values of v^2. At low energies $s \ll m_Z^2$, only the axial coupling is relevant and the effect is expected to be as large as 10% for energies $\sqrt{s} \simeq 40\,\text{GeV}$ characteristic of PEP and PETRA. Near the Z^0 pole, the shape of the asymmetry is sensitive to v^2/a^2. The asymmetry reaches its minimum value

$$A_\mu^{\min} = -\frac{3}{4}\frac{a^2}{2v^2 + a^2} \qquad\qquad 87.$$

for $\rho = -1/(3v^2 + a^2)$, and its maximum value

$$A_\mu^{\max} = \frac{3}{4} \qquad\qquad 88.$$

for $\rho = 1/(a^2 - v^2)$. As seen from Figure 2, the effects are very pronounced and should yield precision measurements of a^2 and v^2. For hadronic final states, asymmetry measurements require some method of

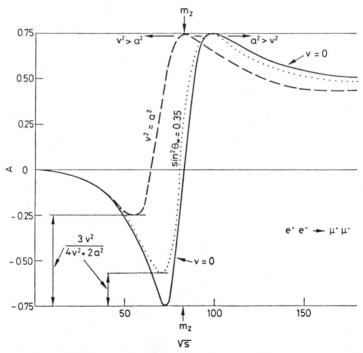

Figure 2   The forward-backward asymmetry A_μ (Equation 86) for $e^+e^- \rightarrow \mu^+\mu^-$, plotted for axial couplings $a_e = a_\mu = 1$, and different values of the e and μ vector coupling.

distinguishing jets of hadrons associated with a primary quark from those of an antiquark. One method might be to measure the total charge of a jet. Then the quantity in Equation 85, with θ defined as the angle óf a positively charged jet with respect to the incident e^- direction, would, to the extent that the jet charge "remembers" its parent quark charge, measure the weighted asymmetry

$$A_{Q_{jet}>0} = \frac{4}{5}\sum A_{q_f=2/3} - \frac{1}{5}\sum A_{q_f=-1/3}.$$ 89.

Another method is to assume that the fastest particle in a jet contains the primary quark; this again entails a weighted average over quark types. It seems plausible that at least consistency checks can be made with regard to the axial and vector couplings predicted by the standard model. When the final-state fermion pair is e^+e^-, Reaction 75 receives contributions from γ and Z^0 exchange in the crossed channel as well. The corresponding cross section and asymmetry formulae are considerably more complicated (see the above cited references) and we do not reproduce them here.

With the assumption of μ-e universality (Equation 83), the absolute magnitude and energy dependence of the cross section ratio R_μ and the asymmetry A_μ depend only on the three parameters v^2, a^2, and m_Z, and their measurement provides a check of universality. However, no information on the sign of v/a can be obtained from these measurements; this requires some polarization-dependent information, which can be obtained either by measuring the final-state lepton polarization (perhaps most realistically, τ polarization) or by comparing cross sections from left- and right-handed polarized incident beams.

In general, for unpolarized incident beams, the helicity of the final-state fermion (averaged over production angle) is given by (again in the narrow width approximation):

$$H^f(s) = -H^{\bar{f}}(s) = \frac{2\rho(s)a_f[q_f v_e - \rho(s)v_f(a_e^2 + v_e^2)]}{q_f^2 - 2q_f\rho(s)v_e v_f + \rho^2(s)(a_e^2 + v_e^2)(a_f^2 + v_f^2)}.$$ 90.

On resonance this beomes

$$H^f(s) = -\frac{2a_f v_f}{(a_f^2 + v_f^2)},$$ 91.

providing a direct measurement of the sign as well as the magnitude of a/v for the final-state fermion. A similar determination of a_e/v_e would be provided by measuring the dependence of the cross-section rate on resonance as a function of the longitudinal polarization of the incident beam. Complete expressions for the angular dependence of the final-state helicity and of the cross section for polarized beams are given in the above-cited references. Within the gauge theory context, the case for demanding polarized beams in the energy range around the Z^0 pole is in fact not very strong. The assumption of lepton universality allows the determination of the common parameter v/a from polarization measurements on a final-state muon or τ. In this case the polarization is maximum in the forward direction, where for example, on resonance, the polarization attains the value

$$H^{\ell^-}(m_Z^2, \cos\theta = 1) = -\frac{4av(a^2 + v^2)}{(a^2 + v^2)^2 + 4av} \simeq 0.32, \qquad 92.$$

even for $v \simeq -0.12$ corresponding to $\sin^2\theta_w = 0.22$ in the standard model, so measurements of $H^\ell(s, \cos\theta)$ can provide a dramatic discriminator. Various suggestions have been made (P-violating momentum correlations and polarization of leading particles of non-zero spin) for probing quark polarization, but they do not appear very convincing as true quark polarimeters. Such measurements, however, are not actually necessary if the general form of the couplings (Equation 58) is accepted, since the possibly more feasible measurements of quark cross-section ratios R_q and asymmetries A_q allow measurements of the signs as well as magnitudes of the quantities $a_q a_e$ and $v_q v_e$, while universality (to be tested in A_{ℓ^-} and R_{ℓ^-}) together with the measurement of H^{ℓ^-} provides the sign of a_e/v_e.

More interesting phenomena may occur if the standard model is incorrect. Figure 3, taken from de Groot et al (1979), shows the shape of the cross section in a model where a second Z^0 is introduced whose couplings to fermions enter only through mixing effects with the standard Z^0 — this leaves the low-energy phenomenology of the standard model intact. In such a model, similar exotic patterns would occur in the energy dependence of asymmetries and polarizations.

3.2 Production of Charged Intermediate Bosons

The obvious way to measure the Yang-Mills couplings in e^+e^- collisions is through the process

$$e^+e^- \to W^+W^- \qquad 93.$$

arising from direct channel Z^0 or photon exchange as in Reaction 75. As this requires a center-of-mass energy above the W^+W^- threshold, it is relevant to consider means of probing these couplings at lower energies.

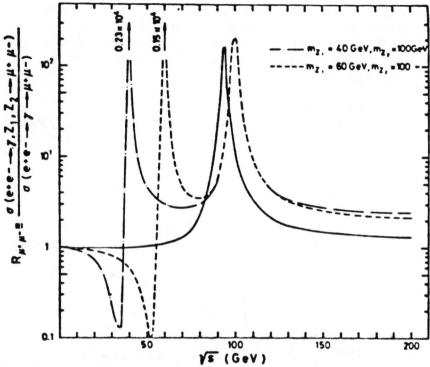

Figure 3 The cross-section ratio R_μ as a function of c.m. energy in a model with two neutral vector bosons and in the standard model $m_Z = 94\,\text{GeV}$ and $\sin^2\theta_w = 0.20$ (*solid line*) (deGroot et al 1979).

Their contribution to the cross section for the processes

$$e^+e^- \rightarrow e^\pm + W^\pm + \nu \qquad\qquad 94.$$

has been calculated and found to be small for energies below the 2W threshold: For example

$$\sigma(e^+e^- \rightarrow We\gamma) \sim 10^{-7}\,\text{nb} \simeq 2 \times 10^{-5}\,\sigma_0 \qquad\qquad 95.$$

in the standard model at $\sqrt{s} \simeq 150\,\text{GeV}$, and the cross section falls rapidly at lower energies. The trilinear boson coupling contribution to

$$e^+e^- \rightarrow \gamma + \nu_e \bar{\nu}_e \qquad\qquad 96.$$

has not been calculated, but is expected to be similar to that for Reaction 94 in order of magnitude. A possible probe of the ZWW vertex is through the decay

$$Z \rightarrow W + f\bar{f}' \qquad\qquad 97.$$

where the fermion-antifermion pair ($\bar{f}f' = e\bar{\nu}_e$, $\bar{u}d$, ...) arises from internal conversion of a virtual W produced at the ZWW vertex. The total branching ratio for processes like Reaction 97 is only about 10^{-7}, giving a cross section (Alles et al 1977, Albert et al 1980)

$$\sigma(e^+e^- \rightarrow W + \bar{f}f') \sim 10^{-8}\,\text{nb} \sim 10^{-6}\,\sigma_0 \qquad\qquad 98.$$

at $\sqrt{s} = m_Z$ where it is maximal.

We are therefore forced to consider Reaction 93 as the only feasible probe of the Yang-Mills couplings. The total cross section and angular

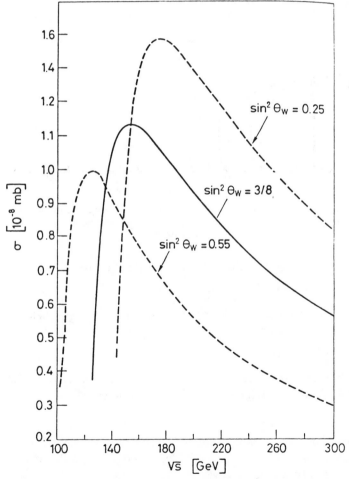

Figure 4　Cross section for $e^+e^- \rightarrow W^+W^-$ for different values of the weak neutral-current angle in the standard model (Alles et al 1977).

distributions for this process were calculated by Alles et al (1977) (see also Sushkov et al 1975) using the standard model. Detailed formulae are given in their paper. Figure 4 shows the cross section as a function of energy for several values of $\sin^2 \theta_w$; with decreasing values of $\sin^2 \theta_w$, the W^+W^- threshold increases (e.g. $\sqrt{s}_{th} = 170\,\text{GeV}$ for $\sin^2 \theta = 0.20$) but the cross section peaks at a higher value. The angular distributions have also been given and show a strong asymmetry in $\cos \theta$ (θ is defined as the W^- angle with respect to the e^- direction), which becomes increasingly pronounced with increasing c.m. energy. This effect is due to crossed-channel ν_e exchange, which is "uninteresting" compared to direct-channel γ and Z exchange contributions because it involves only couplings that are already known from low-energy data. Unfortunately, ν_e exchange gives the dominant contribution for energies not far above threshold, as can be seen from Figure 5 (Alles et al 1977), where different contributions to the cross section for Reaction 93 are plotted separately. However, Figure 5 also illustrates the sensitivity of the total cross section to the delicate cancellation required by a gauge theory:

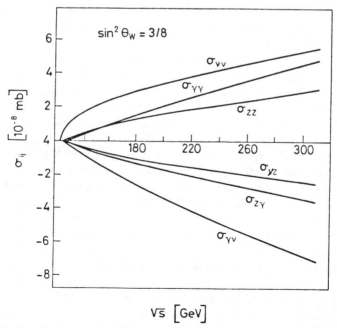

Figure 5 Contributions σ_{aa} from pure a exchange and σ_{ab} from a,b exchange interference to the cross section for $e^+e^- \rightarrow W^+W^-$ calculated in the standard model using $\sin^2 \theta_w = 3/8$ (Alles et al 1977).

each of the interference terms gives a negative contribution comparable in magnitude to the square of each single-exchange contribution.

In view of the importance of probing the Yang-Mills couplings, we must consider ways in which the different contributions may be separated. A useful tool in this respect is the use of longitudinally polarized beams. Since the ν exchange contribution involves only left- (right-) handed electrons (positrons), the γ and Z^0 exchange contributions can be selected by colliding right-handed electrons with left-handed positrons, for example, provided there is no contribution from right-handed ν exchange. In addition, the polarization of the final-state W's, which can be analyzed using lepton angular distributions with respect to the W production axis, is also an effective probe. In particular, the coupling of longitudinally polarized W's would grow rapidly with c.m. energy in the absence of the cancellations assured by the Yang-Mills coupling of Equation 43. These effects were analyzed in detail by Gaemers & Gounaris (1979), who give angular distributions for the different helicity amplitudes and show their dependence on the trilinear coupling parameters. Should polarized beams not be feasible, one must rely on angular distributions to separate the different contributions. At fixed energy, the angular dependence of the lowest-order Yang-Mills coupling is uniquely determined by the requirements of renormalizability, independently of the gauge group or symmetry breaking. If the only other contribution is the t-channel exchange of light $(m^2 \ll m_Z^2, s)$ neutrinos, there are only two independent amplitudes at fixed s, so that contributions to the cross section from s-channel gauge boson exchange, t-channel neutrino exchange, and their interference can be separated by weighted averaging of the data with suitably defined angular functions at each s. The s-dependence of the resulting functions will probe the structure of the gauge interactions. It turns out that the interference term, shown in Figure 6 for three models, is the most sensitive probe. (An analysis of this type could be complicated by an additional contribution from crossed-channel exchange of doubly charged leptons and/or the exchange of appreciably massive fermions.) The error bars in Figure 6 are based on 100 events per energy, which corresponds to about 250 running hours per energy for a beam luminosity of $0.5 \times 10^{32} \text{cm}^{-2}\text{s}^{-1}$ in an experiment that triggers on one leptonic decay mode of the W. Since the dominant decay modes are expected to be hadronic, rates would be dramatically improved if hadronic decays of the W exhibit a clear enough two-jet structure to permit detection above the annihilation background. Perrottet (1978) discussed this possibility in detail and calculated jet angular distributions. Once the threshold for $e^+e^- \to Z^0Z^0$ is also passed, comparison of this reaction (Brown et al

1979), which in the standard model proceeds only via crossed-channel lepton exchange, with Reaction 93 should allow a further probe of the Yang-Mills couplings contributing to the latter.

The higher-order process

$$e^+e^- \rightarrow W^+W^-\gamma \qquad\qquad\qquad 99.$$

receives contributions from quadrilinear vector boson couplings. Their detection may be feasible if c.m. energies well above $200\,\text{GeV}$ can be attained. Contributions from electron bremsstrahlung can be suppressed by excluding small-angle photons, and contributions involving ν exchange with a trilinear boson vertex could be eliminated if longitudinally polarized beams were available at the relevant energy. However, the maximum value of the quadrilinear coupling contribution to the cross section for Reaction 99 occurs at an energy

$$\sqrt{s} = 4m_{\text{W}} - m_Z^2/4m_{\text{W}} = m_{\text{W}}\left(4 - \frac{1}{4\cos^2\theta_{\text{w}}}\right) \approx 300\,\text{GeV} \qquad 100.$$

in the standard model if $\sin^2\theta_{\text{w}} \simeq 0.22$, and is only

$$\sigma_{\max}(\text{quadrilinear}) \simeq 2\times 10^{-7}\,\text{nb} \simeq 2\times 10^{-4}\,\sigma_0. \qquad 101.$$

While experiments of the type envisioned in this subsection will be difficult, it is clear that energies sufficiently high to allow multiple vector

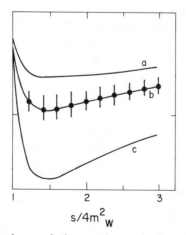

$$s/4m^2_{\text{W}}$$

Figure 6 Energy dependence of the angular projection of the contribution to $\sigma(e^+e^- \rightarrow W^+W^-)$ from interference between direct-channel vector exchange and crossed-channel neutrino exchange for (*a*) photon and ν exchange only, (*b*) the standard model with $\sin^2\theta_{\text{w}} = 0.25$, and (*c*) the standard model including an $(e_{\text{R}}^-, \nu_{\text{R}})$ weak doublet. Error bars represent a typical LEP experiment with a leptonic W-decay trigger.

boson production can provide a unique and invaluable probe of the underlying electroweak theory.

4.   LEPTON-HADRON COLLISIONS

We now turn to the manifestations of the weak vector bosons in very high energy lepton-hadron collisions. One can imagine experiments using any of three different types of leptons: neutrinos, muons, or electrons. High-energy neutrino-hadron collisions could be studied using cosmic-ray neutrino interactions in a very large underwater detector such as DUMAND (1982). Very high energy muon-hadron collisions could be studied using secondary muon beams obtained from a very high energy proton synchrotron such as the proposed VBA (Amaldi 1980). However, the best prospects for studying very high energy lepton-hadron interactions seem to be with electron-proton colliding-beam machines, of which several are now being proposed (see Table 2). Weak vector bosons may manifest themselves in these reactions either indirectly through their propagator effects on cross sections and on the interferences between neutral weak and electromagnetic interactions, or directly through production of the weak vector bosons in the final state. Since the cross sections for the production of the $W^{\pm}$ and Z^0 are not very high, much of the interest in their signatures in lepton-hadron collisions centers on their indirect propagator effects, to which we now turn.

Low-energy weak interaction amplitudes are proportional to the Fermi constant G_F, whose relation to the masses and couplings of the weak vector bosons we have already seen:

$$\frac{G_F}{\sqrt{2}} = \frac{g^2}{8m_W^2} = \frac{g^2 + g'^2}{8m_Z^2}.$$

102.

Table 2   Comparison of ep colliding-beam projects

Name	Location	Maximum center-of-mass energy (GeV)	Expected luminosity $(cm^{-2}s^{-1})$
TRISTAN	KEK, Japan	173	1.8×10^{31}
HERA	DESY, Germany	314	6×10^{31}
Proposal 659	FNAL, USA	200	4×10^{31}
CHEER	FNAL, USA	200	$(1.7 \text{ to } 2.7) \times 10^{31}$
SPS/LEP	CERN, Switz.	249	1.3×10^{31}
ISABELLE	Brookhaven, USA	126	2×10^{31}

At high energies the four-momentum q_μ transferred between the incoming and outgoing leptons may no longer be negligible compared with the vector boson mass. In this case the weak interaction amplitudes are proportional to

$$\frac{1}{m_{W^\pm}^2 + Q^2} \quad \text{or} \quad \frac{1}{m_{Z^0}^2 + Q^2}, \qquad \qquad 103.$$

where $Q^2 \equiv -q^2$ is the modulus of the space-like four-momentum transfer squared. This means that charged-current weak interaction cross sections will no longer rise like the square of the center-of-mass energy $s \equiv (p_\ell + p_h)^2$, as they do at low energies:

$$\sigma(\ell + h \to \ell' + X) \propto G_F^2 s, \qquad \qquad 104.$$

when $s \ll m_{W^\pm}^2$ if one assumes that strong interaction effects are negligible at large momentum transfers. One does not in fact expect these effects to be entirely absent, and the standard asymptotically free gauge theory of the strong interactions, namely QCD, leads one to expect logarithmic corrections to Expression 104 at high-momentum transfers Q^2. These are to be contrasted with the power-law Q^2 dependences to be expected from the vector boson propagators, Expression 103. Furthermore, these propagator effects are independent of parameters of the hadronic kinematics, whereas the QCD scaling violations are not universal functions of Q^2. For these reasons, it is in principle possible to disentangle vector boson propagator effects from conventional QCD effects.

Attempts have been made using present-day accelerator data to establish a lower limit on the mass of the $W^\pm$, or even to determine that it is finite. These experiments are very difficult since the values of Q^2 ($\leq 400\,\text{GeV}^2$) presently available are much lower than the expected $W^\pm$ boson mass squared, and so far they have not yielded any positive results. The best published limit (Barish 1978) on the W boson mass is

$$m_{W^\pm} > 36\ \text{GeV}, \qquad \qquad 105.$$

while a preliminary analysis (Wahl 1981) of CERN-Dortmand-Heidelberg-Saclay (CDHS) collaboration data suggests

$$m_{W^\pm} > 100\ \text{GeV} \pm \text{systematic effects}. \qquad \qquad 106.$$

This result should not yet be construed as conflicting with the expected mass of order 80 GeV, since so far the analysis only takes into account statistical errors in the data, but does not yet take into account systematic errors. One can expect that the lower bound (Expression 106) will be decreased when these are included. It is possible that $W^\pm$

propagator effects may be detected in neutrino-hadron scattering data to be obtained at the FNAL Tevatron in a few years time. However, the main hope for detecting these effects rests with ultra-high-energy cosmis-ray neutrinos and electron-proton colliding-ring machines.

Figure 7, taken from an analysis made for the DUMAND (1982) project, indicates how one may be able to detect $W^\pm$ propagator effects in ultra-high-energy (of order 10 TeV) cosmic-ray neutrino interactions.

The upcoming muons originate from neutrinos that have passed through the Earth. Their number is diminished by absorption, which is less significant if the neutrino cross section rises more slowly than linearly (Expression 104) at high energies, as would be the case if $m_W \approx 80$ GeV. Stronger absorption occurs if there are no W propagator effects (Expression 103), as would be true if $m_W \geqslant 80$ GeV. In principle, measurement of the reaction $e^\pm + p \rightarrow {}^{(\bar{\nu})} + X$ with a high-energy colliding-beam machine can provide better-determined kinematical conditions in a comparable range of center-of-mass energies. As an example, colliding 30 GeV electrons with 820 GeV protons as proposed with HERA (ECFA & DESY 1979) provides a center-of-mass energy $\sqrt{s} = 314$ GeV, which corresponds to a lepton beam of 52 TeV striking a fixed target. Figure 8 shows the number of events expected per day at HERA of the reaction $e_L^- + p \rightarrow \nu_e + X$, binned in different ranges of Q^2. The event rates for a conventional $W^\pm$ mass of about 80 GeV and

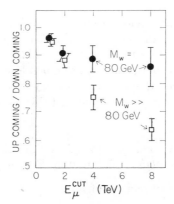

Figure 7 The ratio between up-coming and down-coming muons from ultra-high-energy cosmic-ray neutrino interactions is sensitive to the amount of absorption of neutrinos passing through the Earth. The ratio would be unity in the absence of absorption, while a linearly rising neutrino cross section (corresponding to $m_W \geqslant 80$ GeV) entails more absorption than the slower-rising cross section found if $m_W = 80$ GeV. The error bars indicate the experimental results that would be obtained from DUMAND (1982).

for an infinite $W^{\pm}$ mass are clearly very different, and it is believed that one could measure indirectly the $W^{\pm}$ mass to an accuracy of a few percent. It is also possible to look beyond the conventional $W^{\pm}$ to see if there are any more charged weak bosons $W^{\pm\prime}$. For example, if the simple $W^{\pm}$ propagator (Expression 103) were replaced by

$$\frac{1}{m^2_{W^{\pm}} + Q^2} \to \frac{1}{2}\left[\frac{1}{m^2_{W^{\pm}} + Q^2} + \left(\frac{m^2_{W^{\pm\prime}}}{m^2_{W^{\pm}}}\right)\left(\frac{1}{m^2_{W^{\pm\prime}} + Q^2}\right)\right],$$ 107.

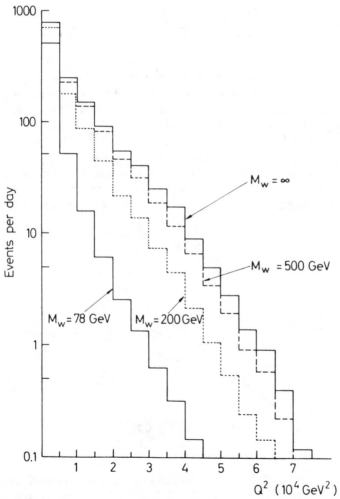

Figure 8 Rate of events per day expected at HERA for $e^- + p \to \nu + X$ in Q^2 bins of 5000 GeV².

so that one half of the low-Q^2 amplitude were due to $W^{\pm\prime}$ exchange, then experiments with a machine such as HERA could tell the difference between $m_{W^{\pm\prime}} = 500$ GeV and infinity (ECFA & DESY 1979).

One can also look for Z^0 propagator effects in high-energy neutral-current reactions. Here measurements are simplest in high-energy electron and muon-hadron scattering, where one can easily know the kinematics of both the incoming and outgoing leptons. The picture is, however, complicated by the dominance of photon exchanges at low Q^2 and by important interference effects:

$$\sigma(\gamma + Z^0) \propto |A|^2 = \left| \frac{\hat{A}_{\mathrm{em}}}{Q^2} + \frac{\hat{A}_{\mathrm{weak}}}{m_{Z^0}^2 + Q^2} \right|^2 . \qquad 108.$$

At low $Q^2 \ll m_Z^2$ the interference effects are of order $10^{-5} Q^2$ (GeV^2) and hence very small. They have been measured in two experiments: the classic SLAC experiment (Prescott et al 1978) demonstrating parity violation in deep inelastic ep scattering at $Q^2 = O(1)$ GeV^2, and more recently a CERN experiment (Bollini et al 1981) indicating a charge asymmetry in deep inelastic $\mu^{\pm} N$ scattering at $Q^2 = O(10^2)$ GeV. Both these experiments are at Q^2 much too small to offer a hope of seeing Z^0 propagator effects, which would cause a deviation from linearity with Q^2 in the interference effects. At high-energy electron-proton colliding-beam machines, the electromagnetic and weak contributions (Expres-

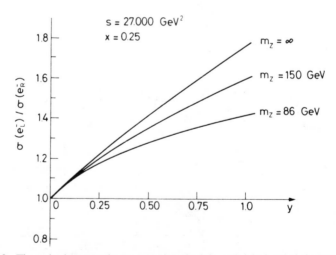

Figure 9 The ratios between the cross sections for left- and right-handed electron-proton scattering at a center-of-mass (energy)^2 = 27,000 GeV^2, for different values of the Z^0 mass: x and y are the standard kinematical variables (Barish 1978) for lepton-nucleon scattering.

sion 108) to the deep inelastic scattering cross sections are comparable; dramatic interference effects should be seen and the Z^0 propagator clearly identifiable. Figure 9 (Ellis et al 1978) shows the ratios between the cross sections for different electron polarization states, both with and without the finite-mass Z^0 propagator. As in the case of the $W^\pm$, measurements with high-energy ep colliding beams should be able to determine indirectly the mass of the Z^0 to within a few percent, and look for the indirect effects of further Z^0's with masses of a few hundred GeV.

Looking for indirect effects of the $W^\pm$ and Z^0 through their propagators is all very well, but it would be much more satisfying to see them produced directly. Unfortunately, the cross sections expected in high-energy ep collisions are discouragingly small, although the cleanliness of many of the final states relative to those encountered in hadronic collisions alleviates some of the detection problems. The most important Feynman diagrams for the production of $W^\pm$ and Z^0 are shown in Figure 10. The diagrams of classes (a) and (b) where the $W^\pm$ or Z^0 are

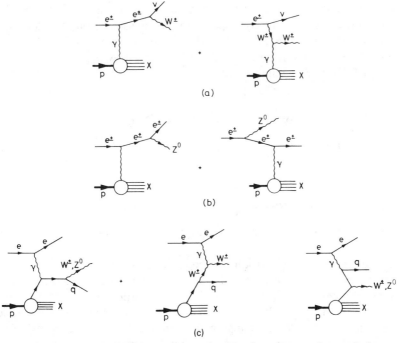

Figure 10 Feynman diagrams for direct production of weak vector bosons in lepton-hadron collisions: (a) for $W^\pm$ production from the leptonic vertex, (b) for Z^0 production from the leptonic vertex, (c) for $W^\pm$ or Z^0 production from the hadronic vertex.

emitted from the leptonic vertex are expected to be accompanied by a relatively simple hadronic system that will often be just a proton or a nucleon resonance. This enables one to plan searches for all decay modes of the $W^{\pm}$ and Z^0, and not just their relatively disfavored (Equations 53 and 63) leptonic decay modes. On the other hand, the "hadronic" diagrams of class (c) can be expected to yield quite a complicated hadronic system, among which only leptonic decays of the $W^{\pm}$ and Z^0 may be discernible.

The cross sections for leptoproduction of $W^{\pm}$ and Z^0 do not have simple analytic forms, and complete calculations of their production with ep colliding beams have not in fact been performed. The production cross sections have been estimated (Llewellyn Smith & Wiik 1977) by scaling up old calculations for producing light $W^{\pm}$ and Z^0 at lower-energy accelerators, or alternatively (Kamal et al 1981) using the Weizsäcker-Williams approximation to treat the photon exchanged in Figure 10. This latter method gives somewhat larger cross sections than the former method, but it is not clear whether the Weizsäcker-Williams approximation is appropriate for this calculation so we prefer to be conservative and quote the old rescaled calculations. The cross sections for the relatively clean events corresponding to emission from the leptonic vertices of Figure 10a, b are plotted in Figure 11 as a function of $s/m_{W^{\pm}}^2$ and $s/m_{Z^0}^2$. Taking the values of $s/m_{W^{\pm}}^2 = 15$ and $s/m_{Z^0}^2 = 12$ appropriate to HERA we estimate (Llewellyn Smith & Wiik 1977, Ellis et al 1978)

$$\sigma(e^- + p \rightarrow \nu + W^- + p)_{\text{leptonic}} \approx 2 \times 10^{-38} \text{ cm}^2$$
$$\sigma(e^- + p \rightarrow \nu + W^- + X \neq p)_{\text{leptonic}} \approx 4 \times 10^{-38} \text{ cm}^2$$

109.

while

$$\sigma(e^- + p \rightarrow e^- + Z^0 + p)_{\text{leptonic}} \approx 3 \times 10^{-37} \text{ cm}^2$$
$$\sigma(e^- + p \rightarrow e^- + Z^0 + X \neq p)_{\text{leptonic}} \approx 2 \times 10^{-37} \text{ cm}^2.$$

110.

It would probably be possible to detect the Z^0 with the cross sections in Equations 110, but detecting the $W^{\pm}$ with the cross sections in Equations 109 seems a rather more doubtful proposition. However, perhaps it is worth noting that the $W^{\pm}$ cross sections are sensitive to the magnetic moment κ of the $W^{\pm}$. Changing the canonical gauge theory value of $\kappa = +1$, derived from Equation 43, to $\kappa = 0$ decreases the expected cross section by 10%, while changing to -1 increases the cross section by 30%. In principle, observations of the reaction $e^+e^- \rightarrow W^+W^-$ have a

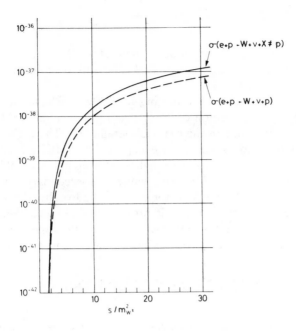

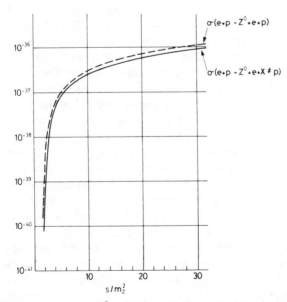

Figure 11 The cross section for Z^0 and $W^\pm$ production at the leptonic vertex estimated by scaling up low-energy results. These events are generally relatively clean, with either a proton or a simple hadronic system accompanying the vector boson. Events where the production takes place at the hadronic vertex are likely to contain a much more complicated hadronic system.

greater sensitivity to the magnetic moment of the $W^{\pm}$, but it is only with a second higher-energy phase of LEP that this reaction will become accessible.

We close this section by returning to Table 2, which lists different high-energy electron-proton colliding-beam projects that are known to us, together with some of their crucial parameters. The highest center-of-mass energy and luminosity are expected for HERA. This is not surprising since HERA is a device dedicated to ep physics, unlike all the other projects except TRISTAN, which are secondary adjuncts to accelerators primarily intended for e^+e^- or hadron-hadron collisions. On the other hand, these parasitic projects are all much cheaper then HERA. At present neither HERA nor any of the other projects has been approved, and their prospects are somewhat uncertain.

5.   HADRON-HADRON COLLISIONS

Hadronic collisions are not as clean as e^+e^- or even ep collisions for making a thorough study of the predicted weak bosons. They will, however, provide higher energies at which these heavy objects should be produced more easily. Unlike e^+e^- collisions, the initial-state interaction is not simple, since hadrons are made of quarks and gluons that interact strongly, presumably according to today's orthodoxy of QCD. Therefore all the calculations of weak boson production will call upon a model of the hadrons in terms of its constituents, which play the role of the incoming e^+e^- to create the weak bosons. Once these are produced, they will decay in the usual way as dictated by the electroweak theory. Unfortunately, these decays are just a very tiny part of the possible final states in hadronic collisions. In the haystack of all produced states we have to look for the characteristic "needle" signatures that are evidence for weak boson production. In recent years a large amount of literature has been devoted to this subject in connection with the various high-energy pp and p$\bar{\text{p}}$ collider projects listed in Table 3: at Brookhaven National Lab, a proton-proton collider, ISABELLE, with center-of-mass energy $\sqrt{s} = 400-800\,\text{GeV}$; at FERMILAB, a proton-antiproton machine using the Tevatron with $\sqrt{s} = 2000\,\text{GeV}$. In Europe, the SPS p$\bar{\text{p}}$ collider ($\sqrt{s} = 540\,\text{GeV}$) is now operating and the first collisions have been observed, which may make the discovery of the weak bosons, if they exist, very imminent. In this section we give an overview of the physics of the weak bosons in pp and p$\bar{\text{p}}$ collisions without going into more technical details, which can be found in the literature by the interested reader.

Table 3 Comparison of hadron-hadron colliding-beam projects

Name	Location	Maximum center-of-mass energy (GeV)	Expected luminosity $(\text{cm}^{-2}\text{s}^{-1})$	Status
$\bar{p}p$ collider	CERN, Switzerland	540	10^{30}	In operation
ISABELLE (pp)	Brookhaven, USA	800	10^{32}	Construction/R & D
Tevatron I($\bar{p}p$)	FNAL, USA	2000	10^{30}	Approved, construction begun

5.1 Total Cross Sections for the Production of Weak Bosons

The starting point for weak boson production is the Drell-Yan model (Drell & Yan 1971, Yan 1976), which was originally intended to describe lepton-pair production in hadronic collisions. A quark from one colliding hadron and an antiquark from the other fuse into a virtual photon, which in turn decays into a lepton pair (see Figure 12). In the naive version of the model (Feynman 1972) quarks and antiquarks are supposed to be essentially free objects inside the hadrons. However, one knows that strong-interaction corrections estimated using QCD are to be included in the calculations. Nevertheless, we use the naive model to present rough estimates and comment on the influence of QCD corrections on this Born approximation when necessary. For production of weak bosons we can follow the same physical picture as in lepton-pair production via a virtual photon; the only change is that we consider different elementary fusion subprocesses. At the constituent level the relevant subreactions are the following (Peierls et al 1977, Quigg 1977):

$$u\bar{u}, d\bar{d}, s\bar{s} \ldots \rightarrow Z_0; \quad u\bar{d}, u\bar{s} \ldots \rightarrow W^+; \quad \bar{u}d, \bar{u}s \ldots \rightarrow W^- \qquad 111.$$

if we neglect the small contributions of heavier quarks in the proton and the antiproton, which are expected to be relatively rare. The hadronic cross section for production of a weak vector boson $\mathcal{W}(W^+, W^-, Z^0)$ is

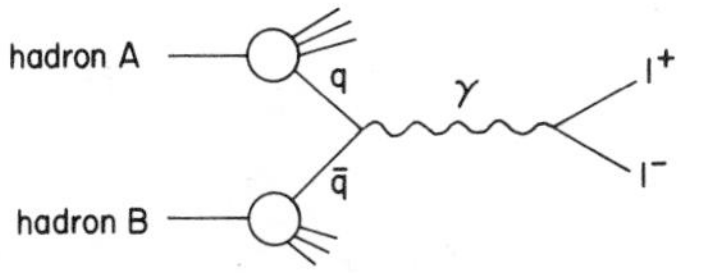

Figure 12 The original Drell-Yan mechanism.

obtained as an incoherent sum of all possible fermion subprocesses weighted by the respective probability distributions of the incoming constituents (see Figure 13 for the kinematics):

$$\frac{d\sigma}{ds}(A + B \rightarrow W + X)$$

$$= \sum_{a,b} \int_0^1 dx_a \int_0^1 dx_b\, \delta(x_a x_b s - \hat{s}) f_{a/A}(x_a) f_{b/B}(x_b) \hat{\sigma}(a + b \rightarrow W) \quad 112.$$

where careted variables refer to the subprocess and $f_{a/A}(x_a)$ is the probability of finding the constituent a inside the hadron A carrying the fraction x_a of the parent hadron's longitudinal momentum. From Equation 112 we can study any kind of distribution for a definite final state, provided we replace $\hat{\sigma}(a + b \rightarrow W)$ by the relevant subprocess distribution. For instance, the total cross section for producing a $W^\pm$ is easily calculated, knowing that

$$\hat{\sigma}(a + b \rightarrow W^+) = \sqrt{2}\, G_F \pi \tau s\, \delta(\hat{s} - m_W^2), \quad\quad\quad 113.$$

which when substituted into Equation 112 yields

$$\sigma(A + B \rightarrow W^+ + X) = \sqrt{2}\, \pi G_F \tau \int_\tau^1 \frac{dx}{x}\, \mathscr{L}^+(x, \tau/x) \quad\quad 114.$$

where

$$\mathscr{L}^+(x, \tau/x) = \frac{1}{3} \{[u_A(x)\bar{d}_B(\tau/x) + \bar{d}_A(x)u_B(\tau/x)]\cos^2\theta_c$$
$$+ [u_A(x)\bar{s}_B(\tau/x) + \bar{s}_A(x)u_B(\tau/x)]\sin^2\theta_c\}. \quad\quad 115.$$

In Equation 115 we have denoted the distribution function of a given quark by its own symbol. We remark that the total cross section only depends on the dimensionless variable $\tau = m_W^2/s$. For W^- we interchange quark and antiquark densities, whereas for Z^0 production, the corresponding function $\mathscr{L}^0(x, \tau/x)$ is slightly more complicated because

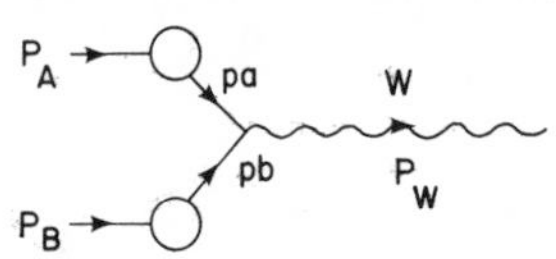

$$P_A = \left(\frac{\sqrt{s}}{2}, \vec{P}\right) \ i \ P_B = \left(\frac{\sqrt{s}}{2}, -\vec{P}\right)$$

$$Pa = x_a P_A \ i \ P_b = x_b P_B \ i \ P_W = \left(\frac{x_a + x_b}{2}\right)\sqrt{s} \ i \ P_\perp \sim 0 \ i \ P_\parallel = \frac{\sqrt{s}}{2}(x_a - x_b)$$

$$\hat{s} = (p_a + p_b)^2 = x_a x_b\, s = \tau s$$

Figure 13   Kinematics of weak boson (W) production.

of the weak electromagnetic mixing:

$$\mathscr{L}^0(x, \tau/x) = \frac{1}{3}\left\{[u_A(x)\bar{u}_B(\tau/x) + \bar{u}_A(x)u_B(\tau/x)]\left[\frac{1}{4} - \frac{2}{3}\sin^2\theta_w + \frac{8}{9}\sin^4\theta_w\right]\right.$$

$$+ [d_A(x)\bar{d}_B(\tau/x) + \bar{d}_A(x)d_B(\tau/x) + s_A(x)\bar{s}_B(\tau/x) + \bar{s}_A(x)s_B(\tau/x)]$$

$$\left.\times\left[\frac{1}{4} - \frac{\sin^2\theta_w}{3} + \frac{2}{9}\sin^4\theta_w\right]\right\}.   \qquad 116.$$

The factors of $\frac{1}{3}$ in Equations 115 and 116 account for the color degree of freedom. Charge invariance tells as that quark (antiquark) densities in a proton are the same as antiquark (quark) densities in an antiproton. In this simple picture, p$\bar{\text{p}}$ collisions generally offer more constituent luminosity than pp collisions because the three "valence" quarks (antiquarks) carrying the p($\bar{\text{p}}$) quantum numbers also carry most of its momentum, but this enhancement does not apply at very low values of τ where the "sea" of q$\bar{\text{q}}$ pairs (and gluons) is important. For the planned colliders, the larger luminosity expected in proton-proton collisions may overcome this drawback and provide a larger absolute event rate than the proton-antiproton colliders.

A convenient variable for parametrizing the final-state distribution is the rapidity, which measures the longitudinal momentum of the produced weak boson

$$x_a - x_b = 2\sqrt{\tau}\sinh y, \quad \text{i.e.} \quad y = \frac{1}{2}\ln\frac{E_\omega + p_\parallel}{E_\omega - p_\parallel}.   \qquad 117.$$

Distributions in this variable reflect the longitudinal distribution of the constituents, and for instance in pp collisions one expects a maximum away from $y = 0$, since the valence quarks carry more momentum than the sea antiquarks and the $\mathscr{W}$ moves in the direction of the valence quarks. So far we have neglected strong interaction effects, but in fact quarks can radiate gluons, and hadrons contain gluons, and therefore there are additional diagrams to be considered (Reya 1981) as in Figure 14. The main effect of gluon radiation is to soften the quark distribution; that is, the low-x region is more populated and the large-x

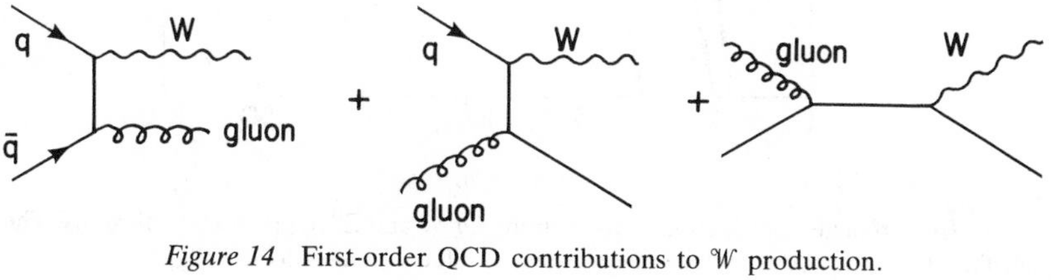

Figure 14   First-order QCD contributions to $\mathscr{W}$ production.

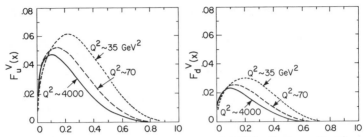

Figure 15 Evolution of the valence quark (u, d) distributions with Q^2.

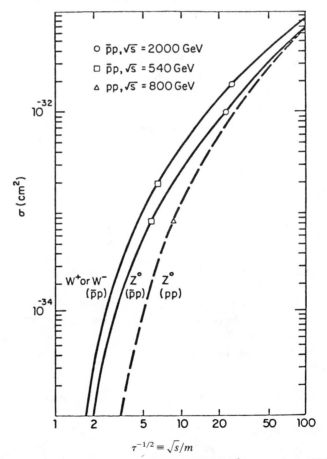

Figure 16 Predicted production cross sections for W and Z^0 in pp and $\bar{p}p$ collisions. The calculated cross-sections include scale-breaking effects (Paige 1979).

region is depressed. This evolution of the distribution is energy-momentum dependent, so that constituent distributions have a logarithmic dependence on the energy scale associated with the process, as shown in Figure 15:

$$f_{a/A}(x) \to f_{a/A}(x, Q^2).$$

To leading order in $1/\ln Q^2$, the momentum scale Q^2 is specified only up to a scale factor of order 1, but it seems physically reasonable to take $Q^2 = \hat{s}$, the energy of the constituent reaction. Another effect of gluon emission is to give the $\mathcal{W}$ a transverse kick much larger than that expected from the Fermi motion of the constituents inside the hadrons. These strong interaction refinements have been included in the calculation of the total cross sections shown in Figure 16 (Paige 1979).

Logarithmic corrections are not important for $\sqrt{s} \sim 500\text{--}1000 \, \text{GeV}$ since there the average fraction of energy carried by the constituents is $\langle x \rangle \sim \sqrt{\tau} \sim 0.1$ or 0.2, a region in which the constituent distribution functions are not greatly affected. However, for larger values of $\sqrt{s}$, as at the FNAL Tevatron, one expects a substantial increase of the cross sections.

As a final remark, let us point out that virtual gluon exchange induces a multiplicative correction factor of order 2 to the Drell-Yan cross section (Altarelli et al 1979). An enhancement of this magnitude has in fact been seen in hadronic lepton-pair production. We have not taken this factor into account in the estimation; its effect would be to increase substantially the present estimates but not to modify the shape of the calculated distributions (Humpert & van Neerven 1980).

5.2 Leptonic Final States

Among all possible decays of the weak bosons, their leptonic modes are the easiest to select for a final-state analysis. As indicated in Section 2, the branching ratios into charged leptons in a three-generation model are

$$B(Z^0 \to e^+ e^-) \sim 3\% \qquad B(W^+ \to \ell^{\pm}(\bar{\nu})_{\ell}) \sim 8\%. \qquad 118.$$
$$\text{or } \mu^+ \mu^-$$

For an integrated luminosity of 10^{36}cm^{-2} this gives for $Z^0 \to \ell^+ \ell^-$ ($W^{\pm} \to \ell^{\pm}$) at the $p\bar{p}$ collider ($\sqrt{s} = 540 \, \text{GeV}$) about 23 events (60 events), an observable signal even in this rather disfavored case. For lepton-pair production by the Z^0, the competing mechanisms are the usual Drell-Yan one via a virtual photon and the leptonic decays of heavy quarks like c and b(t?). Fortunately these backgrounds are much below the nice expected peak in the lepton-pair mass spectrum at the Z^0 mass (Figure 17) (Pakvasa et al 1979). The situation is less favorable for the leptonic modes of the charged weak bosons $W^\pm$ because of the emitted (anti)neutrino that escapes detection. However, if we consider the single charged spectrum, kinematics conspires in our favor. Denoting the momentum of the charged lepton by **l**, we look at the following distribution

$$l^0 \frac{d^3\sigma}{dl} = \sum_{a,b} \int dx_a \, dx_b f_{a/A}(x_a) f_{b/B}(x_b) l^0 \frac{d^3\hat{\sigma}}{dl} (a + b \to W^\pm \to \ell^\pm \nu)$$

$$= \sum_{a,b} \int \frac{d\hat{s}}{\hat{s}} \frac{f_{a/A}(x_a) f_{b/B}(x_b)}{\sqrt{1 - 4l_T^2/\hat{s}}} \frac{l_T^2}{\hat{s}} |M|^2, \qquad\qquad 119.$$

where $|M|^2$ is the squared matrix element of the subprocess containing the W propagator, and l_T is the momentum component perpendicular to the incident beam. We expect that the largest contribution will come from the region $\hat{s} \approx m_W^2$. The denominator in Equation 119 should lead to a sharp peaking of the cross section at $l_T \sim \frac{1}{2} m_W$, which is known as the Jacobian peak and is a very distinctive feature of a heavy object decaying into two light fermions (Figure 18). In the limit of zero width for the $W^\pm$, the l_T spectrum would be cut off at $\frac{1}{2} m_W$, but there are some smearing effects. First the finite width of the $W^\pm$, and secondly the intrinsic transverse momentum of the constituents and the QCD corrections, which can give a rather significant transverse kick to the weak boson (Pakvasa et al 1979, Aurenche & Lindfors 1981). This latter contribution enlarges the available phase space for the lepton and we

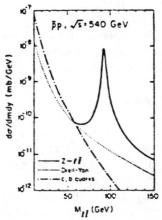

Figure 17   Lepton-pair mass spectrum in p$\bar{\text{p}}$ collisions at $\sqrt{s} = 540 \,\text{GeV}$.

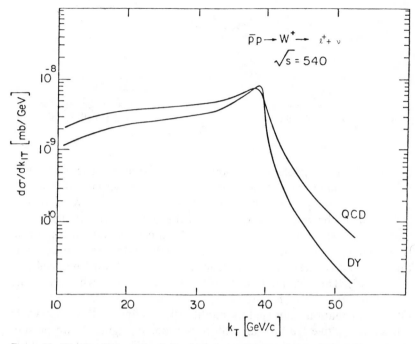

Figure 18   The Jacobian peak of the single-lepton spectrum from W^+ production and decay. The calculation is indicated by "DY," while "QCD" contains the QCD corrections (Aurenche & Lindfors 1981).

expect a smearing of the peak with events at $l_T > \frac{1}{2}m_W$. Fortunately, the Jacobian peak should still emerge from the above-mentioned backgrounds, but it will be difficult to infer the W mass from the position of the maximum.

Having observed peaks in some distributions, one would like to pin-point the specific characteristics of the weak bosons, which include the vector and axial pieces of their couplings introduced in Section 1. In $p\bar{p}$ collisions a clear signal of the axial-vector interference is the forward-backward asymmetry of the lepton spectrum, the forward hemisphere being defined by the proton beam direction. More precisely, the asymmetry can be defined as:

$$A_{FB}(s, \cos\theta) = \frac{\dfrac{d\sigma}{d\cos\theta}(\cos\theta) - \dfrac{d\sigma}{d\cos\theta}(-\cos\theta)}{\dfrac{d\sigma}{d\cos\theta}(\cos\theta) + \dfrac{d\sigma}{d\cos\theta}(-\cos\theta)}.$$ 120.

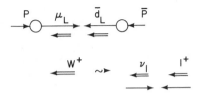

Figure 19 Schematic description of the helicity configuration in the reaction $p\bar{p} \to W^+ \to \ell^+ \nu_\ell$. Single (double) arrows indicates the momentum (helicity) direction.

For the W, one expects this asymmetry to be large because of the pure (V − A) coupling and the qualitative effect can be visualized easily on the basis of helicity arguments. For instance we know that W^+ is mainly produced by the collisions of u_L (from the proton) with $\bar{d}_L$ (from the antiproton), which fixes the helicity of W^+. Since the outgoing neutrino is left-handed one clearly sees that the ℓ^+ will tend to be emitted along the direction of the antiproton beam (into the backward hemisphere) (Figure 19). The asymmetry of ℓ^+ is therefore negative, whereas for the ℓ^- coming from W^- the situation is just the reverse. Sea quarks and QCD contributions can slightly modify the results for this observable but its qualitative features are still preserved. In Figure 20 we present the forward-backward asymmetry (Perrottet 1978, Finjord et al 1981) for W^- as a function of cos θ for various value of $\tau = m_W^2/s$.

For the Z^0 this asymmetry will be more difficult to measure because the vector coupling to the lepton is proportional to $(1 - 4\sin^2 \theta_w)$ and is rather suppressed for today's value of $\sin^2 \theta_w \sim 0.215$. The predicted value of A_{FB} is about 7% and detecting such a low value calls for rather large statistics that will be difficult to achieve in the near future.

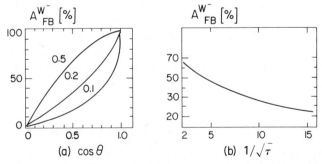

Figure 20 The front-back W^- asymmetry (*a*) versus cos θ for various values of $\sqrt{\tau}$, (*b*) integrated version as a function of $1/\sqrt{\tau}$ (Perrottet 1978, Finjord et al 1981).

5.3   *Hadronic Final States*

As described in Section 2, the weak bosons will decay most of the time into hadronic modes. We therefore expect pairs of hadronic jets at large transverse momentum with a weak origin, characterized by a coupling strength $\alpha_w \simeq 3 \times 10^{-2}$, to be distinguished from the formidable QCD background arising for quark and gluon scattering reactions with a strong coupling constant $\alpha_s(m_W^2) \simeq 0.1$ to 0.2. We expect that the strong interaction two-jet cross section will dominate the weak one (Figure 21)

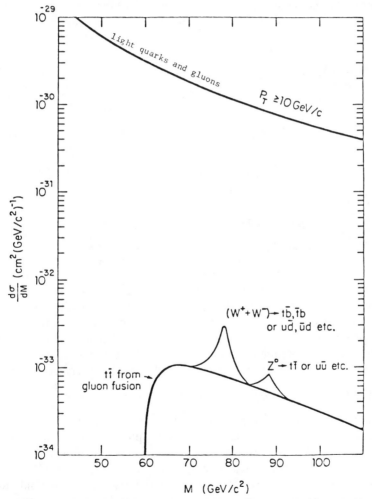

Figure 21   Invariant mass spectrum of heavy quark pairs from W, Z, and gluon fusion. The mass of the top quark is chosen as 30 GeV/c.

by a large factor. The situation looks desperate unless we can devise a clever criterion to disentangle the two types of cross sections. An interesting possibility is to look for heavy quark jets, since the weak bosons are heavy enough to decay in essentially the same way into light or heavy quarks (provided the unseen top quark is not too heavy). This would reduce the QCD background to the extent that we are able to distinguish a heavy quark jet from a light quark jet. In particular, for the W^- decaying into $b\bar{t}$ the only competing QCD mechanism comes for $t\bar{t}$ and $b\bar{b}$ pairs. If we further require the observation of two energetic leptons of the same sign, the W^- signal can emerge from the background (Abud et al 1978, 1979). The decay chains are the following:

$$W^- \to b\bar{t} \begin{array}{l} \longrightarrow \bar{b}\ell^- \bar{\nu}_\ell \\ \\ \longrightarrow c\ell^- \bar{\nu}_\ell \end{array} \qquad\qquad 121.$$

$$\text{QCD} \begin{array}{l} \nearrow t\bar{t} \begin{array}{l} \longrightarrow b\ell^- \bar{\nu}_\ell \\ \\ \longrightarrow b \to c\ell^- \bar{\nu}_\ell \end{array} \\ \\ \searrow b\bar{b} \to \bar{c} \to \bar{s}\ell^- \bar{\nu}_\ell \\ \qquad\quad \longrightarrow c\ell^- \bar{\nu}_\ell \end{array} \qquad 122.$$

The QCD mechanism involves cascade decays and therefore less energetic like-sign leptons than those from the direct decays in the W^- case. Also the W^+ is more often produced forward than the W^-, and we therefore expect an excess of positive dileptons over negative ones in the proton direction. The analysis of this type of signal requires a very careful study of the final state but may provide information on the decay modes of the weak bosons that is not obtainable otherwise.

5.4 Pair Production of Weak Bosons

One of the fundamental properties of the gauge model of electroweak interactions is the presence of trilinear couplings among gauge bosons like WWZ and WWγ, as shown in Section 1. These couplings allow one to produce pairs of weak bosons W^+W^-, $W^\pm Z^0$, $W^\pm \gamma$ (Brown & Mikaelian 1979, Brown et al 1979). As in the case of e^+e^- annihilation, pair production of weak bosons can tell us about the gauge structure and the relevance of the renormalizability of physical theories. It would be particularly interesting to compare the W^+W^-, $W^\pm Z^0$, and Z^0Z^0 cross sections and their dependences on the invariant masses of the boson pairs. At high-energy colliders with $\sqrt{s} \sim 500\text{--}2000$ GeV, the expected cross sections are at the picobarn level (10^{-36} cm^2), and in addition the final-state analysis can be complicated because of the leptons or hadronic jets. A final state with a higher cross section is $W^\pm \gamma$, which directly probes the WWγ coupling and can give a measure of the anomalous moment κ of the W, whose gauge theory value is $+1$. This process may be seen earlier because of a larger rate and the relatively easy indentification of a large transverse momentum photon.

6. HIGGS BOSONS

While the main subjects of this review are the intermediate vector bosons $W^\pm$ and Z^0, for completeness we also add here a few remarks about Higgs bosons. We saw in Section 1 how in the standard $SU(2) \times U(1)$ model the vector bosons acquire their masses through the vacuum expectation value of a complex isodoublet Higgs fields, and how one linear combination H (Equation 30) of electrically neutral components survives the mass generation mechanism to remain as a physical scalar Higgs particle. In more complicated versions of the theory there are more physical Higgs particles, including charged states as well as neutrals. Many physicists find Higgs fields inelegant, and would like to replace elementary spin-zero fields by composites made out of elementary fermions (Bég & Sirlin 1974, Farhi & Susskind 1981). We do not have space here to describe in detail the phenomenology of either the minimal Higgs scenario or the alternatives, but limit ourselves to a few descriptive remarks.

The single physical Higgs particle in the minimal model has an unknown mass. In the notation of Equation 28 in Section 1,

$$m_H^2 = 2\mu^2 + \text{radiative corrections},\tag{123.}$$

where μ^2 is unknown, only the Higgs vacuum expectation value $v^2 = \mu^2/\lambda$ (Equation 29) is phenomenologically determined:

$$v = (\sqrt{2}\,G_F)^{-1/2}.\tag{124.}$$

The radiative corrections to the Higgs boson mass are significant (Coleman & Weinberg 1973) for small values of μ^2, and give a lower bound to m_H corresponding to $\mu^2 = 0$:

$$m_H^2 = \frac{3\alpha^2}{8\sqrt{2}\,G_F}\left[\frac{2 + \sec^4\theta_w}{\sin^4\theta_w} - O\left(\frac{m_f}{m_w}\right)^4\right].\tag{125.}$$

The experimental value of $\sin^2 \theta_w$ leads to a lower bound on the Higgs boson mass of order 10 GeV. Larger masses correspond to larger values of the Higgs self-coupling (Equation 28), which becomes strong unless $m_H < O(1)$ TeV (Lee et al 1977, Veltman 1977). One may therefore expect the Higgs boson mass to be within an order of magnitude of the $W^\pm$ and Z^0 masses.

In contrast to its mass, the couplings of the minimal Higgs particle are completely determined. The couplings to the $W^\pm$ and Z^0 are given in Equation 33; since the Higgs vacuum expectation value gives masses to all the fundamental fermions f, we find

$$g_{H f \bar{f}} = \frac{m_f}{v} = (\sqrt{2} G_F)^{1/2} m_f. \qquad 126.$$

We see from Equations 33 and 126 that the Higgs particle likes to couple to heavy particles such as the top quark, the $W^\pm$, and the Z^0, and these provide the most favorable production mechanisms.

If the Higgs boson weighs significantly less than twice the mass of the top quark, a good place to look for the Higgs boson is in $^3S_1$ toponium decay, where it has been calculated (Wilczek 1977) that

$$\frac{\Gamma(^3S_1(\bar{t}t) \to H + \gamma)}{\Gamma(^3S_1(\bar{t}t) \to \gamma^* \to e^+ e^-)} \approx \frac{G_F m_t^2}{\sqrt{2} \pi \alpha}. \qquad 127.$$

This ratio is $\gtrsim 10\%$ for a top quark of mass $\gtrsim 18$ GeV, the lower limit established by the nonobservation of the top quark at PETRA. When toponium is found, the Higgs boson should be one of the first objects looked for in its decays. Other good mechanisms for producing the Higgs boson involve the Z^0. Figure 22 shows the branching ratios relative to Z^0 decay into $e^+ e^-$ or $\mu^+ \mu^-$ of the decays $Z^0 \to H + (e^+ e^-$ or $\mu^+ \mu^-)$ (Bjorken 1976) and $Z^0 \to H + \gamma$ (Cahn et al 1979). The branching ratios for both decays are greater than 10^{-6} if $m_H \lesssim 50$ GeV, so that the Higgs boson may be visible in $e^+ e^-$ experiments with the expected ten million Z^0 decays. Another good way (Ellis et al 1976, Lee et al 1977) to produce the Higgs boson is in association with the Z^0 through the reaction $e^+ e^- \to Z^0 \to Z^0 + H$ shown in Figure 23. This has a cross section

$$\frac{\sigma(e^+ e^- \to Z^0 + H)}{\sigma(e^+ e^- \to \mu^+ \mu^-)} \gtrsim 0.1 \qquad 128.$$

for $m_H < 100$ GeV at a center-of-mass energy of 200 GeV. Thus a high-energy $e^+ e^-$ machine such as LEP should be able to see a neutral Higgs boson with a mass up to the order of 100 GeV (Camilleri et al 1976).

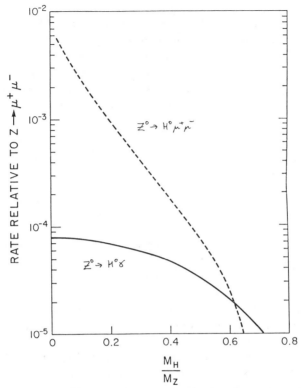

Figure 22   Branching ratios relative to $Z^0 \rightarrow \mu^+\mu^-$ for $Z^0 \rightarrow H^0\mu^+\mu^-$ and $Z^0 \rightarrow H + \gamma$.

The charged Higgs bosons expected in more complicated theories can be pair produced in e^+e^- annihilation, and may also be the dominant decay products of heavier quarks and leptons. Their couplings to other particles are generally correlated with their masses, but this connection is not as clear-cut as it is in the minimal model with just one neutral Higgs boson. While theories of dynamical symmetry breaking avoid the introduction of fundamental spin-zero fields, they nevertheless tend to

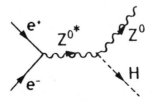

Figure 23   Lowest-order Feynman diagram for the process $e^+e^- \rightarrow Z^0 + H$.

predict many light composite spin-zero bosons with properties analogous to Higgs particles (Bég & Sirlin 1974, Farhi & Susskind 1981). Thus they can be searched for in many of the reactions previously mentioned in connection with Higgs bosons. However, an important difference is that one does not expect large couplings to the $W^\pm$ and Z^0 (Ellis et al 1981), and so copious production of composite spin-zero bosons in association with the intermediate vector bosons is not to be expected.

7. CONCLUSIONS

In this review we have shown how the phenomenological successes of gauge theories of the weak and electromagnetic interactions motivate very strongly the search for charged and neutral vector bosons with masses of the order of 80–90 GeV in the standard model. We have also seen how one can make quite precise predictions for the masses, decays, and production cross sections for these vector bosons in e^+e^-, lepton-hadron, and hadron-hadron collisions. These three different types of experiment are largely complementary in the information they may provide about the vector bosons and their couplings. Electron-positron collisions will not produce very heavy vector bosons in the near future, but can provide us with very detailed information about the Z^0 and provide the best prospects for searching for scalar bosons. Lepton-hadron collisions can provide indirect evidence even for rather heavy vector bosons, but are not the best type of collision to produce them directly. Hadron-hadron collisions can produce even rather heavy vector bosons because of the very high center-of-mass energies they provide, but the complicated nature of the hadronic final states will make it difficult to determine their detailed properties. Hadron-hadron colliders that should be able to produce the $W^\pm$ and Z^0 are in operation and under construction, while one e^+e^- machine capable of producing the Z^0 has been approved. Projects for detecting ultra-high-energy cosmic-ray neutrino collisions and ep colliding rings are being proposed. There is every reason to hope that in a few years the physics of the weak vector bosons will be thoroughly studied in many experiments. We hope and trust they will provide us with exciting surprises.

ACKNOWLEDGMENTS

We thank the many participants in various studies of future high-energy accelerator projects for enjoyable collaborations. In particular, we wish to acknowledge the LEP study contributions of P. Darriulat and

F. Renard, and we especially thank C. H. Llewellyn Smith, B. Richter, and B. H. Wiik for advice and encouragement. This work was supported by the Director, Office of Energy Research, Office of High-Energy and Nuclear Physics, Division of High-Energy Physics of the US Department of Energy under Contract W-7405-ENG-48.

Literature Cited

Abbott, L. F., Farhi, E. 1981. *Nucl. Phys. B* 180:547

Abud, M., Gatto, R., Savoy, C. A. 1978. *Phys. Lett.* 79B:435

Abud, M., Gatto, R., Savoy, C. A. 1979. *Phys. Rev. D* 20:1164

Albert, D., Marciano, W. J., Wyler, D., Parsa, Z. 1980. *Nucl. Phys. B* 119:125

Alles, W., Boyer, C., Buras, A. J. 1977. *Nucl. Phys. B* 119:125

Altarelli, G., Ellis, R. K., Martinelli, G. 1979. *Nucl. Phys. B* 147:461

Amaldi, U., ed. 1980. *Possibilities and Limitations of Accelerators and Detectors. Proc. 2nd ICFA Workshop, Les Diablerets, Switzerland, 4–10 Oct. 1979.* Geneva: CERN, Publ. Group, 1980. 442p.

Asano, Y., Kikutani, E., Kurokawa, S., Miyachi, T., Miyajima, M., Nasashima, Y., Shinkawa, T., Susimoto, S., Yoshimura, Y. 1981. *Phys. Lett.* 107B:159

Aubert, J. J., Becker, U., Biggs, P. J., Burger, J., Chen, M., Everhart, G., Goldhagen, P., Leong, J., McCorrison, T., Rhoades, T. G., Rohde, M., Ting, S. C. C., Wu, S. L., Lee, Y. Y. 1974. *Phys. Rev. Lett.* 33:1404–1406

Augustin, J.-E., Boyarski, A. M., Breidenbach, M., Bulos, F., Dakin, J. T., Feldman, G. J., Fischer, G. E., Fryberger, D., Hanson, G., Jean-Marie, B., Larsen, R. R., Luth, V., Lynch, H. L., Lyon, J., Morehouse, C. C., Paterson, J. M., Perl, M. L., Richter, B., Rapidis, P., Schwitters, R. F., Tanenbaum, W. M., Vannucci, F., Abrams, G. S, Briggs, D., Chinowsky, W., Friedberg, C. E., Goldhaber, G., Hollenbeek, R. J., Kadyk, J. A., Lulu, B., Pierre, F., Trilling, G. H., Whitaker, J. S., Wiss, J., Zipse, J. E. 1974. *Phys. Rev. Lett.* 33:1406–8

Aurenche, P., Lindfors, J. 1981. *Nucl. Phys. B* 185:274–300, 301–17

Bardin, D. Yu., Dokuchaeva, V. A. 1981. On Radiative Corrections To neutrino N → neutrino X Process. *JINR-P2-81-522* (In Russian). Submitted to *Yadernaya Fiz.*

Barish, B. 1978. *Phys. Rep.* 39:279

Bég, M. A. B., Sirlin, A. 1974. *Ann. Rev. Nucl. Sci.* 24:379–449

Bjorken, J. D. 1976. *Proc. 1976 SLAC Summer Inst. Part. Phys. (SLAC-198)*, p. 22. Stanford, Calif: SLAC

Bollini, D., Frabetti, P. L., Heiman, G., Monari, L., Navarria, F. L., Benvenutti, A. C., Bozzo, M., Brun, R., Gennow, M., Goossens, M., Kopp, R., Navach, F., Piemontese, L., Pilcher, J., Schinzer, D., Bardin, D. Yu., Cvach, J., Fadeev, N. G., Golutvin, I. A., Kiryushin, Y. T., Kisselev, V. S., Klein, M., Krivokhizhin, V. G., Kukhtin, V. V., Nowak, W. D., Savin, I. A., Smirnov, G. I., Vesztersombi, G., Volodko, A. G., Zacek, J., Jamnik, D., Meyer-Berkhout, U., Staude, A., Teichert, K. M., Tirler, R., Voss, R., Zupancic, C., Dobrowolski, T., Feltesse, J., Maillard, J., Malasoma, J. M., Milsztajn, A., Renardy, J. F., Sacquin, Y., Smadja, G., Verrecchia, P., Virchaux, M. 1981. *CERN Preprint.* To be published.

Branson, J. G. 1981. *Proc. 1981 Int. Symp. Lepton and Photon Interact. at High Energies*, pp. 279–300. Univ. Bonn, Phys. Inst.

Brown, R. W., Mikaelian, K. O. 1979. *Phys. Rev. D* 19:922–34

Brown, R. W., Sahdev, D., Mikaelian, K. O. 1979. *Phys. Rev. D* 20:1164–74

Cabibbo, N. 1963. *Phys. Rev. Lett.* 10:531

Cahn, R. N., Chanowitz, M. S., Fleishon, N. 1979. *Phys. Lett.* 82B:113–16

Camilleri, L., Cundy, D., Darriulat, P., Ellis, J., Field, J., Fisher, H., Gabathuler, E., Gaillard, M. K., Hoffmann, H., Johnsen, K., Keil, E., Palmonari, R., Preparata, G., Richter, B., Rubbia, C., Steinberger, J., Wiik, B., Willis, W., Winter, K. 1976. *Physics with very high energy e^+e^- colliding beams, CERN Rep. No. 76–18.* Geneva: CERN

Coleman, S., Weinberg, E. 1973. *Phys. Rev. D* 7:1888–1910

Danby, G., Gaillard, J.-M., Goulianos, D., Lederman, L. M., Mistry, N., Schwartz, M., Steinberger, J. 1962. *Phys. Rev, Lett.* 9:36

de Groot, E. H., Gounaris, G. J., Schild-knect, D. 1979. *Phys. Lett.* 85B:399–403

Drell, S. D., Yan, T. M. 1971. *Ann. Phys. NY* 66:578

DUMAND 1982. Project Description prepared by the Hawaii Dumand Center

ECFA, CERN. 1979. *Proc. LEP Summer Study, Les Houches and CERN, Sept. 1978. CERN Rep. No. 79–01,* Vols. 1, 2. Geneva: CERN

ECFA, DESY. 1979. *Proc. ep Facility for Eur., DESY 79/48.* Hamburg: DESY

Ellis, J. 1981. Lectures presented at the 1981 Les Houches Summer School, *LAPP preprint TH-48/CERN TH-3174.* Annecy: L.A.P.P.

Ellis, J., Gaillard, M. K., Nanopoulos, D. V. 1976. *Nucl. Phys. B* 106:292–340

Ellis, J., Gaillard, M. K., Nanopoulos, D. V., Sikivie, P. 1981. *Nucl. Phys. B* 182:529–45

Ellis, J., et al. 1978. *CHEEP: An ep Facility in the SPS. CHEEP Study Groups CERN Rep. No. 78–02.* Geneva: CERN

Farhi, E., Susskind, L. 1981. *Phys. Rep.* 74C:277

Fermi, E. 1933. *La Ricerca Scientifica* 4:P491.

Feynman, R. P. 1972. *Photon Hadron Interactions.* Reading, Mass: Benjamin

Finjord, J., Girardi, G., Perrottet, M., Sorba, P. 1981. *Nucl. Phys. B* 182:427–40

Gaemers, K. J. F., Gounaris, G. J. 1979. *Z. Phys. C* 1:259

Gaillard, M. K., Lee, B. W. 1974. *Phys. Rev D* 10:897–916

Gell-Mann, M., Levy, M. 1960. *Nuovo Cimento* 16:705

Georgi, H., Glashow, S. L. 1972. *Phys. Rev. Lett.* 28:1494

Georgi, H., Weinberg, S. 1978. *Phys. Rev. D* 17:275–79

Glashow, S. L. 1961. *Nucl. Phys.* 22:579–88

Glashow, S. L., Iliopoulos, J., Maiani, L. 1970. *Phys. Rev. D* 2:1285–92

Goldhaber, G., Pierre, F. M., Abrams, G. S., Alam, M. S., Boyarski, A. M., Breidenbach, M., Carithers, W. C., Chinowsky, W., Cooper, S. C., DeVoe, R. G., Dorfan, J. M., Feldman, G. J., Friedberg, C. E., Fryberger, D., Hanson, G., Jaros, J., Johnson, A. D., Kadyk, J. A., Larsen, R. R., Lüke, D., Lüth, V., Lynch, H. L., Madaras, R. J., Morehouse, C. C., Nguyen, H. K., Paterson, J. M., Perl, M. L., Peruzzi, I., Piccolo, M., Pun, T. P., Rapidis, P., Richter, B., Sadoulet, B., Schindler, R. H., Schwitters, R. F., Siegrist, J., Tanenbaun, W., Trilling, G. H.,

Vannucci, F., Whitaker, J., Wiss, J. E. 1976. *Phys. Rev. Lett.* 37:255–59

Goldstone, J. 1961. *Nuovo Cimento* 19:564

Hasert, F. J., Kabe, S., Krenz, W., Von Krogh, J., Lanske, D., Morfin, J., Schultze, K., Weerts, H., Bertrand-Coremans, G. H., Sacton, J., Van Doninck, W., Vilain, P., Camerini, U., Cundy, D. C., Baldi, R., Danilchenko, I., Fry, W. F., Haidt, D., Natali, S., Musset, P., Osculati, B., Palmer, R., Pattison, J. B. M., Perkins, D. H., Pullia, A., Rousset, A., Venus, W., Wachsmuth, H., Brisson, V., Degrange, B., Haguenauer, M., Kluberg, L., Nguyen-Khan, U., Petiau, P., Belotti, E., Bonetti, S., Cavalli, D., Conta, C., Fiorini, E., Rollier, M., Aubert, B., Blum, D., Chounet, L. M., Heusse, P., Lagarrigue, A., Lutz, A. M., Orkin-Lecourtois, A., Vialle, J. P., Bullock, F. W., Esten, M. J., Jones, T. W., McKenzie, J., Michette, A. G., Myatt, G., Scott, W. G. 1973. *Phys. Lett.* 46B:138–40

Herb. S., Hom, D. C., Lederman, L. M., Sens, J. C., Snyder, H. D., Yoh, J. K., Appel, J. A., Brown, B. C., Brown, C. N., Innes, W. R., Ueno, K., Yamanouchi, T., Ito. A. S., Jöstlein, H., Kaplan, D. M., Kephart, R. D. 1977. *Phys. Rev. Lett.* 39:252–55

Higgs, P. W. 1964. *Phys. Lett.* 12:132

Humpert, B., van Neerven, W. L. 1980. *Phys. Lett.* 93B:456

Hung, P. W., Sakurai, J. J. 1981. *Ann. Rev. Nucl. Part. Sci.* 31:375–438

Kamal, A. N., Ng, J. N., Lee, H. C. 1981. *TRIUMF and Univ. Alberta preprint TRI-PP-81-15/Alberta thy-8-91.* Submitted to *Phys. Rev. D*

Kibble, T. W. B. 1967. *Phys. Rev.* 155:1554

Kobayashi, M., Maskawa, K. 1973. *Prog. Theor. Phys.* 49:652–57

Lee, B. W. 1972. *Phys. Rev. D* 6:1188

Lee, B. W., Quigg, C., Thacker, H. B. 1977. *Phys. Rev. D* 16:1519

Lee, B. W., Zinn-Justin, J. 1972. *Phys. Rev. D* 5:3121, 3137, 3155

Llewellyn Smith, C. H., Wheater, J. 1981. *Phys. Lett.* 105B:486

Llewellyn Smith, C. H. Wiik, B. 1977. *DESY preprint 77/38* (Unpubl.)

Marciano, W., Parsa,, Z. 1981. See Peskin & Tye 1981, p. 127

Paige, F. 1979. *Proc. Topical Workshop Product. New Part. Super High Energy Collisions,* ed. V. Barger, F. Halzen. Univ. Wis., Madison

Pakvasa, S., Dechantsreiter, M., Halzen, F., Scott, D. M. 1979. *Phys. Rev. D* 20:2862

Peierls, R. F., Trueman, T. L., Wang, L. L. 1977. *Phys. Rev. D* 16:1397

Perl, M., Abrams, G. S., Boyarski, A. M., Breidenbach, M., Briggs, D. D., Bulos, F., Chinowsky, W., Dakin, J. T., Feldman, G. J., Friedberg, D., Fryberger, D., Goldhaber, G., Hanson, G., Heile, F. B., Jean-Marie, B., Kadyk, J. A., Larsen, R. R., Litke, A. M., Lüke, D., Lulu, B. A., Lüth, V., Lyon, D., Morehouse, C. C., Paterson, J. M., Pierre, F. M., Pun, T. P., Rapidis, P. A., Richter, B., Sadoulet, B., Schwitters, R. F., Tanenbaum, W., Trilling, G. H., Vannucci, F., Whitaker, J. S., Winkelmann F. C., Wiss, J. E. 1975. *Phys. Rev. Lett.* 35:1489–92

Perrottet, M. 1978. *Ann. Phys. NY* 115:107

Peruzzi, I., Piccolo, M., Feldman, G. J., Lecomte, P., Vuillemin, V., Barbaro-Galtieri, A., Dorfan, J. M., Ely, R., Feller, J. M., Fong, A., Gobbi, B., Hanson, G., Jaros, J. A., Kwan, B. P., Litke, A. M., Lüke, D., Madaras, R. J., Martin, J. F., Miller, D. H., Parker, S. I., Perl, M. L., Pun, P. T., Rapidis, P. A., Ronan, M. T., Ross, R. R., Scharre, D. L., Trippe, T. G., Yount, D. E. 1977. *Phys. Rev. Lett.* 39:1301–4

Peskin, M. E., Tye, S.-H. H., eds. 1981. *Proc. Cornell Z^0 Theory Workship. CLNS 81–485.* Ithaca, NY: Cornell Univ. Press

Prentki, J., Zumino, B. 1972. *Nucl. Phys. B* 47:99

Prescott, C. Y., Atwood, W. B., Cottrell, R. L., DeStaebler, H., Garwin, E. L., Gonidec, A., Miller, R. H., Rochester, L. S., Sato, T., Sherden, D. J., Sinclair,

C. K., Stein, S., Taylor, R. E., Clendenin, J. E., Hughes, V. M., Sasao, N. Schüler, K. P., Borghini, M. G., Lübelsmeyer, K., Jentschke, W. 1978. *Phys. Lett.* 77B:347–52

Quigg, C. 1977. *Rev. Mod. Phys.* 49:297–315

Reya, E. 1981. *Phys. Rep.* 69:195

Salam, A. 1968. *Proc. 8th Nobel Symp., Stockholm,* ed. N. Svartholm, p. 367–77. Stockholm: Almquist & Wiksells

Sirlin, A., Marciano, W. J. 1981. *Nucl. Phys. B* 189:442

Steigman, G. 1979. *Ann. Rev. Nucl. Part. Sci.* 29:313–37

Sushkov, O. P., Fambaum, V. V., Khriplovich, I. B. 1975. *Sov. J. Nucl. Phys.* 20:537

't Hooft, G. 1971. *Nucl. Phys. B* 35:167–88

Vainshtein, A. I., Khriplovich, I. B. 1973. *JETP Lett.* 18:83

Veltman, M. 1977. *Acta Phys. Polonica B* 8:475

Veltman, M. 1980. *Phys. Lett.* 91B:95–98

Wahl, H. 1981. Rapporteur Talk at the 1981 E. P. S. Int. Conf. High Energy Phys., Lisbon, to be published in the proceedings

Weinberg, S. 1967. *Phys. Rev. Lett.* 19:1264

Wilczek, F. A. 1977. *Phys. Rev. Lett.* 39:1304

Yan, T. M. 1976. *Ann. Rev. Nucl. Sci.* 26:199–238

Yang, C. N., Mills, R. L. 1954. *Phys. Rev.* 96:191

Yukawa, H. 1935. *Proc. Math. Soc. Jpn.* 17:48

Ann. Rev. Nucl. Part. Sci. 1982. 32:499–573
Copyright © 1982 by Annual Reviews Inc. All rights reserved

CHARGED-CURRENT NEUTRINO INTERACTIONS

H. Eugene Fisk

Fermilab, Batavia, Illinois 60510

Frank Sciulli

Columbia University, New York, New York 10027

CONTENTS

0163-8998/1201-0499$02.00

499

1. INTRODUCTION

1.1 *Deep Inelastic Scattering*

This article is primarily intended to review the status of the deep inelastic, charged-current scattering of neutrinos from nucleons. Processes of the form, $\nu_\mu + N \rightarrow \mu^- + X$, have been an important subject for research since such processes were found to have the profound simplicity referred to as scaling. This high-energy feature was first seen in lower-energy electron and neutrino scattering experiments (Miller et al 1972, Bodek et al 1973, Eichten et al 1973). In this reaction, the target nucleon, N, is at rest in the laboratory, and the final-state hadron system, X, generally consists of an assemblage of at least one nucleon and several mesons.

The neutrino-nucleon reaction is pictured in Figure 1, where the conventional notation is defined. The principle goal of research is to further our understanding of the structure of the nucleon at the bottom vertex. The most physically intuitive framework to describe that structure is in terms of parton constituents (e.g. Feynman 1969, 1974, Bjorken & Paschos 1969). (The evidence strongly indicates that these partons largely have the properties expected of quarks.) The weak interaction is assumed to be well understood and serves as a tool for investigating the elementary neutrino-quark interaction.

Subsequently, in this introduction, we show how the general features of the scaling hypothesis follow from the assumption of nearly free quarks inhabiting the nucleon, and we derive the relevant formulae from a simple, physical picture. (For a more complete and accurate formulation, see Feynman 1972 or Close 1979). We also briefly review the relevant properties of the neutrino, its charged-current interactions

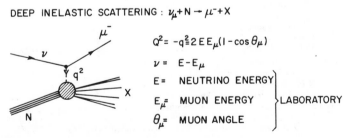

DEEP INELASTIC SCATTERING : $\nu_\mu + N \rightarrow \mu^- + X$

$Q^2 = -q^2 \stackrel{\sim}{=} 2\,E\,E_\mu(1-\cos\theta_\mu)$

$\nu = E - E_\mu$

$E =$ NEUTRINO ENERGY

$E_\mu =$ MUON ENERGY $\Big\}$ LABORATORY

$\theta_\mu =$ MUON ANGLE

Figure 1 Our conception of a neutrino-nucleon reaction, in which the neutrino converts to a μ^- with the emission of a positive intermediate vector boson. This weak virtual particle interacts with the nucleon to produce a hadronic final state consisting of baryons and mesons (X). At high-momentum transfers (large Q^2), the boson wavelength is short enough so that a single constituent of the nucleon interacts.

with other elementary targets, the quark model of nucleon structure, the anticipated form of the inelastic cross section, and the tests and complications of the quark model using deep inelastic data. Readers familiar with the underlying rationale and principles may wish to proceed directly to later sections for a review of the current experimental situation.

1.2 The Neutrino: Intrinsic Properties

The neutrino probe is a unique particle in its own right. As a lepton, it does not partake of the strong (nuclear) forces; because it is uncharged, it does not interact electromagnetically. It has half-integer spin, and fits comfortably into a coherent picture of the world of leptons. Because there are three known charged leptons (e^-, μ^-, τ^-), neutrinos are thought to also come in three varieties (ν_e, ν_μ, ν_τ). Only the first two of these have been directly observed to interact with matter, but it is expected that the ν_τ will be similarly observed within a few years. We anticipate that the ν_τ will, like the others, preserve throughout its interactions its own "tau-ness" just as the interactions of the electron and muon varieties are observed to conserve electron and muon numbers, respectively.

While all neutrinos have well-known charge and spin, our knowledge and understanding of some other important properties of neutrinos are hazy. Lepton number conservation is one example. The evidence seems strong that each charged lepton and its associated neutrino carry their own lepton number, and that these are individually conserved. Some proposed schemes for lepton number assignment and conservation have been ruled out experimentally (Willis et al 1980). There are other schemes consistent with present experimental data, but there is no compelling evidence for any violation of the usual electron, muon, and tau number assignment and conservation (see review by Primakoff & Rosen 1981).

The value of the neutrino rest mass will always be open to experimental tests. There have been hints of nonzero mass for electron neutrinos (Lubimov et al 1980), but these have yet to be corroborated. We know with confidence that the electron neutrino mass is less than about $60\,\mathrm{eV}/c^2$, the muon neutrino mass is less than $600\,\mathrm{keV}/c^2$, and the tau neutrino is less massive than $250\,\mathrm{MeV}/c^2$ (see review by Sciulli 1980). Current experimental work is being undertaken to reduce these limits, because the finiteness of neutrino mass is a fundamental question in both particle physics and cosmology. However, the mass of muon neutrinos is certainly well below where it could kinematically affect results on charged-current scattering from nucleons.

Another topic of current interest involves both nonzero neutrino mass and the violation of lepton number. If neutrinos were massive and conservation of lepton number were violated, spontaneous transitions might occur between neutrino types. This phenomenon, known as neutrino oscillations, was originally proposed by Pontecorvo (1957). Under appropriate conditions, an initially pure beam of, for example, ν_μ would make spontaneous transitions to another neutrino type. The flux of ν_μ would therefore depend on both energy and distance from the production point (for a review, see Baltay 1981 or Trilling 1980). While there have been some favorable hints, there is again no compelling evidence that such phenomena occur.

1.3 The Charged-Current Interactions

Neutrinos have only been observed to interact with matter through the weak force. There exist both charged- and neutral-current forms of the weak interaction; neutrinos partake of both. We confine ourselves in this article to the charged-current reactions, since these give the most direct information on nucleon structure. Neutral-current reactions are studied primarily to determine the properties of the neutral-current force. (For a review of neutral current structure, see Hung & Sakurai 1981.) The charged-current forces are thought to be well understood at present energies to within a few percent (for review, see Sakurai 1981).

The simplest process exemplifying charged-current interactions is the decay of the muon:

$$\mu^- \to e^- + \bar{\nu}_e + \nu_\mu. \tag{1.}$$

In this case, the muon spontaneously disintegrates into a muon type neutrino and an electron with its associated antineutrino. The muon lifetime provides a direct measure of the coupling strength of the weak interactions: the Fermi constant $G = 1.166 \times 10^{-5}\,\text{GeV}^{-2}$. The other observed features of the final-state particles, particularly those of the electron, demostrate that the weak interaction underlying the decay maximally violates parity conservation (i.e. the $V - A$ interaction). In its simplest form, this principle requires that any weakly interacting fermion (including leptons) will interact at high velocity ($v \approx c$) only if the fermion is polarized left-handedly or if the antifermion is polarized right-handedly. In other words, high-velocity right-handed fermions and left-handed antifermions neither interact nor are produced by weak interactions. A direct consequence of this principle is that, in muon decay, the low-mass ν_μ is always left-handed. The $\bar{\nu}_e$ is always polarized right-handedly, and the electron prefers left-handed spin orientation unless it has very low energy. This feature is observed.

Our detailed understanding of muon decay permits us to calculate directly the analogous production cross section involving a beam of muon neutrinos incident on a target of electrons:

$$\nu_\mu + e^- \rightarrow \mu^- + \nu_e. \qquad\qquad 2.$$

This process has been observed and found to comply with expectations (Büsser 1981). For center-of-mass energies ($s^{1/2}$) large compared with the electron or muon mass, the cross section per unit solid angle for the outgoing muon is

$$\frac{d\sigma}{d(\cos\theta)} = \frac{G^2 s}{2\pi}, \qquad\qquad 3.$$

where θ refers to the angle of the final-state muon with respect to the incident neutrino in the center-of-mass system. The dependence on the square of the center-of-mass energy is a consequence of the point-like structure of the weak interaction at presently available energies. Dimensional considerations require the cross section to depend on energy in this way. For energies at which the finite range of the weak interactions are probed, the formula is still appropriate with the modification $G \rightarrow G/(1 + Q^2/M_W^2)^2$, where M_W is the mass of the intermediate vector boson exchanged in the weak interaction.

The angular dependence of this cross section (Equation 3) is a consequence of the particle angular momenta and the $V - A$ character. In the center-of-mass system, the two colliding leptons only interact if they are spinning opposite to their respective directions of motion, as previously stated. Their spins cancel, so that zero total angular momentum characterizes the collision. The angular distribution in the center of mass is therefore isotropic, as reflected in Equation 3.

We see that our understanding of the structure of charged-current interactions permits direct calculation of cross sections between neutrinos and elementary fields such as the electron. It follows that the cross section of neutrinos interacting with other fields, like quarks, are similarly calculable. We return to these calculations after a brief review of the quark model of hadron structure.

1.4 The Quark Model of Nucleon Structure

The idea of elementary spin-$\frac{1}{2}$ fields (quarks), which, in bound states, constitute all hadrons, is now a familiar one. It was originally posited to explain the periodicity in hadron spectroscopy, where it has had renowned success (Gell-Mann 1964, Zweig 1964). The quarks, with fractional charge and baryon number, give hadrons their properties (spin, charge, strangeness, etc). We were faced, however, with a singular problem: the quarks could not be isolated (at least easily) from hadronic matter. Their spin, charge, and other properties could not be measured in the usual way (e.g. passing isolated quarks through electric and magnetic fields). Such measurements clearly require a different approach. The inelastic scattering of leptons from nucleons provides such a technique.

In the quark model, there are strong spectroscopic reasons for assuming the existence of at least four quark and four antiquark types: u, d, s, c. (The newest quark flavor, b, and its presumed partner, t, are not essential to our discussion.) The s and c quarks carry the quanta of strangeness and charm, respectively. The u and c quarks have electric charge equal to $+\frac{2}{3}$, and d and s carry $-\frac{1}{3}$ of the elementary proton charge. Protons and neutrons, the target nucleons, obtain their quantum numbers from the respective valence combinations (uud) and (udd). In general, additional sea pairs of quark-antiquark, e.g. $u\bar{u}$, $d\bar{d}$, etc, could be present. If we picture a nucleon traveling at very high momentum, P, this momentum will be constituted by fractions, $\xi_i P$, carried by the constituents. We might intuitively guess that the largest single fraction will be carried by the valence quarks, a smaller fraction carried by the sea quarks, and an unknown amount carried by nucleon constituents that do not interact directly with neutrinos. Experimentally the noninteracting part is about half of the total, and is likely to originate in the fields responsible for the binding of the quarks (e.g. gluons). The sea constituency may consist primarily of the low-mass variety quark since it is expected that a kinematic penalty exists for including more massive quarks in the nucleon wave function.

1.5   Neutrinos Interacting with (Free) Quarks

The interaction of a neutrino with a free elementary d quark would be similar to the interaction of neutrinos with electrons,

$$\nu_\mu + d \rightarrow \mu^- + u, \qquad\qquad 4.$$

and would therefore give the same cross section as Equation 3. Neutrinos could not interact directly with u quarks, since no charge-$\frac{5}{3}$ quark exists for the final state. However, we can have the process

$$\bar{\nu}_\mu + u \rightarrow \mu^+ + d. \qquad\qquad 5.$$

This reaction is directly analogous to Reaction 4 with one notable exception: the $\bar{\nu}_\mu$ (right-handed) and u (left-handed) give unit total angular momentum, so that the cross section for antineutrinos scattering from quarks is that for neutrinos (Equation 3) multiplied by the square of the angular-momentum-one amplitude, $(1 + \cos \theta)/2$.

Table 1 summarizes in the second column the differential center-of-mass cross sections for the scattering of neutrinos (and antineutrinos) from quarks (and antiquarks). Since, for example, the net angular momentum of $\bar{\nu}_\mu \bar{d}$ is the same as that for $\nu_\mu d$, the cross section is identical. For completeness, we add the scattering of neutrinos from hypothetical spin-zero constituents (k partons), which contains the spin-$\frac{1}{2}$ rotation element, $\cos \theta/2$, and an additional factor-of-two increase over the spin-$\frac{1}{2}$ case, where averaging over target spins reduces the cross section. The three forms of the cross section in the second column are complete. Any higher-spin elementary target will scatter with some linear combinations of these. (This follows from the spin-one nature of the weak force).

The target quarks are, of course, not really free. They are bound in hadrons, which complicates the analysis. However, at high energies the center-of-mass scattering angle, θ, is directly related to the inelasticity of the process:

$$y \equiv 1 - E_\mu/E, \qquad\qquad 6.$$

where E_μ is the outgoing laboratory muon energy and E is the incident neutrino energy. This is demonstrated in Figure 2. This relationship provides a critical interpretative step, so we reiterate: *measurement of laboratory energies only, in an event, permits determination of the center-of-mass scattering angle between the neutrino and the outgoing muon.* Since $1 - y = (1 + \cos \theta)/2$, it follows that the high-energy differential cross sections in y are as given in the third column of Table 1.

The fact that the quarks are moving inside the nucleon, particularly that they have longitudinal momenta, is important to the energy

Table 1 Cross sections for neutrino quark scatterings

Collision	$\dfrac{d\sigma}{d(\cos \theta)}$	$\dfrac{d\sigma}{dy}$
$\nu_\mu d,\ \bar{\nu}_\mu \bar{d}$	$\dfrac{G^2 s}{2\pi}$	$\dfrac{G^2 s}{\pi}$
$\bar{\nu}_\mu u,\ \nu_\mu \bar{u}$	$\dfrac{G^2 s}{2\pi}\left(\dfrac{1+\cos \theta}{2}\right)^2$	$\dfrac{G^2 s}{\pi}(1-y)^2$
$\nu_\mu K,\ \bar{\nu}_\mu K$	$2\dfrac{G^2 s}{2\pi}\cos^2\dfrac{\theta}{2}$	$2\dfrac{G^2 s}{\pi}(1-y)$

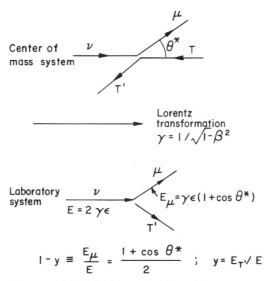

$$1-y \equiv \frac{E_\mu}{E} = \frac{1 + \cos \theta^*}{2} \quad ; \quad y = E_{T'}/E$$

Figure 2 The scaling *y* variable. At high energies where all masses are small compared to particle energies (ε) in the center-of-momentum frame, the collision between the neutrino and an elementary fermion field, T, is essentially elastic. The event scattering angle in this frame (θ^*) is directly calculable from the event inelasticity (y) of the reaction. The *y* value is obtained from the measured laboratory neutrino energy and the final-state muon energy. The spins of the nucleon constituents determine the θ^*, or *y*, dependence of the cross section.

dependence of the cross section. Figure 3 shows the neutrino incident on the target quark in the center of momentum of the neutrino-quark system. In this frame, the proton has total momentum P, and the quark has a fraction ξ of this. The chosen Lorentz frame then defines the neutrino to have the same momentum as the struck quark, ξP. The cross section is proportional to $s_{\nu q}$: the square of the total four-momentum for the neutrino-quark system. This is just $s_{\nu q} = (2\xi P)^2$. The square of the four-momentum for the entire neutrino-nucleon

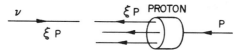

Figure 3 The collision between the neutrino and proton constituent as viewed in the Lorentz frame in which the total momentum between them is zero. In this frame, the proton has (high) momentum, P, and the constituent has momentum ξP. The square of the center-of-momentum energy between neutrino and constituent is $s_{\nu q} = \xi s$, where s is the square of the invariant energy of the neutrino and nucleon (see text). For laboratory neutrino energy, E, and a stationary nucleon target of mass, M, then $s \approx 2ME$ at high energies.

system is $s = (\xi P + P)^2 - (\xi P - P)^2 = 4\xi P^2$. Hence $s_{\nu q} = \xi s$. The differential cross section for neutrinos, for example, scattering from d quarks is then a modified version of the entry in Table 1 (Equation 3):

$$\frac{d\sigma}{dy} = \frac{G^2 \xi s}{\pi} \, \rho(\xi) \, d\xi,$$ 7.

where $\rho(\xi) \, d\xi$ is the probability for finding such a quark in a frame in which the proton has high momentum with the struck quark fraction between ξ and $\xi + d\xi$.

We still need, for the interpretation of data, a direct empirical measurement of this fractional momentum, ξ. One of the most important ideas that has helped us interpret deep inelastic data provides us with an experimental measure of this variable. Refer to Figure 4, where the scattering of the virtual boson from the (nearly massless) quark is shown in the frame in which the (time-like) energy of the boson is zero. By energy conservation, the outgoing massless quark must have energy equal to its incoming energy. The space-like momentum of the propagator must then just equal $\mathbf{q} = 2\xi\mathbf{P}$. It follows that ξ is related to a ratio of Lorentz scalars as

$$\xi = x \equiv \frac{-q^2}{2p \cdot q} = \frac{2EE_\mu (1 - \cos\theta_\mu)}{2M\nu}.$$ 8.

That is, *the fractional momentum carried by the struck quark may be obtained from measured laboratory quantities*: neutrino energy, E; outgoing muon energy, E_μ; with the difference being equal to the laboratory hadron energy, ν; and the outgoing muon angle relative to the incident neutrino direction, θ_μ. These quantities define the scaling variable, x, which approximates the fractional momentum of the struck quark. Of course, this is rigorously correct only in the scaling limit; that is, all energies in the center-of-mass are large compared to all masses, all binding energies, and all internal momenta. This is not and cannot be precisely true. How closely it is approximated is one of the interesting questions to ask of the data.

In a simple scaling model in which a free spin-$\frac{1}{2}$ constituent (q), its antiparticle (q̄), and a spin-0 type (k) inhabit the nucleon, the high-energy cross sections for neutrinos incident on a target of equal numbers of neutrons and protons is

$$\frac{d^2\sigma^\nu}{dx \, dy} = \frac{G^2 s}{2\pi} \left[q^\nu(x) + \bar{q}^\nu(x)(1 - y)^2 + 2k(x)(1 - y) \right]$$ 9a.

$$\frac{d^2\sigma^{\bar\nu}}{dx \, dy} = \frac{G^2 s}{2\pi} \left[q^{\bar\nu}(x)(1 - y)^2 + \bar{q}^{\bar\nu}(x) + 2k(x)(1 - y) \right],$$ 9b.

where $q(x)\,dx = x\rho_q(x)\,dx$ for example is the distribution in fractional momentum x multiplied by x. The normalization of q is chosen by convention such that it is the total number of quarks per nucleon. Since neutrinos interact, on average, with half the quarks (of appropriate charge), the factor of two appears in the denominator. (The normalization of k becomes model dependent at this stage.)

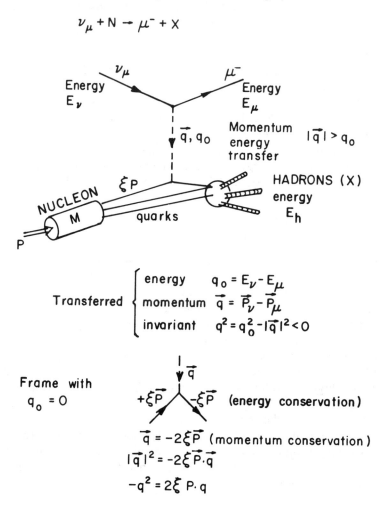

$$\nu_\mu + N \rightarrow \mu^- + X$$

Figure 4 The scaling variable, x. This experimental parameter can be calculated for each event from laboratory energies: E (neutrino), E_μ (muon), and ν (hadron); and from muon angle, θ_μ. The variable x is defined in Equation 8. If all masses, binding energies, and internal momenta can be neglected compared to the longitudinal momenta, x approximates the fraction of nucleon momentum carried by the interacting constituent, ξ.

Some important features predicted with these equations are (a) The total cross section ($E \gg M$) rises linearly with $s \approx 2ME$, where M is the nucleon mass. (b) The momentum distributions of nucleon constituents can be directly extracted, $q(x)$, $\bar{q}(x)$, $k(x)$, and with proper normalization, the integrals of these provide measures of the fractions of momentum carried by the several interacting constituents. The rate of increase of the total cross section with energy is linearly related to these integrals. (c) The fraction of antiquark ($\bar{q}$), and of non-spin-$\frac{1}{2}$ k can also be obtained. (d) Comparison of these experimental cross sections with analogous data taken with electron or muon beams gives a measure of the charges of the interacting constituents, as discussed below.

It comes as no surprise that the electromagnetic scattering of charged leptons with nucleons, under the same assumptions, has a form very similar to the sum of Equations 9a and 9b. The electromagnetic process is pure V, while Equations 9a and 9b behave like $V - A$ and $V + A$, respectively. Also, the zero-mass photon propagator gives a $1/Q^2$ dependence, characteristic of Mott scattering. Finally, the strength of the weak interaction is independent of electric charge, while the electromagnetic process is not. Thus, the $eN \rightarrow eX$ cross section is given by

$$\frac{d^2\sigma^e}{dx\,dy} = \frac{2\pi\alpha^2}{Q^4}\, s\{[r_q^2 q(x) + r_{\bar{q}}^2 \bar{q}(x)][1 + (1-y)^2] + r_k^2 k(x)(1-y)\}. \quad 10.$$

The parameters r_q^2 etc are the mean-square charges of the interacting quark constituents, which measure the relative coupling strengths through the electromagnetic force. Comparison of data from neutrino (weak) scattering with electron or muon (electromagnetic) scattering permits, therefore, measurement of the mean-square charge of interacting constituents.

1.6   Quark Types

Table 2 summarizes the quark constituency probed by neutrino beams. The appearance of the Cabibbo angle, θ_c, is a consequence of the weak hadronic current being a linear combination of strangeness-conserving and strangeness-violating parts (Gell-Mann & Levy 1960, Cabibbo 1963). This is observed in the relative leptonic decay rates of kaons and pions, for example.

In Table 2, it is assumed that u, d, s quark masses can be neglected in comparison to center-of-mass energies. At low Q^2, this may not be the case for charm (c) production, where $m_c \approx 1.5$ GeV. The parameter t_c is

Table 2 Interactions of neutrinos and antineutrinos with various quark types in a four-quark model.[a]

Reaction	Neutron	Proton	Nucleon
$\nu + d \to \mu^- + u$	$d_n \cos^2 \theta_c$	$d_p \cos^2 \theta_c$	$(d + u) \cos^2 \theta_c$
$\to \mu^- + c$	$d_n \sin^2 \theta_c$	$d_p \sin^2 \theta_c$	$(d + u)t_c \sin^2 \theta_c$
$\nu + s \to \mu^- + c$	$s_n \cos^2 \theta_c$	$s_p \cos^2 \theta_c$	$2s\, t_c \cos^2 \theta_c$
$\to \mu^- + u$	$s_n \sin^2 \theta_c$	$s_p \sin^2 \theta_c$	$2s \sin^2 \theta_c$
$\nu + \bar{u} \to \mu^- + \bar{d}$	$\bar{u}_n \cos^2 \theta_c$	$\bar{u}_p \cos^2 \theta_c$	$(\bar{d} + \bar{u}) \cos^2 \theta_c$
$\to \mu^- + \bar{s}$	$\bar{u}_n \sin^2 \theta_c$	$\bar{u}_p \sin^2 \theta_c$	$(\bar{d} + \bar{u}) \sin^2 \theta_c$
$\nu + \bar{c} \to \mu^- + \bar{s}$	$\bar{c}_n \cos^2 \theta_c$	$\bar{c}_p \cos^2 \theta_c$	$2\bar{c} \cos^2 \theta_c$
$\to \mu^- + \bar{d}$	$\bar{c}_n \sin^2 \theta_c$	$\bar{c}_p \sin^2 \theta_c$	$2\bar{c} \sin^2 \theta_c$
$\bar{\nu} + u \to \mu^+ + d$	$u_n \cos^2 \theta_c$	$u_p \cos^2 \theta_c$	$(d + u) \cos^2 \theta_c$
$\to \mu^+ + s$	$u_n \sin^2 \theta_c$	$u_p \sin^2 \theta_c$	$(d + u) \sin^2 \theta_c$
$\bar{\nu} + c \to \mu^+ + s$	$c_n \cos^2 \theta_c$	$c_p \cos^2 \theta_c$	$2c \cos^2 \theta_c$
$\to \mu^+ + d$	$c_n \sin^2 \theta_c$	$c_n \sin^2 \theta_c$	$2c \sin^2 \theta_c$
$\bar{\nu} + \bar{d} \to \mu^+ + \bar{u}$	$\bar{d}_n \cos^2 \theta_c$	$\bar{d}_p \cos^2 \theta_c$	$(\bar{d} + \bar{u}) \cos^2 \theta_c$
$\to \mu^+ + \bar{c}$	$\bar{d}_n \sin^2 \theta_c$	$\bar{d}_p \sin^2 \theta_c$	$(\bar{d} + \bar{u})t_c \sin^2 \theta_c$
$\bar{\nu} + \bar{s} \to \mu^+ + \bar{c}$	$\bar{s}_n \cos^2 \theta_c$	$\bar{s}_p \cos^2 \theta_c$	$2\bar{s}\, t_c \cos^2 \theta_c$
$\to \mu^+ + \bar{u}$	$\bar{s}_n \sin^2 \theta_c$	$\bar{s}_p \sin^2 \theta_c$	$2\bar{s} \sin^2 \theta_c$

[a] θ_c = cabibbo angle; t_c = threshold factor for charm quark production. We assume (a) isospin symmetry (e.g. $d_n = u_p$, etc) and (b) charm and strange quark (antiquark) constituency are independent of neutron-proton isospin (e.g. $s_n = s_p$).

included for this case, where the slow rescaling hypothesis is the most natural way to handle the threshold behavior (see Section 4). For $m_c^2 \ll W^2$, where W is the invariant mass of the final-state hadron system, $t_c \to 1$. We consider this case below.

The net quark and antiquark distributions are defined as: (x dependence assumed)

$$q = u + d + s + c, \qquad\qquad\qquad 10a.$$
$$\bar{q} = \bar{u} + \bar{d} + \bar{s} + \bar{c},$$

where, for example, $u = u(x)$ is the momentum fraction times the probability for finding a u quark with momentum fraction between x and $x + dx$ in the proton. If isospin symmetry is assumed, then the coefficients in Equations 9a and 9b at high energies are

$$q^\nu = q - c + s, \qquad\qquad\qquad 11a.$$
$$\bar{q}^\nu = \bar{q} + \bar{c} - \bar{s}, \qquad\qquad\qquad 11b.$$
$$q^{\bar{\nu}} = q + c - s, \qquad\qquad\qquad 11c.$$
$$\bar{q}^\nu = \bar{q} - \bar{c} + \bar{s}. \qquad\qquad\qquad 11d.$$

We expect, therefore, a slight asymmetry in these coefficients unless the charm content in the nucleon were equal to the strange content. It is expected that the strange content will dominate at present energies, because the strange-quark mass is small in comparison to that of the charm quark.

A measurement of the strange content of the nucleon is obtained from events with two muons in the final state. Such events result from reactions that produce a charm quark, followed by leptonic decay of the charmed particle (e.g. $D^0 \rightarrow K^- \mu^+ \nu_\mu$). Because of the weak selection rules operative in these decays, such events should always have the oppositely charged leptons in the final state, with one from the neutrino collision and the other from the charm decay. Indeed, it is found that the opposite-sign dilepton event rates are at least an order-of-magnitude more intense than same-sign dilepton events. This gives strong evidence that charm production is the dominant mechanism for dilepton production. From Table 2, the contributions to these data in Equations 9a and 9b will be

$$q_{2\mu}^\nu = T[(d + u) \sin^2 \theta_c + 2s \cos^2 \theta_c],$$

$$\bar{q}_{2\mu}^\nu = 0,$$

$$q_{2\mu}^{\bar{\nu}} = 0,$$

$$\bar{q}_{2\mu}^{\bar{\nu}} = T[(\bar{d} + \bar{u}) \sin^2 \theta_c + 2\bar{s} \cos^2 \theta_c]. \tag{12.}$$

Note that neutrinos produce charm from valence quarks, while antineutrinos do not. Here, T includes the threshold dependence for charm production, as well as the total probability for the charmed final state to produce an observable decay lepton. This is discussed at greater length in Section 4.

Events with two leptons of the same electric charge should not occur in the usual four-quark model, since the weak selection rules forbid them. But such events are observed about once in each 2000 charged-current events. Several sources have been hypothesized: (a) $D^0 - \bar{D}^0$ mixing, (b) b-quark constituency of the nucleon, etc. None has provided a universally acceptable explanation.

1.7 General Formulation of Deep Inelastic Scattering

Table 3 summarizes the formulae and notation commonly used in the analysis of deep inelastic scattering data. These are valid to first order in the target nucleon mass and ignore the charged-lepton mass relative to the incident neutrino energy. The small asymmetries already discussed

Table 3 Formulae and notation of deep inelastic scattering

$x \equiv Q^2/2M\nu$ E = beam (neutrino) energy

$y \equiv \nu/E$ E_μ = outgoing lepton (muon) energy

 $\nu \ = E - E_\mu$

 $Q^2 = 2EE_\mu \ (1 - \cos \theta_\mu)$

 $s \ = M^2 + 2ME$

Scaling
two-quark
model

$$\left\{\begin{array}{l} F_2(x, Q^2) = q(x) + \bar{q}(x) + 2k(x) \\ 2xF_1(x, Q^2) = q(x) + \bar{q}(x); \ R(x, Q^2) = 2\dfrac{k(x)}{q(x) + \bar{q}(x)} \\ xF_3(x, Q^2) = q(x) - \bar{q}(x) \end{array}\right.$$

Neutrino: $q(x) \equiv u(x) + d(x) + s(x) + c(x)$

 $\bar{q}(x) \equiv \bar{u}(x) + \bar{d}(x) + \bar{s}(x) + \bar{c}(x)$

Muon-electron $q^{em}(x) = \frac{4}{9}u(x) + \frac{1}{9}d(x) + \frac{1}{9}s(x) + \frac{4}{9}c(x)$

 $\bar{q}^{em}(x) = \frac{4}{9}\bar{u}(x) + \frac{1}{9}\bar{d}(x) + \frac{1}{9}\bar{s}(x) + \frac{4}{9}\bar{c}(x)$

Neutrino

$$\frac{d^2(\sigma^\nu + \sigma^{\bar{\nu}})}{dx\,dy} = \frac{G^2 s}{\pi}\left\{\left(1 - y - \frac{Mxy}{2E}\right)F_2^{wk}(x, Q^2) + \frac{1}{2}y^2[2xF_1^{wk}(x, Q^2)]\right\}$$

$$\frac{d^2(\sigma^\nu - \sigma^{\bar{\nu}})}{dx\,dy} = \frac{G^2 s}{\pi}\left[y(1 - \frac{1}{2}y)xF_3^{wk}(x, Q^2)\right]$$

Muon-electron

$$\frac{d^2\sigma^{em}}{dx\,dy} = \frac{4\pi\alpha^2}{Q^4}s\left\{\left(1 - y - \frac{Mxy}{2E}\right)F_2^{em}(x, Q^2) + \frac{1}{2}y^2[2xF_1^{em}(x, Q^2)]\right\}$$

$$\frac{2xF_1}{F_2} \equiv \left(1 + \frac{4M^2x^2}{Q^2}\right)\bigg/(1 + R); \ 0 \le R \le \infty$$

due to the presence of s quarks (with few c quarks) are typically explicitly corrected to provide F_2, R, and xF_3 as defined in the table. Included for completeness are the analogous formulae for the inelastic scattering of muons or electrons.

From this table, we see the following predictions of the simple quark model.

1. Approximate scaling: $F_i(x, Q^2) \approx F_i(x)$. This requires that the total cross section be linearly dependent on energy.
2. Spin-$\frac{1}{2}$ dominance (Callan-Gross relation): $R^{wk} \approx R^{em} \approx 0$ (Callan & Gross 1969, Bjorken & Paschos 1969).
3. Quark charges: $F_2^{em}/F_2^{wk} \approx \frac{5}{18}$. More precisely, we expect, for negligible charm content,

$$\frac{F_2^{em}}{F_2^{wk}} = \frac{5}{18}\left[1 - \frac{3}{5}\frac{s + \bar{s}}{q + \bar{q}}\right]. \tag{13.}$$

4. Number of valence quarks (Gross–Llewellyn Smith sum rule). Table 2 gives the difference of neutrino-antineutrino scattering, well above charm threshold, as

$$\frac{d^2(\sigma^\nu - \sigma^{\bar{\nu}})}{dx\,dy} = \frac{G^2 s}{\pi} \{[q(x) - \bar{q}(x)][(1 - (1 - y)^2]$$

$$+ \{[s(x) - \bar{s}(x)] + [c(x) - \bar{c}(x)]\}[1 + (1 - y)^2]\}. \qquad 14.$$

We anticipate that $s(x) = \bar{s}(x)$ and $c(x) = \bar{c}(x)$. These equalities are expected but are not required by any general principle. In any case, we must have $\int[s(x)/x]dx = \int[\bar{s}(x)/x]dx$, etc, since the net strangeness (charm) of the nucleon is zero. It follows that

$$\int_0^1 \frac{xF_3(x)}{x}\,dx = \int_0^1 \frac{[q(x) - \bar{q}(x)]}{x}\,dx = 3, \qquad 15.$$

which is the net number of valence quarks in the nucleon. This is referred to as the Gross–Llewellyn Smith (GLS) sum rule (Gross & Llewellyn Smith 1969).

5. Up-down quark difference (Adler sum rule). From Table 2, we see that the difference between scattering from neutrons and protons is, assuming the strange and charm components cancel,

$$\frac{d^2(\sigma^{\nu n} - \sigma^{\nu p})}{dx\,dy} = \frac{G^2 s}{\pi} [(u - d) - (\bar{u} - \bar{d})(1 - y)^2] \qquad 16.$$

and precisely the reverse y dependence for antineutrinos. This gives rise to the prediction (Adler 1966)

$$\int \frac{xF_1^{\nu n} - xF_1^{\nu p}}{x}\,dx = 1. \qquad 17.$$

This reflects the asymmetry of (uud) and (udd) in protons and neutrons, respectively. To test the Adler sum rule requires the use of hydrogen targets.

1.8   Deviations from the Naive Quark Model

From the beginning of work on deep inelastic scattering, it was recognized that scaling should be an approximate phenomenon. Indeed, as it was understood in terms of point-like constituent scattering, it was simultaneously recognized that nonscaling effects should exist at lower energies. The quarks are not, as assumed in the parton model, precisely free with no transverse momentum since they are bound inside hadrons. It is expected that binding and transverse momentum will produce effects that behave approximately as a power of $1/Q^2$.

For example, the Callan-Gross relation must obtain some violation from such effects. If the spin-$\frac{1}{2}$ constituents have a nonzero transverse momentum, p_T, then the collision between quarks is noncolinear at this level, and from this effect alone,

$$R \simeq 4 \, \frac{\langle p_T^2 \rangle}{Q^2}.$$

18.

Similar deviations, in magnitude and in dependence on Q^2, are expected from finite quark mass and binding energies.

The other structure functions should also have some Q^2 dependence from such effects. The very definition of the variable x and its interpretation as fractional momentum requires the assumption of high Q^2. Early interpretations of ep scattering data invoked novel definitions of scaling variables, e.g. $x' = Q^2/(2M\nu + M^2)$, that satisfy the only necessary requirement: $x' \rightarrow x$ as $Q^2 \rightarrow \infty$. With such definitions, the effect on structure functions could be very big, especially at larger x values, e.g. $F_2(x, Q^2) \approx F_2(x)[1 + nM^2x/Q^2(1-x)]$, if $F_2(x') \approx (1-x')^n$ at large x'.

Scale-breaking effects could also come from leptons scattering from bound two-quark systems. These subsystems would give scale breaking similar to that discussed above, and could induce finite R values, particularly at large x (Schmidt & Blankenbecler 1977).

1.9 Quantum Chromodynamics

There has been considerable enthusiasm in recent years for the theory of quantum chromodynamics (QCD) as a viable theory of strong interactions. This theory describes the strong force binding quarks in hadrons as carried by colored gluons, in a manner similar to the way the photon carries the electromagnetic force in atoms (for review, see Appelquist et al 1978). This theory predicts a pattern of scale breaking in deep inelastic scattering that is different at high Q^2 than that anticipated above from bound-state complications. The pattern is dominated by the quark-gluon coupling constant, which, to first order, is $\alpha_s(Q^2) = 12\pi/(33 - 2N_f) \ln(Q^2/\Lambda^2)$. Here N_f is the number of quark flavors, and Λ is a scale thus far undefined by the theory, but it is anticipated to lie in the range of strong interaction binding scales, which is $\sim 10^{-13}$ cm ($0.1 < \Lambda < 0.7$ GeV). (For a recent review of evidence for QCD, see Söding & Wolf 1981.)

The pattern of such perturbative effects is logarithmic in Q^2, rather than being a power law. The greatest experimental difficulty lies in comparing structure function data with predictions that have an unknown part falling like a power of $1/Q^2$ and predictions that have a logarithmic dependence.

The prediction for R, in perturbative QCD, is particularly simple and direct. It is (Altarelli & Martinelli 1978)

$$2xF_1 R(x,\ Q^2) = \frac{\alpha_s(Q^2)}{2\pi}\ x^2 \left[\int_x^1 \frac{dz}{z^3}\ \frac{8}{3}\ F_2(z,\ Q^2)\ dz \right. \qquad 19.$$

$$\left. + \int_x^1 \frac{dz}{z^3}\ 16\left(1 - \frac{x}{z}\right) G(z,\ Q^2)\ dz\right].$$

The first term in the brackets comes from perturbative processes in which the constituent quark creates a bremsstrahlung gluon before collision with the neutrino; the second term comes from pair production of quarks by a gluon, hence the dependence on the gluon distribution, $G(x,\ Q^2)$, with subsequent interaction between a quark and the neutrino. The features predicted by this formula are rather striking, as illustrated in Figure 5. There should be a sharp falloff in x at fixed Q^2 and a slow (logarithmic) dependence on Q^2. Unfortunately the R

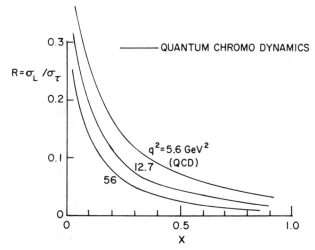

Figure 5 Predictions for R from quantum chromodynamics (Equation 19). Other contributions can come from mechanisms that produce power law dependence on $1/Q^2$.

parameter is very difficult to measure. We review the available data in Section 3.

2   HIGH-ENERGY NEUTRINO EXPERIMENTS

2.1   *Neutrino Beams*

2.1.1   INTRODUCTION   The advent of high-energy accelerators plus the expected linear energy increase in the cross section gave promise for extensive studies of neutrino interactions (Pontecorvo 1951, Schwartz 1960). Such studies have come to pass with a variety of different neutrino beams. Generally, all of these beams are created in long evacuated pipes as the products of π- and K-mesons decaying into their major decay mode: $\mu^- + \nu_\mu$. The parent π and K mesons are themselves produced by machine energy protons striking a target upstream of the decay pipe. Downstream of the decay region, there must be sufficient shielding to remove the high-energy muons. (The muon intensity accompanying a useful neutrino beam is too intense to allow its passage through the neutrino detector.)

The earliest accelerator neutrino experiment used neutrinos from a "bare target" system without any focusing of the meson parents (Danby et al 1962). Nowadays, the secondary mesons are usually focused to obtain more neutrino flux. The focusing has been accomplished with both horn systems and quadrupole magnets. The mesons can also be momentum selected before they enter the decay region. A momentum-selected meson beam creates a narrow-band neutrino beam; without such selection, the neutrino beam is called wide band. The remainder of this section deals with examples of existing beams.

An example of the present narrow-band neutrino beam layout at Fermilab is shown in Figure 6. The simultaneous use by several large and varied detectors, shown here, also exists at CERN. Moreover, different kinds of beams can be utilized by making relatively minor changes in the targeting region. One significant difference between the CERN and Fermilab facilities is that the distance from the midpoint of the decay region to the detectors is shorter at CERN (600 m). The Fermilab layout was optimized for 1000 GeV during its initial construction.

Generally speaking, the bubble chamber experiments with fiducial tonnage of 1 to 10 tons have opted for wide-band beams that maximize neutrino flux by focusing the wide-angle, lower-momentum secondaries in the forward direction. On the other hand, the massive electronic detectors (300 to 1500 tons) have typically requested beams that accentuate high-energy neutrinos, such as narrow-band beams or quadrupole-focused beams.

Since at high energies the laboratory decay angle, θ, of the high-energy neutrinos intercepting the downstream neutrino detector is small, typically less than a few milliradians, the neutrino energy from a pencil beam of hadron secondaries is accurately given by the approximation

$$E_\nu = E_\nu(\max)/(1 + \gamma^2 \theta^2), \qquad\qquad 20.$$

where γ is the usual ratio of beam hadron energy to rest mass, and $E_\nu(\max)$ is the maximum energy neutrino, given in terms of the hadron beam energy by $E_\nu(\max) = E_{\pi,K}[1 - (m_\mu/m_{\pi,K})^2]$. Thus a secondary hadron beam of fixed momentum leads to two distinct energies for zero-degree neutrinos. Which of these energies occurs depends on whether the parent hadron was a K or π meson: $E_\nu^\pi(\max) = 0.42\, E_\pi$ and $E_\nu^K(\max) = 0.95\, E_K$.

2.1.2 NARROW-BAND BEAMS   The recognition of the dichromatic nature of the neutrino's energy spectrum led to proposals for accelerator neutrino beams where the correlation between the neutrino's angle and energy could be exploited to determine the energy of each neutrino interaction in a downstream detector (Peterson 1964, Sciulli et al 1970). A by-product of such a beam is the relative ease of measuring and

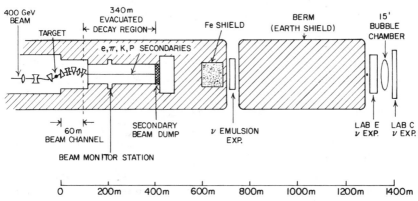

Figure 6 The Fermilab Neutrino Area in 1980. The several detectors may make simultaneous use of the neutrino beam. During narrow-band beam operation, illustrated here, the extracted proton beam is separated from the momentum-selected secondary beam and dumped prior to the decay pipe. This permits monitoring and measurement of the meson secondaries in the decay pipe for purposes of flux normalization. The transverse dimension has been expanded in the figure for clarity.

monitoring flux (see Figure 6). The first example of such a beam had a maximum energy of 230 GeV. It was brought into operation at Fermilab in 1972–1973 (Limon et al 1974). The optics of this beam consisted of point-to-parallel focusing plus momentum selection ($\Delta p/p \simeq \pm 18\%$). At the CERN SPS a more sophisticated dichromatic beam has operated since 1976 with a momentum bite $\Delta p/p$ of $\pm 5\%$ and a maximum energy of 300 GeV. The CERN beam optics (May et al 1977) is point-to-intermediate focus, followed by additional focusing to parallel. A second-generation Fermilab beam, capable of focusing 350-Ge secondaries, was designed (Edwards & Sciulli 1976) and brought into operation in 1978. It has been used in several recent experiments. Figure 7 shows the correlation implied by Equation 20 between the neutrino interaction radius and neutrino energy for data at ten different beam settings from the Columbia/Caltech-Fermilab-Rochester-Rockefeller (CFRR) experiment located in Lab E. The contributions of neutrinos from π and K decay are clearly visible as separate bands.

Momentum bite is only one of the factors determining the average neutrino energy and its spread, σ_E, at a given detector radius. Other factors affecting energy resolution are the secondary beam angular divergence and parallax, i.e. the fact that neutrinos of different energy originating at different points in the decay pipe reach the detector at a common radius. From Equation 20 the characteristic decay angle, θ_o, is γ^{-1}. For secondary pions, where θ_o is less than 1 mrad at high energy, a typical beam angular divergence of ± 0.2 mrad has a substantial effect on the mean neutrino energy and resolution. For neutrinos generated from K mesons, angular divergence does not significantly affect energy resolution. Parallax is important in this case, especially at CERN where the detectors are close to the midpoint of the decay pipe. Although the meson beam momentum bite at CERN is smaller than that at Fermilab, the average energy resolutions are very similar, primarily because of the parallax difference. For the Fermilab beam, tuned to 200 GeV, the $\nu_K(\nu_\pi)$ energy resolution, σ_E/E, varies from 10 (15)% at the center of the detector to 9 (20)% at the edge of the fiducial volume where the radius is 1.25 (0.75) m.

From measurements of the secondary beam intensity, momentum bite, angular divergence, and π, K, p beam composition, the absolute neutrino flux can be calculated. As an example, the flux from the Fermilab beam is shown in Figure 8. In addition to the dichromatic components, there is additional calculable flux at the level of 5 (7)% of the ν_K flux from $K_{\mu_3}(K_{e_3})$ decays as indicated in Figure 8. The K_{μ_3} flux is more of a nuisance for charged-current studies since it produces

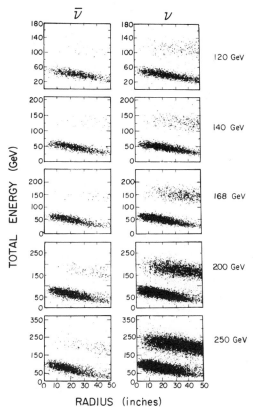

Figure 7 Measured correlation in the narrow-band beam between detected neutrino event energies and the radial position of interaction at the Lab E detector (CFRR group). The settings of the beam central momentum for the selected mesons is shown to the right. Negative and positive charged-meson selection produces $\bar{\nu}$ and ν, respectively. The bands correspond to K decay (*upper band*) and π decay (*lower band*), respectively. The smaller number of K^- mesons relative to K^+ mesons is evident from the comparative paucity of high-energy antineutrino events.

acceptable charged-current data. Unwanted broad-band ν_μ and $\bar{\nu}_\mu$ flux is also generated from decays of mesons before momentum selection. Both the CERN and Fermilab beams minimize these neutrinos by targeting the primary proton beam at a large angle (11 to 14 mrad) relative to the neutrino detector direction. The contribution of this wide-band background to a given neutrino data sample is measured by obtaining neutrino data with the beam entrance to the decay pipe blocked. Since this wide-band flux is dominated by decays of low-momentum mesons, it is most important for pion neutrino data where,

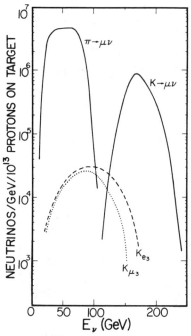

Figure 8   The flux of neutrinos at the Fermilab Lab E detector (radius 1.5 m) with the central hadron momentum selected at 200 GeV with positive electric charge. The dichromatic nature of the beam is clear; the small calculable neutrino contaminations from three-body kaon decays are also shown. Background from low-momentum decays before momentum selection of the meson beam produces additional background primarily below 20 GeV. This can be directly measured by stopping the dichromatic beam before its entry into the decay region.

with a secondary momentum of 200 GeV/c, it averages 5% and is a maximum of 10% at the lowest energies in the Fermilab beam.

2.1.3 WIDE-BAND BEAMS   Charged-current event rates for the 1000-ton detectors with narrow-band beams are typically a few per machine cycle. For smaller-mass targets, like bubble chambers, and for study of rare processes in massive detectors, it was suggested quite early that beams would be needed where wide-angle, low-momentum secondaries would be focused in the forward direction (van der Meer et al 1963, Giesch et al 1963). It has been possible to build these wide-band beams where almost half of the secondary mesons of a given sign are focused into the decay pipe. Focusing is achieved by pulsing high-current (water-cooled) aluminum horn-shaped conductors to create an azimuthal magnetic field. In these "horn" beams, wide-angle (lower-energy) secondaries of one sign are focused while those of opposite sign are defocused.

The quadrupole triplet beam, as the name implies, uses quadrupole focusing to produce a beam that eliminates much of the very-low-energy neutrino flux and relatively enhances the high-energy flux from K decays. Because of the nature of quadrupole focusing, sign selection is relatively poor. This beam, at Fermilab, has been especially useful for investigating charm production by neutrinos.

Figure 9 shows the neutrino flux from three different wide-band beams at Fermilab: horn (Grimson & Mori 1978), quadrupole triplet (Skuja et al 1976), and bare target (Stefanski & White 1976). For comparison, the flux that would be generated by a perfectly focused beam, an ideal situation in which all produced mesons are directed toward the neutrino target, is also shown. In all $\nu\,(\bar{\nu})$ wide-band beams, there is a significant amount of $\bar{\nu}\,(\nu)$ background flux from the secondaries being defocused. When the horn beam focuses negative secondaries to generate antineutrinos, the neutrino flux at energies above 60–80 GeV exceeds the antineutrino flux because of the preponderance of K^+ to K^- secondaries. This has motivated the use of

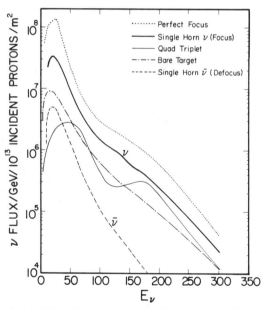

Figure 9 The calculated flux of neutrinos from various broad-band focusing devices used at Fermilab compared to that which would result from a perfect focusing device. The proton beam energy has been taken to be 400 GeV.

plugs in the central region of the secondary beam to increase the $\bar{\nu}/\nu$ flux ratio at high energies.

A schematic physical layout for the Fermilab horn is shown in Figure 10. The horn is powered with a 100-kA half-sine-wave pulse generated from a 2400-μF capacitor bank charged to 6 kV. At both CERN and Fermilab, the current pulse has been lengthened from 40 μs to 1 ms by using a transformer with the capacitor bank on the primary winding and the horn on the secondary winding. This more closely matches the beam extraction period during neutrino operation.

2.2   Beam Flux Measurements

2.2.1   MONITORING   Measuring neutrino total cross sections requires knowledge of the absolute neutrino flux and energy spectrum at a given detector radius. This flux can be reliably calculated only for narrow-band beams where the properties of the secondary beam or of the decay muons, which correlate directly with the ν_μ flux, can be measured. Required are measurements of the absolute numbers of π and K mesons, mean secondary momentum $\langle p \rangle$, momentum spread σ_p, and beam angular divergence.

Two basic and complementary techniques have been developed to monitor the beam. The method used primarily at Fermilab has been to measure the absolute hadron flux in the decay pipe and the fractions of e, μ, π, K, and p ($\bar{p}$). The CERN technique also involves measurement of particle fractions, using a similar method. The absolute intensity measurement initially depended heavily on a beam current transformer in the secondary beam but has more recently relied on measuring the absolute number and profile of muons from π decay, which along with the K/π particle ratio gives the absolute neutrino flux. No method is completely straightforward since the typical narrow-band beam secon-

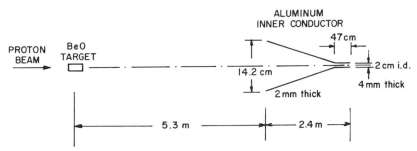

Figure 10 The horn at Fermilab, showing the placement relative to the target position and the location of the cylindrical inner conductor. Particles of one sign that penetrate this conductor are focused; those with electric charge of opposite sign are defocussed.

dary flux is 10^9 to 10^{11} particles during a 1-ms spill, i.e. $\sim 10^{13}$ s^{-1} instantaneous rate. This clearly precludes measuring the total number and kinds of beam particles with standard counting techniques.

2.2.2 PARTICLE RATIOS   Both CERN and Fermilab use short (2 m long) focusing Čerenkov detectors with He gas as the radiator to measure the relative numbers of π's, K's, and protons in the decay pipe. Unlike typical Čerenkov detectors in particle physics, which tag individual particles, these counters integrate all light traversing a small iris at the focal point during the beam spill. Their length implies that a phototube typically observes ~ 0.01 photoelectrons per beam particle. Such short length also means that diffraction effects become important in understanding resolutions. A Čerenkov curve is obtained by recording the light output normalized to beam intensity as a function of Čerenkov pressure, P.

At Fermilab, Čerenkov light with angles between 0.7 and 1.0 mrad is transmitted to a photomultiplier tube. Because of the high beam intensity, only a few stages of phototube amplification are needed to provide adequate current. A sample Čerenkov curve, after background subtraction, is shown in Figure 11. The relative fractions of π, K, and p are proportional to the areas under the peaks. The K/π ratios obtained in the CFRR experiment are shown in Figure 12. Measurements at CERN made in similar fashion (de Groot et al 1979), as well as those made using more traditional techniques (Atherton et al 1980), are in good agreement with the CFRR data (Rapidis et al 1981).

These Čerenkov curves contain a wealth of information in addition to particle fractions. A good approximation governing Čerenkov light in

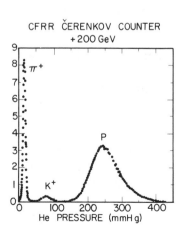

CFRR ČERENKOV COUNTER
+ 200 GeV

Figure 11   The net Čerenkov light per beam particle transmitted through a small (0.7– 1 mrad) iris as a function of helium pressure. The peaks occur at the pressures expected for pions, kaons, and protons at this setting of the dichromatic beam (200 GeV). The areas under the peaks are directly proportional to the relative fractions of these particles in the beam.

K/π RATIOS MEASURED CFRR
E616

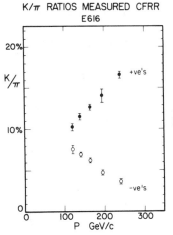

Figure 12   The ratio of kaons to pions in the Fermilab dichromatic beam as a function of beam setting.

this regime is the relation, $\theta_c^2 = 2(n_0 - 1)P/P_0 - (m/p)^2$, between Čerenkov angle (θ_c), pressure (P), index of refraction (n_0) calibrated at known pressure (P_0), mass (m), and momentum (p). The pressure means of the Čerenkov peaks on the curve are directly related to the mean beam momenta. The width of a peak is mainly determined (for pions) by the beam angular divergence and iris opening and (for protons) by the momentum spread. All of these quantities are relevant to the accurate calculation of flux. The index of refraction and counter resolution function are measured with 200-GeV protons extracted from the main accelerator.

2.2.3   ABSOLUTE INTENSITY   A combination of devices has been used to measure the number of secondary particles in the Fermilab decay pipe. At two different locations, helium-filled ionization chambers measured signals proportional to the beam intensity. (The choice of gas was influenced by small dE/dx pion-proton differences and the absence of saturation effects.) The ion chambers were calibrated in four different ways. A 200-GeV beam of protons from the accelerator was transported through the secondary beam line (with the target removed) into the decay pipe where this intensity was simultaneously measured with several devices. The first method consisted of integrating the ion chamber output current while exposing Cu foils that were subsequently counted for Na^{24} content. From the Na^{24} activation cross section, previously measured at CERN (Chapman-Hatchett et al 1979), an ion chamber calibration was obtained. An independent measurement of this cross section was made at Fermilab, and it agreed well with the CERN result. A second calibration used the rectified signal from a 53-MHz (beam structure frequency) rf cavity, located in the decay pipe to measure absolute intensity. The cavity's calibration depends primarily on the directly measured resonant frequency and quality factor. Finally a calibration was accomplished in a low-intensity beam where each particle was counted with standard detectors and circuitry. All of these ion chamber calibrations agree to within 3%, when measured differences of $6\frac{1}{2}\%$ between proton and pion ion chamber response are taken into account. These differences can be understood from calculations showing that secondary beam interactions in the ionization chamber foils are different for pions and protons, and that these interactions produce heavily ionizing nuclear fragments (Rapidis et al 1981).

2.2.4 MUON FLUX   The CERN technique of measuring the muon flux involves the use of small solid-state counters distributed at different depths in the muon shield downstream of the decay pipe (Mount 1981). From these counters the relative π-decay muon intensity is known as a function of radius and depth (Shultze 1977, Wachsmuth 1977). The shapes of the distributions, shown in Figure 13, reflect the dominance of multiple scattering in all but the first gap, where there is some sensitivity

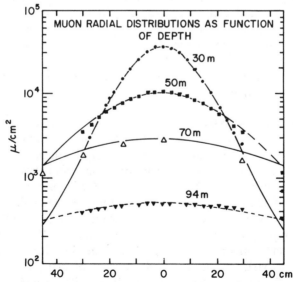

Figure 13 The distribution of muons from the CERN narrow-band beam at several depths in the shield. The peak is due primarily to muons from pion decay in the decay pipe (Shultze 1977).

to the angular divergence of the secondary beam. The solid-state counters seė all forms of ionization, including secondaries from upstream interactions in the shield, cosmic rays, etc. Hence, an absolute calibration is required to obtain the absolute intensity of muons. This was accomplished by counting tracks in nuclear emulsion exposed in the same location as one of the solid-state counters. A plot of the angular distribution of these tracks exhibits a large forward peak ($\theta < 20$ mrad), which, when compared with the background from knock-on electrons and cosmic rays at larger angles ($20 < \theta < 250$ mrad), gives the fraction of the solid-state counter signal that is due to π-decay muons alone (Perkins 1978). The number of ν_π is then calculable, assuming that the signal/background in the counters is independent of radial position and beam setting.

The Fermilab groups are presently trying to measure the muon flux using segmented ion chambers, which can be calibrated in a known-intensity muon beam. This should provide a necessary redundancy for the entire question of neutrino flux measurement.

2.3   Detectors

Our present understanding of the weak interaction and the intent to study charged-current nucleon structure lead to several requirements for a neutrino detector: (a) sizeable mass from cross-section considerations; (b) muon detection for the identification of charged-current events; and (c) energy measurement of the hadrons and muons, and muon angle measurement for determining the values of the relevant variables (x, y, Q^2) for individual events. Both bubble chamber and counter experiments, with appropriate beams, have been used to accumulate ν, $\bar{\nu}$ data.

2.3.1   BUBBLE CHAMBER EXPERIMENTS   The bubble chamber experiments have employed hydrogen, deuterium, Ne-H_2, and propane-freon mixtures as the active target liquids. Typical fiducial mass for the Fermilab 15-ft bubble chamber or BEBC range from about one ton for hydrogen to twelve tons for a 74% molar Ne-H_2 mix in BEBC. The neon mix, in addition to providing extra tonnage, has a short radiation length, which permits efficient detection of electrons. This, for example, has been important in studies of charm production. The detection and identification of muons is accomplished primarily with external muon identifiers (EMI). The BEBC EMI consists of two planes of PWC's following 15 collision lengths of iron. It identifies muons whose momenta are greater than 10 GeV/c with 98% efficiency (Brand et al 1978, 1982, Beuselinck et al 1978). The Fermilab EMI was reconfigured in

1977 into two PWC planes, following 3 to 5 pion absorption lengths. At the same time, an array of drift chambers internal to the bubble chamber vacuum vessel was added and this significantly improved the detection of dimuon events (Cence et al 1976, Stevenson 1978). Muon momentum resolution for the bubble chambers is typically 4–5% for 100 GeV/c muons, while the rms error on hadron energy, E_H, is 20%. The value of E_H for an event is determined by measuring all visible charged particles plus neutrals that interact. A correction for unmeasurable hadron energy of order 20% is applied to all events. The approximate magnitude of this correction is common to the CERN and Fermilab experience. It comes from the inability, in complicated events, to uniquely associate visible gamma rays and neutrons with a primary vertex. Checks on this correction, obtained by the transverse momentum balance technique, have been carried out using π^- beams into the bubble chamber, as well as by comparing the measured event energy with that expected in narrow-band beam experiments (Bosetti et al 1978).

2.4   Counter Experiments

The counter experiments designed to investigate charged-current neutrino physics have three major components: target, calorimeter, and muon spectrometer. The calorimeter provides a measurement of the hadronic energy by sampling the energy deposition throughout the cascade shower downstream of the vertex. For all but the emulsion experiment (discussed in Section 3), the target and calorimeter are combined with tracking chambers to provide the above requirements. In Table 4, the basic properties are given for the CDHS, CHARM, HPWFOR, Fermilab E 594, and CFRR neutrino detectors, all of which have been used to collect deep inelastic neutrino scattering data. We do not discuss here other counter neutrino detectors at Brookhaven, CERN, Fermilab, LAMPF, and various reactors designed specifically to investigate ν_μ and ν_e interactions at energies below the deep inelastic regime.

2.4.1   CFRR DETECTOR   The experimental arrangement for the CFRR detector is shown in Figure 14. The calorimetry (for measurement of hadronic energy, E_H) is accomplished by recording the pulse height from light produced by the hadron shower in liquid scintillation counters (Barish et al 1975). Each counter, 3 m × 3 m × 2.5 cm, is a lucite box containing primary scintillator and wavelength shifter chemicals dissolved in a mineral oil base. Along the four edges of the counter and separated optically from it by an air gap, there exist wavelength shifter bars that collect light, shift it from blue to green, and transmit it to

Table 4 Neutrino counter experiment apparatus summary

Experiment	Tonnage (metric tons) and target material	Calorimetry type/resolution	Tracking θ_μ type/resolution[a] (mrad/P_μ)	P_μ resolution (%)	Average density (g cm^{-3})
CDHS	1400 Iron	Fe + plastic scint. (5-cm sample) $0.70E_H^{-1/2}$	Drift chambers 150	±9	5.2
CHARM	170 Marble	Plastic scint. + proportional counters $0.53E_H^{-1/2}$	Proportional drift tubes 70	±16	1.3
HPWFOR	385 Fe Fe + Scint. Liq. Scint.	Fe + scint. $1.7E_H^{-1/2}$ Liq. scint. $0.7E_H^{-1/2}$	Optical spark chambers Unpublished	±12	7.9 3.0 0.8
Fermilab E 594	340 Sand + Fe Shot	Flash chamber Hits $0.8E_H^{-1/2}$	Flash chambers 70	±12	1.4
CFRR	1000 Fe + Scint.	Liq. scint. $0.89E_H^{-1/2}$ (10-cm sample)	Magnetostrictive spark chambers 35	±11	4.0

[a] Projected angular resolution for tracks outside of hadron showers. Resolution at the event vertex, which is more relevant, is not quoted here.

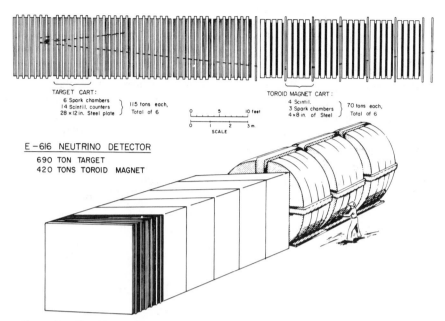

Figure 14 The CFRR detector in Fermilab's Lab E. The steel target region is instrumented with counters and spark chambers to detect the interaction point and to track the muon downstream. The toroids permit measurement of final state muon momenta.

photomultipliers located at each corner of the counter (Barish et al 1978a). In this manner, hadron showers are sampled every 10 cm of Fe. After appropriate addition of all the pulse heights in a shower, E_H is obtained with a fractional error of $0.89E_H^{-1/2}$ (Blair 1982). The individual counters yield ~10–15 photoelectrons for a minimum ionizing particle, which means that a threshold is easily set to signal the passage of a single muon for triggering purposes. In all, there are 82 planes of liquid counters in the CFRR target calorimeter.

Muons from neutrino interactions are identified and their position is determined by 3.2×3.2 m^2 magnetostrictive readout spark chambers. The 37 target spark chambers provide horizontal and vertical coordinates every 20 cm of Fe. To provide good multiple track efficiency, each chamber is individually pulsed in a transmission line mode with a 5-kV, 200-ns pulse across the 1.25-cm gap between plates. The spatial resolution of the chambers is ~0.5 mm. This implies that the projected muon angle error is 35 (mrad)/$P\mu$ (GeV) for muon tracks in which positions are measured in all chambers beginning at the vertex. In operation, hadron showers often mask the position measurement in the first chamber so that the actual angular resolution at the vertex is approximately twice this value $(70/P_\mu)$ in the CFRR experiment. The degree of muon visibility after the vertex, which always compromises angular resolution, depends highly on the experiment.

The muon momentum is measured in a downstream group of instrumented toroids. The toroid spectrometer consists of 24 planes of acrylic scintillator for calorimetry and triggering, 37 spark chambers 1.5×3.0 m, and 4.9 m of magnetized iron for muon momentum determination. In addition, there are three separate banks of scintillators for muon triggering and time-of-flight measurements. Muons traversing the length of the spectrometer have a fractional momentum error of 11%, which is largely determined by multiple scattering.

With the typical errors discussed above for the CFRR experiment, the x resolution can be calculated. Two examples are given in Figure 15, where it is noted that for small x ($x = 0.05$) the x resolution at small and large y is dominated by the error on the muon angle, while the large-x

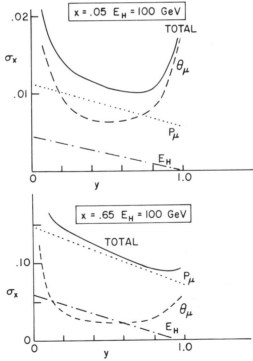

Figure 15 Examples of resolution on the x variable (σ_x) in the CFRR detector. At low x, the resolution on muon angle is most important; at large x, energy resolution is dominant. (Note difference in scale for σ_x.)

resolution is dominated by the muon momentum error. These resolutions produce systematic effects on structure function data that are similar in magnitude to the effects of Fermi motion in an iron nucleus.

2.4.2 OTHER DETECTORS   In the CDHS detector, all the iron is magnetized and thus provides for measurement of the muon momentum throughout the detector (Holder et al 1978). The measurement of track position comes from 19 large planar drift chambers with y, u, v wire orientation (Marel et al 1977). Calorimetry is accomplished by sampling at 5- or 15-cm intervals of Fe (Abramowicz et al 1981). In total, there are 145 planes of NE 110 scintillator (5 mm thick) with each plane consisting of 8 counters and 16 photomultipliers. The resolution achieved (Abramowicz et al 1981a) with 15(5)-cm sampling is $1.35E^{-1/2}(0.70E^{-1/2})$. This represents an improved resolution that is obtained by weighting of the individual counter readouts using a technique previously employed (Dishaw et al 1979) with an iron calorimeter at Fermilab, where the improvement obtained was considerably less dramatic. The CFRR calorimetry data also show very little improvement by employing weighting techniques.

The CHARM detector (Diddens et al 1980) was developed primarily to study neutral-current interactions. It is a fine-grained calorimeter that can measure hadron shower energy and direction. The fiducial mass of the detector is derived from 8-cm thick marble slabs, which give approximately equal lengths for hadronic and electromagnetic showers. Surrounding the marble and scintillator center of the detector is an iron frame magnetized for measurement of muons leaving the sides of the apparatus. Calorimetry for the showers that develop in the iron frame is accomplished by using proportional tubes extending through this frame (Bosio et al 1978). The target calorimeter is followed by toroidal iron magnets for measurement of the momenta of forward muons.

The HPWFOR detector, which has completed its experimental program, consisted of three separate regions of target and calorimeter followed by 3.6- and 7.2-m diameter muon spectrometers (Benvenuti et al 1977, 1978, Heagy et al 1981). The three regions of detector included: (*a*) solid Fe (250 mT), (*b*) liquid scintillator (45 mT), and (*c*) Fe and scintillator target calorimeter (90 mT). Tracking was done with wide- and narrow-gap optical spark chambers.

The Fermilab E 594 detector (Taylor et al 1978, 1980, Bogert et al 1982, Bofill et al 1982, Stutte et al 1981) is a fine-grained calorimeter built to study neutral currents and $\nu_\mu e$ scattering, and consequently has many of the same design objectives as the CHARM detector. It is constructed of 38 modules each of which contain 16 (4 each of u, x, y, x)

flash chambers (5-mm cell size) for a total of 304 x horizontal chambers and 152 u ($+10°$) and 152 y ($-10°$) vertical planes. The mass of the detector is made by filling 1.6-cm thick acrylic extrusions alternately with sand and steel shot. Proportional tubes, which follow each flash chamber module, provide the ability to trigger the apparatus and also give an alternative measure of an event's hadron energy. The gas-filled flash cells that glow discharge through the ionization path after being pulsed with high voltage are read out using magnetostrictive techniques. Hadron energy and energy flow for an event are then determined by the cells that have been hit. Figure 16 shows the response of the detector to a 35-GeV electron and a 35-GeV π^-. Downstream of the flash chamber

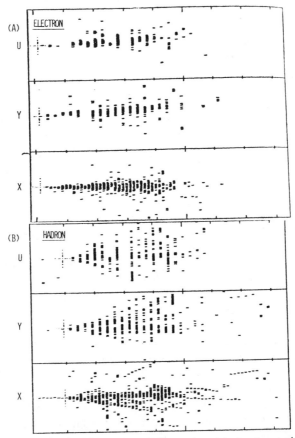

Figure 16 Response of the Fermilab Lab C neutrino detector to an electron and a hadron. The different transverse growth of the showers permits separation of electrons from hadrons in the final state of neutrino collisions.

target calorimeter are the muon spectrometer magnets built by the HPWFOR group, which consist of 3.6- and 7.2-m diameter toroids. The E 594 group has installed planes of proportional tubes for muon momentum measurement.

2.4.3 EVENT TYPES All of the above detectors, using various arrangements for their trigger logic, can trigger on both charged- and neutral-current events. Examples of such triggers are shown in Figure 17.

3 TESTS OF THE QUARK MODEL AND DATA COMPARISONS

3.1 *Total Cross Sections and* y *Dependence*

An important test of the constituent quark model involves the energy dependence of the total cross sections (σ^ν and $\sigma^{\bar\nu}$). These should be nearly linear with respect to incident laboratory neutrino energy. The slope parameters, $\alpha^\nu = \sigma^\nu/E$ and $\alpha^{\bar\nu} = \sigma^{\bar\nu}/E$, are commonly used to refer to cross-section measurements. In the scaling limit, the sum and difference of these slope parameters for a target of equal numbers of neutrons and protons are given by

$$\alpha^+ \equiv \alpha^\nu + \alpha^{\bar\nu} = \frac{4}{3} \frac{G^2 M}{\pi} c' \int F_2(x) \, dx,$$

$$\alpha^- \equiv \alpha^\nu - \alpha^{\bar\nu} = \frac{2}{3} \frac{G^2 M}{\pi} \int x F_3(x) \, dx,$$

21.

where $F_2(x)$ and $xF_3(x)$ are defined in Table 3. The coefficient c' includes a small correction for non-spin-$\frac{1}{2}$ ($R \neq 0$) and for the complications arising from the strange sea:

$$c' \equiv \left[1 + \frac{3}{4} R + \frac{1}{2} \int (s + \bar{s}) \, dx \middle/ \int (q + \bar{q}) \, dx \right] \middle/ (1 + R).$$

For the assumptions $R = 0.1$ (see Section 3.2) and $\int \bar{s} \, dx / \int (q + \bar{q}) \, dx = 0.025$ (see Section 4), these $2\frac{1}{2}\%$ effects exactly cancel ($c' = 1.00$).

The constancy of the cross sections divided by energy is an important test of the point-like constituency of nucleons. This test is obeyed within the accuracy of all measurements. For example, Figure 18 shows recent measurements by the CFRR group (Blair 1982, Rapidis et al 1981, Shaevitz et al 1981) of neutrino and antineutrino cross sections divided by energy ($\sigma^{\nu, \bar{\nu}}/E$), which are expected to be nearly constant. Although these data hint at a gentle rise with energy, the hypothesis of constancy

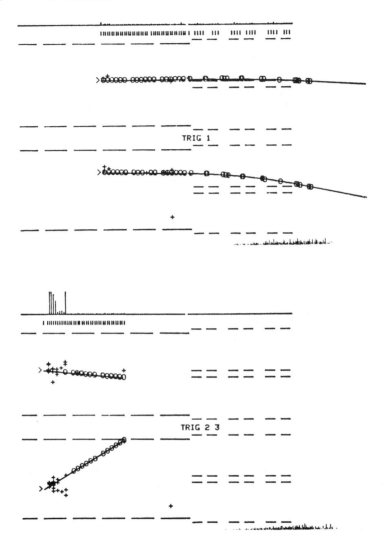

Figure 17 Neutrino charged-current interactions as recorded in the CFRR detector. Side and top views are shown for:
(*a*) event with very little hadronic energy and forward muon,
(*b*) event with very-wide-angle muon and considerable hadronic energy as sensed by the scintillation counters. The energy deposited in each counter is illustrated in the histogram at the top of the figure. The linear scale of energy deposition is calibrated with muons. Calibration of energy deposition vs hadronic energy is accomplished with a pion beam of known energy.

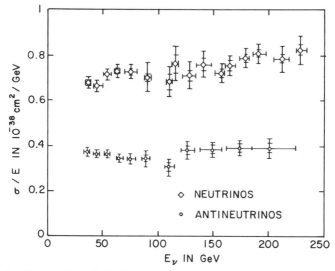

Figure 18 Cross sections divided by neutrino energy from measurements by the CFRR group. This quantity is seen to be independent of energy within present errors. Such cross sections may be converted into integrals over the structure functions.

is acceptable within the errors. Earlier measurements (Barish et al 1977, Bosetti et al 1977, de Groot et al 1979) are also consistent with constant values of α.

By utilizing Equation 21 and the values of α^ν and $\alpha^{\bar\nu}$ shown in Figure 18, one can determine the fraction of momentum carried by interacting constituents. The result is $\int F_2 \, dx = 0.512 \pm 0.018$ and $\int x F_3 \, dx = 0.341 \pm 0.036$. In terms of fractional momentum carried by quark constituents, this is equivalent to $\int q(x) \, dx = 0.403 \pm 0.022$ and $\int \bar q(x) \, dx = 0.062 \pm 0.018$, and, by assumption, the remaining 10% is carried by non-spin-$\frac{1}{2}$ constituents $(R = 0.1)$. One feature, dependent on the ratio of antineutrino to neutrino cross sections, is that the antiquark fraction is approximately 13–17% of the observed total. This number is generally agreed upon. It has been verified by the y dependence of the cross section, where the relative dominance of quark over antiquark fraction should produce a nearly flat y dependence for neutrinos, and an antineutrino cross section dominated by the $(1 - y)^2$ term (see Equation 9). Figure 19 shows this behavior explicitly for the CDHS data (Holder et al 1977, de Groot et al 1979), where the data indicate (assuming $R = 0$) that $\int \bar q \, dx/\int(q + \bar q) \, dx = 0.15 \pm 0.02$. Others observe very similar y dependence, both qualitatively and quantitatively (e.g. Barish et al 1978, Bosetti et al 1978, Heagy et al 1981, Blair 1982). Precise agreement between these two techniques cannot be expected if the

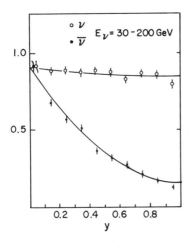

Figure 19 The differential cross section vs *y* for neutrinos and antineutrinos from CDHS data. The qualitative behavior of dominantly flat for neutrinos and dominantly $(1 - y)^2$ for antineutrinos is a feature anticipated in the quark model with valence quarks participating in the major portion of the scattering cross section. Similar behavior is observed in all neutrino experiments.

structure functions show explicit Q^2 dependence (see Section 3.6) or some finite R value exists to change the y dependence (see Section 3.2).

There is not, however, universal quantitative agreement on the magnitudes of the cross sections, and, therefore, on the net fraction of momentum carried by interacting constituents. The CFRR data of Figure 18 imply that the quarks carry 47% of the nucleon momentum, but older data imply less than 40% for this fraction. Table 5 shows a comparison of total cross sections obtained from several measurements. Here we see the discrepancy between the recent measurements of the CFRR group (Lee 1980, Blair 1982, Rapidis et al 1981, Shaevitz et al 1981) and older measurements (Barish et al 1977, Bosetti et al 1977, de Groot et al 1979), which average about 15% in the overall normalization and therefore in the fractional momentum as well. The last entry is a recent reanalysis of the BEBC data, incorporating a new flux calibration that employed muon detectors in the shield. One of these detectors was calibrated using emulsions to separate the muons of in-flight pion decay from extraneous background Section 2.2.3). This neutrino cross section lies between the others. There is agreement with the CFRR (Rapidis et al 1981) value of the integrated xF_3 value, but rather poor agreement on the sea contribution.

The discrepancy in normalizations is of some importance for the reasons discussed above, but also because it is directly related to tests of the mean-square charge of the interacting constituents (see Section 3.3). Since the experiments use different techniques for absolute ν flux normalization, efforts are under way to more precisely monitor flux and to incorporate aspects of the two techniques in all experiments (Rapidis et al 1981).

Table 5 Measurements of cross-section slopes

Reference	Energy range (GeV)	$\alpha^{\nu\,a}$	$\alpha^{\bar{\nu}\,a}$	$\int x F_3 \, \mathrm{d}x$	$\int F_2 \, \mathrm{d}x$	Q	$\bar{Q}$
Astratyan et al 1978	3–30	0.72±0.07	0.32±0.03	0.379±0.072	0.493±0.036	0.413±0.049	0.034±0.027
Barish et al 1977	40–200	0.61±0.03	0.29±0.02	0.303±0.034	0.427±0.017	0.346±0.021	0.042±0.016
Bosetti et al 1977	40–200	0.63±0.05	0.29±0.03	0.322±0.055	0.436±0.028	0.359±0.035	0.037±0.024
deGroot et al 1979	30–200	0.62±0.05	0.30±0.02	0.303±0.051	0.436±0.026	0.350±0.035	0.046±0.019
Jonker et al 1981	25–260	0.60±0.03	0.30±0.02	0.284±0.034	0.427±0.017	0.336±0.021	0.051±0.016
Lee 1980	25–260	0.70±0.03					
Rapidis et al 1981	30–230	0.70±0.03	0.35±0.023	0.331±0.036	0.498±0.018	0.392±0.022	0.061±0.018
Fritze et al 1981	10–200	0.66±0.03	0.30±0.02	0.341±0.034	0.455±0.017	0.377±0.021	0.036±0.016

[a] Measured $\times\ 10^{-38}\ \mathrm{cm}^2\ \mathrm{GeV}^{-1}$.

3.2   *The* R *Parameter*

The prediction obtained by assuming spin-$\frac{1}{2}$ constituents (i.e. $R = 0$) is not expected to be exact, as discussed in the introduction. For example, at low Q^2, R is expected to be nonzero (e.g. $R \sim 4\langle P_T^2\rangle/Q^2$) and perhaps it has logarithmic dependence on Q^2 (from QCD effects) at higher energies. Thus far, measurements have not been very precise because they are exceedingly difficult to make. It requires extracting the R parameter from y distributions at fixed x and Q^2. That is, the data should be divided among three variables, which makes statistical precision difficult. The systematic problems created by resolutions and radiative corrections are also very important.

Table 6 shows the average values of R as obtained from two sets of neutrino data. Also shown are R parameters obtained from (μ, e) inelastic scattering. There is general agreement among all but the last entry (to which we return) that R is small ($\lesssim 0.2$) on the average. The most recent measurement (Abramowicz et al 1981) is shown in Figure 20 as a function of x, averaged over Q^2. The data are not precise enough to be treated in these two variables separately, so they have been averaged over Q^2 at fixed x. This means that the low-x data also are dominated by low Q^2 events. Tentatively then, the R parameter

Table 6   Measurements of R

Group/Reference/Reaction	x Range	Q^2 Range (GeV^2)	R
BEBC/Bosetti et al 1978/νN	0–1.0	0.1–50	0.15±0.10±0.04
CDHS/Abramowicz et al 1981/νN	0–0.6	2.5–75	0.13±0.035±0.04
SLAC/Mestayer 1978/ep	0.1–0.9	2–20	0.21±0.10
CHIO/Gordon et al 1979/μp	0–0.1	1.5–12	$0.52 {}^{0.17}_{0.16}$

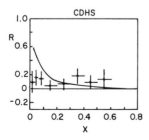

Figure 20   Experimental values of $R = \sigma_s/\sigma_T$ from neutrino scattering. A value of $R = 0.13$ independent of x and Q^2 is consistent with the data (Abramowicz et al 1981).

shows no dramatic dependence on x (or Q^2) and the qualitative features expected from QCD predictions do not appear. It is possible, of course, that the R parameter gets contributions from more than a single source, which may make it difficult to draw definitive conclusions regarding QCD.

The last entry for R, obtained from muon scattering (Gordon et al 1979) on hydrogen, is specifically in the small-x region. The dependence on small-x values is shown in Figure 21a. This is somewhat reminiscent of the behavior expected from QCD, in the sharp falloff with x. It should be noted that the Q^2 in this small-x region are quite low ($\lesssim 1$ GeV^2). The dependence on higher Q^2 is shown in Figure 21b, where the upper dashed curve parametrizes a ln Q^2/Λ^2 dependence, and the other curve has a $1/Q^2$ dependence. The data cannot choose between them. The only neutrino data available with $x < 0.1$ (Figure 20) do not show any hint of large R at small x. More information on R in this small-x region is clearly desirable both from muon scattering and from neutrino scattering. It is important to know whether the scattering is dominantly non-spin-$\frac{1}{2}$, as observed by CHIO, or whether the dominant spin-$\frac{1}{2}$ structure is preserved to very small x. If R is large in this region, it is equally important to determine the Q^2 dependence; does it fall like a power of $1/Q^2$, or is it slower in its falloff and therefore has its origin in perturbative effects?

3.3 F_2 Structure Functions—x Dependence

The deep inelastic structure functions have been measured by several groups, which we designate BEBC (Bosetti et al 1978), CDHS (Eisele

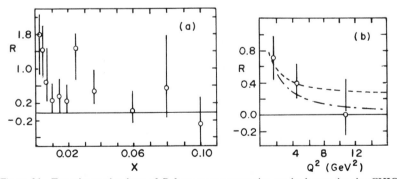

Figure 21 Experimental values of R from muon scattering on hydrogen by the CHIO group. (a) The very-small-x region shows indication of a major enhancement in the non-spin-$\frac{1}{2}$ scattering. (b) The experimental values show a falloff in Q^2, but cannot differentiate between hypotheses of logarithmic falloff and power law falloff, illustrated by the two curves.

et al 1981, de Groot et al 1979, 1980) HPWF (Heagy et al 1981) and GGM (Morfin et al 1981). In this section, we discuss the qualitative features of this neutrino data, and make comparisons among the various data sets. Since an important prediction of the quark model relies upon the mean-square-charge test, i.e. a comparison of electron/muon structure functions (F_2^{eN}) with those from neutrino ($F_2^{\nu N}$), we also proceed to determine the consistency among several measurements of F_2^{eN}. Finally, we remark on the present level of agreement between weak and electromagnetic structure functions in the context of the quark mean-square-charge hypothesis. We postpone discussion of the Q^2 dependence (at fixed x) to a later section.

3.3.1 NEUTRINO STRUCTURE FUNCTIONS   The most statistically significant structure function data are those of the CDHS group (de Groot et al 1979, 1980, Eisele et al 1981). In particular, the latter data (more recent) employs a 120,000-event sample. Other groups use 5–10 times fewer high-energy events. The CFRR group has a high-statistics sample, from which structure functions are expected in 1982. Figure 22 shows the x dependence of the F_2 and xF_3 values from the earlier CDHS data (de Groot et al 1979) averaged over the neutrino energy range, $30 < E_\nu < 200$ GeV. The curve utilizes the functional dependence described below with the exponents best fitted by their data; the overall

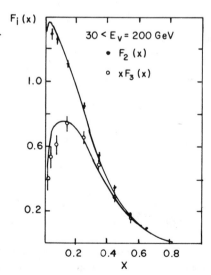

Figure 22   Qualitative features of the x dependence of F_2 and xF_3 as seen in the early CDHS data, averaged over neutino energy. The curves drawn (Equation 22) are a reasonable representation of the qualitative behavior, especially at larger x values.

normalization of the curve has been constrained by the GLS sum rule (see Section 1.7) and by the antiquark fraction, $\int \bar{q} \, dx / \int (q + \bar{q}) \, dx = 0.15$ (see Section 3.1). This parametrization assumes that $R = 0$ and

$$q(x) - \bar{q}(x) = 3\sqrt{x}(1 - x)^{3.5}/\beta(0.5, 4.5) \qquad\qquad 22.$$

$$\bar{q}(x) = 0.478 \, (1 - x)^{6.5}. \qquad\qquad 23.$$

This is a useful functional approximation. It conforms to these data shown for $x > 0.1$, but does not fit very well at lower x values. The newer CDHS data (de Groot et al 1980, Eisele et al 1981) lie even lower at small x.

The other neutrino data sets have been compared directly to the new CDHS data to determine whether there are any significant differences. Figure 23 shows the comparison between BEBC and CDHS $F_2(x)$ in overlapping Q^2 ranges as a function of x. Since both groups obtained

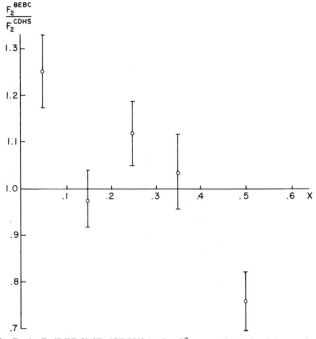

Figure 23   Ratio F_2 (BEBC)$/F_2$ (CDHS) in the Q^2 range where the data overlap for these neutrino experiments. The error bars primarily reflect the statistical limitations in the BEBC data. The ratios at small and large x indicate a qualitative shape difference between the data sets.

very similar total cross-section values, we expect little difference on the average, but we may find some shape differences. The BEBC data appear to be significantly higher at small x $(x < 0.10)$ and to have a smaller value of F_2 at large x $(x \sim 0.5)$. The BEBC structure function appears, therefore, to be more sharply peaked to small x than does that of CDHS. This is a bigger effect than might be anticipated on the basis of Fermi-momentum correction effects between heavy liquid (BEBC) and iron (CDHS) targets (Bodek & Ritchie 1981). The GGM data show a similar, but less significant, behavior relative to CDHS. The HPWF data do not show it. We conclude that the x distributions from neutrino scattering have the qualitative behavior vs x as expressed in Equations 22 and 23; but there are some specific uncertainties: (*a*) the overall normalization has been measured to be different at approximately the 15% level (see Section 3.1); (*b*) the shape as observed in bubble chamber experiments is somewhat "sharper," showing differences of about 25% at small x and at large x; and (*c*) a finite R value affects the overall level of the quark and antiquark functions in proportion to its value. Although measurements of F_2 are largely independent of R, the magnitudes of the q and $\bar{q}$ distributions do depend on R.

3.3.2   ELECTRON/MUON STRUCTURE FUNCTIONS   Several groups have measured electromagnetic structure functions (Bollini et al 1981, Aubert et al 1981, 1981a, Bodek et al 1979, Gordon et al 1979); the groups (targets) are BCDMS (carbon), EMC (H_2), EMC (iron), SLAC (H_2, D_2), and CHIO (H_2), respectively. The data with the highest statistical value and covering the greatest range of x and Q^2 are those of the EMC group. We proceed to compare, where possible, the data of the various groups in the least biased manner. The most direct way to do this is to utilize H_2 data, which does not suffer from differing Fermi-motion corrections.

Figure 24 shows the comparison of F_2 vs x for the data of CHIO (Gordon et al 1979) and EMC (Aubert et al 1981) over the same Q^2 range. The values of F_2 are obtained with the same assumption for R ($R = 0.1$), although the qualitative features shown are not very dependent on this value. On the same graph is a comparison made independently (Smadja 1981) between SLAC (Bodek et al 1979) and EMC data. We perceive that the SLAC data lie systematically higher by about 10%, while the CHIO data are about 8% higher on average. The latter data indicate some x-dependence difference as well. The systematic errors in normalization (3–5%) quoted by these groups are similar to each other. To make the comparison between muon and neutrino scattering, we use the EMC data renormalized by the factor 1.06 ± 0.06.

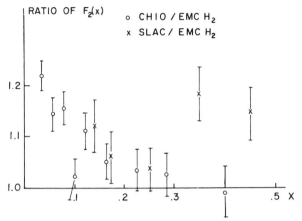

Figure 24 Ratio of F_2 (CHIO)/F_2 (EMC) and F_2(SLAC)/F_2 (EMC) in regions of Q^2 overlap from these electron and muon experiments on hydrogen. The ratios have been obtained with the assumption $R = 0.1$ in the former comparison. The errors primarily reflect the errors quoted by the CHIO and SLAC groups. We assign an empirical renormalization factor, on the basis of these measurements, of 1.06 ± 0.06 to the EMC data for subsequent comparison.

This factor and its assigned error reflects an empirical uncertainty in normalization observed for the three experiments being compared.

3.3.3   TEST OF THE MEAN-SQUARE CHARGE   The νN structure function predicted by quark charges for μN scattering is given by

$$F_2^{\mathrm{pred}}(x) = \frac{\frac{18}{5}CF_2^{\mathrm{EMC}}(x)}{[1 - \frac{3}{5}s^{\mathrm{corr}}(x)]},\qquad 24.$$

where C is the normalization parameter obtained in the last section, $C = 1.06 \pm 0.06$, and $s^{\mathrm{corr}}(x) = [s(x) + \bar{s}(x)/[q(x) + \bar{q}(x)]$ is the correction for strange sea. Here $q(x)$ and $\bar{q}(x)$ have been taken from Equations 22, and $s(x) = \bar{s}(x) = (0.2 \pm 0.1)\bar{q}(x)$, corresponding to $(50\% \pm 25\%)$ of an SU(3) symmetric strange sea $(\bar{s} = \bar{u} = \bar{d})$. The ratio $F_2^{\mathrm{pred}}(x)/F_2^{\mathrm{CDHS}}(s)$ is shown in Figure 25. This ratio should be unity at all x if the hypothesis of the $\frac{5}{18}$ mean-square charge for quarks is to be corroborated by these data. Also shown in the figure is the uncertainty in μN normalization and the uncertainty in the strange-sea correction. At large x ($x > 0.4$), the data are consistent with the 6% normalization uncertainties that we have assigned to the muon data. At smaller x, the discrepancy is greater than 20% for $x < 0.1$. This discrepancy cannot be explained by strange-sea assumptions and muon normalization uncertainties. In these terms, the difference could only be explained if the

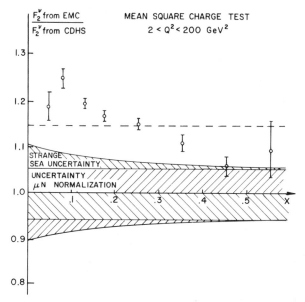

Figure 25   The ratio of F_2^ν predicted by EMC muon data (on the basis of the mean-square-charge relation of Equation 24) divided by the F_2^ν measured by CDHS. Both data sets have the assumption $R = 0.1$. The ratio predicted in the quark model is 1.0, independent of x. Uncertainties on the muon data normalization and on the strange-sea correction are indicated. The dashed line corresponds to the quark model prediction if the neutrino cross section of CDHS were 15% higher.

charm sea were larger than the strange sea; this seems unlikely. The systematic discrepancy (i.e. all x values lie above unity) may be related in part to the total cross-section controversy. If the neutrino data were renormalized upward by 15%, as indicated by the CFRR cross-section measurement, the quark model prediction would lie on the dashed line. This may help to resolve the systematic discrepancy; but there would still remain an x-dependent difference. The muon data show higher values at small x, lower values at large x; in other words, a distribution in x that is considerably more peaked toward small x.

3.4   *The* xF_3 *Structure Function—x Dependence*

The xF_3 structure function is unique to neutrino scattering since it exists only because of the parity violation in the weak interaction. The integral of F_3 is important because it directly measures the valence quark composition of the nucleon. As with the F_2 structure function, both x and Q^2 dependence are observed for xF_3. We concentrate here on the high-statistics data of the CDHS group (Eisele et al 1981). Systematic differences in the x dependence when compared with data of other groups (Bosetti et al 1978, Heagy et al 1981, Morfin et al 1981) are qualitatively similar to those already discussed in Section 3.3.1 for the F_2 structure function.

The Q^2 dependence at fixed x of xF_3 is rather similar to that of F_2, discussed in the next section. Here we concentrate on fixed Q^2, where the data can be satisfactorily fit to a functional form similar to Equations 22 and 23, i.e.

$$xF_3(x) = Cx^B(1-x)^A. \qquad 25.$$

Here A, B, and C will, in general, depend on Q^2.

The qualitative behavior of these exponents has been anticipated from model arguments in the extreme low-x and high-x regions. At small x, the exponent B dominates the behavior. From Regge theory, we expect B will be directly related to the trajectory intercept (α), which is dominant for xF_3: $B = 1 - \alpha$ (Reya 1981). Since xF_3 carries the nucleon's isospin, the highest contributing exchange is the (ρ, ω) trajectory with $\alpha \simeq \frac{1}{2}$. The next lowest trajectory is that of the π with $\alpha \approx 0$, which would put B in the range $\frac{1}{2} < B < 1$. The behavior near $x = 1$ is dominated by the exponent, A, where the predictions are based on counting rules and imply $A \simeq 3$ to 4 (Gunion 1974, Farrar 1974).

Figure 26 shows the values of these exponents from fits to the CDHS data at fixed Q^2. The parameter, A, lies slightly below, but is consistent with the value three over this range of Q^2. The parameter B falls from ≈ 0.65 to ≈ 0.4 over the Q^2 range between 2.5 and 25 GeV^2. A lower value of B signals a more gentle rise with x, which could be an artifact due to the absence of very-small-x data at higher Q^2.

3.5   The Gross–Llewellyn Smith Sum Rule

The GLS sum rule (see Section 1.7) is not expected to be exact, but is predicted to have perturbative QCD corrections at high Q^2 and possibly substantial power law dependence at low Q^2. To first order in perturbative QCD, we expect (Bardeen et al 1978)

$$S \equiv \int_0^1 F_3^{\nu N}(x)\, dx = \int_0^1 \frac{xF_3}{x}\, dx = 3\left[1 - \frac{12}{(33 - 2f)\ln Q^2/\Lambda^2}\right], \qquad 26.$$

where f is the number of active flavors (3 to 4), and Λ is the scale parameter of QCD, $\Lambda \simeq 0.2$ GeV (see Section 3.6). Hence, we might expect a result roughly 5–10% lower than the nominal value at presently available Q^2.

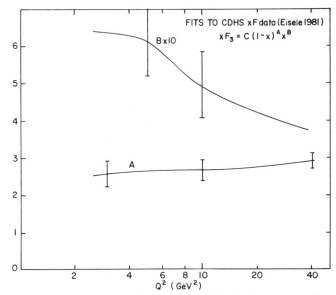

Figure 26 Values of the exponents from fitting CDHS xF_3 data to the indicated form. Expectations are that A would be around 3 to 4 and that B would be between 0.5 and 1.

A determination of S was made several years ago by the CDHS group (de Groot et al 1979). They pointed out that the result depends sensitively on the very-small-x behavior of xF_3. By numerically integrating the data down to $x_{min} = 0.005$, they obtained $S_{x_{min}} = 2.7 \pm 0.3$. Extrapolating from x_{min} to zero, and using the fitted values in Equations 22 and 23, gave $S = 3.2 \pm 0.5$ which dramatically illustrates their point. A similar effect was reported earlier by the BEBC group (Bosetti et al 1978).

Alternatively, we may use the values of A, B, and C at fixed Q^2 found by fitting the newer CDHS data discussed in the last section. From these parameters and their covariance matrix, the integral $\int F_3 \, dx$ and its error were calculated directly. Figure 27 shows the resultant integral over the range $1 < Q^2 < 10$ GeV^2, where there exist data at small x. This technique of finding the integral of F_3 also suffers from insufficient knowledge in the small-x region. Note that the value of the integral is substantially lower than the prediction (Equation 26), which is also shown in the figure.

In any case, in the Q^2 range where data exist at smaller x values ($Q^2 < 10$ GeV^2), the sum rule result is very low. This represents a problem at present. Its resolution may be found in one or more of the

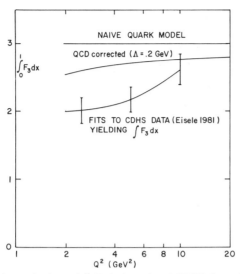

Figure 27 Experimental values of $\int_0^1 F_3 \, dx$ from fits of CDHS data of Equation 25. The values are somewhat lower than expectations.

following explanations: (*a*) The technique applied here is generally inferior to one in which the integration is performed on unbinned data. (*b*) A renormalization of the structure functions upward may occur because of a higher total cross section. (*c*) A shape difference with relatively higher structure function values at small x would raise this integral. (*d*) A serious problem exists with the quark model.

An illustration of points (*b*) and (*c*) lies in a recent evaluation of this sum rule by the BEBC group (Fritze et al 1981). Their new cross-section measurements are such that the integrated value of xF_3 is higher than that from the CDHS data by 13% (see Table 5) and their values of xF_3 at small x are considerably higher (illustrated by the F_2 comparison in Figure 23). Their evaluation of the sum rule is shown in Figure 28, which is by contrast even higher than the QCD corrected prediction. Clearly, this very critical test of the quark model requires some attention.

3.6 *The* Q^2 *Dependence of Structure Functions at Fixed* x

Scaling violations have been observed ever since approximate scaling was discovered (Miller et al 1972, Bodek et al 1973, Eichten et al 1973). Over the low-Q^2 regions ($Q^2 \lesssim 2$ GeV^2), the structure functions at fixed x were found to vary by large factors. A better approximation to scaling was found by using a new scaling variable, e.g. $x' = Q^2/(2M\nu + M^2)$, which approaches x at large Q^2 (Bodek et al 1973). This is very similar in its effect to a $1/Q^2$ dependence of the structure functions.

Theoretical interest in the past few years has focused on perturbative QCD, which predicts that structure functions have an approximate logarithmic dependence on Q^2. Quantitative predictions can be made at any Q^2 by using the quark and gluon structure functions at a single value of $Q^2 = Q_0^2$ as input. The evolution of the distributions, in the leading order of a perturbation series, depends on a single scale parameter, Λ, measured in GeV. Higher-order corrections in the perturbation can also be calculated, but the $1/Q^2$ terms cannot. These depend on the bound-state wave functions of the quarks, which thus far have eluded calculation in the theory.

The principle uncertainty faced in confronting the predictions of QCD with experimental tests is that we do not know *a priori* where the perturbative (logarithmic) behavior begins to dominate over the bound-state effects. At some high Q^2, logarithmic dependence will dominate over power law (higher-twist) dependence. While we might naively presume that the higher-twist effects will disappear near energies corresponding to quark bound states (~ 1 GeV^2), there could be bizarre x dependences that amplify them (see Section 1).

Historically there have been proponents of the ansatz that perturbative behavior begins at very low Q^2. The only trustworthy empirical result that might corroborate the dominance of perturbative QCD behavior would be in the agreement of the parameter Λ over a very large range of Q^2. The agreement of the data with the best-fit structure function curves and the agreement with prediction of the ratios of structure function moments have been touted as verification that the perturbative behavior has been seen. But there is reason to doubt the sensitivity of these tests as we discuss below.

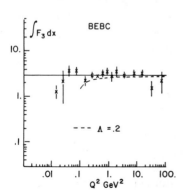

Figure 28   Values of $\int_0^1 F_3 \, dx$ quoted by the BEBC group over a large range of Q^2, using their most recent normalization (Table 5). These values are consistent with 3 at essentially all Q^2. Note the logarithmic scale for the ordinate.

Table 7 shows the values of Λ obtained by various groups using neutrino and muon beams. The results have been obtained using three different techniques: (*a*) moment analysis (Nachtmann 1973), (*b*) the Buras-Gaemers prescription (Buras & Gaemers 1978), (*c*) evolution of the structure functions from a simple parametrization at Q_0^2 using the Altarelli-Parisi equations (Altarelli & Parisi 1977, Gonzalez-Arroya et al 1979, 1980). These procedures are described in detail by others (see, for example, Söding & Wolf 1981). The second technique (*b*) also evolves the structure functions in Q^2, but assumes a specific functional dependence on x at each Q^2.

The first entry in the table was the earliest attempt at confrontation of perturbative QCD with experiment. Data were used that extended to very low Q^2 (~ 0.15 GeV^2). The moment analysis technique seemed appropriate since the ratio of moments obtained from the data agreed with the predictions of QCD. There have been many criticisms of this technique since then (see, for example, Söding & Wolf 1981). However, we see that experiments using that technique or others obtain values of Λ largely independent of the procedure. On the other hand, the ratio of moments test does not appear to be very sensitive to the source of the scaling violations.

The most striking feature from this table is the evolution in values of Λ: from over 700 MeV in 1978 down to 80–250 MeV most recently. The ratios of moments agreed with QCD predictions in all cases; the data were consistent with the theoretical fits in all cases. The change in the value of Λ is primarily associated with the Q^2 range of the data used for the fitting. It seems unlikely, therefore, that the low-Q^2 data are described by the same QCD as that describing the high-Q^2 data.

As an illustration of this point, Figure 29 shows the F_2 data at $x = 0.5$ of BEBC (Bosetti et al 1978) and CDHS (de Groot et al 1980, Eisele et al 1981). For $Q^2 > 5$ GeV^2, the qualitative variation with Q^2 is similar. (The difference in normalization has already been mentioned in the discussion of Figure 23.) For $Q^2 < 5$ GeV^2, the BEBC data show a very pronounced rate of change. This is directly related to the large values of Λ obtained when lower-Q^2 data are included in the fit.

The best and perhaps only test of perturbative QCD rests in the measured values of Λ (in first order) or, more generally, in the value of the quark-gluon coupling, $\alpha_s(Q^2)$. We should obtain consistency for Λ in muon and electron experiments and in e^+e^- production of jets, etc. More importantly, the value of Λ should not depend on the range of Q^2 used in the fit so long as the momentum transfers are high enough.

Values of Λ in the range 100–200 MeV seem very reasonable for a scale of the strong interactions. Higher values have also been deemed

Table 7 Values of Λ (GeV) obtained from fits to inelastic data

Reference	Group	Reaction	Moments	Buras-Gaemers technique	Altarelli-Parisi technique
Bosetti et al 1978	BEBC	νN	0.74 ± 0.05 GeV $(Q^2 > 0.15 \text{ GeV}^2)$		
de Groot et al 1979	CDHS	νN (Fe)		$0.47 \pm 0.1 \pm 0.1$ $(Q^2 > 3 \text{ GeV}^2)$	
Aubert et al 1981	EMC	μN (Fe)			$0.122^{+0.022+0.114}_{-0.020-0.070}$ $(Q^2 > 3.2 \text{ GeV}^2)$
Aubert et al 1981	EMC	μp (H$_2$)			$0.110^{+0.058+0.114}_{-0.020-0.070}$ $(Q^2 > 2.5 \text{ GeV}^2)$
Bollini et al 1981	BCDMS	μN (C)	$0.080^{+0.130+0.100}_{-0.080-0.070}$ $(Q^2 > 25 \text{ GeV}^2)$	$0.136^{+0.050+0.090}_{-0.040-0.080}$ $0.3 < x < 0.7$	$0.085^{+0.060+0.090}_{-0.040-0.080}$
Eisele et al 1981	CDHS	νN (Fe)			0.19 ± 0.08 $(Q^2 > 2 \text{ GeV}^2)$ $(W^2 > 11 \text{ GeV}^2)$
Fritze et al 1981	BEBC	νN	$0.245^{+0.130}_{-0.145}$ $(Q^2 > 1.5 \text{ GeV}^2)$		0.210 ± 0.095 $(Q^2 > 2 \text{ GeV}^2)$

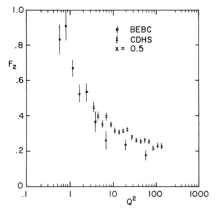

Figure 29 Comparison of BEBC and CDHS F_2 data at $x = 0.5$. The behavior below $Q^2 = 5\,\mathrm{GeV}^2$ in the BEBC data shows very strong dependence on Q^2.

reasonable. One source of concern is whether Λ will continue to grow smaller as data become better and extend to higher Q^2. It will be peculiar if bound states of quarks are characterized by radii of order $(m_\pi c/\hbar)^{-1} \approx (140\,\mathrm{MeV})^{-1}$, and the distance scale of the force were to be significantly longer than this.

For the experimentalist, the problem is a difficult one. The experiments represented in the last five entries of Table 7 generally have values of Λ consistent with 200 MeV, but with some variation around this value. The differences in quoted values of Λ are directly attributable to differences in the behavior of the data, particularly at high x. All techniques, especially the moments analysis and the Altarelli-Parisi evolution, give very consistent answers, so differences in analysis procedure are not likely to be producing different Λ values. As an illustration that the Λ values are closely related to the raw data, Figure 30 shows a slope parameter, $b(x)$, defined by fits to the functional form

$$F_2(x,\ Q^2) = A(x)[1 + b(x)\ \log_{10}(Q^2/10)].\qquad\qquad 27.$$

The slope parameter $b(x)$ is approximately $\Delta F_2(x)/F_2(x)$, or the fractional change in F_2, over a decade in Q^2. The data points are from the BCDMS (μN), EMC (μN), and CDHS (νN) groups, respectively. The curves have been drawn to guide the eye. At small x ($\lesssim 0.2$), the two groups of data show very similar behavior. At large x, where muon and neutrino results should be dominated by the same valence quark distribution, there are systematic differences in the change in F_2 per decade. These are directly correlated with the Λ values obtained by the groups: larger Λ is associated with more negative values of b. It should be noted that the error bars shown for the EMC data are statistical only. At the present level of precision, the experimenters agree that the differences in Λ are consistent within systematic and statistical errors.

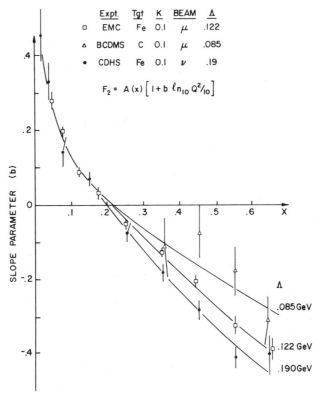

Figure 30 Slope parameter vs x over similar Q^2 ranges from three different experiments. The curves have been drawn to guide the eye only. The best values of Λ obtained in QCD fits are indicated. Differences in Λ are correlated with different systematic logarithmic behavior at large x in the raw data.

With small values of Λ generally agreed upon, there is a critical challenge to be met. This is to obtain high-statistics muon and neutrino inelastic data with the greatest possible precision and over the widest possible range of Q^2, so that the value of Λ can be measured accurately. The extension of the Q^2 range is crucial: if the value of Λ continues to shrink, perturbative QCD will be in question.

4. PRODUCTION OF CHARM AND NEW FLAVORS

4.1 *Introduction*

The four-quark theory of Glashow, Illiopoulos, and Maiani (Altarelli et al 1974, Gaillard 1974, Glashow, Illiopoulos & Maiani 1970) permitted predictions of the rate for deep inelastic neutrino production of charm to be made quite early (Aitarelli et al 1974, Gaillard et al 1974). (For a theoretical review, see Gaillard, Lee & Rosner 1975.) These gave impetus to experimental searches for neutrino-induced charmed particles.

As discussed in Section 1.6, the production of charm with its subsequent semileptonic decay leads to final states with two opposite-sign leptons, and it was in fact the observation of such dimuons early in 1974 that gave the first experimental hint of charm (Rubbia et al 1974).

Owing to the short lifetimes, $\sim 10^{-13}$, of charmed mesons and baryons, the conventional bubble chamber spatial resolution has not permitted the systematic observation of charmed particle tracks before they decay. Bubble chamber groups have studied some specific charmed states, using traditional techniques of invariant mass peaks for final-state particles. These searches look for weakly decaying charmed mesons like D^+ ($c\bar{d}$), D^0 ($c\bar{u}$), F^+ ($c\bar{s}$), and the baryon Λ_c^+ (cud), as well as looking for the strongly decaying higher states.

Since inclusive charm production from neutrino and antineutrino beams is primarily related to the strangeness content of the nucleon (see Section 1.6), the study of dileptons in both bubble chamber and counter experiments provides information on the strength and x dependence of the strange-quark sea.

4.2 Resonant States

In order to definitely demonstrate that a new flavor has been found, it is not sufficient to see resonances in a mass plot. Historically (e.g. with strangeness) the most compelling demonstrations have included (a) a state long-lived enough to see a separation between production and decay, and/or (b) a violation of a selection rule well established for existing flavors. The first explicit observation of charm production and decay, which occurred shortly after the discovery of J/ψ, involved the second of these. The BNL group observed a single event due to the reaction $\nu_\mu p \to \mu^- \Lambda^0 \pi^+ \pi^+ \pi^+ \pi^-$, which violates the sacred $\Delta S = \Delta Q$ rule for semileptonic weak interactions (Cazzoli et al 1975). We now attribute this to the process $\nu_\mu + p \to \mu^- + \Sigma_c^{++}$ and subsequent decays $\Sigma_c^{++} \to \Lambda_c^+ + \pi^+$ and $\Lambda_c^+ \to \Lambda^0 \pi^+ \pi^+ \pi^-$. This discovery was followed by the observation in a Fermilab emulsion experiment of a 182-μm track that very likely was a charmed particle, although the decay could not be uniquely identified (Burhop et al 1976).

Many examples of charmed particle decays have since been observed in bubble chamber experiments, where mass peaks are directly correlated with charmed-particle mass peaks seen from $e^+ e^-$ collisions. In many cases, these neutrino events demonstrate through their violation of strangeness and isospin selection rules that a new flavor is involved. Table 8 shows a summary of these results.

Several experimental groups have recently attempted to use emulsion stacks in neutrino beams to exploit the above-mentioned property (a) of finite distance between production and decay. Besides providing a very clean signature for weakly decaying charmed particles, this promises quantitative determination of their lifetimes (or absolute total decay rates). Six such events were reported by a BEBC group (Angelini et al 1979a, b, Allasia et al 1979) and one apparently unique $F^+ \to \pi^+ \pi^+ \pi^- \pi^0$ by the Fermilab group (Ammar et al 1980).

At Fermilab, the E 531 hybrid emulsion experiment has accumulated the largest existing sample of charmed-particle states, with masses, lifetimes, and decay modes measured (Ushida et al 1980a, b). They used 23 liters of emulsion as target for the detector shown in Figure 31. By searching the emulsion for vertices predicted from the drift chamber tracks, they found 23 charged and 21 neutral multiprong short-lived decays. The emulsion was exposed to neutrinos in a single-horn, wide-band beam produced by 350-GeV protons (Ushida et al 1980b). A photomicrograph of an F^- production event is shown in Figure 32. Kinematic fitting of the E 531 candidates and a single F candidate from another experiment (Ammar et al 1980) leads to the information on the 34 events listed in Table 9.

Table 8   Charmed particles identified in bubble chamber neutrino experiments[a]

Experiment	D	Λ_c^+	Σ_c	D^*
Baltay et al 1978	64/180			
Baltay et al 1979			20/6 + +	
Cnops et al 1979		2/0		
Grassler et al 1981		2/0		
Blietschau et al 1979				2/0 +
Kitagaki et al 1980		19/10		
Kitagaki et al 1982[b]		1/0	2/0	
Armenise et al 1981		51/16[c]		

[a] Number of identified charm events/number of non charm background events is indicated. Charge mode is also indicated where appropriate.

[b] These three events have the decay mode $\Lambda_c^+ \to \Sigma^0 \pi^+$.

[c] Inclusive Λ_c^+, Σ_c^+, and Σ_c^{++} events.

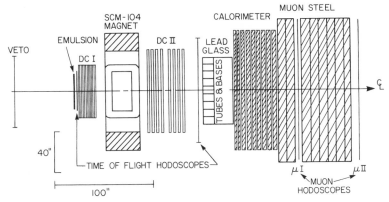

Figure 31 E 531 experiment utilizing 23 liters of emulsion target. The momenta of secondary charged particles are measured using the drift chambers DCI, DCII, and the SCM 104 magnet. The time-of-flight hodoscopes are used for charged-particle identification. There is Pb glass for e and γ identification ($\Delta E = \pm 0.14\ E^{1/2}$) and a hadron calorimeter for identification of neutral hadrons. Muons are identified by their passage through steel.

Note that there are included a single D^-, a single F^-, and three $\bar{D}^0$ events in Table 9 that are accompanied by a μ^+. This implies that they are produced by incident antineutrinos. The $\nu/\bar{\nu}$ beam flux estimates agree with the observed $D/\bar{D}$ production ratio; this is consistent with the absence of short-lived mixing of these charmed mesons, in contrast to the analogous strange-particle case where $K^0/\bar{K}^0$ mixing is complete.

The table also indicates the observed decay modes and aggregate lifetimes and shows the predominance of charmed meson production

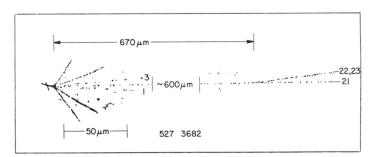

Figure 32 Photomicrograph of an E 531 F^- event. The figure shows the result of an antineutrino interaction in emulsion that produces a 12.8-GeV F^- meson ($\bar{c}s$ quark state). The F^- is accompanied by a 30.3-GeV μ^+ (track 1) and a 5.1-GeV positive track (track 3). The F^- decays after 670 μm to a 5.4-GeV π^- or K^- (track 21), a 1.5-GeV π^- (track 22), a 1.1-GeV π^+ (track 23), and a 4.8-GeV π^0, which has two measured γ's. The two-constraint fitted F^- mass is $2.026 \pm 0.056\,\text{GeV}/c^2$ and the decay time is 3.6×10^{-13} s.

Table 9 E 531 emulsion experiment event data

$\Lambda_c{}^a$	F
$\Lambda_c^+ \to p\bar{K}^0$	$F^- \to \pi^+\pi^-\pi^-\pi^0$
$\quad pK^-\pi^+\pi^0$	$F^+ \to K^+\pi^+\pi^+\pi^-\bar{K}^0$
$\quad p\pi^+\pi^-\bar{K}^0$ $(2)^b$	$\quad \to K^+K^-\pi^+\pi^0$
$\quad \Lambda\pi^+\pi^-\pi^+$ $(3)^b$	$\quad \to \pi^+\pi^+\pi^-\pi^{0c}$
$\quad \Sigma^0\pi^+$	$M_F = 2042 \pm 33$
$M_{\Lambda_c} = 2265 \pm 30$	$\tau_F = 2.0\,{}^{+1.8}_{-0.8} \times 10^{-13}$ s
$\tau_{\Lambda_c} = 1.4\,{}^{+0.8}_{-0.4} \times 10^{-13}$ s	

D^0 or $\bar{D}^0$	D^+ or D^-
$D^0 \to K^-\pi^+\pi^0$	$D^+ \to K^-\pi^+\pi^+\pi^0$
$\quad K^-\pi^+\pi^0\pi^0$	$\quad \to K^-K^+\pi^+\pi^0$
$\quad K^-\pi^+\pi^-\pi^+$ $(2)^b$	$\quad \to K^-\pi^+e^+(\nu)$
$\quad K^-\pi^+\pi^-\pi^+\pi^0$ $(2)^b$	$\quad \to K^-\pi^+\mu^+(\nu)$
$\quad K^-\pi^+\pi^+\pi^+\pi^-\pi^-$	$D^- \to K^+\pi^-e^-(\nu)$
$\quad \bar{K}^0\pi^+\pi^-$ $(2)^b$	$M_{D^\pm} = 1851 \pm 20$
$\quad \bar{K}^0\pi^+\pi^-\pi^0$	$\tau_{\Lambda_{D^\pm}} = 10.3\,{}^{+10.3}_{-4.2} \times 10^{-13}$ s
$\quad \pi^+\pi^+\pi^+\pi^-\pi^-\pi^-\pi^0$	
$\quad K^-e^+\nu$	Summary
$\quad K^-\mu^+\nu$	
$\quad K^-\pi^+\pi^-\mu^+\nu$	8 Λ_c
$\bar{D}^0 \to K^+\pi^-\pi^0$	4 F
$\quad K^+\pi^-\pi^0\pi^0$	5 charged D's
$\quad K^+\pi^-\pi^+\pi^-\pi^0$	17 neutral D's
$M_{D^0} = 1842 \pm 16$	
$\tau_{D^0} = 2.3\,{}^{+0.8}_{-0.5} \times 10^{-3}$ s	

[a] 4 Λ_c^+ are consistent with having either a Σ_c^{++} or Σ_c^0 parent.
[b] Indicates multiple candidates of a given type.
[c] This event is from E 564, Ammar et al (1980).

over that of charmed baryons (Stanton 1981). The D^0, $D^\pm$ lifetime results indicate that the $D^\pm$ lifetime is larger than that of the D^0 (Ushida et al 1982). Indirect evidence to this effect came from a comparison of the relative leptonic branching ratios of these states as measured in e^+e^- collisions (Bacino et al 1980, Schindler et al 1981).

The production characteristics of the charmed mesons and baryons are different. These data and others (Armenise et al 1981) give evidence for some production of charmed baryons with low Q^2. Armenise et al quote a quasi-elastic cross section times branching ratio for the total production of Λ_c^+, Σ_c^+, Σ_c^{*+} to be $\sigma B = 14.3 \pm 7.4 \times 10^{-40}\, cm^2$. Here the branching ratio B refers to the probability to decay into neutral vees: $B = [\Gamma(\Lambda_c^+ \to \Lambda_0 + \dots) + \Gamma(\Lambda_c^+ \to \bar{K}^0 + p + \dots)]/\Gamma(\Lambda_c^+ \to \text{all})$. The number of similar E 531 candidates is consistent with this number. For scale, we compare this value of σB to the asymptotic cross sections for the quasi-elastic processes, $\nu n \to \mu^- p$ and $\nu p \to \mu^- \Delta^{++}$ of $\sim 90 \times 10^{-40}$ and $\sim 65 \times 10^{-40}\, cm^2$, respectively (Mann et al 1973, Barish et al 1979). Theoretical estimates of the quasi-elastic charmed-baryon cross sections depend on how the SU(4) symmetry breaking and quark model with orbital excitations are handled (Finjord & Ravndal 1975, Schrock & Lee 1976, Avilez et al 1978, Amer et al 1979, Avilez & Kobayashi 1979). The precision of the present measurements does not uniquely establish one of the theoretical models, but is generally consistent in magnitude with expectations.

The difference between charmed-meson and -baryon production is clearly seen in Figure 33, which shows the x_F distribution for the E 531 data. Half of the charmed baryons are target fragments while the charmed mesons are primarily identified as current fragments. It would appear that a reasonable fraction, perhaps one third, of the charmed-baryon production is quasi-elastic, and that charmed-baryon production is relatively small at high Q^2. It may thus be justifiably omitted at high Q^2 in the analysis of opposite-sign dimuon data to obtain the fraction of strange sea in the nucleon.

E-531

CHARM PRODUCTION CHARACTERISTICS

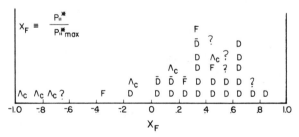

Figure 33 E 531 charm production characteristics. The figure shows the Feynman x distribution for the E 531 charm events produced by neutrinos. X_F is defined in terms of center-of-mass longitudinal momentum as shown in the figure. Note the target-like nature of three of the Λ_c events with X_F near -1.

The relative numbers of identified D^0, D^+, D^{*+} are generally consistent, within present statistics, with the production ratios D^*/D of about three, D^+/D^0 (not through D^*) of about one, and D^{*+}/D^{*0} about the same. These are the expectations from the most naive assumption, i.e. relative population directly proportional to the number of spin states.

4.3 Opposite-Sign Dileptons

Because of the short lifetimes and relatively large semileptonic branching ratios of charmed particles, both counter and bubble chamber experiments have used the large numbers of easily identified, oppositely charged dilepton events to study the physics of deep inelastic charm production (see Figure 34 and Section 1.6). The helicity structure, assuming V–A currents, predicts the y distribution to be flat, aside from the kinematic effect of transforming a light d or s quark to the heavy c quark. This threshold behavior changes the flat y distribution to one peaked at higher y.

To create the high-mass quark m_c, the fraction of nucleon momentum ξ carried by the interacting constituent becomes

$$\xi \approx x' = x + m_c^2/(2ME_\nu y), \qquad\qquad 28.$$

where x is the traditional scaling variable $(Q^2/2M\nu)$. This follows from the kinematics of the propagator four-momentum, q, combining with the quark whose momentum fraction is $x'p$ to give a mass m_c^2: $(q + x'p)^2 = -m_c^2$ (Barnett 1976, Georgi & Politzer 1976). The x' variable then has a bound, $m_c^2/(2ME_\nu y)$, below which charm cannot be produced (Kaplan & Martin 1976).

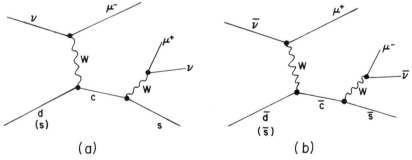

(a)   (b)

Figure 34 Feynman diagrams for production of dimuons through charm quark decay. Valence quarks (d) can only produce charm with neutrino beams. Hence the characteristics of dimuons from neutrino and antineutrino collisions are somewhat different.

For valence quarks, the charm cross section is not affected much in the low-x' region, but for the strange-sea distribution, e.g. $x's(x') = A(1 - x')^7$, which is strongly peaked at small x', there is substantial suppression at low neutrino (antineutrino) energies (Brock 1980, Edwards & Gottschalk 1981). If in addition to the assumption of x' as the appropriate variable, one includes the phase space factor (slow rescaling) for producing a heavy quark in the two-body scattering, $\nu + [^{d}_{s}] \rightarrow \mu^{-} + c$, the charm production differential cross section becomes (Kaplan & Martin 1976, Lai 1978)

$$\frac{d^3\sigma}{dx\,dy\,dz} = \frac{G^2 M E_\nu}{\pi} \left\{ [x'd(x')\sin^2\theta_c + x's(x')\cos^2\theta_c] \right.$$

$$\left. \times \left[1 - \frac{m_c^2}{2ME_\nu x'}\right] D(z) \right\}. \qquad\qquad 29.$$

In this equation the fragmentation of the charm quark is described by the function $D(z)$ where z is the fraction of the energy taken by the charm particle (predominantly D mesons) in the W-boson-nucleon center of mass.

The rate for opposite-sign dileptons is then the cross section (Equation 29) multiplied by the average leptonic branching ratio for the decays of charmed particles. The D-meson leptonic branching ratios measured at SPEAR (Schindler et al 1981, Bacino et al 1980) imply an average charm to muon branching ratio of 0.08 ± 0.03, assuming a D^0/D^+ total production rate of 2 to 3. Since the E 531 data measure the relative charm to single muon cross section to be $7 \pm 2\%$ at 60 GeV, the expected $\mu^+\mu^-/\mu^-$ rate at comparable energy should be ~0.55% with the assumption of the branching ratio equal to 0.08. This is in good agreement with the dimuon rate at the same energy given in the next section and in Figure 35.

4.3.1 CHARM CROSS SECTION VS ENERGY Several neutrino bubble chamber experiments have observed events with a muon and an electron in the final state (μ^-e^+) (von Krogh et al 1976, Blietschau et al 1976, Deden et al 1977, Bosetti et al 1977a, 1978a, Baltay et al 1977, Erriquez et al 1978, Ballagh et al 1981). The phenomena have also been seen in antineutrino exposures (μ^+e^-) (Berge et al 1979, Ballagh et al 1977, 1981). These data come primarily from exposures to wide-band beams where the neutrino energy producing the event is not known *a priori*. The identification of positrons (electrons) is good with this technique for momenta above 0.3 GeV/c. However, only the energy

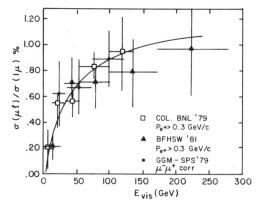

Figure 35 $\mu^- l^+/\mu^-$ cross-section ratio vs E_{vis}. The data are $\mu^- e$ bubble chamber data from Ballagh et al (1981) and Murtagh (1979). The GGM-SPS data are $\mu^- \mu^+/\mu^-$ rates corrected for full muon acceptance (Armenise et al 1979). E_{vis} is the measured energy for the event, $E_{h^+} E_{\mu^-} + E_{l^+}$. The curve shown is a calculation that includes slow rescaling (Brock 1980).

visible to the apparatus (E_{vis}) is measured in an event; final-state neutrinos, for example, are lost. Thus, the $\mu^- e^+/\mu^-$ relative cross sections for the experiments that report energy-dependent ratios are shown in Figure 35 as a function of E_{vis}. This shows that the dilepton cross section rises to 0.6% of the charged-current cross section at 50 GeV and then continues to rise to ~1% at 150 GeV.

There are several mechanisms in the production and decay of charmed mesons that should raise the number of neutral-strange-particle decays (V's) in events with charm. Much of the production involves transformation of an s quark, $s \rightarrow c$, leaving net strangeness behind. The subsequent decay of the charmed mesons decay preferentially to strange particles. Indeed, the number of observed strange particles per event is about four times as large in events with a prompt electron in the final states as in events without charm (Murtagh 1979). The raw numbers of excess neutral strange particles in μe events seem to be generally agreed upon: ~0.3 per event. This is in reasonable agreement with the expectations of charm production (Murtagh 1979, Ballagh et al 1981).

Charm dimuon events are reported by both bubble chamber and counter experiments. However, for dimuons a substantial loss in signal results because the momentum of the second muon, P_{μ_2}, is required to be larger than some minimum value. This cut is needed to eliminate confusion between hadrons and very low-momentum muons of short range as well as low-momentum muon background from the decays of π

and K mesons in the hadronic final state. The Gargamelle (Armenise et al 1979) and Fermilab E 546 (Ballagh et al 1980) collaborations have both measured dimuon rates with $P_{\mu_2} > 2.6$ and $4.0 \, \text{GeV}/c$, respectively. The $\mu^- \mu^+$ data of Gargamelle (117 events) corrected for the P_{μ_2} cut is shown in Figure 35 vs E_{vis} with the μe bubble chamber data.

There are also recently published high-statistics cross-section rates for CDHS, CHARM, and CFRR. The CFRR data (484 ν events with $P_{\mu_2} > 4 \, \text{GeV}/c$) are obtained with the Fermilab dichromatic beam while the CDHS (15,000 events) and CHARM (494 ν, 285 $\bar{\nu}$ events) data come from the CERN wide-band beam. A comparison of the CDHS and CFRR $2\mu/1\mu$ cross-section ratios for $P_{\mu_2} > 6.5 \, \text{GeV}/c$ shows reasonable agreement (Fisk 1981). To obtain the total dimuon rate, this restriction on muon momentum must be corrected with a theoretical model.

There is an additional problem associated with using wide-band beams for obtaining dimuon to single muon production ratios. The opposite-sign events from charm decay involve missing energy as discussed in the last section. For a typical case, e.g. $E_{\text{missing}} \approx 15 \, \text{GeV}$ at 150-GeV incident energy, the dimuon events will be systematically underestimated in energy and compared to single muon events of lower energy, on average. Because wide-band beams provide flux with a substantial fall with energy, this can produce a systematic underestimate of the $2\mu/1\mu$ ratio by as much as 50%.

To find the total charm cross section, the CDHS group has employed a Monte Carlo calculation incorporating a specific charm model that corrects for the P_{μ_2} cut and includes the effects of the missing final-state ν energy. This allows them to establish the $2\mu/1\mu$ cross-section ratio as a function of E_ν, rather than E_{vis} (Knobloch et al 1981). Their data have also had the K- and π-decay background of ~13% for ν and ~6% for $\bar{\nu}$ subtracted. The resulting cross-section points are shown in Figure 36. It should be pointed out that the CDHS antineutrino data derived from the wide-band beam have made a specific definition assigning the lower-energy muon to the charm decay. They have checked the reliability of this definition with a sample of narrow-band beam dimuons where the designation is not so arbitrary.

A comparison of the corrected CDHS neutrino and antineutrino data shows the $\bar{\nu}$ cross-section ratio to exceed the ν data at all energies. This is somewhat unexpected since slow rescaling should suppress the $\bar{\nu}$ rate more than the ν rate at low energies. This comes about because the $\bar{\nu}$ charm dimuons are produced entirely from sea quarks (see Equation 12). A calculation of the dimuon to single muon rate depends sensitively (see Section 1.6) on the valence and sea x distribution, the charm quark

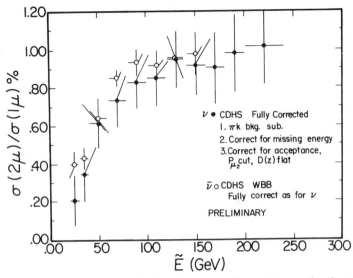

Figure 36 CDHS corrected opposite-sign dimuon rates for neutrinos and antineutrinos. The data, accumulated during wide-band beam operation, have been corrected using a Monte Carlo program for (*a*) the P_{μ_2} cut, (*b*) missing final-state energy, and (*c*) background due to hadronic decays to muons (Knobloch et al 1981). $\tilde{E}$ is the Monte-Carlo-corrected energy, which includes an estimate of the missing final-state neutrino energy.

mass, and the fraction of strange sea in the nucleon (Edwards & Gottschalk 1981).

4.3.2 *x* DISTRIBUTIONS   Figure 37 shows the *x* distributions of dimuon events obtained by CDHS in their wide-band running without corrections for apparatus acceptance, which they have calculated to be negligible (Knobloch et al 1980). Both neutrino and antineutrino data are shown. The quantity histogrammed is x_{vis}, which is usually about 10% larger than the true *x* owing to missing neutrino energy. The mean values of x_{vis} for the neutrino and antineutrino data shown are 0.195 ± 0.01 and 0.095 ± 0.01, respectively. Both CFRR (Fisk 1981) and CHARM (Jonker et al 1981a) experiments are in qualitative agreement with these *x* distributions. The striking difference between the ν and $\bar{\nu}$ *x* distributions clearly indicates the production of charm for antineutrinos almost exclusively from the charm-sea quarks while the neutrino data show production from both sea and valence quarks.

4.3.3   STRANGE-SEA CONTENT OF THE NUCLEON   Two methods have been used to extract the strange-sea content of the nucleon:

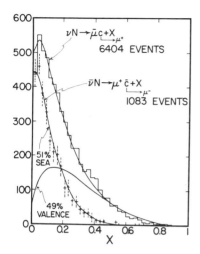

Figure 37 CHDS x distributions for opposite-sign dimuons. The x_{vis} distributions for CDHS wide-band neutrino and antineutrino dimuon data are shown. As expected the antineutrino data are sharply peaked at small x since charm production occurs almost entirely by conversion of an anti-strange-sea quark to an anti-charm quark. The neutrino charm production involves both the valence down quarks and sea strange quarks. The valence distribution alone is also shown (Buras & Gaemers 1978). The neutrino data are fitted with 51% strange sea and 49% valence quarks. x_{vis} is the x value calculated using the visible energy for E_ν.

$\eta_s = 2S/(U + D)$, where S, U, and D refer to the integrals over x of the appropriate momentum fractions, e.g. $U = \int xu(x) \, dx$. In the first method the neutrino dimuon x distribution is fitted using the valence x distributions, $u(x)$ and $d(x)$, obtained from deep inelastic scattering and the strange sea, $s(x)$ from the opposite-sign antineutrino dimuons. For example, CDHS assume a Buras-Gaemers (Buras & Gaemers 1978) valence distribution and $A(1 - x)^7$ for the strange sea, both features being qualitatively consistent with their data in shape, to obtain η_s.

The second or double ratio method is based not on x distributions but on the ratios

$$R_1 = \frac{\sigma_{\bar{\nu}}(1\mu)}{\sigma_\nu(1\mu)}$$

$$R_2 = \frac{\sigma_\nu(2\mu)/\sigma_\nu(1\mu)}{\sigma_{\bar{\nu}}(2\mu)/\sigma_{\bar{\nu}}(1\mu)}$$

to find η_s. Assuming $\eta \equiv \bar{U} + \bar{D}/U + D \approx 0.15$, the fraction of strange sea is given by

$$\eta_s = \frac{\tan^2 \theta_c[1 - \eta R_2/R_1]}{[R_2/R_1 - 1]}.$$

CDHS and CHARM have reported values of η_s. Figure 38 shows the CDHS results, determined using the two methods, as a function of E_ν (Knobloch et al 1981). The overall average value of η_s is 0.046 ± 0.006, not including slow rescaling or systematic errors. Similarly the CHARM group finds $\eta_s = 0.050 \pm 0.015$ again without slow rescaling (Jonker et al

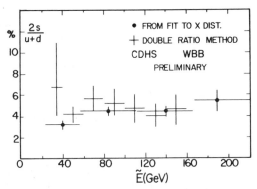

Figure 38 CDHS determination of the strange-sea fractional momentum in the nucleon (Knobloch et al 1981). Two methods are used to determine the strange-sea fraction, $\eta_s \equiv 2S/(U+D)$. The first technique (*solid points*) uses the neutrino x distribution as fitted in Figure 37 to obtain η_s, while the second method, discussed in Section 4.3, employs the double ratio technique. Slow rescaling effects have not been taken into account. Preliminarly indications suggest they would approximately double η_s (Knobloch et al 1981).

1981a). The effects of slow rescaling could double the value of η_s (Knobloch et al 1981, Edwards & Gottschalk 1981), making it $\sim 10\%$, which is still less than is expected for an SU(3) symmetric strange sea, i.e. $\eta_s = \eta = 12–15\%$. It should be emphasized that quantitative results on the magnitude of the strange sea have only become available recently and the effects of slow rescaling have not been reported by the groups who have sizable data samples.

4.3.4   CHARM FRAGMENTATION   Since the dimuon experiments do not measure the energy of the charmed particle, the charm fragmentation function is difficult to determine. Based on models of charm production that include the assumption of production of only D or D* mesons, one can fit the $z_{\mu 2}$ distribution, i.e. the fraction of energy carried off by the charm-produced second muon, to determine the fragmentation function. Generally the experiments seem to agree that the z distribution is flat or peaked toward larger z, although most of the data on this subject is still preliminary. The hybrid emulsion experiment should give some insight into this problem albeit with small statistics and at lower energies.

4.3.5   SUMMARY   In summary, the qualitative features of charm production by neutrinos (antineutrinos) appears to agree with the basic GIM model, which predicts that the charm production should rise to a value determined by the valence d and sea-strange quarks. The x distributions indicate the important role taken by the strange-sea quarks, and the energy dependence of the charm cross section exhibits threshold dependence that may be accounted for by slow rescaling. The momentum fraction of strange sea in the nucleon is 4–5% and could be double that with slow rescaling. This is an important point especially because knowledge of the magnitude of the strange sea is essential in comparing deep inelastic muon and neutrino scattering (see Section 3.3.3).

4.4 Like-Sign Dimuons

Several experimental groups have reported the observation of like-sign dimuons, although the level of the signals and backgrounds have not always agreed. All experiments conclude that the second muon origin-ates at the hadron vertex. This follows from the fact that the ϕ distribution is peaked at 180°, where ϕ is the angle between the two muon tracks projected on a plane perpendicular to the incident neutrino (de Groot et al 1979a, Nishikawa et al 1981, Benvenuti et al 1978a). Mechanisms suggested as prompt sources for these signals include (a) associated charm production, (b) bottom meson or baryon produc-tion, and (c) $D^0 - \bar{D}^0$ mixing.

4.4.1 ENERGY DEPENDENCE OF THE LIKE-SIGN RATE   A summary of recent experimental $\mu^- \mu^- / \mu^-$ rates is shown in Figure 39 (Knobloch et al 1981, Jonker et al 1981a, Holder et al 1977, Trinko et al 1981, Shaevitz et al 1980). To substantially reduce the background from second muons coming from the decay of K and π mesons, a $P_{\mu_2} > 9$-GeV cut has been imposed on most of the data. Note that the results come from experiments that use narrow-band, horn, and quadrupole triplet beams. As mentioned earlier, in the discussion of opposite-sign dimuon rates, for data taken in horn beams this can lead to an underestimate of the cross-section ratio that cannot be compensated by calculation, in this case, since the mechanism producing the like-sign dimuon is unknown. Narrow-band beam data do not suffer from this problem but are typically less statistically significant.

4.4.2 BACKGROUND CALCULATIONS   Although the P_{μ_2} cut eliminates most of the background, calculations must still be made to determine an absolute background level. Such a calculation, which is based directly on experimental data, is discussed by Shaevitz et al (1980). The anticipated background sources are: (a) decay of primary π's or K's produced at the hadron vertex in ordinary single-muon charged-current

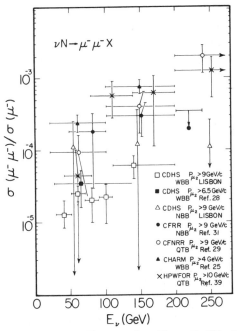

Figure 39 Like-sign dimuon rates. The rates for $\nu N \rightarrow \mu^- \mu^- X$ relative to the single muon charged-current cross section are shown. The data include the momentum cut for the second muon as shown. References from the top to the bottom of the list are Knobloch et al 1981, Holder et al 1977, de Groot et al 1979, Shaevitz et al 1980, Nishikawa et al 1981, Jonker et al 1981, and Trinko et al 1981. The CDHS open triangles and the CFRR data are taken with the narrow-band beam. The CFNRR and HPWFOR data were obtained with the quadrupole triplet beam and the remaining CDHS and CHARM data result from wide-band beam experiments. The one-standard-deviation upper limit shown at 220 GeV is from the CFRR experiment.

interactions, and (*b*) the production of either a prompt or nonprompt second muon from the interaction of the primary hadrons in the hadron shower. The rate from decays of the primary hadrons is calculated two ways. The π and K inclusive spectra and multiplicity are determined either from Field-Feynman quarks jets based on νNe data obtained with a wide-band beam or they are taken directly from high-energy νNe data. The secondary particle cascade calculation makes use of the hadrons just described and the measured prompt and nonprompt muon production by hadrons in the Fermilab experiment E 379/595 variable density target. This yields the probability, as a function of hadron energy E_H, for producing a muon with momentum greater than a particular cutoff value. An example of the calculation is given in Figure 40.

The calculated numbers of background events is obtained from the convolution of the probability curve with the observed hadron energy distribution for single-muon events. In the first experiment (CFNRR) definitively to see a signal, there were 12 events with 1.3 ± 0.7 background (Nishikawa et al 1981). In the more recent CFRR narrow-band data there are 13 $\mu^-\mu^-$ events and the calculated background is 3.5 events. Calculations that roughly agree with this background rate have been made by CDHS and HPWFOR (de Groot et al 1979a, Trinko et al 1981). The HPWFOR and CFNRR groups have checked their background calculation by comparing it with the $\mu^-\mu^+/\mu^-$ rates they obtain in different target densities.

The CHARM collaboration group with $P_{\mu_2} > 4$ GeV/c did not make a background subtraction but instead extrapolated both muons back to the event vertex, where they can find the projected horizontal and vertical distances between the two muons. A comparison of this distribution with those for prompt and decay muons generated by

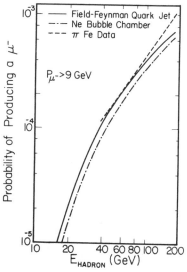

Figure 40 Muon production probability vs hadron energy. The curves are the probability vs hadron energy that a μ^- with momentum greater than 9 GeV/c will be produced and observed in the CFRR detector from π or K decay. The solid curve uses the Field-Feynman quark jets to generate the hadrons. The dashed curve, given for comparison to the solid curve, is the measured probability with pions on the CFRR detector (Shaevitz et al 1980). The dot-dashed curve utilizes neon bubble chamber data on hadronic final states from neutrino collisions to calculate the first-generation decays and the πFe data for subsequent generations.

Monte Carlo techniques gives them the fraction of prompt events in their sample.

Although the data shown in Figure 39 do not all agree, there is good agreement among the narrow-band and quadrupole triplet data. The above discussion does leave some room for differences in normalization and background subtractions. In any case, a level of 10^{-4} at 200 GeV is difficult to obtain with present theoretical models.

4.4.3 POSSIBLE SOURCES OF LIKE-SIGN DIMUONS   The earliest predictions of like-sign dimuons resulted from considering the associated production of charmed pairs, $c\bar{c}$ as shown in Figure 41a (Goldberg 1977, Young et al 1978, Kane et al 1979). These authors concluded that for the same cuts on the theory as experiment, the predicted like-sign rate is low compared to the data by one to two orders of magnitude. As an aside comment it should be pointed out that theoretical predictions for

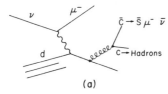

(a)

Gluon Bremsstrahlung

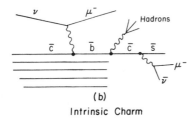

(b)

Intrinsic Charm

Figure 41 Possible sources for like-sign dimuons. (*a*) Associated charm production by gluon bremsstrahlung (Goldberg 1977, Young et al 1978, Kane et al 1979). (*b*) b production with intrinsic charm (Brodsky et al 1981). (*c*) b production via gluon fusion (Barger et al 1981).

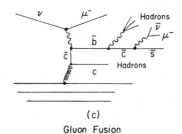

(c)

Gluon Fusion

hadronic production of bare charm pairs by gluon bremsstrahlung are also lower than the observed rate by one or two orders of magnitude.

As an example of b (bottom quark) production, Brodsky, Peterson & Sakai (Brodsky et al 1981) proposed intrinsic charm within the nucleon (Figure 41b) to explain neutrino like-sign dimuon rates and the large diffractive cross section claimed for pp → Λ_c^+ at the ISR (Giboni et al 1979). With 1% intrinsic charm and $p_{\mu_2} > 9$ GeV/c, the model gives too small a rate. The rate is severely limited partly because of the V − A coupling of c to b quarks, which gives a $(1 - y)^2$ factor in the cross section. They point out that intrinsic charm would give a like-sign dimuon x distribution whose mean value is significantly larger than is the case for gluon bremsstrahlung.

Barger, Keung & Phillips (Barger et al 1981) investigated a b-production model with gluon fusion, as shown in Figure 41c. They find the cross section for this process to be smaller than the intrinsic charm model and conclude that of the three models discussed, the intrinsic charm model with full strength right-handed coupling comes closest to fitting the data. If the rate is to be explained with left-handed (V − A) coupling, the intrinsic charm in the nucleon would have to be more than 1% and this is inconsistent with the large-x dimuon cross section observed in muon scattering by the EMC group (Strovink 1981).

If there were sufficient $D^0 − \bar{D}^0$ mixing, then same-sign dimuons would follow from the same mechanisms producing opposite-sign events. For example, if all the like-sign dimuons in the CFRR experiment came from $D^0 − \bar{D}^0$ mixing, the D^0/D^+ production ratio plus $D^0 \to \mu$ inclusive branching ratio would imply $D^0 − \bar{D}^0$ mixing at about 20%. At the present time, there is no evidence for mixing with the measured limits (90% confidence level) of less than 11% (Avery et al 1980) and (95% confidence level) of less than 16% (Feldman et al 1977). And there is evidence from hadronic interactions that the mixing limit is even smaller than this (Bodek et al 1982). This strongly suggests that the fraction of like-sign events due to mixing is negligible.

Thus, we conclude that several experiments see evidence for like-sign dimuons, but there is at present no single source to account for them.

5. SUMMARY

The study of charged-current neutrino interactions with nucleons has confirmed our understanding of the V − A nature of the charged weak current. It has also provided, through the measurement of nucleon structure functions, corroboration of the quark model. The predominantly spin-$\frac{1}{2}$ nature of the quarks is shown in the charged-current neutrino and antineutrino y distributions. The approximate correctness of scaling is demonstrated by the linear rise of the neutrino cross section with energy. Measurements of neutrino and antineutrino cross sections give integrated values of the structure function $F_2(x)$ implying that only half of the nucleon's longitudinal momentum is carried by the quarks. The remaining momentum is presumably taken by the gluons that carry the strong forces between quarks. Comparison of the structure function F_2, as measured separately with muons and neutrinos, convincingly demonstrates the existence of fractional charge for the quark partons. That the number of valence quarks in the nucleon is three has also been approximately verified with the measurement of the integral of $F_3(x)$ (Gross-Llewellyn Smith sum rule). Any disagreement in these fundamental tests ($<20\%$) is probably within present experimental precision among all the various experiments.

The flavor-changing nature of the weak hadronic current permits the observation of neutrino-induced opposite-sign dimuon events. Here the second muon results from the semileptonic decay of a charmed hadron. In particular, the antineutrino dimuon data, that come primarily from the conversion of an $\bar{s}$ to a $\bar{c}$ quark, allow measurement of the strange sea quark x distribution. Experimentally the neutrino-induced charm states, as observed in emulsion, have provided lifetimes for the D^{+-}, D^0, F^{+-}, and Λ_c states, and as more data become available will give information on charm fragmentation. The origin of like-sign dimuons observed at a rate about 10% of the opposite-sign dimuons is not well understood but may be due to associated charm production.

Deviations from the simple quark model and scaling hypothesis occur in the structure functions at fixed x as Q^2 is varied. The theory of perturbative quantum chromodynamics (QCD) predicts such nonscaling behavior. How much of the observed Q^2 variation is related to perturbative QCD remains to be determined. If the Q^2 dependence is dominated by perturbative QCD diagrams, the dependence of the structure functions should be approximately logarithmic with Q^2. The amount of variation permits a measurement of the quark-gluon coupling constant, α_s, or alternatively the scale parameter of the strong interactions, Λ. Values of Λ from muon and neutrino structure function measurements are presently in the vicinity of 100 to 250 $(\text{MeV}/c)^2$. An alternative determination of α_s may result from the measurement of the R parameter (σ_L/σ_T). Present data that reach a maximum Q^2 of about 200 $(\text{GeV}/c)^2$, along with new data at larger Q^2 which will be available at Tevatron energies, should provide more conclusive tests of QCD. This presumes, of course, that there are no new surprises to be uncovered in charged-current phenomena in that energy region.

Literature Cited

Abramowicz, H., et al. 1981. *Phys. Lett.* 107B:141–44
Abramowicz, H., et al. 1981a. *Nucl. Instrum. Methods* 180:429
Adler, S. L. 1966. *Phys. Rev.* 143:1144
Allasia, D., et al. 1979. *Phys. Lett.* 87B:287
Altarelli, G. Martinelli, G. 1978. *Phys. Lett.* 76B:89
Altarelli, G., et al. 1974. *Phys. Lett.* 48B:435
Altarelli, G., Parisi, G. 1977. *Nucl. Phys.* B126:298
Amer, A., et al. 1979. *Phys. Lett.* 81B:48
Ammar, R., et al. 1980. *Phys. Lett.* 94B:118
Angelini, C., et al. 1979a. *Phys. Lett.* 80B:428
Angelini, C., et al. 1979b. *Phys. Lett.* 84B:150
Appelquist, T., Barnett, R. M., Lane, K. 1978. *Ann. Rev. Nucl. Part. Sci.* 28:387–499
Armenise, N., et al. 1979. *Phys. Lett.* 86B:115
Armenise N., et al. 1981. *Phys. Lett.* 104B:409
Astratyan, A. E., et al. 1978. *Phys. Lett.* 105B:315
Atherton, H. W., et al. 1980. *CERN Preprint 80–07* (unpublished)
Aubert, J. J., et al. 1981. *Phys. Lett.* 105B:315
Aubert, J. J., et al. 1981a. *Phys. Lett.* 105B:322
Avery, P., et al. 1980. *Phys. Rev. Lett.* 44:1309
Avilez, C., Kobayashi, T. 1979. *Phys. Rev.* D19:3448
Avilez, C., et al. 1978. *Phys. Rev.* D17:709
Bacino, W., et al. 1980. *Phys. Rev. Lett.* 45:329
Ballagh, H. C., et al. 1977. *Phys. Rev. Lett.* 39:1650
Ballagh, H. C., et al. 1980. *Phys. Rev.* D21:569
Ballagh, H. C., et al. 1981. *Phys. Rev.* D24:7
Baltay, C. 1981. *Proc. Neutrino '81*, ed. R. J. Cence, E. Ma, A. Roberts, 2:195–328. (Maui, Hawaii)
Baltay, C., et al. 1977. *Phys. Rev. Lett.* 39:62
Baltay, C., et al. 1978, *Phys. Rev. Lett.* 41:73
Baltay, C., et al. 1979. *Phys. Rev. Lett.* 42:1721
Bardeen, W. A., et al. 1978. *Phys. Rev.* D18:3998
Barger, V., et al. 1981. *Phys. Rev.*

D24:244
Barish, B. C., et al. 1975. *Nucl. Instrum. Methods* 130:49
Barish, B. C., et al. 1977. *Phys. Rev. Lett.* 39:1595
Barish, B. C., et al. 1978. *Phys. Rev. Lett.* 40:1414
Barish, B. C., et al. 1978a. *IEEE Trans. Nucl. Sci.* NS-25:532
Barish, S. J., et al. 1979. *Phys. Rev.* D19:2521
Barnett, R. M. 1976. *Phys. Rev. Lett.* 36:1163
Benvenuti, A., et al. 1977. *Phys. Rev. Lett.* 38:1110
Benvenuti, A., et al. 1978. *Phys. Rev. Lett.* 40:488
Benvenuti, A., et al. 1978a. *Phys. Rev. Lett.* 41:725
Berge, J. P., et al. 1979. *Phys. Lett.* 81B:89
Beuselinck, R., et al. 1978. *Nucl. Instrum. Methods* 154:445
Bjorken, J. D., Paschos, E. A., 1969. *Phys. Rev.* 185:1975
Blair, R. E. 1982. PhD thesis. *A Total Cross Section and y-Distribution Measurement for Muon-Type Neutrinos and Anti-Neutrinos on Iron*, Caltech, Pasadena
Blietschau, J., et al. 1976. *Phys. Lett.* 60B:207
Blietschau, J., et al. 1979. *Phys. Lett.* 86B:108
Bodek, A., et al. 1973. *Phys. Rev. Lett.* 30:1087
Bodek, A., et al. 1979. *Phys. Rev.* D20:1471
Bodek, A., Ritchie, J. L. 1981. *Phys. Rev.* D23:1070; D24:1400
Bodek, A., et al. 1982. *Phys. Lett.* 113B:82
Bofill, J., et al. 1982. *IEEE Trans Nucl. Sci.* NS-29 1:400
Bogert, D., et al. 1982. *IEEE Trans. on Nucl. Sci.* NS-29 1:363
Bollini, D., et al. 1981. *Phys. Lett.* 104B:403
Bosetti, P.C., et al. 1977. *Phys. Lett.* 70B:273
Bosetti, P. C., et al. 1977a. *Phys. Rev. Lett.* 38:1248
Bosetti, P. C., et al. 1978. *Nucl. Phys.* B142:1–28
Bosetti, P. C., et al. 1978a. *Phys. Lett.* 73B:380
Bosio, C., et al. 1978. *Nucl. Instrum. Methods* 157:35
Brand, C., et al. 1978. *Nucl. Instrum. Methods* 136:485
Brand, C., et al. 1982. *IEEE Trans. Nucl. Sci.* NS-29 1:272

Brock, R. 1980. *Phys. Rev. Lett.* 44:1027
Brodsky, S. J., et al. 1981. *Phys. Rev.* D23:2745
Buras, A. J., Gaemers, K. J. F. 1978. *Nucl. Phys.* B132:249
Burhop, E. H. S., et al. 1976. *Phys. Lett.* 65B:299
Büsser, F. W. 1981. See Baltay 1981, p. 351
Cabibbo, N. 1963. *Phys. Rev. Lett.* 10:531
Callan, C. G., Gross, D. J. 1969. *Phys. Rev. Lett.* 22:156
Cazzoli, E. G., et al. 1975. *Phys. Rev. Lett.* 34:1125
Cence, R. J., et al. 1976. *Nucl. Instrum. Methods* 138:245
Chapman-Hatchett, A., et al. 1979. *CERN Rep. SPS/ABT/Int. 79-1* (unpublished)
Close, F. E. 1979. *An Introduction to Quarks and Partons.* New York: Academic
Cnops, A. M., et al. 1979. *Phys. Rev. Lett.* 42:197
Danby, G., et al. 1962. *Phys. Rev. Lett.* 9:36
Deden, H., et al. 1977. *Phys. Lett.* 67B:474
de Groot, J. G. H., et al. 1979. *Z. Phys.* C1:143
de Groot, J. G. H., et al. 1979a. *Phys. Lett.* 86B:103
de Groot, J. G. H., et al. 1980. *Proc. 20th Int. Conf. High Energy Phys.*
Diddens, A. N., et al. 1980. *Nucl. Instrum. Methods* 178:27
Dishaw, J. P., et al. 1979. *Phys. Lett.* 85B:142
Edwards, B. J., Gottschalk, T. D. 1981. *Nucl Phys.* B186:309
Edwards, D. A., Sciulli, F. J. 1976. A Second Generation Narrow Band Neutrino Beam. *FERMILAB TM-660* (unpublished)
Eichten, T., et al. 1973. *Phys Lett.* 46B:274
Eisele, F., et al. 1981. See Baltay 1981, 1:297–310; also, private communication with R. Turlay
Erriquez, O., et al. 1978. *Phys. Lett.* 77B:227
Farrar, G. 1974. *Nucl Phys.* B77:429–42
Feldman, G. J., et al. 1977. *Phys. Rev. Lett.* 38:1313
Feynman, R. P. 1969. *Phys. Rev. Lett.* 23:1415
Feynman, R. P. 1972. *Photon-Hadron Interactions.* Reading, Mass: Benjamin
Feynman, R. P. 1974. *Proc. Neutrino '74*, ed. C. Baltay, pp. 299–327. New York: Am. Inst. Phys.
Finjord, J., Ravndal, F. 1975. *Phys. Lett.* 58B:61
Fisk, H. E. 1981. *Proc. Int. Symp. Leptons*

Photons High Energies, ed. W. Pfeil, p. 703. Univ. Bonn Phys. Inst.
Fritze, P., et al. 1981. See Baltay 1981, 1:344–48
Gaillard, M. K. 1974. See Feynman 1974, p. 65
Gaillard, M. K., et al. 1975. *Rev. Mod. Phys.* 47:277
Gell-Mann, M. 1964. *Phys. Lett.* 8:214
Gell-Mann, M., Levy, M. 1960. *Nuovo Cimento* 16:705
Georgi, H. Politzer, H.D.. 1976. *Phys. Rev.* D14:1829
Giboni, K. L., et al. 1979. *Phys. Lett.* 85B:437
Giesch, M., et al. 1963. *Nucl. Instrum. Methods* 20:58
Glashow, S., et al. 1970. *Phys. Rev.* D2:1285
Goldberg, H. 1977. *Phys. Rev. Lett.* 39:1598
Gonzalez-Arroyo, H. L., et al. 1979. *Nucl. Phys.* B153:161; B159:512
Gonzalez-Arroyo, H. L., et al. 1980. *Nucl. Phys.* B166:429
Gordon, B. A., et al. 1979. *Phys. Rev.* D20:2645–91
Grassler, H., et al. 1981. *Phys. Lett.* 99B:159
Grimson, J., Mori, S. 1978. New Single Horn System. *FERMILAB TM-824* (unpublished)
Gross, D. J., Llewellyn Smith, C. H. 1969. *Nucl. Phys.* B14:337
Gunion, J. F. 1974. *Phys. Rev.* D10:242–50
Heagy, S. M., et al. 1981. *Phys. Rev.* D23:1045–69
Holder, M., et al. 1977. *Phys. Rev. Lett.* 39:433
Holder, M., et al. 1978. *Nucl. Instrum. Methods* 148:235; 151:69
Hung, P. Q., Sakurai, J. J. 1981. *Ann. Rev. Nucl. Part. Sci.* 31:375–438
Jonker, M., et al. 1981. *Phys Lett.* 99B:265; Erratum 100B:520
Jonker, M., et al. 1981a. *Phys. Lett.* 107B:241
Kane, G. L., et al. 1979. *Phys. Rev.* D19:1978
Kaplan, J., Martin, F. 1976. *Nucl. Phys.* B115:333
Kitagaki, T., et al. 1980. *Phys. Rev. Lett.* 45:955
Kitagaki, T., et al. 1982. *Phys. Rev. Lett.* 48:299
Knobloch, J., et al. 1980. *Proc. 20th Int. Conf. High Energy Phys. Madison, Wis.*, ed. L. Durand, L. Pondrom, p. 769. New York: Am. Inst. Phys.
Knobloch, J., et al. 1981. See Baltay 1981, 1:421
Lai, C. H. 1978. *Phys. Rev.* D18:1422

Lee, J. R. 1980. PhD thesis. *Measurements of νN Charged Current Cross Sections from 25 to 260 GeV*, Caltech, Pasadena (Unpublished)
Limon, P., et al. 1974. *Nucl. Instrum. Methods* 116:317
Lubimov, V. A., et al. 1980. *Phys. Lett.* 94B:266
Mann, A. W., et al. 1973. *Phys. Rev. Lett.* 31:844
Marel, G., et al. 1977. *Nucl. Instrum. Methods* 141:43
May, J., et al. 1977. Status Report on Narrow Band Neutrino Beam Running and Monitoring, *CERN Internal Rep.* 26.4.7 (unpublished)
Mestayer, M. D. 1978. PhD thesis. *SLAC Rep. No. 214* (unpublished)
Miller, G. et al. 1972. *Phys. Rev.* D5:6528
Morfin, J., et al. 1981. *Phys. Lett.* 104B:235
Mount, R. P. 1981. *Nucl. Instrum. Methods* 187:401
Murtagh, M. J. 1979. *Proc. 1979 Int. Symp. Lepton Photon Interact. High Energies*, ed. T. B. W. Kirk, H. D. I. Arbarbane p. 277. Batavia, Ill: Fermi: Natl. Lab
Nachtmann, O. 1973. *Nucl. Phys.* B63:237
Nishikawa, K., et al. 1981. *Phys. Rev. Lett.* 46:1555
Perkins, D. H. 1978. *Proc. Summer Inst. Part. Phys.* p. 1. Stanford, Calif: SLAC
Peterson, V. Z. 1964. A Monochromatic Neutrino Beam from a High Intensity 200-GeV Proton Synchrontron. *LRL Rep. UCID-10028* (unpublished)
Pontecorvo, B. 1951. *J. Exp. Theor. Phys. (USSR)* 37:1751 (Transl. 1960. *Sov. Phys. JETP* 10:1236)
Pontecorvo, B. 1957. *Sov. Phys. JETP* 6:429, 1972
Primakoff, H., Rosen, S. P. 1981. *Ann. Rev. Nucl. Part. Sci.* 31:145–92
Rapidis, P., et al, 1981. *Proc. Summer Inst. Part. Phys.*, ed. A. Mosher, p. 641. Stanford, Calif: SLAC
Reya, E. 1981. *Phys. Rep.* 69:195–333
Rubbia, C., et al. 1974. *Proc. 17th Int. Conf. High Energy Phys., London*, ed. J. R. Smith, p. IV-117. Chilton: Sci. Res. Counc. Rutherford Lab.
Sakurai, J. J. 1981. See Baltay 1981, 2:457–91
Schindler, R. H., et al. 1981. *Phys. Rev.* 24:78
Schmidt, I. A., Blankenbecler, R. 1977. *Phys. Rev.* D16:1318
Schrock, R. E., Lee, B. W. 1976. *Phys. Rev.* D13:2539
Schwartz, M. 1960. *Phys. Rev. Lett.* 4:306

Sciulli, F. 1980. *Proc. Summer Inst. Part. Phys.* ed. A. Mosher, p. 29. Stanford, Calif: SLAC
Sciulli, F., et al. 1970. Neutrino Physics at Very High Energies, *Fermilab Proposal 21* (unpublished)
Shaevitz, M. H., et al. 1980. See Sciulli 1980, p. 475
Shaevitz, M., et al. 1981. See Balty 1981, 1:311
Shultze, K. 1977. *Proc. 1977 Int. Symp. Lepton Photon Interact. High Energy, Hamburg, Germany,* p. 359. Hamburg: DESY
Skuja, A., et al. 1976. The 300-GeV Triplet Train. *Fermilab TM-646* (unpublished)
Smadja, G. 1981. See Fisk 1981, p. 444
Söding, P., Wolf, G. 1981. *Ann. Rev. Nucl. Part. Sci.* 31:231–93
Stanton, N. R. 1981. See Baltay 1981, 1:491
Stefanski, R., White, H. 1976. Neutrino Flux Distributions. *Fermilab FN-292* (unpublished)
Stevenson, M. L. 1978. *Proc. Topical Conf. Neutrino Phys. Accel., Oxford,* ed. A. G. Michette, P. B. Renton, p. 362. Chilton, England: Rutherford Lab.
Strovink, M. 1981. See Fisk 1981, p. 594
Stutte, L., et al. 1981. See Baltay 1981, 1:377
Taylor, F. E., et al. 1978. *IEEE Trans. Nucl. Sci.* NS-25:31
Taylor, F. E., et al. 1980. *IEEE Trans. Nucl. Sci.* NS-27:30
Trilling, G. H. 1980. See Knobloch et al 1980, p. 1140
Trinko, T., et al. 1981. *Phys.. Rev.* D23:1889
Ushida, N., et al. 1980a. *Phys. Rev. Lett.* 45:1049
Ushida, N., et al. 1980b. *Phys. Rev. Lett.* 45:1053
Ushida, N., et al. 1982. *Phys. Rev. Lett.* 48:844
van der Meer, S., et al. 1963. *Proc. Siena Int. Conf. Elementary Part.* 1:536, ed. G. Bernardin:, G. P. Puppi. Bologna: Soc. Ital. Fis.
von Krogh, J., et al. 1976. *Phys. Rev. Lett.* 36:710
Wachsmuth, H. 1977. *CERN Rep. EP/PHYS 77-43* (unpublished)
Willis, S. E., et al. 1980. *Phys. Rev. Lett.* 45:1370
Young, B. L., et al. 1978. *Phys. Lett.* 74B:111
Zweig, G. 1964. *CERN Reps. 8182/TH 401; 8419/TH 412* (unpublished)

AUTHOR INDEX

(Names appearing in capital letters indicate authors of chapters in this volume.)

CUMULATIVE INDEXES

CONTRIBUTING AUTHORS VOLUMES 23–32

CHAPTER TITLES, VOLUMES 23–32

PARTICLE INTERACTIONS AT HIGH ENERGIES

NEW BOOKS
FROM
ANNUAL REVIEWS INC.

NOW YOU CAN
CHARGE THEM
TO

VISA *MasterCard*

ORDER FORM

A NONPROFIT SCIENTIFIC PUBLISHER

Annual Reviews Inc.

4139 EL CAMINO WAY • PALO ALTO, CA 94306 USA • (415) 493-4400

Please list the volumes you wish to order by volume number. If you wish a standing order (the latest volume sent to you automatically each year), indicate volume number to begin order. Volumes not yet published will be shipped in month and year indicated. Specified back volumes may be purchased at lower prices if ordered **prior to January 1, 1983.** See bracketed volume and price information listed below. All prices subject to change without notice.

ANNUAL REVIEW SERIES

Annual Review of **ANTHROPOLOGY**

		Prices Postpaid per volume USA/elsewhere	Regular Order Please send:	Standing Order Begin with:
			Vol. number	Vol. number

[Vols. 1-8 $17.00/$17.50
Price effective through 12/31/82]

Vols. 1-10	(1972-1981)	$20.00/$21.00		
Vol. 11	(1982)	$22.00/$25.00		
Vol. 12	(avail. Oct. 1983)	$27.00/$30.00	Vol(s). _____	Vol. _____

Annual Review of **ASTRONOMY AND ASTROPHYSICS**

[Vols. 1-17 $17.00/$17.50
Price effective through 12/31/82]

Vols. 1-19	(1963-1981)	$20.00/$21.00		
Vol. 20	(1982)	$22.00/$25.00		
Vol. 21	(avail. Sept. 1983)	$44.00/$47.00	Vol(s). _____	Vol. _____

Annual Review of **BIOCHEMISTRY**

[Vols. 28-48 $18.00/$18.50
Price effective through 12/31/82]

Vols. 28-50	(1959-1981)	$21.00/$22.00		
Vol. 51	(1982)	$23.00/$26.00		
Vol. 52	(avail. July 1983)	$29.00/$32.00	Vol(s). _____	Vol. _____

Annual Review of **BIOPHYSICS AND BIOENGINEERING**

[Vols. 1-9 $17.00/$17.50
Price effective through 12/31/82]

Vols. 1-10	(1972-1981)	$20.00/$21.00		
Vol. 11	(1982)	$22.00/$25.00		
Vol. 12	(avail. June 1983)	$47.00/$50.00	Vol(s). _____	Vol. _____

Annual Review of **EARTH AND PLANETARY SCIENCES**

[Vols. 1-8 $17.00/$17.50
Price effective through 12/31/82]

Vols. 1-9	(1973-1981)	$20.00/$21.00		
Vol. 10	(1982)	$22.00/$25.00		
Vol. 11	(avail. May 1983)	$44.00/$47.00	Vol(s). _____	Vol. _____

Annual Review of **ECOLOGY AND SYSTEMATICS**

[Vols. 1-10 $17.00/$17.50
Price effective through 12/31/82]

Vols. 1-12	(1970-1981)	$20.00/$21.00		
Vol. 13	(1982)	$22.00/$25.00		
Vol. 14	(avail. Nov. 1983)	$27.00/$30.00	Vol(s). _____	Vol. _____

SEE ORDERING INFORMATION ON PAGE 4.

Annual Review of **ENERGY**	Prices Postpaid per volume USA/elsewhere	Regular Order Please send: Vol. number	Standing Order Begin with: Vol. number

Annual Review of **ENERGY**

[Vols. 1-4 $17.00/$17.50
Price effective through 12/31/82]

Vols. 1-6	(1976-1981) .... $20.00/$21.00		
Vol. 7	(1982) .... $22.00/$25.00		
Vol. 8	(avail. Oct. 1983) .... $56.00/$59.00	Vol(s). _____	Vol. _____

Annual Review of **ENTOMOLOGY**

[Vols. 7-25 $17.00/$17.50
Price effective through 12/31/82]

Vols. 7-26	(1962-1981) .... $20.00/$21.00		
Vol. 27	(1982) .... $22.00/$25.00		
Vol. 28	(avail. Jan. 1983) .... $27.00/$30.00	Vol(s). _____	Vol. _____

Annual Review of **FLUID MECHANICS**

[Vols. 1-12 $17.00/$17.50
Price effective through 12/31/82]

Vols. 1-13	(1969-1981) .... $20.00/$21.00		
Vol. 14	(1982) .... $22.00/$25.00		
Vol. 15	(avail. Jan. 1983) .... $28.00/$31.00	Vol(s). _____	Vol. _____

Annual Review of **GENETICS**

[Vols. 1-13 $17.00/$17.50
Price effective through 12/31/82]

Vols. 1-15	(1967-1981) .... $20.00/$21.00		
Vol. 16	(1982) .... $22.00/$25.00		
Vol. 17	(avail. Dec. 1983) .... $27.00/$30.00	Vol(s). _____	Vol. _____

Annual Review of **IMMUNOLOGY — New Series 1983**

Vol. 1	(avail. April 1983) .... $27.00/$30.00	Vol(s). _____	Vol. _____

Annual Review of **MATERIALS SCIENCE**

[Vols. 1-9 $17.00/$17.50
Price effective through 12/31/82]

Vols. 1-11	(1971-1981) .... $20.00/$21.00		
Vol. 12	(1982) .... $22.00/$25.00		
Vol. 13	(avail. Aug. 1983) .... $64.00/$67.00	Vol(s). _____	Vol. _____

Annual Review of **MEDICINE: Selected Topics in the Clinical Sciences**

[Vols. 1-3, 5-15, 17-31 $17.00/$17.50
Price effective through 12/31/82]

Vols. 1-3, 5-15	(1950-1952; 1954-1964) .... $20.00/$21.00		
Vols. 17-32	(1966-1981) .... $20.00/$21.00		
Vol. 33	(1982) .... $22.00/$25.00		
Vol. 34	(avail. April 1983) .... $27.00/$30.00	Vol(s). _____	Vol. _____

Annual Review of **MICROBIOLOGY**

[Vols. 15-33 $17.00/$17.50
Price effective through 12/31/82]

Vols. 15-35	(1961-1981) .... $20.00/$21.00		
Vol. 36	(1982) .... $22.00/$25.00		
Vol. 37	(avail. Oct. 1983) .... $27.00/$30.00	Vol(s). _____	Vol. _____

Annual Review of **NEUROSCIENCE**

[Vols. 1-3 $17.00/$17.50
Price effective through 12/31/82]

Vols. 1-4	(1978-1981) .... $20.00/$21.00		
Vol. 5	(1982) .... $22.00/$25.00		
Vol. 6	(avail. March 1983) .... $27.00/$30.00	Vol(s). _____	Vol. _____

SEE ORDERING INFORMATION ON PAGE 4.

FROM

NAME —————————————

ADDRESS —————————————

————— ZIP CODE —————

Annual Reviews Inc. A NONPROFIT
SCIENTIFIC PUBLISHER
4139 EL CAMINO WAY
PALO ALTO, CALIFORNIA 94306, USA

PLACE
STAMP
HERE

Annual Review of **SOCIOLOGY**

⌈Vols. 1-5 $17.00/$17.50⌉
⌊Price effective through 12/31/82⌋

	Prices Postpaid per volume USA/elsewhere	Regular Order Please send: Vol. number	Standing Order Begin with: Vol. number
Vols. 1-7	(1975-1981) .... $20.00/$21.00		
Vol. 8	(1982) .... $22.00/$25.00		
Vol. 9	(avail. Aug. 1983) .... $27.00/$30.00	Vol(s). _____	Vol. _____

SPECIAL PUBLICATIONS	Prices Postpaid per volume USA/elsewhere	Regular Order Please send:

Annual Reviews Reprints: **Cell Membranes**, 1975-1977

(published 1978)   Softcover .... $12.00/$12.50   _____ Copy(ies).

Annual Reviews Reprints: **Cell Membranes**, 1978-1980

(published 1981)   Hardcover .... $28.00/$29.00   _____ Copy(ies).

Annual Reviews Reprints: **Immunology**, 1977-1979

(published 1980)   Softcover .... $12.00/$12.50   _____ Copy(ies).

History of Entomology

(published 1973)   Clothbound .... $10.00/$10.50   _____ Copy(ies).

**Intelligence and Affectivity:
Their Relationship During Child Development**, by Jean Piaget

(published 1981)   Hardcover .... $8.00/$9.00   _____ Copy(ies).

Telescopes for the 1980s

(published 1981)   Hardcover .... $27.00/$28.00   _____ Copy(ies).

The Excitement and Fascination of Science, Volume 1

(published 1965)   Clothbound .... $6.50/$7.00   _____ Copy(ies).

The Excitement and Fascination of Science, Volume 2

(published 1978)   Hardcover .... $12.00/$12.50
                     Softcover .... $10.00/$10.50   _____ Copy(ies).

To: ANNUAL REVIEWS INC. 4139 El Camino Way, Palo Alto, CA 94306-9981 USA   Tel. 415-493-4400

Please enter my order for the publications checked above.

Institutional purchase order No. _____

Amount of remittance enclosed $_____

Charge my account ☐ MasterCard ☐ Visa   Acct. No._____

Exp. Date _____

Individuals: Prepayment required, or charge account below.

California residents, please add applicable sales tax.
Prices subject to change without notice.

_____
Signature

Name _____
        Please print

Address _____
          Please print

_____ Zip Code _____   ☐ Please send free copy of current *Prospectus*